Pulmonomics: Omics Approaches for Understanding Pulmonary Diseases

Sudipto Saha • Sreyashi Majumdar •
Parthasarathi Bhattacharyya

Pulmonomics: Omics Approaches for Understanding Pulmonary Diseases

Springer

Sudipto Saha
Department of Biological Sciences
Bose Institute
Kolkata, West Bengal, India

Sreyashi Majumdar
Department of Biological Sciences
Bose Institute
Kolkata, West Bengal, India

Parthasarathi Bhattacharyya
Institute of Pulmocare & Research
Kolkata, West Bengal, India

ISBN 978-981-99-3504-8 ISBN 978-981-99-3505-5 (eBook)
https://doi.org/10.1007/978-981-99-3505-5

This Springer imprint is published by the registered company Springer Nature Singapore Pte Ltd.
The registered company address is: 152 Beach Road, #21-01/04 Gateway East, Singapore 189721, Singapore

Paper in this product is recyclable.

Preface

Pulmonary diseases are a global burden and cause morbidity, mortality, and disability. The number of patients with lung diseases is increasing due to present lifestyle, environmental pollution, and genetic factors. Lungs work in association with the heart and their proper functioning is essential for healthy living. Physicians often encounter difficulties in accurately diagnosing pulmonary diseases in a non-invasive way since the symptoms (like cough and shortness of breath) are similar in different types of pulmonary diseases. In addition, there are overlaps between different lung diseases like Asthma-COPD Overlap Syndrome (ACOS). The pathogenesis and diagnosis for various pulmonary diseases are different and so is the associated treatment regimen. There are a few clinical biomarkers from blood and urine samples available in practice for liver, kidney, and other infectious diseases, but not in the case of lung diseases. So with the advent of OMICS approaches, researchers studied different lung diseases including asthma, COPD, IPF, COVID-19, and lung cancer to identify potential biomarkers and drug targets. Recently, several OMICS approaches like genomics, proteomics, transcriptomics, and metabolomics have enhanced the understanding of these diseases at molecular and systems levels. Besides, X-ray and HRCT images of the lungs added a new dimension to the diagnosis of lung diseases. Integration of the multi-omics approaches using bioinformatics tools and mathematical modeling provides a better understanding of these complex lung diseases thus, leading to a better diagnosis and treatment plan to improve and maintain lung health. In this book, pulmonary diseases have been classified into five major categories, namely, obstructive diseases (COPD, asthma, Bronchiectasis), restrictive diseases (Idiopathic Pulmonary Fibrosis, Sarcoidosis), pulmonary vascular diseases (Pulmonary Hypertension), infectious pulmonary diseases (Tuberculosis, COVID-19, Pneumonia), and lung cancer. Each book chapter has been dedicated to specific pulmonary disease-associated clinical features and disease trends across the world, different biomarkers identified using genomics, transcriptomics, proteomics, metabolomics, metagenomics, and radiomics approaches. Also, how these OMICS approaches were integrated using machine learning techniques to predict the biomarkers across different molecular levels. Besides these, in each chapter, details on present therapeutics including chemotherapy, biologics, and vaccines are included. Each chapter ends with future research goals, research problems, and opportunities associated with pulmonary disease.

The purpose of this book is to provide students and researchers in the field of medical and life sciences with a detailed description of the application of state-of-the-art OMICS techniques in pulmonary diseases for the identification of potential biomarkers. We hope that this book will be useful for new-generation physicians cum researchers and students/fellows working directly or indirectly with pulmonary diseases.

Kolkata, India — Sudipto Saha
Kolkata, India — Sreyashi Majumdar
Kolkata, India — Parthasarathi Bhattacharyya

Contents

About the Authors

Sudipto Saha is an Associate Professor at the Department of Biological Sciences, Bose Institute, Kolkata, India. He joined Bose Institute as a DBT-Ramalingaswami fellow in 2012. He has earlier worked as a post-doctoral research fellow at Indiana University, School of Informatics, Indianapolis, USA (2007–2008) and at Case Western Reserve University, School of Medicine, Cleveland, USA (2008–2012). He did his Ph.D. (Bioinformatics) in 2007 from Jawaharlal Nehru University (work done at CSIR-Institute of Microbial Technology, Chandigarh). His research interests are in pulmonary diseases like asthma/COPD and tuberculosis and has a specialization in bioinformatics and OMICS data analysis. He has more than 22 years of research experience in Immunoinformatics, Proteomics, Bioinformatics, and Machine Learning Techniques. He has published more than 85 research articles in peer-reviewed international journals and authored several book chapters.

Sreyashi Majumdar has obtained her Ph.D. degree from the University of Calcutta, Kolkata, India (2022). She pursued her Ph.D. research at Bioinformatics Centre (Department of Biological Sciences), Bose Institute, Kolkata, India under the guidance of Dr. Sudipto Saha. Her research interest encompasses the systemic discovery of asthma biomarkers for therapeutics. She has published several research articles in peer-reviewed international journals and has also authored a book chapter. She has several awards accredited to her.

Parthasarathi Bhattacharyya did his post-doctoral fellowship in pulmonary medicine from PGIMER, Chandigarh and presently working as the lead pulmonologist attached to the Institute of Pulmocare and Research, Kolkata, India. He has been engaged to serve the motto of the institute as research, education, and patient care for over two decades. He keeps a keen interest in research, teaching, and innovative work. He has published more than 75 research articles in peer-reviewed national and international journals.

1 Introduction to Pulmonary Diseases and OMICS Approaches

Abstract

Incidence of diseases affecting the respiratory tract and lungs have increased in the past few decades. Exposure to harmful environmental triggers, infections, genetic factors, and aging contributes to progressive deterioration of lung health. Identification of disease biomarkers and drug targets can enable improved disease diagnosis and management. Different omics approaches, like genomics, epigenomics, transcriptomics, proteomics, metabolomics, metagenomics, and radiomics have brought about revolutionizing change in biomarker research. Exploring different biological samples like blood, exhaled breath condensate, sputum, broncho alveolar lavage fluid, bronchial tissue biopsies and urine using the modern state-of-art omics approaches, followed by integration of multi-omics data using bioinformatics tools can provide deeper mechanistic insight into pulmonary disease biology and provide clue about potential biomarkers and drug targets.

Keywords

Pulmonary diseases · Biomarkers · Genomics · Transcriptomics · Proteomics · Metagenomics · Bioinformatics · Multi-omics

1.1 Pulmonary Diseases

Group of diseases affecting airways, lungs, and other parts of the respiratory system are collectively referred to as pulmonary diseases. The prevalence of pulmonary diseases has increased substantially in the past few decades (Labaki and Han 2020). Several factors, including genetic predisposition, exposure to environmental triggers (like allergens, pollutants, microbes), lifestyle habits (like smoking, lack of physical exercise, alcohol consumption, obesity) contribute to gradual deterioration of lung

S. Saha et al., *Pulmonomics: Omics Approaches for Understanding Pulmonary Diseases*, https://doi.org/10.1007/978-981-99-3505-5_1

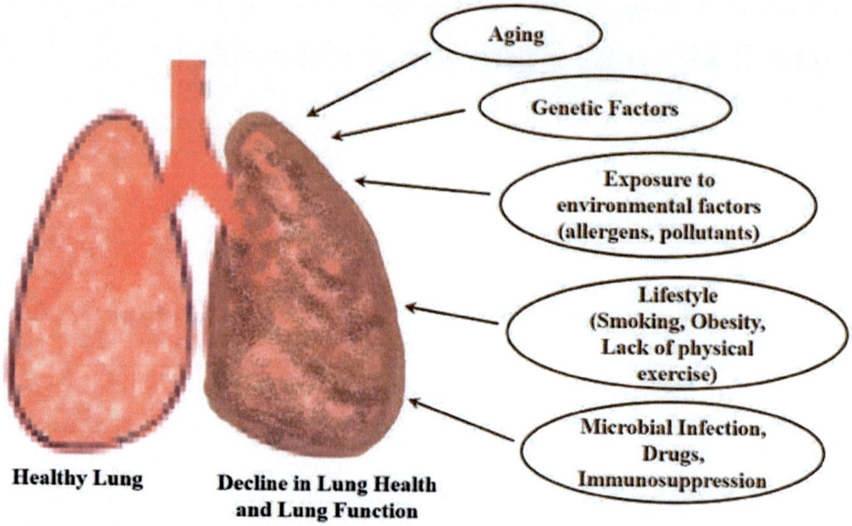

Fig. 1.1 Different factors leading to decline in lung health and lung function

health and pulmonary immune system, thereby paving the way for increased incidences of respiratory diseases (as shown in Fig. 1.1).

Aging also has an effect on lung health. Aging is associated with certain structural changes in lungs and extrapulmonary regions that results in reduced lung volume, decreased volume of thoracic cavity, increased alveolar dead space, and reduced inspiratory and expiratory efficiency. All these changes cumulatively lead to progressive decline in lung function. Reduced lung function along with decline in pulmonary immune response results in predisposition of aged individuals to pulmonary diseases (Schneider et al. 2021; Lowery et al. 2013; Skloot 2017).

1.2 Types of Pulmonary Diseases

Pulmonary diseases can be broadly classified into the five major groups (as shown in Fig. 1.2):

(i) Obstructive pulmonary diseases
(ii) Restrictive pulmonary diseases
(iii) Vascular pulmonary diseases
(iv) Infectious pulmonary disorders
(v) Neoplastic diseases or lung cancer

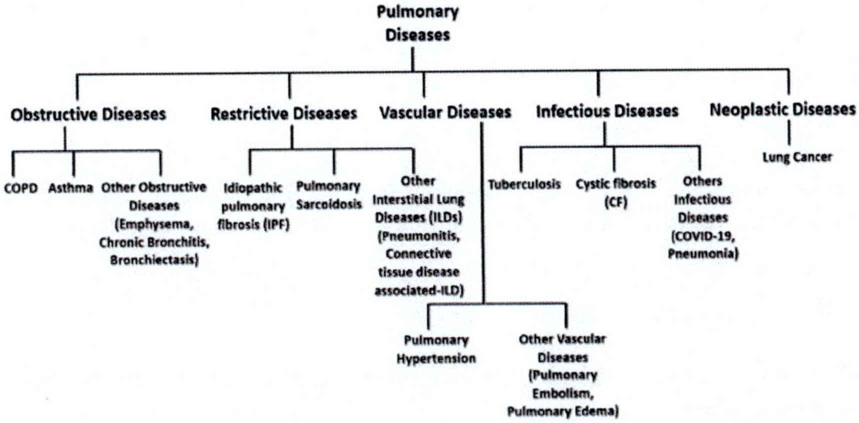

Fig. 1.2 Classification of pulmonary diseases. Pulmonary diseases have been classified into five major categories: Obstructive diseases, restrictive diseases, vascular disease, infectious diseases, and neoplastic diseases

1.3 Need and Significance of Biomarkers in Pulmonary Diseases

The increasing global burden of pulmonary diseases and the high mortality rate demands early disease diagnosis, proper staging and classification of the diseases, and designing novel, effective therapeutics for them. Biomarkers not only lie central to proper disease diagnosis, monitoring disease severity/activity, and impact of therapeutic interventions but also provide clue toward novel drug targets. Biomarkers may be defined as measurable indicators of normal biological processes or diseased conditions (Califf 2018). Bio-molecules that undergo cellular, biochemical, or molecular alterations in diseased state can serve as disease biomarkers. Biomarkers can be grouped into the following categories:

Diagnostic Biomarkers

These biomarkers aid in effective diagnosis and confirmation of presence of a diseased condition. Such biomarkers may also enable disease stratification and categorization of disease severity.

Monitoring Biomarkers

Such biomarkers measured repeatedly over time to monitor the course of disease progression or change in disease severity. They can also be used to evaluate improvement or changes in disease parameters after treatment with therapeutic agents.

Prognostic Biomarkers

These biomarkers help to determine the chances or risk of disease recurrence or disease progression, after primary therapeutic invention in a diseased patient.

Risk or Susceptibility Biomarkers

These biomarkers are indicative of the chances or potential of developing a particular diseased condition.

Pharmacodynamic or Drug Response Biomarkers

The level of these biomarkers changes upon exposure to particular therapeutic agent. These biomarkers are of immense utility in clinical practice during the early phases of drug development.

Predictive Biomarkers

Any change in the profile of these biomarkers or the mere presence of these biomarkers in a group of individuals predicts which individuals would respond favourably or unfavourably to a given therapy. Predictive biomarkers help in observing the effect or impact of therapeutic intervention.

The different types of biomarkers can be of immense utility in proper detection of pulmonary diseases. These biomarkers may serve as stepping stones for designing novel precision medicines for better management of respiratory disorders. However, till date there are a very few validated biomarkers for respiratory disorders (Cazzola 2021).

1.4 Omics Approaches as Tools for Identification of Biomarkers for Pulmonary Diseases

The advent and technical advancement in omics approaches can be exploited at large for the identification and validation of biomarkers for pulmonary diseases. Omics may be defined as a rapidly emerging, multidisciplinary, unbiased technologies that provides better insights into different aspects of normal biological state as well as diseased condition (Kan et al. 2017). Omics approaches have improved the understanding of pulmonary diseases—aided in disease staging, subtyping of diseases, unveiling mechanisms associated with disease phenotypes and drug responsiveness, identifying novel drug targets, and designing potential therapeutics.

1.5 Types of Omics Approaches

Multi-omics and systems biology approaches are being widely exploited to study and address unresolved problems associated with respiratory disorders (as shown in Fig. 1.3). The different omics approaches that find wide application in pulmonary medicine include:

Genomics

The most mature omics approach till date is genomics. Genomics finds enormous application in the field of clinical and disease biology, for highlighting genome sequence and copy number variations associated with various diseases. Modern day genomics employs the use of high-throughput techniques like microarray, next generation sequencing (NGS) for whole genome sequencing or exome sequencing.

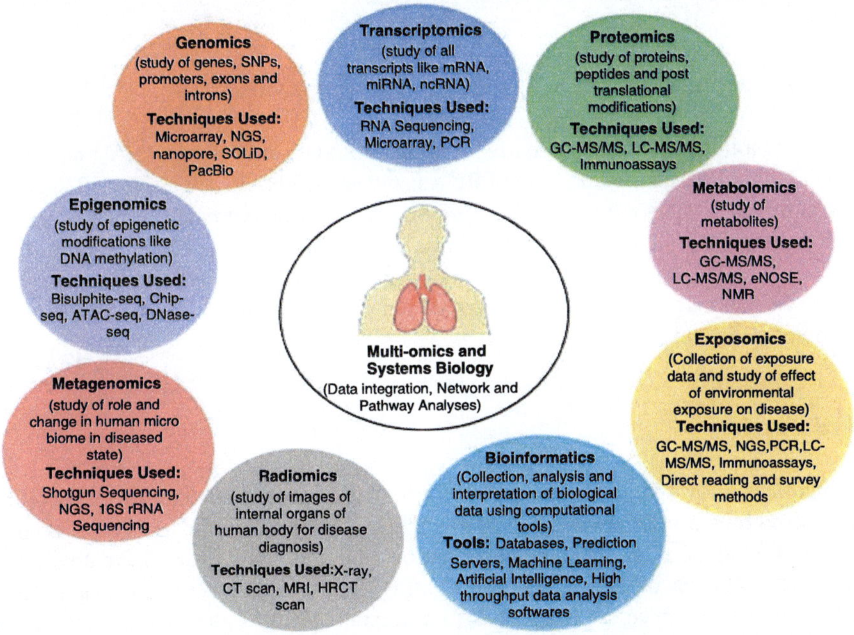

Fig. 1.3 Multi-omics approaches used to study pulmonary disease biology. Different omics approaches like genomics, epigenomics, transcriptomics, proteomics, metabolomics, metagenomics, medical imaging, and bioinformatics finds wide application in study of human diseases. The state-of-art techniques widely used in these omics studies have also been highlighted

High-throughput genomics aids in identifying copy number variants (CNVs), single nucleotide polymorphisms (SNPs), genomic rearrangements, rare variants, and loss of heterozygosity (LOH) variants that are linked to increased disease susceptibility (Misra et al. 2018). Till date, several SNPs and genetic variations identified from blood, tissue biopsies, and broncho alveolar lavage fluids have been strongly correlated with different types of respiratory diseases.

Epigenomics

Epigenomics provides clue about reversible modifications of DNA or DNA-associated proteins (including DNA methylation and histone modification) that regulate chromatin accessibility, transcription factor binding, and gene transcription. Alterations in such covalent modifications of DNA and histone proteins are often associated with disease development including pulmonary diseases (Hasin et al. 2017). Bisulfite sequencing, chromatin immunoprecipitation followed by sequencing (ChIP-seq), methylated DNA immunoprecipitation (Me-DIP), DNase Sequencing (DNase-seq), and ATAC-seq are the different epigenomics technologies widely used for identification of disease associated epigenetic signatures (Ogulur et al. 2021). Mostly blood and tissue samples are explored for identifying epigenomics signatures of pulmonary diseases.

Transcriptomics

Transcriptomics involves qualitative and quantitative study of all RNA transcripts (including mRNA, long non-coding RNAs, and small RNAs such as microRNAs) in a cell or tissue. It serves as a connecting bridge between genomics and proteomics. Microarray, RNA sequencing, and polymerase chain reaction (PCR) are commonly used to identify and validate differentially expressed gene biomarkers associated with the pathobiology of pulmonary diseases from blood and other biological samples (Hasin et al. 2017).

Proteomics

Proteomics allows large scale analysis of the whole proteome (total of all proteins expressed by the genome) in a cell or given tissue. High-resolution mass spectrometry (MS) coupled with different fractionation techniques (like gas chromatography and liquid chromatography), western blotting, and immunoassays are extensively used to study changes in protein expression profile and post-translational modifications (PTMs like phosphorylation, glycosylation, ubiquitination, and nitrosylation) associated with particular disease phenotypes. Apart from its contribution in identification of protein-based disease biomarkers and potential drug targets, proteomics also enables the study of important protein–protein interactions associated with normal biological processes and diseased conditions, using techniques like Affinity Purification Mass Spectrometry (AP-MS) (Hasin et al. 2017). Proteomic tools have enabled the study of biological samples (like blood, sputum, broncho alveolar lavage, and bronchial tissues) for identification of important protein biomarkers and PTMs involved in pulmonary disease biology.

Metabolomics

Metabolomics enables both relative and absolute quantification of metabolites like sugars, steroids, lipids, nucleotides, amino acids, and organic acids. Nuclear magnetic resonance (NMR) and high-resolution mass spectrometry (MS) coupled with various separation techniques (like gas chromatography and liquid chromatography) are commonly used for studying the metabolome and for determining disease associated metabolite variations (Misra et al. 2018). Advances in metabolomics have paved the way for identification of pulmonary disease associated metabolic signatures from serum, blood plasma, exhaled breath, urine, and broncho alveolar lavage samples.

Exposomics

Exposomics, the study of the exposome is a relatively newer concept in the field of omics. Exposome refers to the total non-genetic environmental exposure encountered by an individual during the entire lifetime starting from prenatal period. Exposomics makes effective use of omics tools and techniques to collect exposure data and correlate such exposure data with biochemical and molecular changes associated with diseased conditions (https://www.cdc.gov/niosh/topics/exposome/default.html; Price et al. 2022). Functional exposomics further potentiates our understanding of the impact or contribution of environmental factors in disease biology. Exposomics is of prime significance for pulmonary diseases that involves complex interaction between genetic and environmental factors, including obstructive airway diseases like COPD, asthma; restrictive diseases like silicosis, asbestosis, pneumoconiosis; infectious diseases like tuberculosis and malignancies like lung cancer.

Metagenomics

This relatively newer omics approach allows to study the diverse microbial flora from the microbial genome present in different biological samples including human derived bio-specimens. Metagenomics aids in understanding the role of normal human microbial flora in restoring human health and also highlights alterations in microbiome pattern under diseased state (Martin et al. 2014). NGS, shotgun sequencing, and 16SrRNA sequencing are the routinely used platforms for high-throughput metagenomics studies (Hasin et al. 2017). Most of the metagenomics studies for pulmonary diseases are conducted using broncho alveolar lavage fluid or sputum samples.

Radiomics

The branch of omics involved in dealing with medical imaging is referred to as radiomics. It involves acquisition, extraction, and subsequent analysis of medical images for different pulmonary diseases using X-ray imaging, computed tomography (CT) scans, high resolution computed tomography (HRCT) scans, positron emission tomography (PET), and magnetic resonance imaging (MRI). Radiomics makes efficient use of artificial intelligence, deep learning techniques, and neural networks for constructing descriptive and predictive models; and for establishing correlation between pulmonary disease related clinical characteristics, gene expression profile with features of medical images (Liu et al. 2021).

Bioinformatics

Bioinformatics uses computational tools for efficient collection, compilation, storage, management, analysis, and interpretation of large sets of biological data including clinical data. Several databases, web browsers, webservers, and prediction tools have been constructed for different pulmonary diseases using bioinformatics. Bioinformatics tools like deep learning, artificial neural networks, and artificial intelligence are widely used to generate descriptive and predictive models, which finds enormous application in clinical biology including respiratory disease biology. Bioinformatics also allows analysis of high-throughput data generated by genomics, metagenomics, transcriptomics, epigenomics, proteomics, and metabolomics and enables efficient integration of multi-omics data for obtaining better insight into airway disease pathobiology (Bayat 2002).

Details of the state-of-art omics approaches along with the different platforms, software, and repositories available for each omics technique are enlisted in Table 1.1.

Integrative Omics

Integration of diverse datasets derived from different omics approaches is referred to as integrative omics or multi-omics. Although such integration involves computationally complex processes, it provides a clearer insight into the complex biomolecular crosstalk involved in disease development. Such crosstalks cannot be deduced from individual omics studies. Integrative omics makes use of network-based approaches for representation of higher order interaction. Several network-based integrative omics approaches have been applied for effective identification of asthma

Table 1.1 The state-of-art omics approaches

Sl. no.	Omics approach	Type of studies	Input	Platform	Software	Repository
1.	Genomics	Identification of SNPs or genetic variations, eQTL analysis	DNA	Genome wide association studies (GWAS)	Galaxy pipelines (https://usegalaxy.org/), R/Bioconductor packages (freely available) (https://www.bioconductor.org/), GWASpro (freely available) (Kim et al. 2019), PLINK (free) (Purcell et al. 2007; Chang et al. 2015), SNPTEST (free) (Marchini and Howie 2010), SurvivalGWAS_SV (open access) (Syed et al. 2017)	dbSNP (https://www.ncbi.nlm.nih.gov/snp/), dbVar (https://www.ncbi.nlm.nih.gov/dbvar/), ClinVar (https://www.ncbi.nlm.nih.gov/clinvar/), European Variation Archive (EVA) (https://www.ebi.ac.uk/eva/), Genome Sequence Archive for Human (GSA-Human) (https://ngdc.cncb.ac.cn/gsa-human/), GWAS Catalog (https://www.ebi.ac.uk/gwas/), Genomic Expression Archive (GEA) (https://www.ddbj.nig.ac.jp/gea/index-e.html), The European Genome-phenome Archive (EGA) (https://ega-archive.org/)
		Genome organization and genome wide interaction maps	DNA	Hi C (Belton et al. 2012), 3C-Seq, capture-C [Illumina]	HiCeekR (Di Filippo et al. 2019), R/Bioconductor packages (freely available) (https://www.bioconductor.org/)	4D Nucleome Data Portal (Reiff et al. 2022), ENCODE portal (https://www.encodeproject.org/), NCBI GEO (https://www.ncbi.nlm.nih.gov/geo/), EMBL-EBI ArrayExpress (https://www.ebi.ac.uk/arrayexpress/)

(continued)

Table 1.1 (continued)

Sl. no.	Omics approach	Type of studies	Input	Platform	Software	Repository
2.	Epigenomics	Epigenetic studies, DNA methylation	DNA	Bisulfite sequencing, DNase sequencing, ATAC sequencing, ChiP-seq	DMRichR (https://www.benlaufer.com/DMRichR/articles/DMRichR.html), R/Bioconductor packages (freely available) (https://www.bioconductor.org/), BioWardrobe (Kartashov and Barski 2015)	ENCODE (https://www.encodeproject.org/), NIH Roadmap Epigenomics (http://www.roadmapepigenomics.org/), IHEC Data Portal (https://epigenomesportal.ca/ihec/), DeepBlue (https://deepblue.mpi-inf.mpg.de/)
3.	Transcriptomics	Gene expression profile	Transcripts (mRNA, miRNA, ncRNA)	RNA sequencing microarray real-time PCR	GIANT: Galaxy-based tool (open source) (Vandel et al. 2020), SMAGEXP (freely available) (Blanck 2019), TRAPID (freely available) (Van Bel et al. 2013), ABioTrans (freely available) (Zou et al. 2019), Tools4miRs (free) (Lukasik et al. 2016), Qlucore omics explorer (licensed) (https://qlucore.com/omics-explorer)	NCBI GEO (https://www.ncbi.nlm.nih.gov/geo/), GEA (https://www.ddbj.nig.ac.jp/gea/index-e.html), Sequence Read Archive (SRA) (https://www.ncbi.nlm.nih.gov/sra), EMBL-EBI ArrayExpress (https://www.ebi.ac.uk/arrayexpress/)
4.	Proteomics	Protein identification and quantitation (protein expression changes)	Protein	LC-MS/MS, GC-MS/MS, Immunoassays (Chip assay, Western blotting)	Mascot (licensed software) (https://www.matrixscience.com/server.html), Scaffold (licensed software) (https://www.proteomesoftware.com/products/scaffold-5), skyline (free software) (Pino et al. 2020), ImageJ software	PRIDE (https://www.ebi.ac.uk/pride/), MassIVE (https://massive.ucsd.edu/ProteoSAFe/static/massive.jsp), PeptideAtlas (http://www.peptideatlas.org/), Panorama Public (https://

					(free software) (Schroeder et al. 2021)	panoramaweb.org/home/project-begin.view)
5.	Metabolomics	Metabolite expression change quantification	Small molecule metabolites from body fluids or exhaled breath (short fatty acids, small volatile fatty acids)	NMR, LC-MS, GC-MS, HR-MS, CE-MS, eNOSE	Metabolomics Workbench (Sud et al. 2016), Skyline for Small Molecules (open source software) (Adams et al. 2020), XCMS (open source software) (Domingo-Almenara and Siuzdak 2020), MZmine(open source software) (Pluskal et al. 2010), MS-DIAL(open source software) (Tsugawa et al. 2015)	Metabolomics Workbench (Sud et al. 2016), NIH Common Fund's National Metabolomics Data Repository (NMDR) (https://www.metabolomicsworkbench.org/data/index.php), EMBL-EBI MetaboLights (Haug et al. 2020), Panorama Public (https://panoramaweb.org/home/project-begin.view)
6.	Metagenomics	Identification and comparison of microbiome under disease condition	DNA, RNA	NGS, shotgun sequencing, 16S rRNA sequencing	MGnify (freely available) (Mitchell et al. 2020), MetaWRAP (open source software) (Uritskiy et al. 2018), HOME-BIO (open source) (Ferravante et al. 2021), ASaiM or galaxy (open source) (Mehta et al. 2021), UPARSE (freely available) (Edgar 2013), bioOTU (free) (Chen et al. 2016), Mothur (free) (Schloss et al. 2009), QIIME (open source) (Fung et al. 2021)	NCBI Sequence Read Archive (SRA) (https://www.ncbi.nlm.nih.gov/sra), MG-RAST (https://www.mg-rast.org/), EMBL-EBI Metagenomics portal (https://www.ebi.ac.uk/metagenomics/), European Nucleotide Archive (ENA) (https://www.ebi.ac.uk/ena/browser/home)

This table enlists the various omics approaches that finds application in disease biology and clinical studies along with the commonly used platforms, software used for data analyses, and major data repositories and COPD subtypes (Kan et al. 2017). Integrative analysis of different layers of omics data not only helps to explain heterogeneity of disease manifestations but also aids in identification of disease biomarker and potential drug targets, for better disease diagnosis and management (Karczewski and Snyder 2018; Khan et al. 2020).

1.6 Different Biological Samples Explored by Omics Approaches for Discovery of Biomarkers for Pulmonary Diseases

The choice or selection of biological samples for identification of biomarkers for pulmonary diseases using omics approaches is of prime significance. Clinical feasibility of sample collection and sample processing should be taken into account for selecting biological samples. The state-of-art is to use non-invasive or minimally invasive biological samples for biomarker-based studies (Liu et al. 2021). The different biological samples widely and commonly explored for pulmonary diseases' biomarker research include blood, sputum, exhaled breath, broncho alveolar lavage fluid (BALF), urine and bronchial tissue biopsies. The advantages and disadvantages of each of these biological samples along with examples of a few biomarkers of different pulmonary diseases identified using these bio-samples have been tabulated in Table 1.2.

Blood

Circulating blood contains different blood cells and tissue leakage products including lipids, inflammatory proteins, metabolites, and other bio-molecules. Therefore, blood can serve as an exemplary source of biomarkers for normal physiological processes as well as diseased conditions (Saha et al. 2009; Thambisetty and Lovestone 2010). It involves micro-invasive sample collection. Blood plasma and serum have been widely explored for biomarker identification studies. However, expression profile of blood borne biomarkers may get affected in presence of other co-morbid conditions or infections (Tiotiu 2018).

Sputum

Sputum serves as a rich source of cellular and molecular biomarkers, especially for respiratory inflammation and infections. Sputum cellular counts and protein expression trends can aid in determining phenotypes and diagnosing severity of pulmonary diseases (Sagel et al. 2007; Dasgupta et al. 2021). Induction of sputum is a non-invasive technique that involves inhalation of hypertonic saline solution. Sample collection requires expertise as salivary contamination of sputum may mask the exact sputum cell count or precise sputum protein expression profile (Simpson et al. 2004).

Table 1.2 Major biological samples used for identification of biomarkers for pulmonary diseases using omics approach

Sl. no.	Biological samples	Advantages	Disadvantages	Some biomarkers along with pulmonary diseases
1.	Blood (blood plasma or serum)	• Cost effective • Ease of collection with minimal patient effort • Micro-invasive • Simple measurement procedures • Source of several biomarkers • Enables sub phenotyping of diseases (Wadsworth et al. 2011) • Very commonly used	• Micro-invasive (procedure may be difficult and painful in a few patients) • Blood biomarker level may be affected by presence of other co-morbidities (Tiotiu 2018; Wadsworth et al. 2011)	• Increased Eosinophil Cationic Protein (ECP)—Cystic fibrosis, asthma, and viral bronchiolitis (Dosanjh et al. 2009) • Increased serum IgE—Allergic asthma (Froidure et al. 2016) • Increased serum periostin—Asthma (Scichilone et al. 2016) • Procalcitonin and CRP -for distinguishing pneumonia from asthma and COPD exacerbations (Bafadhel et al. 2011) • Plasma matrix metalloproteinase (MMP1, MMP8, and MMP9) -associated with COPD severity (Koo et al. 2016) • Plasma brain natriuretic peptide (BNP) - pulmonary hypertension in patients with chronic lung disease (Leuchte et al. 2006) • Surfactant proteins SPA and SPD—Diagnosis of idiopathic pulmonary fibrosis (Ley et al. 2014)
2.	Sputum	• Non-invasive • Reliable measure of airway inflammation	• Expertise and specialized training needed • Collection process may be difficult and uncomfortable • Patients may fail to provide adequate samples • Often have reproducibility issues	• Differential sputum cell (eosinophil, neutrophil) count—T2 high and T2 low asthma and response to corticosteroids (Pavlidis et al. 2019; Kyriakopoulos et al. 2021) • Increased Myeloperoxidase

(continued)

Table 1.2 (continued)

Sl. no.	Biological samples	Advantages	Disadvantages	Some biomarkers along with pulmonary diseases
			• Salivary contamination • Not recommended for children below 8 years (Wadsworth et al. 2011)	(MPO)—COPD (Zhu et al. 2014) • Calprotectin—Cystic fibrosis (Jain 2017)
3.	Exhaled breath	• Non-invasive • Ease of sample collection and measurement • Involves negligible patient effort • Source of several biomarkers • Enables sub phenotyping of diseases (Wadsworth et al. 2011)	• Requires specialized instruments • Expensive process • Salivary contamination • Collection procedure severely impacts protein quality • Reproducibility issues • Unproven clinical effectiveness (Wadsworth et al. 2011)	• Fractional nitric oxide in the exhaled breath (FeNO)—measure of airway inflammation (Tiotiu 2018) • H2O2, F2-isoprostanes in exhaled breath condensate (EBC)—Oxidative stress in pulmonary diseases (Jain 2017)
4.	Urine	• Non-invasive • Ease of sample collection • Standard processing • Sensitive and specific • Source of several biomarkers. • Enables sub phenotyping of diseases.	• Requires expensive NMR facility (Wadsworth et al. 2011) • Clinical effectiveness yet to be proved	• Increased urinary leukotriene E4 (uLTE4)— Childhood allergic asthma and aspirin exacerbated respiratory disease (AERD) in adults (Tiotiu 2018) • Increased urinary NO—improved survival in acute respiratory distress syndrome (ARDS) (Jain 2017)
5.	Broncho alveolar lavage fluid (BALF)	• Minimally invasive procedure • Enables sampling of lower respiratory tract (Davidson et al. 2020) • Provides better diagnosis • Reflects the predominant types of inflammatory cells • Provides idea about microbial flora (Meyer and Raghu 2011)	• Trained and expert professionals needed for suction and sample collection after instillation with saline • BALF has a large range of normal values, which serves as a major clinical limitation of BALF (Radha et al. 2014)	• Elevated prostaglandin D2 (PGD2) is a measure of disease severity in asthma (Balzar et al. 2011; Fajt et al. 2013) • S100A9 in BALF serves as a biomarker for idiopathic pulmonary fibrosis (IPF) (Hara et al. 2012)

(continued)

Table 1.2 (continued)

Sl. no.	Biological samples	Advantages	Disadvantages	Some biomarkers along with pulmonary diseases
6.	Bronchial samples and bronchial biopsies	• Provides definitive assessment of airway inflammation (Wadsworth et al. 2011) • Provides better and earlier diagnosis • Aids in better therapeutic decision	• Most invasive technique • Time consuming process • Risky process may be associated with several complications • Trained and expert professionals needed • Specialized medical equipment and facilities needed	• Increased IL-22+ and IL-23+ immunoreactive cells in the bronchial epithelium - Stable COPD (Di Stefano et al. 2009) • Sub-mucosal eosinophilic and lymphocytic infiltration, Epithelial shedding, goblet cell hyperplasia, sub epithelial collagen deposition - Asthma (Bergeron et al. 2010) • Immuno histochemical examination of p40 and cytokeratin 5/6 - squamous cell lung cancer (Affandi et al. 2018)

This table enlists the different clinical samples employed for identifying pulmonary diseases' biomarkers along with the respective advantage and disadvantage of each biological sample

Exhaled Breath

In recent past, exhaled breath condensate is being increasing explored as a non-invasive and promising source of biomarkers (mostly metabolite and protein biomarkers) for pulmonary diseases. However, the low concentration range of biomarkers present in exhaled breath makes detection difficult. Advancement in proteomics and metabolomics techniques enables better assessment of exhaled breath biomarkers and may aid in improved diagnosis of pulmonary diseases (Liang et al. 2012).

Urine

Urine is often explored as a valuable, alternative, non-invasive source of protein and metabolic biomarkers of lung diseases, including lung cancer, asthma, idiopathic pulmonary fibrosis, and infectious diseases like tuberculosis (Tiotiu 2018; Gasparri et al. 2021; Horikiri et al. 2017; Isa et al. 2018).

Broncho Alveolar Lavage Fluid (BALF)

BALF enables studying and sampling of lower respiratory tract for biomarker identification (Davidson et al. 2020). BALF aids in identifying the normal microbial flora of the lower respiratory tract and helps in monitoring its alteration in diseased state. It also gives an idea about the predominant inflammatory cells present under diseased condition (Meyer and Raghu 2011).

Bronchial Tissue Biopsies

Tissue biopsies provide the most accurate detection of airway inflammation, carcinoma, and other lung disorders. However, its use as a biomarker reservoir is largely limited by the invasive nature of sample acquisition by bronchoscopy (Wadsworth et al. 2011).

Extensive omics-based studies using one or more of these biological samples can prove to be crucial in diagnosing, staging of different pulmonary diseases, and also in effectively distinguishing between pulmonary diseases having overlapping symptoms.

1.7 Conclusion

The major challenges associated with use of omics approaches for disease biology include high false discovery rate with fewer sample size, large scale sample processing, bioinformatics challenges for integration of omics datasets, and interpretation of integrative omics data (Hemnes 2018). Nonetheless, pulmonomics—the application of omics approaches for pulmonary diseases hold huge promise. Omics-based study of human biological samples and integration of multi-omics data for pulmonary diseases shall provide better understanding of disease biology and aid in explaining the heterogeneity of pulmonary diseases. Novel molecular biomarkers and drug targets identified through these omics approaches can improve the present therapeutic status of respiratory diseases and pave the way for implementation of precision medicine in the field of pulmonary diseases.

References

Adams KJ, Pratt B, Bose N, Dubois LG, St John-Williams L, Perrott KM, Ky K, Kapahi P, Sharma V, MacCoss MJ et al (2020) Skyline for small molecules: a unifying software package for quantitative metabolomics. J Proteome Res 19(4):1447–1458

Affandi KA, Tizen NMS, Mustangin M, Zin R (2018) p40 immunohistochemistry is an excellent marker in primary lung squamous cell carcinoma. J Pathol Transl Med 52(5):283–289

Bafadhel M, Clark TW, Reid C, Medina MJ, Batham S, Barer MR, Nicholson KG, Brightling CE (2011) Procalcitonin and C-reactive protein in hospitalized adult patients with community-acquired pneumonia or exacerbation of asthma or COPD. Chest 139(6):1410–1418

Balzar S, Fajt ML, Comhair SA, Erzurum SC, Bleecker E, Busse WW, Castro M, Gaston B, Israel E, Schwartz LB et al (2011) Mast cell phenotype, location, and activation in severe asthma. Data from the severe asthma research program. Am J Respir Crit Care Med 183(3):299–309

Bayat A (2002) Science, medicine, and the future: bioinformatics. BMJ 324(7344):1018–1022

Belton JM, McCord RP, Gibcus JH, Naumova N, Zhan Y, Dekker J (2012) Hi-C: a comprehensive technique to capture the conformation of genomes. Methods 58(3):268–276

Bergeron C, Tulic MK, Hamid Q (2010) Airway remodelling in asthma: from benchside to clinical practice. Can Respir J 17(4):e85–e93

Blanck S (2019) Marot G: SMAGEXP: a galaxy tool suite for transcriptomics data meta-analysis. Gigascience 8(2)

Califf RM (2018) Biomarker definitions and their applications. Exp Biol Med (Maywood) 243(3):213–221

Cazzola M (2021) Moving to a personalized approach in respiratory medicine. From Academic Research to Regulatory Intervention. Front Drug Saf Regul

Chang CC, Chow CC, Tellier LC, Vattikuti S, Purcell SM, Lee JJ (2015) Second-generation PLINK: rising to the challenge of larger and richer datasets. Gigascience 4:7

Chen SY, Deng F, Huang Y, Jia X, Liu YP, Lai SJ (2016) bioOTU: an improved method for simultaneous taxonomic assignments and operational taxonomic units clustering of 16s rRNA gene sequences. J Comput Biol 23(4):229–238

Dasgupta A, Chakraborty R, Saha B, Suri H, Singh P, Raj A, Taneja B, Dash D, Sengupta S, Agrawal A (2021) Sputum protein biomarkers in airway diseases: a pilot study. Int J Chron Obstruct Pulmon Dis 16:2203–2215

Davidson KR, Ha DM, Schwarz MI, Chan ED (2020) Bronchoalveolar lavage as a diagnostic procedure: a review of known cellular and molecular findings in various lung diseases. J Thorac Dis 12(9):4991–5019

Di Filippo L, Righelli D, Gagliardi M, Matarazzo MR, Angelini C (2019) HiCeekR: a novel shiny app for hi-C data analysis. Front Genet 10:1079

Di Stefano A, Caramori G, Gnemmi I, Contoli M, Vicari C, Capelli A, Magno F, D'Anna SE, Zanini A, Brun P et al (2009) T helper type 17-related cytokine expression is increased in the bronchial mucosa of stable chronic obstructive pulmonary disease patients. Clin Exp Immunol 157(2):316–324

Domingo-Almenara X, Siuzdak G (2020) Metabolomics data processing using XCMS. Methods Mol Biol 2104:11–24

Dosanjh A, Gamst A, Phillipson J, Broughton A (2009) Elevated serum eosinophil cationic protein levels in cystic fibrosis, pediatric asthma, and bronchiolitis. Pediatr Asthma Allergy Immunol 10(4):169–173

Edgar RC (2013) UPARSE: highly accurate OTU sequences from microbial amplicon reads. Nat Methods 10(10):996–998

Fajt ML, Gelhaus SL, Freeman B, Uvalle CE, Trudeau JB, Holguin F, Wenzel SE (2013) Prostaglandin D(2) pathway upregulation: relation to asthma severity, control, and TH2 inflammation. J Allergy Clin Immunol 131(6):1504–1512

Ferravante C, Memoli D, Palumbo D, Ciaramella P, Di Loria A, D'Agostino Y, Nassa G, Rizzo F, Tarallo R, Weisz A et al (2021) HOME-BIO (sHOtgun MEtagenomic analysis of BIOlogical entities): a specific and comprehensive pipeline for metagenomic shotgun sequencing data analysis. BMC Bioinform 22(Suppl 7):106

Froidure A, Mouthuy J, Durham SR, Chanez P, Sibille Y, Pilette C (2016) Asthma phenotypes and IgE responses. Eur Respir J 47(1):304–319

Fung C, Rusling M, Lampeter T, Love C, Karim A, Bongiorno C, Yuan LL (2021) Automation of QIIME2 metagenomic analysis platform. Curr Protoc 1(9):e254

Gasparri R, Sedda G, Caminiti V, Maisonneuve P, Prisciandaro E, Spaggiari L (2021) Urinary biomarkers for early diagnosis of lung cancer. J Clin Med 10(8)

Hara A, Sakamoto N, Ishimatsu Y, Kakugawa T, Nakashima S, Hara S, Adachi M, Fujita H, Mukae H, Kohno S (2012) S100A9 in BALF is a candidate biomarker of idiopathic pulmonary fibrosis. Respir Med 106(4):571–580

Hasin Y, Seldin M, Lusis A (2017) Multi-omics approaches to disease. Genome Biol 18(1):83

Haug K, Cochrane K, Nainala VC, Williams M, Chang J, Jayaseelan KV, O'Donovan C (2020) MetaboLights: a resource evolving in response to the needs of its scientific community. Nucleic Acids Res 48(D1):D440–D444

Hemnes AR (2018) Using omics to understand and treat pulmonary vascular disease. Front Med (Lausanne) 5:157

Horikiri T, Hara H, Saito N, Araya J, Takasaka N, Utsumi H, Yanagisawa H, Hashimoto M, Yoshii Y, Wakui H et al (2017) Increased levels of prostaglandin E-major urinary metabolite (PGE-MUM) in chronic fibrosing interstitial pneumonia. Respir Med 122:43–50

Isa F, Collins S, Lee MH, Decome D, Dorvil N, Joseph P, Smith L, Salerno S, Wells MT, Fischer S et al (2018) Mass spectrometric identification of urinary biomarkers of pulmonary tuberculosis. EBioMedicine 31:157–165

Jain KK (2017) Biomarkers of pulmonary diseases. Springer, The Handbook of Biomarkers, pp 673–388

Kan M, Shumyatcher M, Himes BE (2017) Using omics approaches to understand pulmonary diseases. Respir Res 18(1):149

Karczewski KJ, Snyder MP (2018) Integrative omics for health and disease. Nat Rev Genet 19(5): 299–310

Kartashov AV, Barski A (2015) BioWardrobe: an integrated platform for analysis of epigenomics and transcriptomics data. Genome Biol 16:158

Khan S, Ince-Dunn G, Suomalainen A, Elo LL (2020) Integrative omics approaches provide biological and clinical insights: examples from mitochondrial diseases. J Clin Invest 130(1): 20–28

Kim B, Dai X, Zhang W, Zhuang Z, Sanchez DL, Lubberstedt T, Kang Y, Udvardi MK, Beavis WD, Xu S et al (2019) GWASpro: a high-performance genome-wide association analysis server. Bioinformatics 35(14):2512–2514

Koo HK, Hong Y, Lim MN, Yim JJ, Kim WJ (2016) Relationship between plasma matrix metalloproteinase levels, pulmonary function, bronchodilator response, and emphysema severity. Int J Chron Obstruct Pulmon Dis 11:1129–1137

Kyriakopoulos C, Gogali A, Bartziokas K, Kostikas K (2021) Identification and treatment of T2-low asthma in the era of biologics. ERJ Open Res 7(2)

Labaki WW, Han MK (2020) Chronic respiratory diseases: a global view. Lancet Respir Med 8(6): 531–533

Leuchte HH, Baumgartner RA, Nounou ME, Vogeser M, Neurohr C, Trautnitz M, Behr J (2006) Brain natriuretic peptide is a prognostic parameter in chronic lung disease. Am J Respir Crit Care Med 173(7):744–750

Ley B, Brown KK, Collard HR (2014) Molecular biomarkers in idiopathic pulmonary fibrosis. Am J Physiol Lung Cell Mol Physiol 307(9):L681–L691

Liang Y, Yeligar SM, Brown LA (2012) Exhaled breath condensate: a promising source for biomarkers of lung disease. ScientificWorldJournal 2012:217518

Liu X, Cui B, Wang Q, Ma Y, Li L, Chen Z (2021) Biomarkers for respiratory diseases: present applications and future discoveries. Clin Transl Disc

Lowery EM, Brubaker AL, Kuhlmann E, Kovacs EJ (2013) The aging lung. Clin Interv Aging 8: 1489–1496

Lukasik A, Wojcikowski M, Zielenkiewicz P (2016) Tools4miRs – one place to gather all the tools for miRNA analysis. Bioinformatics 32(17):2722–2724

Marchini J, Howie B (2010) Genotype imputation for genome-wide association studies. Nat Rev Genet 11(7):499–511

Martin R, Miquel S, Langella P, Bermudez-Humaran LG (2014) The role of metagenomics in understanding the human microbiome in health and disease. Virulence 5(3):413–423

Mehta S, Crane M, Leith E, Batut B, Hiltemann S, Arntzen MO, Kunath BJ, Pope PB, Delogu F, Sajulga R et al (2021) ASaiM-MT: a validated and optimized ASaiM workflow for metatranscriptomics analysis within galaxy framework. F1000Res 10:103

Meyer KC, Raghu G (2011) Bronchoalveolar lavage for the evaluation of interstitial lung disease: is it clinically useful? Eur Respir J 38(4):761–769

Misra BB, Langefeld CD, Olivier M, Cox LA (2018) Integrated omics: tools, advances, and future approaches. J Mol Endocrinol

Mitchell AL, Almeida A, Beracochea M, Boland M, Burgin J, Cochrane G, Crusoe MR, Kale V, Potter SC, Richardson LJ et al (2020) MGnify: the microbiome analysis resource in 2020. Nucleic Acids Res 48(D1):D570–D578

Ogulur I, Pat Y, Ardicli O, Barletta E, Cevhertas L, Fernandez-Santamaria R, Huang M, Bel Imam M, Koch J, Ma S et al (2021) Advances and highlights in biomarkers of allergic diseases. Allergy 76(12):3659–3686

Pavlidis S, Takahashi K, Ng Kee Kwong F, Xie J, Hoda U, Sun K, Elyasigomari V, Agapow P, Loza M, Baribaud F et al (2019) "T2-high" in severe asthma related to blood eosinophil, exhaled nitric oxide and serum periostin. Eur Respir J 53(1)

Pino LK, Searle BC, Bollinger JG, Nunn B, MacLean B, MacCoss MJ (2020) The skyline ecosystem: informatics for quantitative mass spectrometry proteomics. Mass Spectrom Rev 39(3):229–244

Pluskal T, Castillo S, Villar-Briones A, Oresic M (2010) MZmine 2: modular framework for processing, visualizing, and analyzing mass spectrometry-based molecular profile data. BMC Bioinform 11:395

Price EJ, Vitale CM, Miller GW, David A, Barouki R, Audouze K, Walker DI, Antignac JP, Coumoul X, Bessonneau V et al (2022) Merging the exposome into an integrated framework for "omics" sciences. iScience 25(3):103976

Purcell S, Neale B, Todd-Brown K, Thomas L, Ferreira MA, Bender D, Maller J, Sklar P, de Bakker PI, Daly MJ et al (2007) PLINK: a tool set for whole-genome association and population-based linkage analyses. Am J Hum Genet 81(3):559–575

Radha S, Afroz T, Prasad S, Ravindra N (2014) Diagnostic utility of bronchoalveolar lavage. J Cytol 31(3):136–138

Reiff SB, Schroeder AJ, Kirli K, Cosolo A, Bakker C, Lee S, Veit AD, Balashov AK, Vitzthum C, Ronchetti W et al (2022) The 4D Nucleome data portal as a resource for searching and visualizing curated nucleomics data. Nat Commun 13(1):2365

Sagel SD, Chmiel JF, Konstan MW (2007) Sputum biomarkers of inflammation in cystic fibrosis lung disease. Proc Am Thorac Soc 4(4):406–417

Saha S, Harrison SH, Chen JY (2009) Dissecting the human plasma proteome and inflammatory response biomarkers. Proteomics 9(2):470–484

Schloss PD, Westcott SL, Ryabin T, Hall JR, Hartmann M, Hollister EB, Lesniewski RA, Oakley BB, Parks DH, Robinson CJ et al (2009) Introducing mothur: open-source, platform-independent, community-supported software for describing and comparing microbial communities. Appl Environ Microbiol 75(23):7537–7541

Schneider JL, Rowe JH, Garcia-de-Alba C, Kim CF, Sharpe AH, Haigis MC (2021) The aging lung: physiology, disease, and immunity. Cell 184(8):1990–2019

Schroeder AB, Dobson ETA, Rueden CT, Tomancak P, Jug F, Eliceiri KW (2021) The ImageJ ecosystem: open-source software for image visualization, processing, and analysis. Protein Sci 30(1):234–249

Scichilone N, Crimi C, Benfante A, Battaglia S, Iemmolo M, Spatafora M, Crimi N (2016) Higher serum levels of periostin and the risk of exacerbations in moderate asthmatics. Asthma Res Pract 2:1

Simpson JL, Timmins NL, Fakes K, Talbot PI, Gibson PG (2004) Effect of saliva contamination on induced sputum cell counts, IL-8 and eosinophil cationic protein levels. Eur Respir J 23(5): 759–762

Skloot GS (2017) The effects of aging on lung structure and function. Clin Geriatr Med 33(4): 447–457

Sud M, Fahy E, Cotter D, Azam K, Vadivelu I, Burant C, Edison A, Fiehn O, Higashi R, Nair KS et al (2016) Metabolomics workbench: an international repository for metabolomics data and metadata, metabolite standards, protocols, tutorials and training, and analysis tools. Nucleic Acids Res 44(D1):D463–D470

Syed H, Jorgensen AL, Morris AP (2017) SurvivalGWAS_SV: software for the analysis of genome-wide association studies of imputed genotypes with "time-to-event" outcomes. BMC Bioinformatics 18(1):265

Thambisetty M, Lovestone S (2010) Blood-based biomarkers of Alzheimer's disease: challenging but feasible. Biomark Med 4(1):65–79

Tiotiu A (2018) Biomarkers in asthma: state of the art. Asthma Res Pract 4:10

Tsugawa H, Cajka T, Kind T, Ma Y, Higgins B, Ikeda K, Kanazawa M, VanderGheynst J, Fiehn O, Arita M (2015) MS-DIAL: data-independent MS/MS deconvolution for comprehensive metabolome analysis. Nat Methods 12(6):523–526

Uritskiy GV, DiRuggiero J, Taylor J (2018) MetaWRAP-a flexible pipeline for genome-resolved metagenomic data analysis. Microbiome 6(1):158

Van Bel M, Proost S, Van Neste C, Deforce D, Van de Peer Y, Vandepoele K (2013) TRAPID: an efficient online tool for the functional and comparative analysis of de novo RNA-Seq transcriptomes. Genome Biol 14(12):R134

Vandel J, Gheeraert C, Staels B, Eeckhoute J, Lefebvre P, Dubois-Chevalier J (2020) GIANT: galaxy-based tool for interactive analysis of transcriptomic data. Sci Rep 10(1):19835

Wadsworth S, Sin D, Dorscheid D (2011) Clinical update on the use of biomarkers of airway inflammation in the management of asthma. J Asthma Allergy 4:77–86

Zhu A, Ge D, Zhang J, Teng Y, Yuan C, Huang M, Adcock IM, Barnes PJ, Yao X (2014) Sputum myeloperoxidase in chronic obstructive pulmonary disease. Eur J Med Res 19:12

Zou Y, Bui TT, Selvarajoo K (2019) ABioTrans: a biostatistical tool for Transcriptomics analysis. Front Genet 10:499

Obstructive Airway Diseases 2

Abstract

Asthma, COPD (including chronic bronchitis and emphysema), and bronchiectasis are the commonly encountered obstructive airway diseases. Symptomatic similarity across the different obstructive airway diseases makes disease diagnosis and effective management difficult. Spirometry and imaging are widely used for disease diagnosis. The mainstay therapy for obstructive airway disorders includes bronchodilators and anti-inflammatory medications such as corticosteroids that can exert significant side effects in dose-dependent manner upon prolonged usage. Prudent use of modern omics technologies can enable identification of important biomarkers for efficient disease endotyping and for subsequent development of precision medicine for obstructive airway diseases.

Keywords

Obstructive airway diseases · Airflow obstruction · Dyspnea · Pulmonary Function Tests (PFTs) · Imaging · Precision medicine

2.1 Introduction

Obstructive pulmonary diseases can be described as a group of respiratory disorders, characterized by obstruction of airways and the presence of airflow limitation. These diseases are associated with defective pulmonary function, as detected by spirometry or pulmonary function tests (PFTs) (Drummond 2014). These obstructive diseases have a high global prevalence and exert considerable impact on morbidity and mortality. The most common types of obstructive pulmonary disorders include chronic obstructive pulmonary disease (COPD), asthma, chronic bronchitis, emphysema, and bronchiectasis (Webb 1997). Sometimes, cystic fibrosis is also included in

S. Saha et al., *Pulmonomics: Omics Approaches for Understanding Pulmonary Diseases*, https://doi.org/10.1007/978-981-99-3505-5_2

this group (Restrepo 2007). However, cystic fibrosis should be treated as infectious disease on the basis of its pathophysiology (Sibley et al. 2006).

2.2 Clinical Features

Increased resistance to airflow in obstructive lung diseases may be due to one or more of the following factors:

(i) Inflammation and swelling of the airways
(ii) Excessive mucous secretion in the airways and mucous accumulation in the bronchial lumen
(iii) Contraction of the bronchial smooth muscle cells
(iv) Destruction of lung parenchyma and walls of the alveolar air sacs

Each of these conditions leads to consequent narrowing of the airways and increased airflow obstruction (West 2013).

Exposure to environmental irritants, occupational hazards, and genetic predisposition often serves as common risk factors for obstructive pulmonary diseases (Osadnik and Singh 2019). Active smoking, secondhand exposure to smoke, dust, allergens, infectious microorganisms, industrial fumes, and chemicals are the common environmental triggers leading to increased mucous production and airflow obstruction in obstructive lung diseases (Herzog 1989).

Shortness of breath or dyspnea is the primary symptom of obstructive pulmonary disorder. Other symptoms include wheezing, coughing, excessive mucus secretion, and chest tightness. Individuals with obstructive airway disorders are also more prone to respiratory infections. Mild symptoms are noted at an initial stage of obstructive disorder, these symptoms worsen and become more severe with subsequent disease progression (https://www.medicalnewstoday.com/articles/324406).

2.3 Different Types of Obstructive Pulmonary Diseases

Obstructive pulmonary diseases can be of different types, namely chronic bronchitis, emphysema, chronic obstructive pulmonary disease (COPD), asthma, bronchiectasis, and bronchiolitis obliterans (Garibaldi et al. 2012). The major regions of the respiratory tract affected and the symptoms for each of these diseases have been tabulated in Table 2.1.

Table 2.1 Different types of obstructive pulmonary disorders

Obstructive pulmonary diseases	Major affected regions	Symptoms
Chronic bronchitis	Conducting airways (Herzog 1989)	Shortness of breath, chronic cough, phlegm (Mejza et al. 2017)
Emphysema	Pulmonary parenchyma and elastic connective tissues (Herzog 1989)	Shortness of breath, cough with or without sputum (https://www.ncbi.nlm.nih.gov/books/NBK482217/)
COPD	Airways, pulmonary vasculature, and lung parenchyma (https://www.ncbi.nlm.nih.gov/books/NBK559281/)	Shortness of breath, cough, sputum production, wheezing, chest tightness, chest congestion (Miravitlles and Ribera 2017)
Asthma	Conducting airways (Herzog 1989)	Shortness of breath, wheezing, coughing, chest tightness (Papi et al. 2018)
Bronchiectasis	Airway dilation and scarring (Smith 2017)	Shortness of breath, wheezing, cough with mucopurulent sputum, chest pain (Smith 2017)
Bronchiolitis	Small airways (https://www.ncbi.nlm.nih.gov/books/NBK441865/)	Shortness of breath, wheezing, dry cough, fatigue (https://rarediseases.info.nih.gov/diseases/9551/bronchiolitis-obliterans)

This table enlists the different obstructive pulmonary diseases along with the major parts of respiratory tract affected and the major symptoms

2.4 Diagnosis of Obstructive Pulmonary Diseases

Symptomatic overlap among the different obstructive pulmonary diseases (as evident from Table 2.1) makes diagnosis difficult on the basis of disease symptoms alone. The standard diagnostic tools used for detection of different obstructive pulmonary diseases include pulmonary function tests and imaging tests (Webb 1997).

Pulmonary Function Tests (PFTs)

Spirometry is the most common pulmonary function test used for detecting the different obstructive pulmonary diseases and distinguishing them from restrictive pulmonary diseases. This simple, non-invasive technique also helps in distinguishing between the different types of obstructive airway diseases and determining the degree of disease severity. Spirometry assesses lung function and response to bronchodilators based on important parameters, namely forced vital capacity (FVC), forced expiratory volume in the first second (FEV1) and FEV1/FVC ratio (Rivero-Yeverino 2019). Forced vital capacity (FVC) is defined as the amount of air forcibly exhaled after a maximum inspiration to the total lung capacity

Table 2.2 Changes in important lung function parameters in obstructive pulmonary disorders

Sl. no.	Important lung function parameters	Trends in obstructive pulmonary diseases
1.	Forced vital capacity (FVC)	Normal or reduced
2.	Forced expiratory volume in 1 second (FEV1)	Decreased
3.	FEV1/FVC ratio	Reduced
4.	Post bronchodilator FEV1	Increase ≥200 mL and ≥ 12% from baseline—reversible airflow obstruction (asthma) Increase <200 mL or < 12%—irreversible airflow obstruction (COPD)
5.	Diffusing capacity of lung for carbon monoxide (D_{LCO})	Reduced—Emphysema Normal or increased—Asthma
6.	Total lung capacity (TLC)	Normal or increased
7.	Functional residual capacity (FRC)	Increased (e.g., emphysema)
8.	Residual volume (RV)	Increased

This table enlists the trends of vital lung function parameters in obstructive pulmonary disorders, as observed from pulmonary function tests (PFTs)

(TLC), only leaving the residual volume (RV). Forced expiratory volume in 1 s or FEV1, on the other hand, is the measure of the maximum amount of air expelled forcibly in the first second. The ratio of FEV1 to FVC (FEV1/FVC), also referred as the Tiffeneau-Pinelli index, is an important indicator of airflow obstruction (Dempsey and Scanlon 2018). In case of obstructive pulmonary disorders, FEV1 and FEV1/FVC ratio undergoes reduction, whereas FVC is unaffected or relatively less affected in patients with air trapping, as mentioned in Table 2.2. The extent of reduction in FEV1 determines the severity of obstructive pulmonary diseases. Spirometry also helps to assess the reversibility of airflow obstruction in response to bronchodilators like albuterol. A post-bronchodilator increase in FEV1 by greater than or equal to 200 mL and greater than or equal to 12% from baseline denotes reversibility of airflow obstruction and helps in distinguishing asthma, a disease of reversible airflow obstruction from an usually relatively irreversible (poorly reversible) airflow obstruction as COPD (Ye et al. 2018; Agmy et al. 2021). The graphical representation of the spirometry results in the form of a flow volume curve or flow volume loop (FVL) and time volume curve depicts the presence of abnormalities of airflow (obstruction) in the airways. It also provides an idea of the airflow in small airways and a measure of the quality, reproducibility, and acceptability of the test conducted (Heckman and O'Connor 2015). PulmoPred, a machine learning-based web application, has been designed to accurately predict obstructive airway diseases (like COPD) and non-obstructive pulmonary diseases (like DPLD) based on spirometry data (Bhattacharjee et al. 2022).

The second most common type of PFT aims to determine the diffusing capacity of lung for carbon monoxide (D_{LCO}). $D_{LCO,}$ also known as transfer factor for carbon monoxide (T_{LCO}) determines the efficiency of gaseous exchange (oxygen and carbon dioxide) between the lung alveoli and the bloodstream. D_{LCO} gets adversely altered with reduction in lung volume or in presence of pulmonary parenchymal disorders. D_{LCO} is reduced in some obstructive pulmonary diseases like emphysema due to alveolar destruction and reduction in effective surface area for gaseous exchange. D_{LCO} either remains normal or gets increased in case of asthma due to increased lung volumes. The measurement of D_{LCO} also includes a collateral measurement of the lung volumes and a relative increase in the approximate residual volume (RV) alone as well as in relation to reduced approximate TLC (total lung capacity) ratio. This gives an indirect idea of air trapping in chronic long-standing asthma and advanced COPD (Tiwari et al. 2017; Shin et al. 2015). The extent of change in D_{LCO} is a measure of disease severity and denotes increased morbidity and mortality in patients (Enright 2016; https://www.ncbi.nlm.nih.gov/books/NBK556149/).

Another important PFT is lung plethysmography, which is a gold standard for accurate measurements of lung volumes in terms of total lung capacity (TLC), functional residual capacity (FRC), and residual volume (RV). TLC is the total volume of air in the lungs after maximal inspiration. FRC provides a measure of the amount of air in the lungs after normal and passive expiration, while RV is the amount of air present in the lungs after maximal exhalation (Heckman and O'Connor 2015). Obstructive respiratory diseases are characterized by increased TLC, increased FRC, increased RV or RV/TLC ratio due to hyperinflation or air trapping in lungs. However, use of plethysmography for lung volume estimation in obstructive respiratory diseases is largely limited, since, most of the clinical diagnosis is made on the basis of spirometry and imaging studies (Dempsey and Scanlon 2018).

6-min Walk Test & 2-Chair Test

Several other tests, namely 6-min walk test, stair climbing, shuttle-walk test, are also employed for assessing functional capacity or aerobic exercise capacity in patients with obstructive pulmonary disorders and other cardiopulmonary diseases (Laboratories ATSCoPSfCPF 2002; Hernandes et al. 2011; Troosters et al. 2002). 6-min walking test (6MWT) is an inexpensive, simple test that gives a global picture of the cumulative response of pulmonary and cardiovascular system during exercise in terms of the walking distance. 6MWT serves as a predictor of morbidity and mortality in patients with obstructive respiratory diseases like COPD. Two chair test (2CT) has evolved as an effective and reproducible alternative of 6MWT for monitoring the status of cardiopulmonary recovery response after a small but defined exercise. 2CT provides a measure of the changes in pulse rate and oxygen saturation (SpO2) after exercise. Prospect of 2CT has been assessed for COPD and pulmonary hypertension, but, it requires further validation in other respiratory diseases (Bhattacharyya et al. 2020).

Imaging Techniques

Chest X-ray radiography and computed tomography (CT) scan of the chest are the most commonly explored imaging techniques for pulmonary diseases. Chest X-ray is an inexpensive technique associated with minimal exposure to radiation (Washko 2010). Chest radiography finds application in diagnosing abnormalities (like tracheal neoplasm) and complications (like pneumonia) or pneumothorax in patients with obstructive pulmonary diseases (Webb 1997). Computed Tomography (CT) scans provide measures of airway diseases in terms of mural thickening and changes in the airway lumen. High resolution CT (HRCT) scan provides well resolved structural details for better detection of morphologic abnormalities associated with obstructive pulmonary diseases. It makes the diagnosis of emphysema and bronchiectasis, helps to assess the presence of small airway diseases and air trapping, allows the measurement of the bronchial wall thickening, identifies the presence of complications as pneumothorax and pulmonary hypertension, and can help measurement of total lung capacity (Labaki et al. 2017). The use of Magnetic Resonance Imaging (MRI) for respiratory diseases is largely limited by considerably high air to lung tissue ratio. A relatively newer imaging technique used for pulmonary diseases is Optical Coherence Tomography (OCT). OCT provides details of mural structures at a resolution lesser than 5 μm. Clinical application of OCT has revealed details of airway mural remodelling in COPD patients (Washko 2010).

Thus, these different pulmonary function tests together with the imaging tools are of tremendous importance in the clinical diagnosis of airflow obstruction.

2.5 Treatment of Obstructive Pulmonary Diseases

The most common medications used for obstructive pulmonary diseases are bronchodilators. Bronchodilators reduce airflow limitation and bronchial obstruction by relaxing airway smooth muscles. Bronchodilators can be short acting or long acting. Presently, three classes of bronchodilators are available for use—beta 2 agonists, antimuscarinic or anticholinergic agents, and methylxanthines (Cazzola and Matera 2014). Inhaled corticosteroids (ICS) and systemic corticosteroids are widely used to control airway inflammation associated with obstructive respiratory diseases. Other medications that may be used along with bronchodilators and corticosteroids include phosphodiesterase-4 (PDE4) inhibitors, mucolytics, macrolides, mast cell stabilizers, and leukotriene antagonists (Barnes 2006). Oxygen supplemental therapy and pulmonary rehabilitation may be given to patients with severe obstructive pulmonary diseases (van Haarst et al. 2019). Although these medications improve the quality of life in most cases, they might not be equally effective for all disease phenotypes. Besides, prolonged usage of these drugs are often associated with local and systemic side effects (Barnes 2010).

2.6 Dissecting Different Dimensions of Obstructive Pulmonary Diseases: A Step Toward Precision Medicine

The pace of therapeutic innovation needs to be expedited to overcome the limitations associated with the present day treatment strategies available for obstructive pulmonary diseases. The revolutionizing concept of precision medicine was introduced in 2008 to enhance the overall efficacy and minimize the side effects of therapeutics (Ginsburg and Phillips 2018). The term precision medicine may be defined as targeted treatment approach based on genetic variability, molecular biomarkers, phenotypic signatures, environmental exposures, and lifestyle habits that allow classification of patients with similar clinical presentation into distinct groups or endotypes. Such endotyping improves therapeutic responses and reduces unnecessary side effects (Leung et al. 2019; Khan et al. 2022). The development of precision medicine for obstructive respiratory diseases is largely hindered by dearth of adequate disease endotyping based on molecular, cellular, and phenotypic heterogeneity among patients (Tejera et al. 2021). Biomarkers, especially, predictive, response, and prognostic biomarkers are of prime importance in designing precision medicine (Leung et al. 2019). Dissecting different dimensions or levels (namely molecular, cellular, tissue, organ, and patient phenotypes) for each obstructive pulmonary diseases holds the key to biomarker identification as a step toward development of precision medicine (as shown in Fig. 2.1). The advent of omics approaches is of prime importance in biomarker identification and evolution of precision medicine.

2.7 Conclusion

Precision medicine depends on integrated application of different advanced technologies for better understanding of disease biology, precise diagnosis, classification, and treatment of diseases. The ambitious target of developing precision medicine for obstructive airway diseases is in its naïve stage. The marked improvement in electronic medical records systems has enabled efficient collection of medical data and subsequent correlation of these data with biological findings. Such advances in clinical information technology have provided enormous technical support for accelerating the process of development of precision medicine (Liu et al. 2019; Goetz and Schork 2018). Besides, recent advancement in the field of large-scale omics techniques like genomics, epigenomics, transcriptomics, proteomics, lipidomics, exposomics, metabolomics, metagenomics, radiomics, and bioinformatics enable rapid collection and analysis of high-throughput data for efficient identification of biomarkers of obstructive airway diseases. Such biomarkers identified by modern, state of art omics approaches may pave the way for a future era of precision medicine for obstructive pulmonary diseases.

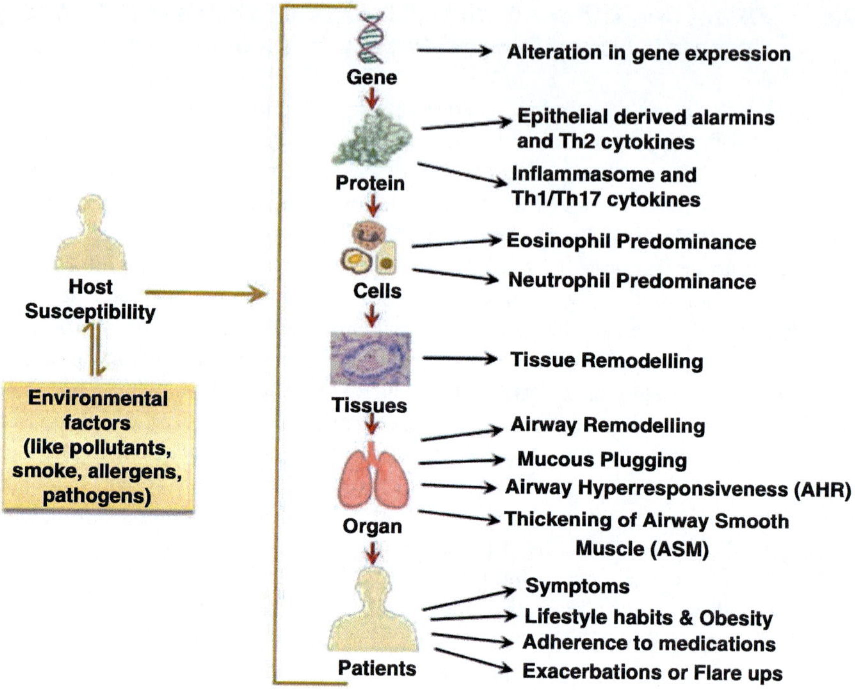

Fig. 2.1 Dissecting and targeting the different dimensions of obstructive airway diseases—a step toward precision medicine

References

Agmy GA, Mahmoud M, ABE A, Adam M (2021) Diagnostic performance of lung volumes in assessment of reversibility in chronic obstructive pulmonary disease. Egypt J Bronchol:15

Barnes PJ (2006) Drugs for asthma. Br J Pharmacol 147(Suppl 1):S297–S303

Barnes PJ (2010) Inhaled corticosteroids. Pharmaceuticals (Basel) 3(3):514–540

Bhattacharjee S, Saha B, Bhattacharyya P, Saha S (2022) Classification of obstructive and non-obstructive pulmonary diseases on the basis of spirometry using machine learning techniques. J Comput Sci:63

Bhattacharyya P, Saha D, Paul M, Ganguly D, Mukherjee B, Roy Chowdhury S, RoyChoudhury S, Agarwal P, Halder I, Ghosh Roy D et al (2020) Two chair test: a substitute of 6 min walk test appear cardiopulmonary reserve specific. BMJ Open Respir Res 7(1)

Cazzola M, Matera MG (2014) Bronchodilators: current and future. Clin Chest Med 35(1):191–201

Dempsey TM, Scanlon PD (2018) Pulmonary function tests for the generalist: a brief review. Mayo Clin Proc 93(6):763–771

Drummond MB (2014) Obstructive airway disease in urban populations. Curr Opin Pulm Med 20(2):180–185

Enright MDP (2016) Office-based DLCO tests help pulmonologists to make important clinical decisions. Respir Investig 54(5):305–311

Garibaldi BT, Illei P, Danoff SK (2012) Bronchiolitis. Immunol Allergy Clin N Am 32(4):601–619

Ginsburg GS, Phillips KA (2018) Precision medicine: from science to value. Health Aff (Millwood) 37(5):694–701
Goetz LH, Schork NJ (2018) Personalized medicine: motivation, challenges, and progress. Fertil Steril 109(6):952–963
Heckman EJ, O'Connor GT (2015) Pulmonary function tests for diagnosing lung disease. JAMA 313(22):2278–2279
Hernandes NA, Wouters EF, Meijer K, Annegarn J, Pitta F, Spruit MA (2011) Reproducibility of 6-minute walking test in patients with COPD. Eur Respir J 38(2):261–267
Herzog H (1989) Obstructive lung disease. Arzneimittelforschung 39(8A):961–971. discussion 971–962
Khan M, Cole AG, Mannino DM (2022) Precision medicine in chronic obstructive pulmonary disease: how far have we come? Curr Opin Pulm Med 28(2):115–120
Labaki WW, Martinez CH, Martinez FJ, Galban CJ, Ross BD, Washko GR, Barr RG, Regan EA, Coxson HO, Hoffman EA et al (2017) The role of chest computed tomography in the evaluation and management of the Patient with chronic obstructive pulmonary disease. Am J Respir Crit Care Med 196(11):1372–1379
Laboratories ATSCoPSfCPF (2002) ATS statement: guidelines for the six-minute walk test. Am J Respir Crit Care Med 166(1):111–117
Leung JM, Obeidat M, Sadatsafavi M, Sin DD (2019) Introduction to precision medicine in COPD. Eur Respir J 53(4)
Liu X, Luo X, Jiang C, Zhao H (2019) Difficulties and challenges in the development of precision medicine. Clin Genet 95(5):569–574
Mejza F, Gnatiuc L, Buist AS, Vollmer WM, Lamprecht B, Obaseki DO, Nastalek P, Nizankowska-Mogilnicka E, Burney PGJ, collaborators B et al (2017) Prevalence and burden of chronic bronchitis symptoms: results from the BOLD study. Eur Respir J 50(5)
Miravitlles M, Ribera A (2017) Understanding the impact of symptoms on the burden of COPD. Respir Res 18(1):67
Osadnik CR, Singh S (2019) Pulmonary rehabilitation for obstructive lung disease. Respirology 24(9):871–878
Papi A, Brightling C, Pedersen SE, Reddel HK (2018) Asthma. Lancet 391(10122):783–800
Restrepo RD (2007) Inhaled adrenergics and anticholinergics in obstructive lung disease: do they enhance mucociliary clearance? Respir Care 52(9):1159–1173. discussion 1173–1155
Rivero-Yeverino D (2019) Spirometry: basic concepts. Rev Alerg Mex 66(1):76–84
Shin TR, Oh YM, Park JH, Lee KS, Oh S, Kang DR, Sheen S, Seo JB, Yoo KH, Lee JH et al (2015) The prognostic value of residual volume/Total lung capacity in patients with chronic obstructive pulmonary disease. J Korean Med Sci 30(10):1459–1465
Sibley CD, Rabin H, Surette MG (2006) Cystic fibrosis: a polymicrobial infectious disease. Future Microbiol 1(1):53–61
Smith MP (2017) Diagnosis and management of bronchiectasis. CMAJ 189(24):E828–E835
Tejera P, Jardim J, Laucho-Contreras ME (2021) Editorial: precision medicine in pulmonary diseases-same principles. New Approach Front Med (Lausanne) 8:821013
Tiwari A, Rahman K, Abejie B, Jain VV, Vempilly JJ (2017) Longer duration of asthma is significantly associated with increased RV/TLC ratio. Respir Med 124:44–48
Troosters T, Vilaro J, Rabinovich R, Casas A, Barbera JA, Rodriguez-Roisin R, Roca J (2002) Physiological responses to the 6-min walk test in patients with chronic obstructive pulmonary disease. Eur Respir J 20(3):564–569
van Haarst A, McGarvey L, Paglialunga S (2019) Review of drug development guidance to treat chronic obstructive pulmonary disease: US and EU perspectives. Clin Pharmacol Ther 106(6):1222–1235
Washko GR (2010) Diagnostic imaging in COPD. Semin Respir Crit Care Med 31(3):276–285

Webb WR (1997) Radiology of obstructive pulmonary disease. AJR Am J Roentgenol 169(3): 637–647
West JB (2013) Pulmonary pathophysiology: the essentials
Ye Q, Liao A, D'Urzo A (2018) FEV1 reversibility for asthma diagnosis: a critical evaluation. Expert Rev Respir Med 12(4):265–267

Check for updates

Chronic Obstructive Pulmonary Disease (COPD) 3

Abstract

Effective COPD diagnosis and disease stratification is largely hindered due to presence of symptomatic similarity with other lung diseases and presence of comorbidities. Several probable biomarker candidates have been identified from different biological samples using various omics tool to alleviate such challenges. Alteration in host metabolomic profile and microbial dysbiosis have also been elaborated for COPD. Integration of multi-omics datasets using computational tools has been explored to enable sub-phenotyping of COPD and to pave the way for identification of novel drug targets. Although present therapeutics are mainly restricted to bronchodilators, steroids, and mucolytics, several biologics and novel small molecules that are being clinically investigated for their efficacy in COPD have been highlighted. Finally, the scientific gaps and future research areas of COPD to be explored using multi-omics approaches have been addressed.

Keywords

Irreversible airflow obstruction · Bronchodilators · Pink puffer · Blue-bloater · GOLD classification · Smokers · Comorbidities

3.1 Introduction

Chronic Obstructive Pulmonary Disease (COPD) is characterized by chronic airway inflammation and airflow limitation that is poorly or partially reversible compared to that of asthma. This chronic, progressive inflammatory airway disorder has long been thought comprise of two different conditions with or without overlaps, namely chronic bronchitis (inflammation of small airways with hyper-secretion) and emphysema (degeneration of lung parenchyma and alveolar membranes), leading to

S. Saha et al., *Pulmonomics: Omics Approaches for Understanding Pulmonary Diseases*, https://doi.org/10.1007/978-981-99-3505-5_3

reduced lung function (Wang et al. 2020a). The major symptoms of COPD include shortness of breath, cough, sputum production, chest tightness, and wheezing (Canas et al. 2020). The symptomatic course of COPD can be superimposed with recurrent episodes of acute exacerbations characterized by increase in symptoms demanding extra-medications than usual. These events of acute exacerbations can lead to rapid decline in lung function. COPD is one of the leading causes of morbidity and mortality all over the world (Christenson et al. 2022).

3.2 Clinical Features

Two major clinical phenotypes of COPD have been recognized as (i) Type A or "pink puffer" phenotype and (ii) Type B or "blue-bloater" phenotype. Phenotype A or a pink puffer is characterized by dyspnoea as the primary symptom and radiological evidences of emphysema. They have active respiratory centre to stimulate breathing or ventilation in response to hypoxemia or low oxygen content in blood. These pink puffer patients have increased lung volumes and impaired diffusing capacity for carbon monoxide (D_{LCO}). Type B, on the other hand, is mostly associated with chronic inflammation and increased mucus secretion with a lazy respiratory centre to respond to hypoxemia. Therefore, instead of dyspnoea, they have usually profuse expectoration and relatively higher risk of retention of carbon dioxide. Lung volumes and D_{LCO} remain unaltered in chronic bronchitis. Oxyhaemoglobin desaturation together with erythrocytosis leads to cyanosis in patients, a condition often referred to as blue-bloater (Candela et al. 2019). Airflow obstruction in COPD may be contributed by dynamic obstruction in COPD of emphysematous variety. However, both the conditions frequently overlap, and the relative contribution of emphysema or chronic bronchitis in airflow obstruction in COPD is difficult to determine under clinical setup.

In general, COPD can be diagnosed by different pulmonary function tests and imaging techniques. Irreversible airflow obstruction in COPD is primarily diagnosed by spirometry and demonstrated by the presence of post bronchodilator FEV1/FVC of less than 70%. The extent of airflow obstruction and severity of COPD are measured in terms of predicted FEV1 (%). The Global Initiative for Chronic Obstructive Lung Disease (GOLD) has classified COPD into four stages or grades, based on predicted FEV1 reflecting the degree or severity of airflow limitations in spirometry analysis. Stage 1 to Stage 4 is associated with a progressive reduction in FEV1 (predicted) and a gradual increase in symptomatic severity, as highlighted in Fig. 3.1 (Manian 2019; Rabe et al. 2007). The American Thoracic Society/European Respiratory Society (ATS/ERS) has also stratified COPD into five stages of disease severity based on predicted FEV1 values, as shown in Fig. 3.1 (Burkhardt and Pankow 2014).

Imaging techniques like CT and HRCT scanning find enormous application in diagnosing and predicting disease prognosis in COPD. Decreased airway count observed in computed tomography (CT) is significantly linked to decline in lung function in COPD patients (Kirby et al. 2018). High-resolution computed

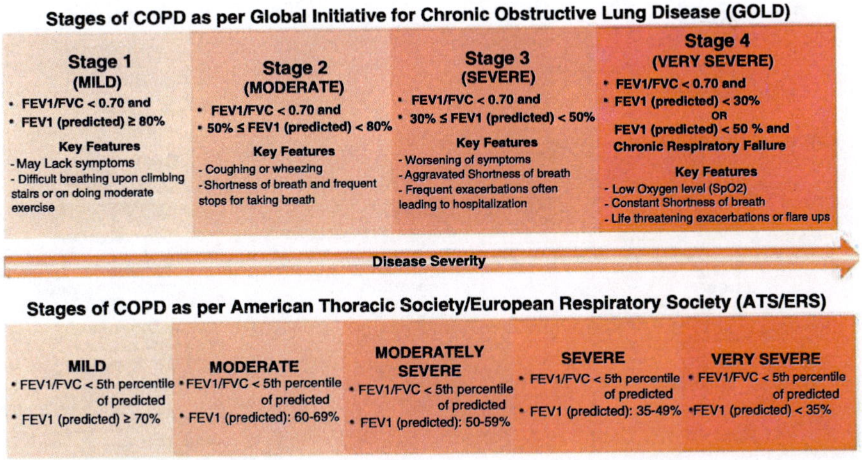

Fig. 3.1 Different stages of COPD based on GOLD standards and ATS/ERS criteria. COPD has been classified into different stages based on the degree of disease severity and airflow obstruction. GOLD standards and ATS/ERS criteria have been obtained from (Manian 2019; Rabe et al. 2007)

tomography (HRCT), on the other hand, enables COPD phenotyping based on thickness of bronchial wall and percentage of emphysema. The emphysema predominant phenotype is characterized by less than 1.75 mm thickness of bronchial wall and greater than 35% of emphysema, whereas the bronchitis predominant phenotype has lesser than 35% emphysema and greater than 1.75 mm thickness of bronchial wall. Increased thickness of bronchial wall and greater emphysema detected by HRCT are related to increased incidences of acute exacerbations in COPD patients (Han et al. 2011).

The changes in lung parenchyma and thickness of bronchial airways observed in COPD can be triggered by multiple predisposing risk factors (as shown in Fig. 3.2). The most important aetiological factor for emphysema and COPD is smoking. Smoking leads to a gradual decline in lung function. Smokers have higher mortality rate from COPD as compared to non-smokers (Hogea et al. 2020). Apart from cigarette or tobacco smoke, other common predisposing factors include biomass fuel smoke, chronic asthma, particulate matter pollutants, dust particles, occupational exposures, and exposure to second-hand smoke (Rabe and Watz 2017). Aging and physical inactivity are also associated with progressive deterioration in lung function parameters. Genetic factors like alpha1-antitrypsin deficiency are also associated with an increased risk of COPD. Childhood respiratory infections may eventually lead to COPD in adulthood (Hayden et al. 2015). Besides, COPD patients are also more prone to microbial infections. Acute respiratory infections in COPD patients are associated with rapid decline in lung function, frequent acute exacerbations, and increased mortality (Silverman and Speizer 1996). Presence of multiple comorbidities, referred to as multi-morbidity is common in COPD patients (Raherison et al. 2018). COPD increases the chance of developing lung cancer and

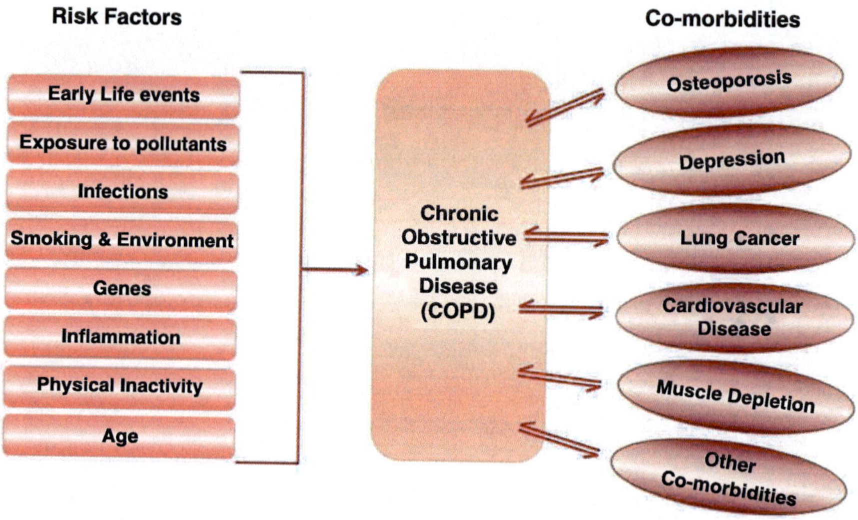

Fig. 3.2 Association of risk factors and comorbidities with COPD. The various risk factors are vital for predisposition to COPD. COPD, on the other hand, increases the chance of developing various other chronic diseases. The presence of other comorbidities alongside COPD further deteriorates the quality of life and affects the morbidity and mortality of COPD

secondary pulmonary hypertension. Common extra-pulmonary comorbidities associated with COPD include coronary artery diseases, hypertension, osteoporosis, diabetes mellitus, muscle weakness, sleep disorders, and depression (Balbirsingh et al. 2022; Chatila et al. 2008). COPD increases the risk of developing one or more of the above-mentioned comorbidities. Likewise, the presence of comorbid conditions makes diagnosis and management of COPD difficult. Such comorbidities often lead to deterioration in the quality of life, increase in disease severity, frequent hospitalizations, and high mortality rate in COPD patients (Cavailles et al. 2013).

3.3 Biomarkers of COPD

Effective diagnosis and management of COPD depend on biomarker discovery. Biomarkers shall not only enable efficient disease diagnosis, but shall also aid in distinguishing between the different disease phenotypes, stratifying the disease based on degree of severity and monitoring COPD prognosis (Ambade et al. 2015; Zemans et al. 2017). Omics approaches serve as promising tools for measuring systemic biomarkers and correlating them with COPD pathophysiology. Some of the biomarkers of COPD identified by different omics tools have been tabulated in Table 3.1.

Table 3.1 Different biomarkers of COPD

Omics approach	Biomarkers	Biological samples	Comments
Genomics	Alpha-1-antitrypsin (AAT)	Serum	Alpha-1-antitrypsin deficiency (AATD), caused by common AAT polymorphism [PiZZ (Glu342Lys) homozygous condition] speeds up emphysema onset and serves as a genetic risk factor for COPD (Green et al. 2015; Blanco et al. 2017; Denden et al. 2010)
	Soluble receptor for advanced glycation end products (sRAGE)		A SNP in AGER encoding for sRAGE (rs2070600) is associated with lower plasma levels of sRAGE in COPD (Regan et al. 2019)
	FGFR2	Blood	A allele of rs2420915 located near FGFR2 increased the risk of developing COPD in Chinese population rs1907240 and rs2257129 increased the risk of COPD in non-smokers (Li et al. 2021)
	MGAT5		rs6430491 near MGAT5 was linked with low risk of developing COPD G allele of rs2593704 of MGAT5 was associated with less risk of COPD (Li et al. 2021)
Transcriptomics	RP9, NAPE-PLD	Peripheral blood, lung tissues	Decreased in COPD patients (Jain 2017)
	miR-20a, miR-28-3p, miR-34c-5p, miR-100, miR-7	Serum	Decreased expression of miR-20a, miR-28-3p, miR-34c-5p, and miR-100, along with increased expression of miR-7 was noted in COPD. These miRNAs can aid in diagnosing COPD (Akbas et al. 2012)
	miR-181a, miR-21		miR-181a significantly reduced while miR-21 significantly increased in COPD patients (Xie et al. 2014)

(continued)

Table 3.1 Different biomarkers of COPD (continued)

Omics approach	Biomarkers	Biological samples	Comments
	miR-125b	Plasma	Upregulation of this miRNA is associated with inflammation, severity and acute exacerbations in COPD (Hu et al. 2017)
	miR-181c	Lung Tissues	Downregulation of this miRNA is related to COPD development (Du et al. 2017)
	miR-223, miR-15b, miR1274a		Upregulation of these miRNAs is associated with COPD pathogenesis (Ezzie et al. 2012)
	miR-183		Related to pathogenesis and severity of COPD (Cao et al. 2014)
	miR-218-5p		Significant reduction is associated with COPD pathogenesis (Conickx et al. 2017)
Epigenomics	PTEN, NFE2L2 (Nrf2)	Bronchial brushings (Bronchial small airway epithelial cells)	Hypermethylation of PTEN and NFE2L2 (Nrf2) genes is linked with inflammatory response and airway remodelling in COPD (Zhang et al. 2021a)
	CDKN2A, MGMT	Sputum	Hypermethylation could serve as early diagnostic tool for COPD (He et al. 2020)
	P16, GATA4		Methylation of P16 and GATA4 promoters was assumed as early diagnostic biomarkers for COPD (He et al. 2020)
	HDAC6	Lung tissues	Hypomethylation of HDAC6 is associated with cigarette smoke induced in epithelial dysfunction in COPD patients (Lam et al. 2020)
	NOS1AP, BID, GABRB1		Hypermethylation serves as epigenetic signatures of COPD patients and smokers (Sundar et al. 2017)

(continued)

Table 3.1 Different biomarkers of COPD (continued)

Omics approach	Biomarkers	Biological samples	Comments
Proteomics	Fibrinogen	Blood	Increased fibrinogen expression is associated with increased risk of exacerbations and increased mortality rate in COPD patients (Fermont et al. 2019)
	IL-6		Elevated IL-6 level is associated with increased incidences of hospitalization in COPD patients (Fermont et al. 2019)
	Soluble receptor for advanced glycation end products (sRAGE)		Reduced sRAGE levels are associated with reduced lung function (FEV1) and acute COPD exacerbations (Pratte et al. 2021)
	Alpha-1-antitrypsin (AAT)	Serum	Deficiency is associated with increased risk of COPD (Jain 2017)
	Copeptin, the precursor of vasopressin		Serves as a prognostic biomarker. Indicates poor prognosis in exacerbation of hospitalized COPD patients (Stolz et al. 2007)
	C-Reactive Protein (CRP)		Increased in acute COPD exacerbations and decline in lung health (Jain 2017; Fermont et al. 2019)
	Procalcitonin (PCT)		Can distinguish COPD exacerbations caused by bacterial infections from exacerbations caused by other factors. Can guide antibiotic administration and dosage in COPD exacerbations caused by bacterial infections (Mathioudakis et al. 2017)
	Surfactant Protein D (SP-D)		Elevated in COPD patients. Serves as prognostic biomarker of disease severity and COPD exacerbation rate (Moon et al. 2018)

(continued)

Table 3.1 Different biomarkers of COPD (continued)

Omics approach	Biomarkers	Biological samples	Comments
	Immunoglobulin G (IgG)		Reduced serum IgG is associated with acute COPD exacerbations and hospitalization (Leitao Filho et al. 2020)
	VEGF		Increased in mild to severe COPD patients (Farid Hosseini et al. 2014)
	Parathyroid hormone (PTH)		Serves as a COPD biomarker. High levels predict COPD exacerbations (Amado et al. 2021)
	Osteoprotegerin (OPG)		Elevated in COPD (Jain 2017)
	Serum amyloid A (SAA)		Associated with COPD exacerbations caused by respiratory tract infections (Jain 2017)
	Pro-forms of collagen type III (Pro-C3) and type VI (Pro-C6)		Low levels of Pro-C3 and Pro-C6 are associated with air trapping, severe airflow limitation, and emphysema in COPD patients (Jain 2017)
	High-sensitivity troponin I (hs-TnI)		hs-TnI is a predictor of all-cause mortality in patients with stable COPD (Waschki et al. 2020)
	PIGF	Serum, Broncho alveolar Lavage fluid (BALF)	Overexpression of PIGF serves as a COPD biomarker (Jain 2017)
	IL-8	Broncho alveolar Lavage fluid (BALF)	Elevated in COPD patients compared to asthmatics (Nocker et al. 1996)
	Neutrophil elastase (NE)	Sputum	Increased sputum NE level showed positive correlation with severity of acute COPD exacerbations (Moon et al. 2018)
	Neutrophil gelatinase associated lipocalin (NGAL)		Helps in stratifying ACOS, COPD, and asthma (Moon et al. 2018)
	Lipocalin 1 (Lp1)		Reduced in COPD Correlates with FEV1/FVC significantly (Moon et al. 2018)

(continued)

Table 3.1 Different biomarkers of COPD (continued)

Omics approach	Biomarkers	Biological samples	Comments
	BKCαβ1	Lung Tissues	BKCαβ1 downregulation is associated with COPD severity (Cao et al. 2014)
Metabolomics	GlycA	Blood	Elevated levels of GlycA are associated with chronic inflammation. Elevated blood level of GlycA serves as a biomarker of COPD inflammatory response even before the clinical onset of COPD symptoms (Prokic et al. 2020)
	Uric acid	Serum	Elevated uric acid level serves as a biomarker of early mortality in COPD patients (Jain 2017)
	Vitamin D	Serum	Hypovitaminosis D in COPD patients is associated with greater risk of exacerbations and hospitalizations (Amado et al. 2021)
	L-homocysteic acid, octadecanoylcarnitine, n-undecanoylglycine, LysoPE (lysophosphatidyl ethanolamine) (18:0), LysoPC (lysophosphatidyl choline) (20:4), MG (monoglycerides) (18:0), PC (phosphatidylcholine) (32:0), PC (34:0), PE (phosphatidyl ethanolamine) (40:7), PI (phosphatidylinositol) (38:5)	Blood Plasma, Broncho alveolar Lavage fluid (BALF)	Expression of these metabolites positively correlated with COPD (Regan et al. 2019)
Metagenomics	*Streptococcus* spp.	Oral	Decreased abundance of *Streptococcus* spp. noted in COPD patients (Tiew et al. 2021)
	Veillonella spp.		*Veillonella* spp. increased in COPD patients (Tiew et al. 2021)
	Haemophilus spp., *Moraxella* spp., *Streptococcus* spp.	Sputum	Increased abundance of *Haemophilus* spp., *Moraxella* spp. and decreased abundance of

(continued)

Table 3.1 Different biomarkers of COPD (continued)

Omics approach	Biomarkers	Biological samples	Comments
			Streptococcus spp. noted during acute COPD exacerbations (Wang et al. 2016)
	Haemophilus spp., *Pseudomonas* spp., *Prevotella* spp.	Broncho alveolar Lavage fluid (BALF)	Increased *Haemophilus* spp. and *Pseudomonas* spp. noted in COPD patients. Decreased *Prevotella* spp. observed in BAL of COPD patients (Tiew et al. 2021)
	Proteobacteria Bacteroidetes	Lung Biopsies	Increased Proteobacteria and decreased Bacteroidetes noted in COPD (Tiew et al. 2021)

This table enlists various important biomarkers of COPD identified using various omics approaches

3.4 Genomics

Several internationally conducted studies, including the ECLIPSE (Evaluation of COPD Longitudinally to Identify Predictive Surrogate End-points) study, the ICGN (International COPD Genetics Network) study, and the COPDGene cohort study have highlighted several genetic abnormalities and single nucleotide polymorphisms (SNPs) that are associated with increased susceptibility to COPD (Faner et al. 2014). Genome-wide association studies (GWAS) have identified significant association of several novel loci with COPD predisposition and airflow obstruction in COPD. SNPs of genes involved in inflammatory pathways, oxidative pathways, and proteases-anti protease pathway have been linked to COPD (Kumar et al. 2013). SNPs in fibroblast growth factor-7 (rs12591300 and rs4480740), surfactant protein D (rs2245121, rs911887, rs6413520, and rs721917), human hedgehog interacting protein (rs13147758), family with sequence similarity 13 member A gene (rs7671167), tumor necrosis factor (rs1800629 and rs361525), and matrix metalloproteinase-9 (rs3918242) are associated with increased susceptibility to COPD (Faner et al. 2014; Ortega-Martinez et al. 2020; Xie et al. 2015; Resendiz-Hernandez et al. 2018; Zhao et al. 2020). Several SNPs in genes like interleukin-6 (rs1800795), telomerase reverse transcriptase (rs61748181), and NLR family pyrin domain containing 1 (rs12150220) are related to impaired lung function in COPD patients (He et al. 2009; Xu et al. 2021; Ozretic et al. 2019). A few other COPD-related SNPs have been documented in Table 3.1.

3.5 Epigenomics

Global disruption of DNA methylation pattern in COPD has been assessed by several epigenomics studies. Changes in DNA methylation have also been associated with SNPs and alteration in gene expression profile in COPD. Increased methylation is associated with repressed gene expression and vice versa. Aberrant DNA methylation patterns have been identified in hundreds of genes and have been correlated with reduced lung function in COPD. Hyper-methylation of cholinergic receptors genes (CHRNB1, CHRNB2, and CHRND), glutathione S-transferase genes (GSTP1, GSTM1, and GSTT1), and other genes (GPR126, HTR4, TFF3, CKB, PTEN, CSNK2A2, GREM1, and EPHX1) have been documented in COPD. Similarly, hypo-methylation of several genes (MUC13, KSR1, SERPINA1, FUT7, DAPP1, AHRR, CXCR1, CHRM3, CCL20, CCL22, and CYP1B1) have also been noted in COPD patients (Vucic et al. 2014; He et al. 2020; Casas-Recasens et al. 2021). Some of these altered methylation patterns may serve as early diagnostic tool for COPD (as shown in Table 3.1).

3.6 Transcriptomics

Transcriptomics studies have highlighted several differentially expressed genes and miRNAs that drive the molecular machinery leading to COPD. Increased expression of hypoxia-inducible factor 1A (HIF1A) may be related to reduced lung function and poor quality of life in COPD patients (Paci et al. 2020). Likewise, heparin-binding EGF-like growth factor (HBEGF) gene expression was elevated in lung tissues of COPD patients and was related to FEV1 and D_{LCO} (Yao et al. 2019). Increased lung tissue expression of several transcripts like CYR61, TNFSF10, TGFβ-1, IL-1R, CX3CL1, and CTGF was associated with increase in COPD disease severity (Chen et al. 2008). Some of the differentially expressed miRNAs associated with COPD have been listed in Table 3.1.

3.7 Proteomics

Protein biomarkers are of prime significance as they not only aid in disease diagnosis but also may be explored as drug targets for designing novel drugs. Blood borne biomarkers like CRP, fibrinogen, interleukin-8, interleukin-6, and lung-derived biomarkers like surfactant protein-D (SP-D), club cell protein 16 (CC16), and pulmonary and activation-regulated chemokine (PARC) have been identified for COPD. CRP, fibrinogen, IL-6, and adiponectin were elevated in the serum of COPD patients. The COPD Biomarker Qualification Consortium (CBQC) has considered to evaluate plasma fibrinogen as a prognostic biomarker of COPD (Faner et al. 2014). Keratin-17 (KRT17) and Dehydrogenase/Reductase 9 (DHRS9) serve as molecular signatures for Type A emphysematous phenotype of COPD (Bai et al. 2021). Several other proteins are also dysregulated in COPD like haptoglobin, ceruloplasmin, prothrombin, serotransferrin, vitamin D binding protein, complement 4B, interleukin-33, caveolin 1, and caveolae-associated protein 1 (and those listed in Table 3.1) which needs to be investigated further to evaluate their potency as biomarker candidates or potential drug targets for COPD (Baralla et al. 2018; Zhang et al. 2021b).

3.8 Metabolomics

Alteration in metabolomics profile in COPD patients has been assessed using nuclear magnetic resonance spectroscopy and liquid chromatography with tandem mass spectrometry (LC-MS/ MS). Serum metabolic profile revealed an increase in phenylalanine, glutamine, 3-methylhistidine, and ketone bodies in COPD patients along with a concomitant decrease in *N,N*-dimethyl glycine, branched chain amino acids, and lipoproteins (Faner et al. 2014; Ubhi et al. 2012). Serum metabolic profiling also enables efficient distinction between smoke-induced COPD and tuberculosis-associated COPD (Kim et al. 2021). Exhaled breath condensate (EBC) has also been screened as a prospective source of COPD specific metabolic signatures. Adenosine and AMP levels were elevated in EBC from COPD patients (Faner et al. 2014). The pH of EBC provides a measure of endogenous airway acidification and serves as a potential biomarker of COPD disease severity (Papaioannou et al. 2011). 8-Iso-Prostaglandin F2α level in induced sputum enables prediction of COPD severity with high sensitivity and specificity (Zhu et al. 2020). Other metabolic markers of COPD identified from serum, blood plasma, and broncho alveolar lavage fluid have been listed in Table 3.1.

3.9 Metagenomics

Alterations in airway and lung microbiome are associated with COPD (as shown in Table 3.1). 16S rRNA gene sequencing of sputum samples identified *Prevotella* spp., *Veillonella* spp., *Rothia* spp., *Neisseria* spp., *Porphyromonas* spp., *Streptococcus* spp., and *Fusobacterium* spp. to be differentially abundant in severe COPD patients as compared to moderate COPD patients (Lee et al. 2016). Increased abundance of *Moraxella* spp. and *Streptococcus* spp. and decreased abundance of *Prevotella* spp. were noted in bronchial brushing samples from COPD patients. Decreased abundance of *Prevotella* spp. was associated with reduced expression of genes involved in epithelial defence. Increased abundance of *Moraxella* spp. on the other hand leads to increased expression of pro-inflammatory genes (Ramsheh et al. 2021). COPD airway microbiome may also influence the host metabolome expression pattern (Wang et al. 2020b). The complex relationship between change in microbiome, host metabolic profile, and host immune response needs to be explored further for designing novel therapeutics for COPD.

3.10 Medical Imaging or Radiomics

Medical imaging like chest X-rays and HRCT scans are not the first-line methods for the diagnosis of COPD. But these provide information to the physicians about the disease phenotyping and assessment of response to treatment (Myc et al. 2019). Recently, medical imaging is used for monitoring COPD patients since there is a link between COPD and lung cancer. Quantitative CT (QCT) is used to quantify airway abnormalities, air trappings, and texture analysis in assessing the stages of COPD (Lynch and Al-Qaisi 2013; Fernandes et al. 2016).

3.11 Bioinformatics

COPD-related transcriptomics data were compiled in PulmonDB, and it is an online freely available database (http://pulmondb.liigh.unam.mx/). The microarray and RNA-Seq COPD data can also be assessed from the R package (https://github.com/AnaBVA/pulmondb; Villasenor-Altamirano et al. 2020). This database allows heatmaps visualization and gene set enrichment analysis and therefore can serve as a useful resource for analysing transcriptomics (mRNAs and miRNAs) data for identifying potential COPD biomarkers. Chest medical imaging of COPD patients is stored in databases and is available for further analysis and development of prediction models using artificial intelligence. PadChest database provides chest X-ray images of several lung diseases including COPD (Bustos et al. 2020).

3.12 Multi-Omics and Data Integration

Integration of multi-omics datasets allows molecular sub-phenotyping of COPD diagnosis. In a small cohort study of different classes of COPD patients, multi-omics integration based on similarity network fusion (SNF) of transcriptomics (miRNA and mRNA), proteomics, and metabolomics data from multiple anatomical locations such as airway epithelium, lung resident immune cells, airway exudates, exosomes, and serum have shown to predict the correct classification of COPD (Li et al. 2018). The R-package SNFtool was used to integrate multi-omics data for COPD subject-based accurate clustering (Li et al. 2018; Wang et al. 2014). A large-scale COPD patient study, multi-omics data based on whole blood transcriptomics, plasma proteome, and plasma metabolomes were integrated machine learning techniques that enables distinct clinical phenotyping (Gillenwater et al. 2021). These multi-omics studies show that they are better in classified of subtyping of COPD than single OMICs data.

3.13 Current Applications

The present-day therapy for COPD primarily relies on the use of bronchodilators and steroids. The common bronchodilators widely used for COPD include beta2-agonists and antimuscarinic agents. Methylxanthines are also used as weak bronchodilators. Combination therapy comprising of a bronchodilator and inhaled corticosteroid is widely used for treatment of COPD patients especially with recurrent exacerbations and hospitalization and in those with increased peripheral blood eosinophil count (Cheng 2018). Blood eosinophil count and the fraction of exhaled nitric oxide (FeNO) have allowed clinicians to choose the responders to inhaled corticosteroids (Annangi et al. 2022). Roflumilast, a PDE4 inhibitor, is also used as maintenance therapy for COPD. Mucolytics have been used to reduce sputum viscosity and promote expectoration in COPD patients (Ejiofor and Turner 2013). Macrolide antibiotics are often used to treat respiratory infections and control acute exacerbations in COPD patients (Barnes and Stockley 2005). The details of each of these therapeutics along with the mechanism of action, effect in COPD patients, and adverse effects have been documented in Table 3.2.

Oxygen therapy is often administered to COPD patients with resting hypoxia developing mostly from secondary pulmonary hypertension related ventilation-perfusion mismatch. Interventional therapy like bronchoscopic interventions and lung volume reduction surgery are used to alleviate lung hyperinflation in advanced stage COPD patients (Rabe and Watz 2017).

Although the current therapeutic regime has provided relief to COPD patients to a large extent, there still remains an urgent need to continue the search for new therapeutic strategies. Several probable drug candidates, small molecule inhibitors, and monoclonal antibodies are in different phases of clinical trials for evaluation of their safety and efficacy in COPD patients (as listed in Table 3.3).

Further research employing omics approaches shall provide better mechanistic understanding of COPD disease pathobiology for the identification of biomarkers, drug targets, and the development of new therapeutic strategies.

3.14 Future Perspectives

COPD, a leading non-infective, chronic and progressive respiratory disorder, is of great concern. Even though, smoking is a primary risk factor for COPD, approximately, 25–45% of the COPD cases are observed in non-smokers. So, risk factors other than smoking needs to be explored further to determine how these factors modify the course and clinical manifestation of COPD (Bakke et al. 2011). Exposomics studies shall aid in assessing the impact of these risk factors for COPD (Benjdir et al. 2021). Also, there is a huge dearth of knowledge regarding the underlying immuno-regulatory mechanism, pathophysiology, and prognosis of COPD in non-smokers (Alvarado 2018). Different state-of-art omics approaches together with multi-omics data integration can help in improving our understanding of pathogenesis and long-term complications of COPD in non-smokers, thereby

Table 3.2 Drugs commonly used to treat COPD

Sl. no.	Class of drugs	Commonly used drugs	Mode of action	Effects in COPD patients	Side effects
1.	Long-acting beta2-agonists (LABA)	Salmeterol xinafoate, Formoterol fumarate, arformoterol tartrate, indacaterol maleate, vilanterol, olodaterol hydrochloride	Binds to beta2 adrenergic receptor in airway smooth muscle cells and triggers production of cellular cAMP and activation of Protein Kinase A (PKA). PKA mediated phosphorylation of several targets leads to relaxation of airway smooth muscles. Important non-bronchodilator effects include reduced release of mast cell mediators, decreased neutrophil recruitment, and enhanced mucociliary transport	Used as maintenance therapy Leads to bronchodilation, improved FEV1 and lung function Enhances exercise tolerance in COPD patients Reduce airway micro-vascular leakage and neutrophil respiratory burst in COPD Combination of LABA with ICS or LAMA or both improves therapeutic efficacy and results in better bronchodilation, more improvement in lung function and FEV1, than LABA mono therapy (Ejiofor and Turner 2013; Barnes and Stockley 2005; Rossi et al. 2008)	Mostly well tolerated Cardiovascular effects of LABA like ventricular arrhythmia, palpitations can at times be troublesome due to presence of cardiac comorbidities in COPD patients. Other rare adverse effects include headache, nausea, and tremors
2.	Short-acting beta2-agonists (SABA)	Salbutamol (albuterol sulphate), Levalbuterol hydrochloride	Binds to beta2 adrenergic receptor in airway smooth muscle cells and triggers production of cellular cAMP and activation of Protein Kinase A (PKA). PKA mediated phosphorylation of several targets leads to relaxation of airway smooth	Used for immediate relief in both chronic and acute COPD exacerbations. Leads to improvement in post bronchodilator lung function and reduction in dyspnoea in moderate to severe COPD patients. More improvement in FEV1 and greater	Overuse of SABA may lead to more severe forms of COPD and may also increase the risk for cardiac side effects

(continued)

Table 3.2 (continued)

Sl. no.	Class of drugs	Commonly used drugs	Mode of action	Effects in COPD patients	Side effects
			muscles. Action starts within 3 min and peaks in 2.5 h (Total duration of action is about 4–6 h). Other non-bronchodilator effects include reduced release of mast cell mediators, decreased neutrophil recruitment and enhanced mucociliary transport	bronchodilation when used with SAMA (Ejiofor and Turner 2013; Fan et al. 2016; Kopsaftis et al. 2018)	
3.	Short-acting antimuscarinic agents (SAMA)	Ipratropium bromide	Anti-cholinergic agent. Blocks muscarinic receptors M1 and M3 on bronchial and airway smooth muscle cells and thereby reduces bronchoconstriction. Action starts within minutes and peaks in 1–2 h (Total duration of action is ~4 h).	Used for the management of both chronic and acute COPD exacerbations. Induce bronchodilation in COPD patients. Better bronchodilation and improved FEV1 observed when used with SABA (Ejiofor and Turner 2013; Kopsaftis et al. 2018)	Minimal side effects. Side effects include headache, nausea, sore throat, sedation, tachycardia, urinary retention and painful urination, enlarged prostate gland, back pain, acute angle closure glaucoma
4.	Long-acting antimuscarinic agents (LAMA)	Tiotropium bromide, aclidinium bromide, glycopyrronium bromide, umeclidinium bromide	Anti-cholinergic agent. Selectively blocks muscarinic receptors M1 and M3 on bronchial and airway smooth muscle cells, leading to inhibition of acetylcholine release and inhibition of parasympathetic pathways;	Used as maintenance therapy Superior to LABA. Leads to improvement in lung function (FEV1), exercise tolerance, and quality of life. Reduces dyspnoea, exacerbations, and hospitalization. More effective when used in	Most common side effect is dry mouth. Other side effects include constipation, urinary retention, and nasopharyngitis

			thereby reducing bronchoconstriction. Sustains this bronchodilating effect for about 24 h. May also exert anti-inflammatory effects	combination with LABA or ICS or both (Ejiofor and Turner 2013)	
5.	Inhaled Corticosteroid (ICS)	Budesonide, Fluticasone, Flunisolide, Beclomethasone dipropionate, Mometasone, Triamcinolone acetonide	Exert anti-inflammatory effects, possibly because in presence of corticosteroids, HDAC reduces transcription of inflammatory cytokines	Often used in combination with LABA or both LABA and LAMA to reduce airway inflammation, improve lung function and bronchodilation Leads to improvement in FEV1 and slows down disease progression in stable COPD patients. Decreases exacerbation rates and increases response to bronchodilators in severe COPD patients (Ejiofor and Turner 2013; Falk et al. 2008)	Significantly increases risk of pneumonia. Local adverse effects like headache, dysphonia, oral candidiasis, adrenal suppression, skin bruising, loss of bone mass, increased risk of fracture and glaucoma
6.	Systemic glucocorticoids	Prednisolone, prednisone, methylprednisolone	Exert anti-inflammatory effects, possibly because in presence of corticosteroids, HDAC reduces transcription of inflammatory cytokines	No long-term benefit noted in stable COPD. Marked improvement in lung function, reduction in dyspnoea, shortening of hospital stay and reduced relapse rate was noted in patients with acute COPD exacerbations (Falk et al. 2008)	Not recommended for long-term use. Dose and duration dependent adverse reactions noted. Adverse reactions include weight gain, insomnia, anxiety, depression, hypertension, hyperglycaemia, myopathy, reduction in serum osteocalcin, osteoporosis, and adrenal insufficiency

(continued)

Table 3.2 (continued)

Sl. no.	Class of drugs	Commonly used drugs	Mode of action	Effects in COPD patients	Side effects
7.	Methylxanthines	Theophylline	Causes airway smooth muscle relaxation and mild bronchodilation by increasing cellular cAMP (by blocking phosphodiesterase PDE3/4) and cGMP (by inhibiting PDE5). Exerts weak exert anti-inflammatory effect via direct HDAC activation (effective activation under oxidative stress)	Used in maintenance therapy as additional relief with LABA or LAMA Provides significant improvement in FEV1 when used with LABA or LAMA HDAC activity is increased and restored to more than normal in COPD alveolar macrophages by low concentrations of theophylline (Barnes and Stockley 2005; https://www.ncbi.nlm.nih.gov/books/NBK559281/)	Adverse reactions may be severe. Side effects include fever, allergy, unusual heart rate or breathing, fainting (syncope), seizures
8.	Macrolides	Azithromycin	Has anti-bacterial effect. May exert anti-inflammatory effect by inhibiting NF-κβ and activator protein 1 (noted in vitro)	Effective in controlling COPD exacerbations (Taylor et al. 2015)	May lead to antibiotic overuse and development of antibiotic resistance
9.	PDE4 inhibitor	Roflumilast	Inhibits phosphodiesterase-4 (PDE4), leading to increased cellular cAMP, reduced micro-vascular leakage, and decreased release of chemokines and cytokines	Reduces dyspnoea and frequent attacks in COPD patients Used as a maintenance therapy in case of severe COPD (Ejiofor and Turner 2013)	Use should be avoided in case of liver diseases, depression, and allergic reactions. Side effects include weight loss, diarrhoea, nausea, and vomiting

10.	Mucolytics	*N*-acetylcysteine (NAC), carbocysteine, erdosteine, ambroxol	Serves as mucolytic agent and glutathione-generating anti-oxidant	Decreases exacerbation rates and improves airway functioning in COPD patients when used as a supplement (Ejiofor and Turner 2013)	Side effects are rare. Side effects include diarrhoea, nausea, rashes, drowsiness, irritation in respiratory tract

Table 3.3 Drugs and Biologics that are under clinical investigation for use in COPD

Sl. no.	Drugs/ biologics	Description	Target	Major effects	Status
1.	GSK-256066	PDE4 inhibitor	PDE4	Lead to increase in FEV1 upon bronchodilator challenge and caused reduction in residual volume	Investigational (Clinical Trial Phase II) (Watz et al. 2013a)
2.	Ensifentrine (RPL554)	Dual inhibitor of phosphodiesterase-3 and phosphodiesterase-4	PDE3, PDE4	Leads to effective bronchodilation in dose dependent manner (Ferguson et al. 2021)	Investigational (Clinical Trial Phase III) (https://clinicaltrials.gov/ct2/show/NCT04542057)
3.	CHF6001	PDE4 inhibitor	PDE4	Exert anti-inflammatory effects	Investigational (Clinical Trial Phase II) (Singh et al. 2019)
4.	Mepolizumab	Humanized anti-interleukin-5 (IL-5) monoclonal antibody	IL-5	Reduced annual rate of moderate/severe exacerbations in COPD patients having eosinophilic phenotype	Investigational (Clinical Trial Phase III) (Cazzola et al. 2021)
5.	Itepekimab	Anti-interleukin-33 (IL-33) monoclonal antibody	IL-33	Decreased exacerbation rate and improved lung function in erstwhile smokers with COPD	Investigational (Clinical Trial Phase III) (Cazzola et al. 2021)
6.	Navarixin (MK-7123)	Cysteine-X-cysteine chemokine receptor 2 (CXCR2) antagonist	CXCR2	Improved FEV1 significantly in COPD patients	Investigational (Clinical Trial Phase II) (Rennard et al. 2015)
7.	AZD2423	Chemokine (C-C motif) receptor 2 (CCR2) antagonist	CCR2	Reduced monocytes in peripheral blood	Investigational (Clinical Trial Phase II) (Baker and Donnelly 2021)
8.	SB-681323	p38 MAPK inhibitor	p38 MAPK	Inhibited TNF-α production. Serves as more potent inhibitor of p38 MAPK than prednisolone Suppressed inflammation in COPD patients	Investigational (Clinical Trial Phase I) (Singh et al. 2010)
9.	Losmapimod (GW856553)	MAPK inhibitor	p38 α/β MAPK	Reduction in exacerbations in COPD patients with blood eosinophil level ≤2% (Marks-Konczalik et al. 2015)	Investigational (Clinical Trial Phase II) (Watz et al. 2014)

10.	PH-797804	p38 MAPK inhibitor	p38 MAPK	Improved lung function and dyspnoea score in moderate or severe COPD patients treated with short-acting bronchodilators	Investigational (Clinical Trial Phase II) (MacNee et al. 2013)
11.	Etanercept	TNF (Tumor Necrosis Factor) inhibitor	TNF-α	Etanercept did not exhibit better effectiveness than prednisone in controlling acute exacerbations of COPD	Investigational (Clinical Trial Phase II/III) (Aaron et al. 2013)
12.	Zileuton	5-Lipoxygenase inhibitor	5-LO	Lowers urinary leukotriene E4 (LTE4) levels in COPD patients requiring hospitalization, but it did not shorten the duration of hospital stay	Investigational (Clinical Trial Phase III) (Woodruff et al. 2011)
13.	ABX-CXCL8	Monoclonal antibody against CXCL8	CXCL8	Improved dyspnoea in COPD patients	Investigational (Phase II Pilot Study) (Mahler et al. 2004)
14.	Bimosiamose	Inhaled small molecule pan-selectin antagonist	L-selectin, P-selectin, E-selectin	Reduced airway inflammation in COPD patients	Investigational (Clinical Trial Phase II) (Watz et al. 2013b)

aiding in improved therapeutic improvisation. In general, spirometry provides effective diagnosis, but, other diagnostic tools based on imaging and biomarkers should be extensively explored for diagnosing COPD at an initial stage (Bakke et al. 2011). Besides, molecular understanding of COPD with phenotypes and endotypes needs to be upgraded using multi-omics approaches to enable better appreciation of the pathophysiology including the clinical heterogeneity of the disease in near future (Han et al. 2010). COPD research has been largely handicapped for long due to the lack of suitable animal model (Croxton et al. 2002). Although significant advances have been made for designing suitable animal model of COPD, no individual model is known to reproduce all the features of the human disease. Careful selection of animal model for COPD is an important criterion for subsequent mechanistic, pre-clinical, and translational studies (Tanner and Single 2020). Integrating the influence of comorbid conditions in COPD models shall enable better understanding of COPD in aged population and help in improving clinical treatment parameters (Lopez-Campos et al. 2019). Biomedical therapies need to be improved and personalized to maximize efficacy and minimize adverse effects (Gruss et al. 2021). Alongside medication, role of vaccination, stem cell therapy, and pulmonary rehabilitation in controlling COPD exacerbations need to be studied (Lopez-Campos et al. 2019; Chen et al. 2021). COPD patients are often at a higher risk of developing other complications like lung cancer. However, the exact mechanism responsible for such increased instances of lung cancer in COPD patients is unknown and requires elucidation at molecular levels (Durham and Adcock 2015). Leading edge omics and systems biology approaches can help in improving our understanding of such relationship between COPD and lung cancer. All these above-mentioned avenues can be explored in future to improve the overall diagnosis and set the stage for development of precision medication for better control of COPD.

References

Aaron SD, Vandemheen KL, Maltais F, Field SK, Sin DD, Bourbeau J, Marciniuk DD, FitzGerald JM, Nair P, Mallick R (2013) TNFalpha antagonists for acute exacerbations of COPD: a randomised double-blind controlled trial. Thorax 68(2):142–148

Akbas F, Coskunpinar E, Aynaci E, Oltulu YM, Yildiz P (2012) Analysis of serum micro-RNAs as potential biomarker in chronic obstructive pulmonary disease. Exp Lung Res 38(6):286–294

Alvarado A (2018) Chronic obstructive pulmonary disease in non-smokers: an update. Clin Res Trials

Amado CA, Munoz P, Garcia-Unzueta M, Aguero J, Tello S, Fueyo P, Vega C, Lavin BA, Guerra RA, Casanova C (2021) High parathyroid hormone predicts exacerbations in COPD patients with hypovitaminosis D. Respir Med 182:106416

Ambade VN, Sontakke AN, Barthwal MS, Tyagi R, Basannar DR (2015) Diagnostic utility of biomarkers in COPD. Respir Care 60(12):1729–1742

Annangi S, Nutalapati S, Sturgill J, Flenaugh E, Foreman M (2022) Eosinophilia and fractional exhaled nitric oxide levels in chronic obstructive lung disease. Thorax 77(4):351–356

Bai S, Ye R, Wang C, Sun P, Wang D, Yue Y, Wang H, Wu S, Yu M, Xi S et al (2021) Identification of proteomic signatures in chronic obstructive pulmonary disease emphysematous phenotype. Front Mol Biosci 8:650604

Baker JR, Donnelly LE (2021) Leukocyte Function in COPD: clinical relevance and potential for drug therapy. Int J Chron Obstruct Pulmon Dis 16:2227–2242

Bakke PS, Ronmark E, Eagan T, Pistelli F, Annesi-Maesano I, Maly M, Meren M, Vermeire Dagger P, Vestbo J, Viegi G et al (2011) Recommendations for epidemiological studies on COPD. Eur Respir J 38(6):1261–1277

Balbirsingh V, Mohammed AS, Turner AM, Newnham M (2022) Cardiovascular disease in chronic obstructive pulmonary disease: a narrative review. Thorax

Baralla A, Fois AG, Sotgiu E, Zinellu E, Mangoni AA, Sotgia S, Zinellu A, Pirina P, Carru C (2018) Plasma proteomic signatures in early chronic obstructive pulmonary disease. Proteomics Clin Appl 12(3):e1700088

Barnes PJ, Stockley RA (2005) COPD: current therapeutic interventions and future approaches. Eur Respir J 25(6):1084–1106

Benjdir M, Audureau E, Beresniak A, Coll P, Epaud R, Fiedler K, Jacquemin B, Niddam L, Pandis SN, Pohlmann G et al (2021) Assessing the impact of exposome on the course of chronic obstructive pulmonary disease and cystic fibrosis: the REMEDIA European Project Approach. Environ Epidemiol 5(4):e165

Blanco I, Bueno P, Diego I, Perez-Holanda S, Casas-Maldonado F, Esquinas C, Miravitlles M (2017) Alpha-1 antitrypsin Pi*Z gene frequency and Pi*ZZ genotype numbers worldwide: an update. Int J Chron Obstruct Pulmon Dis 12:561–569

Burkhardt R, Pankow W (2014) The diagnosis of chronic obstructive pulmonary disease. Dtsch Arztebl Int 111(49):834–845. quiz 846

Bustos A, Pertusa A, Salinas JM, de la Iglesia-Vaya M (2020) PadChest: a large chest X-ray image dataset with multi-label annotated reports. Med Image Anal 66:101797

Canas JA, Rodrigo-Munoz JM, Sastre B, Gil-Martinez M, Redondo N, Del Pozo V (2020) MicroRNAs as potential regulators of immune response networks in asthma and chronic obstructive pulmonary disease. Front Immunol 11:608666

Candela M, Costorella R, Stassaldi A, Maestrini V, Curradi G (2019) Treatment of COPD: the simplicity is a resolved complexity. Multidiscip Respir Med 14:18

Cao Z, Zhang N, Lou T, Jin Y, Wu Y, Ye Z, Pan J (2014) microRNA-183 down-regulates the expression of BKCabeta1 protein that is related to the severity of chronic obstructive pulmonary disease. Hippokratia 18(4):328–332

Casas-Recasens S, Noell G, Mendoza N, Lopez-Giraldo A, Garcia T, Guirao A, Agusti A, Faner R (2021) Lung DNA methylation in chronic obstructive pulmonary disease: relationship with smoking status and airflow limitation severity. Am J Respir Crit Care Med 203(1):129–134

Cavailles A, Brinchault-Rabin G, Dixmier A, Goupil F, Gut-Gobert C, Marchand-Adam S, Meurice JC, Morel H, Person-Tacnet C, Leroyer C et al (2013) Comorbidities of COPD. Eur Respir Rev 22(130):454–475

Cazzola M, Ora J, Cavalli F, Rogliani P, Matera MG (2021) An overview of the safety and efficacy of monoclonal antibodies for the chronic obstructive pulmonary disease. Biologics 15:363–374

Chatila WM, Thomashow BM, Minai OA, Criner GJ, Make BJ (2008) Comorbidities in chronic obstructive pulmonary disease. Proc Am Thorac Soc 5(4):549–555

Chen ZH, Kim HP, Ryter SW, Choi AM (2008) Identifying targets for COPD treatment through gene expression analyses. Int J Chron Obstruct Pulmon Dis 3(3):359–370

Chen YT, Miao K, Zhou L, Xiong WN (2021) Stem cell therapy for chronic obstructive pulmonary disease. Chin Med J 134(13):1535–1545

Cheng SL (2018) Blood eosinophils and inhaled corticosteroids in patients with COPD: systematic review and meta-analysis. Int J Chron Obstruct Pulmon Dis 13:2775–2784

Christenson SA, Smith BM, Bafadhel M, Putcha N (2022) Chronic obstructive pulmonary disease. Lancet 399(10342):2227–2242

Conickx G, Mestdagh P, Avila Cobos F, Verhamme FM, Maes T, Vanaudenaerde BM, Seys LJ, Lahousse L, Kim RY, Hsu AC et al (2017) MicroRNA profiling reveals a role for microRNA-218-5p in the pathogenesis of chronic obstructive pulmonary disease. Am J Respir Crit Care Med 195(1):43–56

Croxton TL, Weinmann GG, Senior RM, Hoidal JR (2002) Future research directions in chronic obstructive pulmonary disease. Am J Respir Crit Care Med 165(6):838–844
Denden S, Khelil AH, Knani J, Lakhdar R, Perrin P, Lefranc G, Chibani JB (2010) Alpha-1 antitrypsin gene polymorphism in Chronic Obstructive Pulmonary Disease (COPD). Genet Mol Biol 33(1):23–26
Du Y, Ding Y, Chen X, Mei Z, Ding H, Wu Y, Jie Z (2017) MicroRNA-181c inhibits cigarette smoke-induced chronic obstructive pulmonary disease by regulating CCN1 expression. Respir Res 18(1):155
Durham AL, Adcock IM (2015) The relationship between COPD and lung cancer. Lung Cancer 90(2):121–127
Ejiofor S, Turner AM (2013) Pharmacotherapies for COPD. Clin Med Insights Circ Respir Pulm Med 7:17–34
Ezzie ME, Crawford M, Cho JH, Orellana R, Zhang S, Gelinas R, Batte K, Yu L, Nuovo G, Galas D et al (2012) Gene expression networks in COPD: microRNA and mRNA regulation. Thorax 67(2):122–131
Falk JA, Minai OA, Mosenifar Z (2008) Inhaled and systemic corticosteroids in chronic obstructive pulmonary disease. Proc Am Thorac Soc 5(4):506–512
Fan VS, Gylys-Colwell I, Locke E, Sumino K, Nguyen HQ, Thomas RM, Magzamen S (2016) Overuse of short-acting beta-agonist bronchodilators in COPD during periods of clinical stability. Respir Med 116:100–106
Faner R, Tal-Singer R, Riley JH, Celli B, Vestbo J, MacNee W, Bakke P, Calverley PM, Coxson H, Crim C et al (2014) Lessons from ECLIPSE: a review of COPD biomarkers. Thorax 69(7): 666–672
Farid Hosseini R, Jabbari Azad F, Yousefzadeh H, Rafatpanah H, Hafizi S, Tehrani H, Khani M (2014) Serum levels of vascular endothelial growth factor in chronic obstructive pulmonary disease. Med J Islam Repub Iran 28:85
Ferguson GT, Kerwin EM, Rheault T, Bengtsson T, Rickard K (2021) A dose-ranging study of the novel inhaled dual PDE 3 and 4 inhibitor ensifentrine in patients with COPD receiving maintenance Tiotropium therapy. Int J Chron Obstruct Pulmon Dis 16:1137–1148
Fermont JM, Masconi KL, Jensen MT, Ferrari R, Di Lorenzo VAP, Marott JM, Schuetz P, Watz H, Waschki B, Mullerova H et al (2019) Biomarkers and clinical outcomes in COPD: a systematic review and meta-analysis. Thorax 74(5):439–446
Fernandes L, Fernandes Y, Mesquita AM (2016) Quantitative computed tomography imaging in chronic obstructive pulmonary disease. Lung India 33(6):646–652
Gillenwater LA, Helmi S, Stene E, Pratte KA, Zhuang Y, Schuyler RP, Lange L, Castaldi PJ, Hersh CP, Banaei-Kashani F et al (2021) Multi-omics subtyping pipeline for chronic obstructive pulmonary disease. PLoS One 16(8):e0255337
Green CE, Vayalapra S, Hampson JA, Mukherjee D, Stockley RA, Turner AM (2015) PiSZ alpha-1 antitrypsin deficiency (AATD): pulmonary phenotype and prognosis relative to PiZZ AATD and PiMM COPD. Thorax 70(10):939–945
Gruss I, McCreary GM, Ivlev I, Houlihan ME, Yawn BP, Pasquale C, Clark W, Mularski RA (2021) Developing a patient-driven chronic obstructive pulmonary disease (COPD) research agenda in the U.S. J Patient Rep Outcomes 5(1):126
Han MK, Agusti A, Calverley PM, Celli BR, Criner G, Curtis JL, Fabbri LM, Goldin JG, Jones PW, Macnee W et al (2010) Chronic obstructive pulmonary disease phenotypes: the future of COPD. Am J Respir Crit Care Med 182(5):598–604
Han MK, Kazerooni EA, Lynch DA, Liu LX, Murray S, Curtis JL, Criner GJ, Kim V, Bowler RP, Hanania NA et al (2011) Chronic obstructive pulmonary disease exacerbations in the COPDGene study: associated radiologic phenotypes. Radiology 261(1):274–282
Hayden LP, Hobbs BD, Cohen RT, Wise RA, Checkley W, Crapo JD, Hersh CP, Investigators CO (2015) Childhood pneumonia increases risk for chronic obstructive pulmonary disease: the COPDGene study. Respir Res 16:115

He JQ, Foreman MG, Shumansky K, Zhang X, Akhabir L, Sin DD, Man SF, DeMeo DL, Litonjua AA, Silverman EK et al (2009) Associations of IL6 polymorphisms with lung function decline and COPD. Thorax 64(8):698–704

He LX, Tang ZH, Huang QS, Li WH (2020) DNA methylation: a potential biomarker of chronic obstructive pulmonary disease. Front Cell Dev Biol 8:585

Hogea SP, Tudorache E, Fildan AP, Fira-Mladinescu O, Marc M, Oancea C (2020) Risk factors of chronic obstructive pulmonary disease exacerbations. Clin Respir J 14(3):183–197

Hu HL, Nie ZQ, Lu Y, Yang X, Song C, Chen H, Zhu S, Chen BB, Huang J, Geng S et al (2017) Circulating miR-125b but not miR-125a correlates with acute exacerbations of chronic obstructive pulmonary disease and the expressions of inflammatory cytokines. Medicine (Baltimore) 96(51):e9059

Jain KK (2017) Biomarkers of pulmonary diseases. Springer, The Handbook of Biomarkers, pp 673–388

Kim DJ, Oh JY, Rhee CK, Park SJ, Shim JJ, Cho JY (2021) Metabolic fingerprinting uncovers the distinction between the phenotypes of tuberculosis associated COPD and smoking-induced COPD. Front Med (Lausanne) 8:619077

Kirby M, Tanabe N, Tan WC, Zhou G, Obeidat M, Hague CJ, Leipsic J, Bourbeau J, Sin DD, Hogg JC et al (2018) Total airway count on computed tomography and the risk of chronic obstructive pulmonary disease progression. Findings from a population-based study. Am J Respir Crit Care Med 197(1):56–65

Kopsaftis ZA, Sulaiman NS, Mountain OD, Carson-Chahhoud KV, Phillips PA, Smith BJ (2018) Short-acting bronchodilators for the management of acute exacerbations of chronic obstructive pulmonary disease in the hospital setting: systematic review. Syst Rev 7(1):213

Kumar M, Phougat N, Ruhil S, Dhankhar S, Balhara M, Chhillar AK (2013) Genomics of Chronic Obstructive Pulmonary Disease (COPD); exploring the SNPs of protease-antiprotease pathway. Curr Genomics 14(3):204–213

Lam HC, Cloonan SM, Bhashyam AR, Haspel JA, Singh A, Sathirapongsasuti JF, Cervo M, Yao H, Chung AL, Mizumura K et al (2020) Histone deacetylase 6-mediated selective autophagy regulates COPD-associated cilia dysfunction. J Clin Invest 130(11):6189

Lee SW, Kuan CS, Wu LS, Weng JT (2016) Metagenome and metatranscriptome profiling of moderate and severe COPD sputum in Taiwanese han males. PLoS One 11(7):e0159066

Leitao Filho FS, Mattman A, Schellenberg R, Criner GJ, Woodruff P, Lazarus SC, Albert RK, Connett J, Han MK, Gay SE et al (2020) Serum IgG levels and risk of COPD hospitalization: a pooled meta-analysis. Chest 158(4):1420–1430

Li CX, Wheelock CE, Skold CM, Wheelock AM (2018) Integration of multi-omics datasets enables molecular classification of COPD. Eur Respir J 51(5)

Li X, Zhou G, Tian X, Chen F, Li G, Ding Y (2021) The polymorphisms of FGFR2 and MGAT5 affect the susceptibility to COPD in the Chinese people. BMC Pulm Med 21(1):129

Lopez-Campos JL, Rodriguez DA, Quintana-Gallego E, Martinez-Llorens J, Carrasco Hernandez L, Barreiro E (2019) Ten research questions for improving COPD care in the next decade. COPD 16(5–6):311–320

Lynch DA, Al-Qaisi MA (2013) Quantitative computed tomography in chronic obstructive pulmonary disease. J Thorac Imaging 28(5):284–290

MacNee W, Allan RJ, Jones I, De Salvo MC, Tan LF (2013) Efficacy and safety of the oral p38 inhibitor PH-797804 in chronic obstructive pulmonary disease: a randomised clinical trial. Thorax 68(8):738–745

Mahler DA, Huang S, Tabrizi M, Bell GM (2004) Efficacy and safety of a monoclonal antibody recognizing interleukin-8 in COPD: a pilot study. Chest 126(3):926–934

Manian P (2019) Chronic obstructive pulmonary disease classification, phenotypes and risk assessment. J Thorac Dis 11(Suppl 14):S1761–S1766

Marks-Konczalik J, Costa M, Robertson J, McKie E, Yang S, Pascoe S (2015) A post-hoc subgroup analysis of data from a six month clinical trial comparing the efficacy and safety of losmapimod in moderate-severe COPD patients with </=2% and >2% blood eosinophils. Respir Med 109(7):860–869

Mathioudakis AG, Chatzimavridou-Grigoriadou V, Corlateanu A, Vestbo J (2017) Procalcitonin to guide antibiotic administration in COPD exacerbations: a meta-analysis. Eur Respir Rev 26(143)

Moon JY, Leitao Filho FS, Shahangian K, Takiguchi H, Sin DD (2018) Blood and sputum protein biomarkers for chronic obstructive pulmonary disease (COPD). Expert Rev Proteomics 15(11): 923–935

Myc LA, Shim YM, Laubach VE, Dimastromatteo J (2019) Role of medical and molecular imaging in COPD. Clin Transl Med 8(1):12

Nocker RE, Schoonbrood DF, van de Graaf EA, Hack CE, Lutter R, Jansen HM, Out TA (1996) Interleukin-8 in airway inflammation in patients with asthma and chronic obstructive pulmonary disease. Int Arch Allergy Immunol 109(2):183–191

Ortega-Martinez A, Perez-Rubio G, Ambrocio-Ortiz E, Nava-Quiroz KJ, Hernandez-Zenteno RJ, Abarca-Rojano E, Rodriguez-Llamazares S, Hernandez-Perez A, Garcia-Gomez L, Ramirez-Venegas A et al (2020) The SNP rs13147758 in the HHIP gene is associated with COPD susceptibility, serum, and sputum protein levels in smokers. Front Genet 11:882

Ozretic P, da Silva Filho MI, Catalano C, Sokolovic I, Vukic-Dugac A, Sutic M, Kurtovic M, Bubanovic G, Popovic-Grle S, Skrinjaric-Cincar S et al: Association of NLRP1 coding polymorphism with lung function and serum IL-1beta concentration in patients diagnosed with chronic obstructive pulmonary disease (COPD). Genes (Basel) 2019, 10(10)

Paci P, Fiscon G, Conte F, Licursi V, Morrow J, Hersh C, Cho M, Castaldi P, Glass K, Silverman EK et al (2020) Integrated transcriptomic correlation network analysis identifies COPD molecular determinants. Sci Rep 10(1):3361

Papaioannou AI, Loukides S, Minas M, Kontogianni K, Bakakos P, Gourgoulianis KI, Alchanatis M, Papiris S, Kostikas K (2011) Exhaled breath condensate pH as a biomarker of COPD severity in ex-smokers. Respir Res 12:67

Pratte KA, Curtis JL, Kechris K, Couper D, Cho MH, Silverman EK, DeMeo DL, Sciurba FC, Zhang Y, Ortega VE et al (2021) Soluble receptor for advanced glycation end products (sRAGE) as a biomarker of COPD. Respir Res 22(1):127

Prokic I, Lahousse L, de Vries M, Liu J, Kalaoja M, Vonk JM, van der Plaat DA, van Diemen CC, van der Spek A, Zhernakova A et al (2020) A cross-omics integrative study of metabolic signatures of chronic obstructive pulmonary disease. BMC Pulm Med 20(1):193

Rabe KF, Watz H (2017) Chronic obstructive pulmonary disease. Lancet 389(10082):1931–1940

Rabe KF, Hurd S, Anzueto A, Barnes PJ, Buist SA, Calverley P, Fukuchi Y, Jenkins C, Rodriguez-Roisin R, van Weel C et al (2007) Global strategy for the diagnosis, management, and prevention of chronic obstructive pulmonary disease: GOLD executive summary. Am J Respir Crit Care Med 176(6):532–555

Raherison C, Ouaalaya EH, Bernady A, Casteigt J, Nocent-Eijnani C, Falque L, Le Guillou F, Nguyen L, Ozier A, Molimard M (2018) Comorbidities and COPD severity in a clinic-based cohort. BMC Pulm Med 18(1):117

Ramsheh MY, Haldar K, Esteve-Codina A, Purser LF, Richardson M, Muller-Quernheim J, Greulich T, Nowinski A, Barta I, Stendardo M et al (2021) Lung microbiome composition and bronchial epithelial gene expression in patients with COPD versus healthy individuals: a bacterial 16S rRNA gene sequencing and host transcriptomic analysis. Lancet Microbe 2(7): e300–e310

Regan EA, Hersh CP, Castaldi PJ, DeMeo DL, Silverman EK, Crapo JD, Bowler RP (2019) Omics and the search for blood biomarkers in chronic obstructive pulmonary disease. Insights from COPDGene. Am J Respir Cell Mol Biol 61(2):143–149

Rennard SI, Dale DC, Donohue JF, Kanniess F, Magnussen H, Sutherland ER, Watz H, Lu S, Stryszak P, Rosenberg E et al (2015) CXCR2 antagonist MK-7123. A phase 2 proof-of-concept trial for chronic obstructive pulmonary disease. Am J Respir Crit Care Med 191(9):1001–1011

Resendiz-Hernandez JM, Ambrocio-Ortiz E, Perez-Rubio G, Lopez-Flores LA, Abarca-Rojano E, Pavon-Romero GF, Flores-Trujillo F, de Jesus H-ZR, Camarena A, Perez-Rodriguez M et al (2018) TNF promoter polymorphisms are associated with genetic susceptibility in COPD secondary to tobacco smoking and biomass burning. Int J Chron Obstruct Pulmon Dis 13: 627–637

Rossi A, Khirani S, Cazzola M (2008) Long-acting beta2-agonists (LABA) in chronic obstructive pulmonary disease: efficacy and safety. Int J Chron Obstruct Pulmon Dis 3(4):521–529

Silverman EK, Speizer FE (1996) Risk factors for the development of chronic obstructive pulmonary disease. Med Clin North Am 80(3):501–522

Singh D, Smyth L, Borrill Z, Sweeney L, Tal-Singer R (2010) A randomized, placebo-controlled study of the effects of the p38 MAPK inhibitor SB-681323 on blood biomarkers of inflammation in COPD patients. J Clin Pharmacol 50(1):94–100

Singh D, Beeh KM, Colgan B, Kornmann O, Leaker B, Watz H, Lucci G, Geraci S, Emirova A, Govoni M et al (2019) Effect of the inhaled PDE4 inhibitor CHF6001 on biomarkers of inflammation in COPD. Respir Res 20(1):180

Stolz D, Christ-Crain M, Morgenthaler NG, Leuppi J, Miedinger D, Bingisser R, Muller C, Struck J, Muller B, Tamm M (2007) Copeptin, C-reactive protein, and procalcitonin as prognostic biomarkers in acute exacerbation of COPD. Chest 131(4):1058–1067

Sundar IK, Yin Q, Baier BS, Yan L, Mazur W, Li D, Susiarjo M, Rahman I (2017) DNA methylation profiling in peripheral lung tissues of smokers and patients with COPD. Clin Epigenetics 9:38

Tanner L, Single AB (2020) Animal models reflecting chronic obstructive pulmonary disease and related respiratory disorders: translating pre-clinical data into clinical relevance. J Innate Immun 12(3):203–225

Taylor SP, Sellers E, Taylor BT (2015) Azithromycin for the prevention of COPD exacerbations: the good, bad, and ugly. Am J Med 128(12):1362:e1361–e1366

Tiew PY, Jaggi TK, Chan LLY, Chotirmall SH (2021) The airway microbiome in COPD, bronchiectasis and bronchiectasis-COPD overlap. Clin Respir J 15(2):123–133

Ubhi BK, Riley JH, Shaw PA, Lomas DA, Tal-Singer R, MacNee W, Griffin JL, Connor SC (2012) Metabolic profiling detects biomarkers of protein degradation in COPD patients. Eur Respir J 40(2):345–355

Villasenor-Altamirano AB, Moretto M, Maldonado M, Zayas-Del Moral A, Munguia-Reyes A, Romero Y, Garcia-Sotelo JS, Aguilar LA, Aldana-Assad O, Engelen K et al (2020) PulmonDB: a curated lung disease gene expression database. Sci Rep 10(1):514

Vucic EA, Chari R, Thu KL, Wilson IM, Cotton AM, Kennett JY, Zhang M, Lonergan KM, Steiling K, Brown CJ et al (2014) DNA methylation is globally disrupted and associated with expression changes in chronic obstructive pulmonary disease small airways. Am J Respir Cell Mol Biol 50(5):912–922

Wang B, Mezlini AM, Demir F, Fiume M, Tu Z, Brudno M, Haibe-Kains B, Goldenberg A (2014) Similarity network fusion for aggregating data types on a genomic scale. Nat Methods 11(3): 333–337

Wang Z, Bafadhel M, Haldar K, Spivak A, Mayhew D, Miller BE, Tal-Singer R, Johnston SL, Ramsheh MY, Barer MR et al (2016) Lung microbiome dynamics in COPD exacerbations. Eur Respir J 47(4):1082–1092

Wang N, Wang Q, Du T, Gabriel ANA, Wang X, Sun L, Li X, Xu K, Jiang X, Zhang Y (2020a) The potential roles of exosomes in chronic obstructive pulmonary disease. Front Med (Lausanne) 7: 618506

Wang Z, Yang Y, Yan Z, Liu H, Chen B, Liang Z, Wang F, Miller BE, Tal-Singer R, Yi X et al (2020b) Multi-omic meta-analysis identifies functional signatures of airway microbiome in chronic obstructive pulmonary disease. ISME J 14(11):2748–2765

Waschki B, Alter P, Zeller T, Magnussen C, Neumann JT, Twerenbold R, Sinning C, Herr C, Kahnert K, Fahndrich S et al (2020) High-sensitivity troponin I and all-cause mortality in patients with stable COPD: an analysis of the COSYCONET study. Eur Respir J 55(2)

Watz H, Mistry SJ, Lazaar AL (2013a) investigators IPC: safety and tolerability of the inhaled phosphodiesterase 4 inhibitor GSK256066 in moderate COPD. Pulm Pharmacol Ther 26(5): 588–595

Watz H, Bock D, Meyer M, Schierhorn K, Vollhardt K, Woischwill C, Pedersen F, Kirsten A, Beeh KM, Meyer-Sabellek W et al (2013b) Inhaled pan-selectin antagonist Bimosiamose attenuates airway inflammation in COPD. Pulm Pharmacol Ther 26(2):265–270

Watz H, Barnacle H, Hartley BF, Chan R (2014) Efficacy and safety of the p38 MAPK inhibitor losmapimod for patients with chronic obstructive pulmonary disease: a randomised, double-blind, placebo-controlled trial. Lancet Respir Med 2(1):63–72

Woodruff PG, Albert RK, Bailey WC, Casaburi R, Connett JE, Cooper JA Jr, Criner GJ, Curtis JL, Dransfield MT, Han MK et al (2011) Randomized trial of zileuton for treatment of COPD exacerbations requiring hospitalization. COPD 8(1):21–29

Xie L, Wu M, Lin H, Liu C, Yang H, Zhan J, Sun S (2014) An increased ratio of serum miR-21 to miR-181a levels is associated with the early pathogenic process of chronic obstructive pulmonary disease in asymptomatic heavy smokers. Mol BioSyst 10(5):1072–1081

Xie J, Wu H, Xu Y, Wu X, Liu X, Shang J, Zhao J, Zhao J, Wang J, Dela Cruz CS et al (2015) Gene susceptibility identification in a longitudinal study confirms new loci in the development of chronic obstructive pulmonary disease and influences lung function decline. Respir Res 16:49

Xu J, de Oliveira DM, Trudeau MA, Yang Y, Chin JJY, Sin DD, Sandford AJ, Wong JMY (2021) Mild catalytic defects of tert rs61748181 polymorphism affect the clinical presentation of chronic obstructive pulmonary disease. Sci Rep 11(1):4333

Yao Y, Gu Y, Yang M, Cao D, Wu F (2019) The gene expression biomarkers for chronic obstructive pulmonary disease and interstitial lung disease. Front Genet 10:1154

Zemans RL, Jacobson S, Keene J, Kechris K, Miller BE, Tal-Singer R, Bowler RP (2017) Multiple biomarkers predict disease severity, progression and mortality in COPD. Respir Res 18(1):117

Zhang Z, Fu C, Liu J, Sai X, Qin C, Di T, Yang Y, Wu Y, Bian T (2021a) Hypermethylation of the Nrf2 promoter induces ferroptosis by inhibiting the Nrf2-GPX4 axis in COPD. Int J Chron Obstruct Pulmon Dis 16:3347–3362

Zhang YH, Hoopmann MR, Castaldi PJ, Simonsen KA, Midha MK, Cho MH, Criner GJ, Bueno R, Liu J, Moritz RL et al (2021b) Lung proteomic biomarkers associated with chronic obstructive pulmonary disease. Am J Physiol Lung Cell Mol Physiol 321(6):L1119–L1130

Zhao R, Zhou H, Zhu J (2020) MMP-9-C1562T polymorphism and susceptibility to chronic obstructive pulmonary disease: a meta-analysis. Medicine (Baltimore) 99(31):e21479

Zhu T, Li S, Wang J, Liu C, Gao L, Zeng Y, Mao R, Cui B, Ji H, Chen Z (2020) Induced sputum metabolomic profiles and oxidative stress are associated with chronic obstructive pulmonary disease (COPD) severity: potential use for predictive, preventive, and personalized medicine. EPMA J 11(4):645–659

Asthma

4

Abstract

Cutting edge omics approaches have led to phenotyping and sub-phenotyping of asthma based on underlying molecular endotypes. T2-high and T2-low asthma have been widely recognized. Differential airway immune responses contribute to such heterogeneity in asthma phenotypes and variable therapeutic responses. Several pathobiological drivers, biomarkers, and therapeutic targets of asthma identified using different advanced omics tools have been documented. Databases on asthma biomarkers also have been developed for identification of novel drug target, downstream network analyses, and comparative analysis of atopic asthma with other allergic diseases. The contribution of multi-omics data analyses in asthma classification, risk prediction, and drug target identification has also been discussed. Clinical investigations of several potential drug candidates that might aid in revolutionizing asthma therapy have also been highlighted in this chapter. Knowledge gaps, scientific challenges, and further research directions for asthma have been described in the end.

Keywords

Reversible airflow obstruction · FeNO · Allergy · IgE · IL-5 · T2-high asthma · T2-low asthma · omalizumab

4.1 Introduction

Asthma is a complex airway disorder characterized by chronic airway inflammation, airway hyperresponsiveness, reversible airflow obstruction, and airway remodelling. Asthma affects individuals of different age groups—from children to adults. Symptoms like shortness of breath, chest tightness, coughing, wheezing, breathlessness are mostly noted in asthmatics (He et al. 2020). Symptomatic overlap of asthma

S. Saha et al., *Pulmonomics: Omics Approaches for Understanding Pulmonary Diseases*, https://doi.org/10.1007/978-981-99-3505-5_4

with other respiratory diseases makes asthma diagnosis difficult. Besides, asthma has heterogeneous clinical presentation. Distinct underlying molecular mechanisms define each of the different phenotypes and endotypes of asthma and guide their response to present day therapeutics. In the past decade, asthma has emerged as a global socio-economic burden due to the increased disease prevalence and lack of cure for asthma (Loverdos et al. 2019).

4.2 Clinical Features

Clinically, asthma has certain qualities. As an obstructive airway disease, it has symptoms common to all other diseases of airflow limitations as cough, wheeze, shortness of breath, and expectoration. Further, it has certain characteristics in the dynamics of presentation and the course as a disease of "reversible" airflow obstruction (Ye et al. 2018). There are "variability" and "reversibility" in symptomatic presentations (Yawn 2009). Since the disease is genetically predisposed and expressed in environmentally favourable situations. These qualities as variability and reversibility are partly explained. Again, asthma symptoms are precipitated by certain trigger or triggers that are often specific to a certain individual asthmatic. Only those specific exposures, often unknown or uncontrollable, can worsen the disease and, thus, make it "unpredictable" in symptomatic presentation (Kaplan 2021). Therefore, a symptomatic patient may quickly become asymptomatic with medication for the quality of "reversibility", and an asthmatic can suddenly become sick (exacerbation) from asymptomatic state for its "variability" (Quirt et al. 2018). Clinical examination has limitations to appreciate the severity or even presence of asthma; thus, the disease may be clinically "un-assessable".

Pathophysical classification makes asthma grouped as allergic or atopic asthma and non-allergic asthma (Kim et al. 2017a). Whatever be the type, the symptomatology is similar. However, asthma is marked by clinical and pathophysiological heterogeneity.

Classification of Asthma

The classification of allergic and non-allergic asthma is globally accepted. Allergic asthma involves allergenic sensitization, whereas non-allergic lacks involvement of allergens in the pathogenesis of asthma (Kim et al. 2017a; Majumdar et al. 2018). Modern omics-based studies have enabled phenotyping and sub-phenotyping of asthma based on distinct underlying molecular pathways (endotypes). Presently, asthma can be broadly classified as T2-high and T2-low or non-T2 asthma based on the presence of Th2 inflammatory responses (Murphy et al. 2021; Kyriakopoulos et al. 2021). T2-high asthma, characterized by the increased release of Th2 cytokines (like IL-4, IL-5, and IL-13) and responsiveness to corticosteroids can be further dichotomized into allergic asthma and late onset eosinophilic asthma. T2-low asthma lacks Th2 driven inflammation and shows resistance to corticosteroid

Asthma			
T2-High Asthma		**T2-Low Asthma**	
• Mostly Early Onset asthma • Mild to severe forms of asthma (<40% cases are severe) • High T2 inflammatory response • Responsive to inhaled corticosteroids (ICS) and inhibitors of Type 2 inflammation • Variable asthma control • Lower mortality		• Late onset Asthma (common in females) • More severe form of asthma with worse pulmonary function • Associated with obesity (high BMI), smoking, pollutants, viral or bacterial infections, advanced age • Low or No T2 inflammation • Poor response to inhaled corticosteroids (ICS) [Corticosteroid Resistance] and inhibitors of Type 2 inflammation • Poor asthma control • Higher mortality	
Allergic Asthma (AA)	**Eosinophilic Asthma (EA)**	**Neutrophilic Asthma (NA)**	**Paucigranulocytic Asthma (PGA)**
• Early onset asthma • Allergen sensitization • Total IgE>150KIU/L • Overexpression of alarmins and Th2 dependent cytokines	• Late onset asthma • No evidence of atopy • Increased sputum eosinophil count (>1.9-3%) • Patients often develop chronic rhinosinusitis with nasal polyps (CRSwNP)	• Increased neutrophil proportion (≥64%) in induced sputum • Less responsive to ICS • Poor Lung function	• Sputum eosinophil (<2%) and neutrophil (<64%) levels are normal • Less responsive to ICS • Related to better lung function

Fig. 4.1 Classification of asthma. Asthma can be classified broadly as T2-high and T2-low asthma based on the type of airway inflammation. Allergic asthma and eosinophilic asthma belong to the class of T2-high asthma, whereas neutrophilic asthma (NA) and paucigranulocytic asthma (PGA) belong to T2-low asthma group

therapy. T2-low asthma can be grouped into neutrophilic asthma or paucigranulocytic asthma based on the percentage of neutrophils in the sputum. The detailed features of the different types of asthma have been summarized in Fig. 4.1. Another type of asthma has also been recognized called the mixed granulocytic asthma (Ricciardolo et al. 2021). The age of onset and severity of each of these phenotypes of asthma varies as shown in Fig. 4.2 (Wenzel 2012). The severity of asthma is measured in terms of predicted FEV1 (%) as determined by spirometry. Based on the degree of severity, asthma can be classified as mild intermittent (FEV1 predicted ≥80%), mild persistent (FEV1 predicted ≥80%), moderate persistent (60% < FEV1 predicted <80%), and severe persistent asthma (FEV1 predicted ≤60%) (Stout et al. 2006).

The basic clinical diagnosis of asthma primarily relies on the use of several tests, namely spirometry, broncho-provocation test, and peak flow test. Specific tests like total IgE count in blood, blood eosinophil count, sputum cell counts, and FeNO (Fractional exhaled nitric oxide) test are conducted for determining specific phenotype of asthma (Murphy et al. 2021). The detailed scheme of the approaches used for diagnosis and classification of asthma into major phenotypes has been represented in Fig. 4.3.

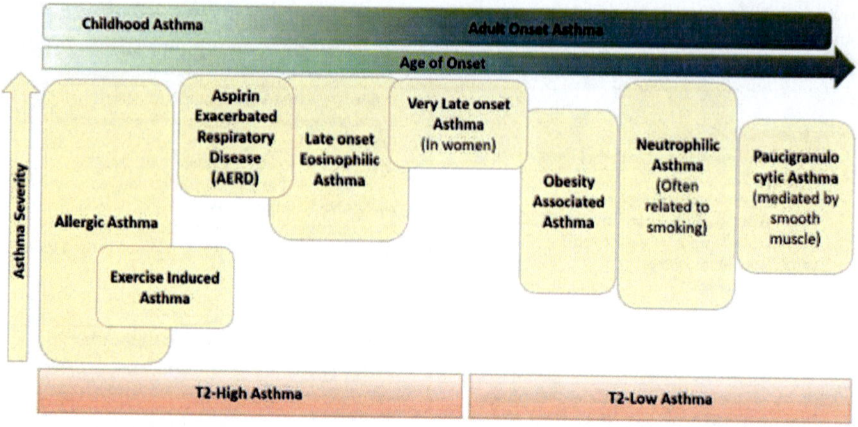

Fig. 4.2 Heterogeneity of asthma and the age of onset. Different types of asthma have different age of onset and different degree of severity (Wenzel 2012)

4.3 Biomarkers of Asthma

In spite of the advancement in diagnostic techniques, the heterogeneous clinical presentation of asthma and presence of symptomatic similarity with other obstructive pulmonary diseases like COPD make effective diagnosis and management of asthma difficult (Bhowmik et al. 2019). Identification and validation of biomarkers for asthma using omics approaches shall allow efficient distinction of asthma from other obstructive respiratory diseases, proper phenotyping and endotyping of asthma and pave the way for designing novel therapeutics with minimal adverse effects. Several omics-based studies for identification of asthma biomarkers have been conducted under the initiatives of programs like Severe Asthma Research Program (SARP) and Unbiased Biomarkers for the Prediction of Respiratory Disease Outcomes (U-BIOPRED) (Lefaudeux et al. 2017; Brasier et al. 2010). Different potential biomarker candidates identified by the various state of art omics tools have been tabulated in Table 4.1.

4.4 Genomics

Genome-wide association studies (GWAS) and linkage analysis have elaborated the contribution of genetic variations towards asthma predisposition and treatment response. Gene clusters including GSDMB and ORMDL3 located on chromosome 17q21 have been strongly related to asthma in GWAS studies (Das et al. 2017). Several intronic and intergenic variants in the ZPBP2-GSDMB region have been associated with both childhood and adult-onset asthma. rs11557467 in the promoter region of ZPBP2 was strongly related to asthma in children and adults. Presence of G

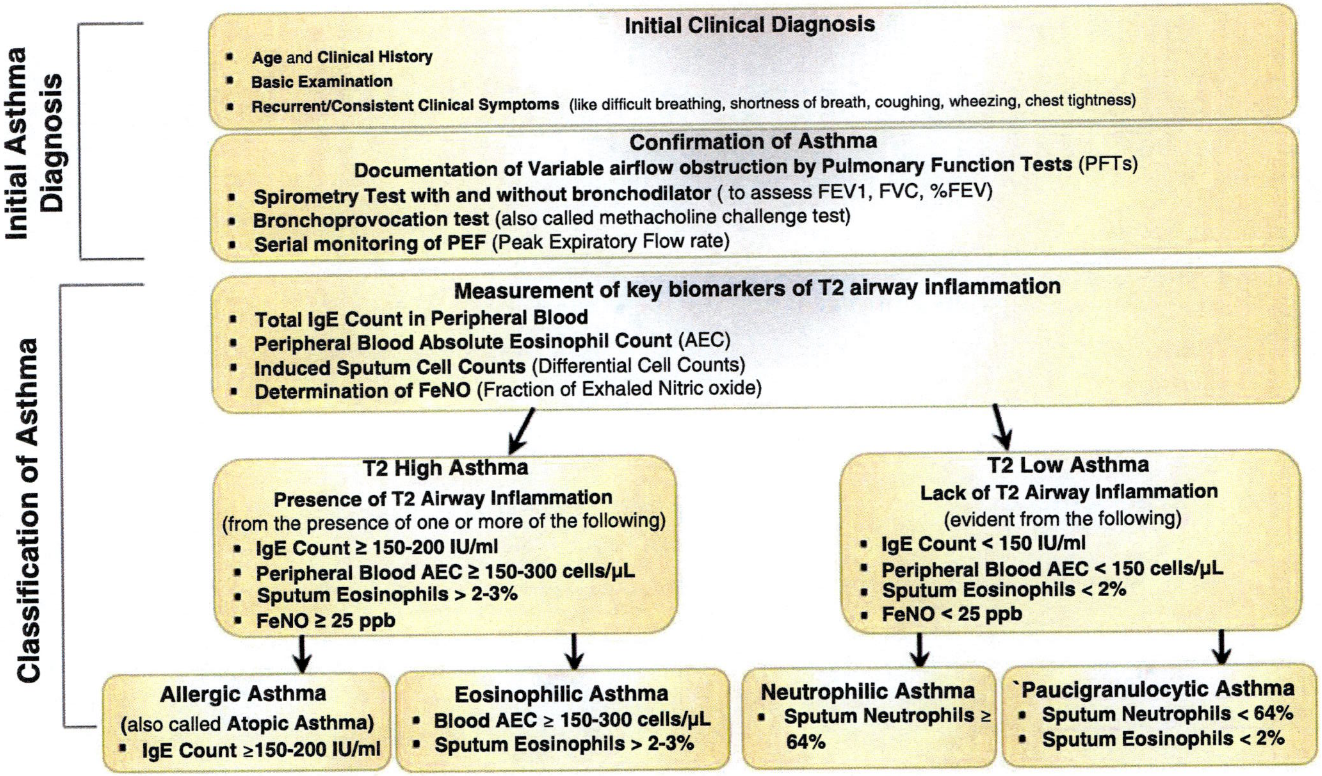

Fig. 4.3 Approaches for diagnosis and classification of asthma as T2-high and T2-low. Various clinical tests and pulmonary function tests aid in detecting asthma and further categorizing it into T2-high asthma and T2-low (or non T2-high) asthma

Table 4.1 Different biomarkers of asthma

Omics approach	Biomarkers	Biological sample	Types of asthma	Comments
Genomics	miR-196a	Blood	Eosinophilic asthma	rs11614913 (CT/CC genotype) of miR-196a is associated with eosinophilic asthma and increased sputum eosinophil count (Specjalski and Jassem 2019)
	ADAM33		Childhood asthma	rs511898 of ADAM33 was associated with increased risk of childhood asthma (Li et al. 2019)
	IL-6		Adult-onset atopic asthma	rs1800797 of IL-6 was associated with increased risk of adult onset asthma (Lajunen et al. 2016)
	IL5RA		Childhood Asthma	rs71058675 and rs1153462 were associated with the risk of childhood asthma. rs6773701 was related to eosinophil counts (Forno et al. 2017)
	CHI3L1		Late onset Asthma	The C allele of rs946261 was associated with the development of late onset asthma (Kanazawa et al. 2019)
	GSDMB		Aspirin-exacerbated respiratory disease (AERD)	rs870830 and rs7216389 of GSDMB were significantly associated with AERD (Kim et al. 2017b)
Transcriptomics		Serum		Increased DPP-4 levels are

(continued)

Table 4.1 (continued)

Omics approach	Biomarkers	Biological sample	Types of asthma	Comments
	Dipeptidyl Peptidase 4 (DPP-4)		Asthma (responsiveness to therapy)	indicative of response to IL-13 therapy (Brightling et al. 2015)
	miR-199a-5p	Plasma, sputum	Neutrophilic asthma	Significantly elevated in neutrophilic asthma and is associated with poor lung function (Huang et al. 2018)
	miR-629-3p, miR-223-3p, miR-142-3p	Sputum	Severe neutrophilic asthma	These miRNAs were significantly upregulated in severe asthma and were associated with neutrophilic inflammation (Maes et al. 2016).
	NLRP3, Caspase-1, IL-1β		Neutrophilic asthma	Increased expression is noted in neutrophilic asthma patients (Simpson et al. 2014)
	Periostin (POSTN), CLCA1, SERPINB2	Bronchial Tissue	T2-high asthma	Increased expression serves as markers of Th2-driven airway inflammation (Southworth et al. 2021)
	miR-26a-5p	Blood, sputum, bronchial biopsies	Eosinophilic asthma	Reduced expression was associated with severe eosinophilic airway inflammation (Roffel et al. 2022)
Epigenomics	IL-13, RUNX3	Peripheral Blood Mononuclear Cells (PBMCs)	Childhood asthma	Hypomethylation of IL-13 and RUNX3 is associated with T2

(continued)

Table 4.1 (continued)

Omics approach	Biomarkers	Biological sample	Types of asthma	Comments
				inflammatory response in childhood asthma (Yang et al. 2015)
	FCER2, TGFB1		Childhood obesity associated asthma	Hypermethylated in obese asthmatics (Rastogi et al. 2013)
	CCL5, IL2RA, TBX21			Decreased promoter methylation is noted in obese asthma patients (Rastogi et al. 2013)
	ASCL3	Umbilical cord white blood cells	Childhood asthma	Methylation of the ASCL3 5′-CpG island(s) is associated with risk of childhood asthma (Perera et al. 2009)
Proteomics	IgE	Serum	Allergic asthma	Elevated (>150 IU/mL) in allergic asthma (Lama et al. 2013)
	Periostin (POSTN)		Eosinophilic asthma	Increased significantly in the T2 high eosinophilic subtype of severe asthma (Jia et al. 2012)
	IP-10		Virus induced acute asthma exacerbations	Elevated levels indicate deterioration of lung function and excessive airway inflammation (Wark et al. 2007)
	IL-17		Severe asthma	Increased in severe asthma as compared to mild/moderate asthma (Agache et al. 2010)

(continued)

Table 4.1 (continued)

Omics approach	Biomarkers	Biological sample	Types of asthma	Comments
	S100A9		Neutrophilic asthma	Overexpressed in neutrophilic asthma and can serve as a potential biomarker (Quoc et al. 2021)
	YKL-40		Neutrophilic asthma	Increased in neutrophilic asthma. May serve as a promising biomarker (Lee et al. 2021)
	IL-33, APOE	Blood plasma	Treatment naïve adult-onset atopic asthma	IL-33 was significantly upregulated while Apolipoprotein E (APOE) was significantly downregulated. IL-33 and APOE can serve as may serve as two-protein classifier-based biomarker of atopic asthma (Bhowmik et al. 2019)
	Endothelin-1 (ET-1)	Exhaled breath	Allergic asthma, exercise induced asthma	Significantly higher in patients with unstable asthma as compared to that in stable asthma (Zietkowski et al. 2008)
Metabolomics	Valine, isobutyric acid, acetic acid, tyrosine, isovalerate, histidine	Blood	Allergen sensitized childhood asthma	Valine, isobutyric acid, and acetic acid are associated with highly sensitized form of asthma, while tyrosine, isovalerate, and histidine are related to less sensitized asthma (Chiu et al. 2021)

(continued)

Table 4.1 (continued)

Omics approach	Biomarkers	Biological sample	Types of asthma	Comments
	Ceramide/ Sphingosine – 1-phosphate (SP-1)	Serum	Uncontrolled Neutrophilic Asthma	Associated with increased neutrophil count in blood (Kim et al. 2020)
	Nitric oxide	Exhaled breath	T2-high asthma	Elevated FeNO (≥30 ppb) levels noted (Ricciardolo et al. 2021)
	Bromotyrosine	Urine	allergic asthma	Elevated levels predict airflow limitation, increased exacerbation rate, and poorly controlled asthma (Tiotiu 2018)
	Leukotriene E4 (LTE4)		Aspirin Exacerbated Respiratory Disease (AERD)	Increased LTE4 level serves as predictive biomarker for aspirin intolerance and future exacerbations (Hagan et al. 2017)
	Prostaglandin D2 (PGD2)	Broncho Alveolar Lavage Fluid (BALF)	Severe asthma	Serves as measure of disease severity. Elevated in severe asthma as compared to mild asthma (Fajt et al. 2013)
Metagenomics	*Haemophilus influenzae, Moraxella catarrhalis*	Sputum	Poorly controlled neutrophilic asthma	Increased prevalence of *Haemophilus influenzae* and *Moraxella catarrhalis* was noted in neutrophilic asthma (Simpson et al. 2016; Abdel-Aziz et al. 2019)
	Tropheryma whipplei		Eosinophilic Asthma	Abundance of *Tropheryma whipplei* was noted in

(continued)

Table 4.1 (continued)

Omics approach	Biomarkers	Biological sample	Types of asthma	Comments
				eosinophilic asthma (Simpson et al. 2016)
	Pasteurella spp., *Fusobacterium* spp.	Broncho Alveolar Lavage (BAL)	Corticosteroid sensitive asthma	Increased prevalence of *Pasteurella* spp. and *Fusobacterium* spp. was noted (Goleva et al. 2013).
	Neisseria spp., *Haemophilus* spp., *Tropheryma* spp.		Corticosteroid Resistant Asthma	Abundance of *Neisseria* spp., *Haemophilus* spp., and *Tropheryma* spp. was observed (Goleva et al. 2013)
	Lactobacillus spp.	Broncho Alveolar Lavage (BAL), Endobronchial brushings	Severe Asthma	Increased prevalence of *Lactobacillus* spp. was related to severely impaired lung function (Denner et al. 2016)
	Prevotella spp., *Veillonella* spp., *Gemella* spp.			Reduced prevalence of these genera was associated with poor lung function (FEV1) in asthmatics (Denner et al. 2016)

This table enlists various important biomarkers of asthma identified using different omics approaches

allele of rs2952156 in ERBB2 conferred increased protection from asthma in ethnically diverse group of individuals (Hernandez-Pacheco et al. 2019). Genetic variants in various genes including ADAM33, IL-4, IL-6, IL-13, CSF2, MUC7, TNF, ADRB2, and RANTES have been associated with asthma susceptibility and can be explored further as predictive biomarkers of asthma. Few of these SNP signatures specific to asthma have been tabulated in Table 4.1. Several SNPs like rs73650726 (SPATA31D1-RASEF), rs7903366, rs7081864, and rs7070958 (PRKG1) have been associated with bronchodilator response in African American asthmatics (Spear et al. 2019). SNPs in CRHR1 (rs242941 and rs1876828) and GLCC1 (rs37973) have been correlated with increased response to corticosteroids in asthma patients (Cazzola et al. 2020; McGeachie et al. 2013).

4.5 Epigenomics

Epigenetic studies have provided evidences for association of epigenetic regulation of gene expression, especially DNA methylation with the onset of asthma. Alteration in DNA methylation pattern of genes involved in epithelial barrier function and inflammation effectively distinguishes asthma from non-asthmatic condition. Hyper-methylation of protocadherin-20 (PCDH20) has been linked to asthma in smokers (Sood et al. 2012). Hypo-methylation of IL5RA, RUNX3, IL-13, and TIGIT has also been correlated with asthma (Gomez 2019; DeVries and Vercelli 2016). Besides, decreased methylation of genes involved in surfactant production (PPT2, ACOT7), nitric oxide synthesis (ACP5), phagocytosis (SERPINC1), and airway remodelling (FOXP1, COL15A1, CCDC19, and RB1) have also been observed in asthma patients (Ntontsi et al. 2021). Specific methylation profiles of CDHR3, CDH26, NTRK1, FBXL7, and SLC9A3 have been associated with atopic asthma in children (Forno et al. 2019). Hypo-methylation and reduced expression of Vanin-1 (VNN1) have been associated with poor therapeutic response to inhaled corticosteroids (Agache et al. 2021). Some of the epigenetic signatures of asthma have been enlisted in Table 4.1. Epigenetic modifications in asthma needs to explored further to evaluate their diagnostic and therapeutic potential.

4.6 Transcriptomics

Transcriptomics studies using microarray and RNA sequencing (RNA-seq) have enabled deeper mechanistic understanding of asthma. Transcriptomics enables effective stratification of asthma into T2-high and T2-low asthma based on the expression profile of Th2 inflammation associated genes. Overexpression of Th2 signatures like IL-4, IL-5, and IL-13 are observed only in T2-high asthma (Singh et al. 2020). Reduced expressions of CCL15, CA8, CSH1, ERBB2, SUCNR1, MMP16, SYT13, FOXA2, HPN, OSBPL6, and PROS1 were associated with increased asthma severity. On the other hand, the expression of FUT3, FUT5, TLE1, FERMT1, DPYSL3, NMU, KCNK6, PTPRH, and HOMER2 was positively correlated with severity of asthma (Modena et al. 2017). Differential expression of miRNAs and transcription factors have been noted in asthma and associated with asthma specific differentially expressed genes (DEGs) via Feed Forward Loop (FFL) (Singh et al. 2020). Persistent airflow limitation in severe asthmatics has been associated with transcriptomics-derived distinct underlying gene regulatory networks (Hekking et al. 2017). Some other asthma-associated DEGs and differentially expressed miRNAs as identified by transcriptomics studies have been tabulated in Table 4.1.

4.7 Proteomics

The emergence and advancement of proteomics technologies have enabled identification of biomarkers and novel drug targets for asthma. These biomarkers often enable phenotyping and sub-phenotyping of asthma. Increased serum periostin expression serves as a biomarker of eosinophilic asthma (Takahashi et al. 2019). Increased serum levels of complement factors (C3 and C4) and cyclooxygenase have been associated with lung function (FEV1) in patients with aspirin-induced asthma (Xu et al. 2020). Sputum alpha-1-antitrypsin and transthyretin were elevated in eosinophilic asthmatics, whereas S100A9, azurocidin, myeloperoxidase, and neutrophil gelatinase-associated lipocalin were significantly increased in neutrophilic asthma patients (Schofield et al. 2019). Increased sputum expression of CFS2 and AGR2 protein has been noted in current smokers with severe asthma (CSA) and aid in distinguishing CSA from non-smoker asthmatics (Takahashi et al. 2018). Galectin 3 has been identified as a predictive tool for determining airway remodelling modulation in severe asthmatics in response to omalizumab (Anti-IgE monoclonal antibody) therapy (Mauri et al. 2014). Other potential asthma biomarker candidates derived from proteomics study have been summarized in Table 4.1.

4.8 Metabolomics

Systemic metabolomics analyses using blood, sputum, exhaled breath condensate, BALF, and urine identified dysregulation of several metabolites to be associated with hypermethylation, hypoxia, and altered immune response in asthma pathogenesis. Arginine and levels of exhaled Nitric Oxide(NO) were related to Th2 inflammatory response in asthma (Sim et al. 2021). Fractional exhaled NO (FeNO) is widely used as a diagnostic biomarker of Type 2 inflammation in severe asthma (Chung 2021). Elevated urinary levels of alanine, acetylcarnitine, carnitine, and trimethylamine-N-oxide were observed in patients with severe asthma. Acetate, malonate, citrate, dimethylglycine, phenylacetylglutamine, and hippurate were significantly reduced in urine of severe asthmatics (Loureiro et al. 2014). Besides changes in amino acids, alteration in lipid metabolites has been linked with lung dysfunction and inflammatory response in asthma. Prostaglandins, leukotrienes, lipoxins, thromboxanes, and hydroxyeicosatetraenoic acids are significantly altered in asthma. Urinary leukotriene E4 (LTE4) levels are commonly used as diagnostic indicators of aspirin-exacerbated respiratory disease (AERD) (Miyata and Arita 2015). Furthermore, increased sphingosine-1-phosphate (S1P) levels have been correlated with asthma severity (Kim et al. 2020). Other metabolic signatures of asthma derived from different biological samples have been listed in Table 4.1.

4.9 Metagenomics

Airway microbial dysbiosis is noted in asthma as compared to normal healthy state. Variations in microbiome are noted amongst the different phenotypes and sub-phenotypes of asthma. Several metagenomics studies have been conducted to evaluate the changes in the microbiome across the gut–lung axis in asthma. Decreased airway abundance of *Gemella* spp. and *Porphyromonas* spp. together with an increased relative abundance of *Haemophilus* spp., *Moraxella* spp., and *Klebsiella* spp. has been associated with Type 2 asthma. Similarly, increased abundance of *Porphyromonas* spp., *Fusobacterium* spp., and *Tropheryma whipplei* has been related to non-type 2 asthma (Barcik et al. 2020). Different other alterations in airway microbiome in asthma have been listed in Table 4.1. Gut bacterial dysbiosis has also been correlated to asthma. Reduced gut microbial diversity was associated with increased risk of asthma in children. Gut *Clostridium difficile* abundance was associated with risk of asthma. Reduced relative abundance of *Lachnospira* spp., *Veillonella* spp., *Faecalibacterium* spp., and *Rothia* spp. in gut was significantly related to increased susceptibility of asthma in children (Hufnagl et al. 2020). These metagenomics observations thus highlight the importance of cross talk between gut–lung microbiome in regulating immune homeostasis and relate microbial dysbiosis with asthma risk.

4.10 Exposomics

Environmental exposures (i.e. the exposome) are major risk factors for asthma. Exposure may include allergens, microbes, air pollutants, radiation, diet, and drugs/chemicals. Second-hand smoke and oxidative stress were found to interact with genetic factors and are strongly associated with asthma risk (Johansson et al. 2021; Wild 2012). Several large collaborative studies were performed to explore the relationship between exposome and allergic diseases like asthma including the European Human Early-Life Exposome Study (HELIX) and the Kingston Allergy Birth Cohort (KABC) (Maitre et al. 2018; North et al. 2017). Exposome research is at an early stage and characterizing all the exposomes and identifying their direct role in asthma is a future research goal.

4.11 Bioinformatics

A few databases of asthma biomarkers are available for end users. Collection and manual curation of asthma biomarkers from the literature was done in the Database of Allergy and Asthma Biomarkers (DAAB) (Sircar et al. 2015; Majumdar et al. 2021). DAAB-v2 is freely available at http://bicresources.jcbose.ac.in/ssaha4/daab-v2/, and it allows users to access genes/proteins that are differentially expressed in allergic asthma, SNPs, and drug candidates. Another database called AllerGAtlas 1.0 also harbours information on biomarkers of allergic asthma (Liu et al. 2018).

Besides, two projects, namely like U-BIOPRED and MeDALL have also been implemented for efficient identification of biomarkers of respiratory and allergic diseases including severe asthma (Lefaudeux et al. 2017; Simpson et al. 2019; Bousquet et al. 2016). There is a database of biomarkers of exposure to environmental pollutants and dietary factors along with their concentration, named Exposome Explorer (Neveu et al. 2017). These databases allow users to search and browse the biomarkers data and classify the disease risk and its phenotypes.

4.12 Medical Imaging

Chest X-ray and Computerized Tomography (CT) scans are performed in severe asthmatics to study airway structure and perform functional analysis. The role of imaging in asthma is primarily in excluding other lung diseases. High-resolution computed tomography (HRCT) and magnetic resonance imaging (MRI) are also used to create distinct clusters of asthmatics (Ash and Diaz 2017).

4.13 Multi-omics and Data Integration

Integrating multi-omics data in asthma has been performed for predicting asthma risk in patients and molecular classification of asthma. Some success stories of integrative omics include the identification of endotypes in infants with rhinovirus bronchiolitis; identification of genes like IL5RA as risk factors using genome-wide (GW) genotype data, GW methylation, GW expression profiling, and cytokine levels (Forno et al. 2017; Raita et al. 2021). Machine learning can be a useful tool in omics data integration that could lead to the development of drug targets for asthma control and prevention (Gautam et al. 2022).

4.14 Current Applications

The mainstay therapy for asthma includes beta-2 agonists, muscarinic antagonists and corticosteroids. Often combination therapies are used to achieve better relief from asthma symptoms. Mast cell stabilizers and leukotriene modifiers are also used in case of T2-high asthma (like allergic asthma and eosinophilic asthma) to minimize airway inflammation and airway hyperresponsiveness (Sharma et al. 2022). T2-low asthma appears to be resistant to corticosteroids. Macrolide antibiotics are used to provide relief from acute exacerbations in case of refractory neutrophilic asthma (Kyriakopoulos et al. 2021). The details of the different drugs widely used for controlling asthma symptoms and exacerbations have been listed in Table 4.2.

Several monoclonal antibodies like omalizumab, reslizumab, mepolizumab, benralizumab, and dupilumab have been approved for use in asthma (Lee et al. 2021). Various other biologics and small molecule inhibitors are being clinically

Table 4.2 Drugs used for controlling asthma exacerbations

Class of drugs		Drugs	Mode of action	Types of asthma	Comments
Beta-2 agonists	Long-acting beta-2 agonists (LABA)	Salmeterol, formoterol, arformoterol, indacaterol, vilanterol	Binds to beta-2 adrenergic receptor and triggers protein kinase A (PKA) mediated phosphorylation of several targets, leading to airway smooth muscle relaxation	Moderate asthma, severe asthma	Monotherapy not given, used with ICS. Provide optimal control of asthma (Sharma et al. 2022; Mauer and Taliercio 2020)
	Short-acting beta-2 agonists (SABA)	Albuterol, levalbuterol		Acute asthma broncho-constriction, exercise-induced asthma	Relieve acute broncho-constriction. Prevents exercise induced asthma. Not used as maintenance therapy for chronic asthma (Sharma et al. 2022; Mauer and Taliercio 2020)
Anticholinergics (Antimuscarinics)	Long-acting muscarinic antagonists (LAMA)	Tiotropium, glycopyrronium, umeclidinium	Blocks muscarinic receptors M1 and M3 on bronchial and airway smooth muscle cells and thereby reduces broncho-constriction	T2-low severe asthma	Improves lung function (FEV1) (Murphy et al. 2021; Kyriakopoulos et al. 2021; Sharma et al. 2022)
	Short-acting muscarinic antagonists (SAMA)	Ipratropium		Acute asthma exacerbations	Used as adjunctive therapy for emergency cases of acute asthma exacerbation. Relieves symptoms of asthma (Sharma et al. 2022; https://www.msdmanuals.com/professional/pulmonary-disorders/asthma-and-related-disorders/drug-treatment-of-asthma)
Corticosteroids		Beclomethasone, budesonide, flunisolide,	Exerts anti-inflammatory effects by promoting	Moderate asthma, severe asthma	Reduce airway inflammation. Control

	Inhaled corticosteroids (ICS)	fluticasone propionate/ fluticasone furoate, mometasone, ciclesonide	HDAC-mediated reduction in the transcription of inflammatory cytokines		asthma symptoms. Used along with LABA as first-line maintenance therapy–this further improves asthma control (Sharma et al. 2022; Mauer and Taliercio 2020)
	Systemic or oral corticosteroids (OCS)	Methyl-prednisolone, prednisolone, prednisone		Severe asthma exacerbations	Resolve asthma exacerbation symptoms rapidly. Decrease hospitalization and relapse rate (Papi et al. 2020)
Combination Therapy	SAMA + SABA	Ipratropium and albuterol (salbutamol)	Acts synergistically to relax airway smooth muscles and reduce broncho-constriction	Acute asthma exacerbations	Rapid relief from dyspnoea and acute asthma symptoms (Xu et al. 2021)
	ICS + LABA	Budesonide and formoterol, mometasone and formoterol, fluticasone and salmeterol, fluticasone and vilanterol	Acts synergistically to improve efficacy of controlling airway inflammation and airway remodelling	Moderate to severe persistent asthma	Improve lung function (FEV1). Better control of asthma symptoms (Papi et al. 2020)
	LABA + LAMA+ ICS	Indacaterol, glycopyrronium bromide, and mometasone furoate	Synergistic interaction between LABA, LAMA, and ICS enhances efficacy	Poorly controlled asthma, severe asthma	Improve lung function and control asthma symptoms. Superior than ICS + LABA therapy (Rogliani et al. 2022)
Methylxanthines		Theophylline	Leads to airway smooth muscle relaxation and weak bronchodilation by increasing cellular cAMP and cGMP (by inhibiting	Nocturnal Asthma	Relax airway passage and improve breathing. Not recommended for asthma by GINA 2018 due to its adverse effects (Sharma et al. 2022; https://www.

(continued)

Table 4.2 (continued)

Class of drugs	Drugs	Mode of action	Types of asthma	Comments
		phosphodiesterases – PDE3/4/5)		msdmanuals.com/professional/pulmonary-disorders/asthma-and-related-disorders/drug-treatment-of-asthma)
Leukotriene modifiers	Montelukast, zafirlukast	Leukotriene receptor antagonist—serves as competitive inhibitor of leukotrienes D4 and E4	Mild to severe persistent asthma, allergic asthma, exercise induced asthma	Improve baseline lung function. Reduce airway hyperresponsiveness (Sharma et al. 2022; Mauer and Taliercio 2020; https://www.msdmanuals.com/professional/pulmonary-disorders/asthma-and-related-disorders/drug-treatment-of-asthma)
	Zileuton	Inhibits 5-Lipoxygenase (5-LO) (Sharma et al. 2022; Bouchette and Preuss 2022)		
Mast cell stabilizers	Cromolyn sodium	Inhibits histamine and leukotriene (slow reacting substance of anaphylaxis) release from mast cells	Allergic asthma, eosinophilic asthma	Reduce airway hyperresponsiveness. Prevent early and late responses to allergens (Sharma et al. 2022; https://www.msdmanuals.com/professional/pulmonary-disorders/asthma-and-related-disorders/drug-treatment-of-asthma)
Macrolides	Azithromycin	Has anti-bacterial effect. Exert anti-inflammatory and immunomodulatory effects by inhibiting activation of	Childhood asthma, eosinophilic asthma, refractory neutrophilic asthma	Improve asthma exacerbations and quality of life significantly in eosinophilic and

		NF-κβ and activator protein 1		non-eosinophilic asthma (Murphy et al. 2021; Kyriakopoulos et al. 2021; Wong et al. 2014)
	Erythromycin		Refractory neutrophilic asthma	Reduce severity of bronchial responsiveness in bronchial asthma (Kyriakopoulos et al. 2021; Korematsu et al. 2010)
	Clarithromycin		Neutrophilic Asthma	Reduce neutrophilic inflammation. Improve asthma control (Kyriakopoulos et al. 2021; Wong et al. 2014)

investigated for their efficacy in different phenotypes and sub-phenotypes of asthma (as shown in Table 4.3).

A non-medical intervention termed as bronchial thermoplasty has been approved for patients with severe uncontrolled persistent asthma. This technique makes use of therapeutic radiofrequency to destroy damaged bronchial tissues, thereby reducing broncho-constriction (Sharma et al. 2022).

Differential response of the different sub-phenotypes of asthma to the currently available therapeutics demands better understanding of asthma disease biology; and multi-omics mediated identification of biomarkers and drug targets as a step towards the quest for novel therapeutics and precision medicine for asthma.

4.15 Future Perspectives

A large part of the clinical, pathogenic, and molecular heterogeneity of asthma has been brought to our arena of understanding through the existing and evolving multi-omics knowledge. It has already resulted in effective phenotyping and endotyping of the disease in terms of understanding the pathogenesis and deciding the therapeutic strategy. However, several areas of asthma research are yet to be explored to ensure further improvements in our understanding of asthma pathobiology. There is an urgent need for identification of biomarkers for efficient stratification of paediatric and adult asthma. Translation of biomarker discovery into clinical studies could help in designing novel biologics for paediatric asthma (Golebski et al. 2020). Apart from this, the treatment of Type 2 low neutrophilic asthma is a huge challenge, since the underlying mechanism is not completely elucidated (Kyriakopoulos et al. 2021). Similarly, obesity is a vital risk factor and disease modifier for asthma, but, the exact mechanism by which obesity modulates airway immune response and triggers asthma pathogenesis is still shrouded in mystery (Peters et al. 2018). Besides, the role of mitochondrial dysfunction in the pathogenesis of asthma needs to be explored (Qian et al. 2022). In recent past, asthma-COPD overlap syndrome (ACOS) has evolved as a huge cause of concern due to increasing disease burden. Biomarkers for efficient diagnosis and distinction form asthma are a need of the hour (Tu et al. 2021). Multi-omics and systems biology-based approaches shall be employed to improve our understanding of neutrophilic asthma, obese asthma, and ACOS. Microbial dysbiosis and perturbation in metabolic profile have been linked with asthma, however, the exact role of microbiome and metabolism in regulating airway immune response in asthma needs to be dissected. The link between altered metabolome and perturbed microbial flora should be explored as potential mediators of asthma (Kozik et al. 2022). Several challenges are encountered including lack of complete metabolic database for reliable identification of metabolites, lack of accurate determination of the relative contribution of host and microbes in the altered metabolomic profile, and lack of suitable animal models, which can harbour human airway microbes. It is also important to evaluate whether reversal of microbial dysbiosis would lead to restoration of normal metabolomic profile or bring about reduction in asthma exacerbations. Besides, gut microbiome has also been known to

Table 4.3 Biologics for Asthma

Sl. no.	Biologics	Description	Target	Types of asthma	Major effects	Status
1.	Omalizumab	Anti-IgE monoclonal antibody	IgE	Allergic asthma	Reduced serum IgE. Decrease in exacerbation rates and use of oral corticosteroids. Little effect on FEV1 and other symptoms of asthma	Approved (Lee et al. 2021; Papi et al. 2018)
2.	Reslizumab	Anti-human IL-5 monoclonal antibody	IL-5	Severe eosinophilic asthma	Lead to improvement in lung function (FEV1) and asthma control. Reduced asthma exacerbations and sputum eosinophil count	Approved (Lee et al. 2021; Papi et al. 2018)
3.	Mepolizumab	Anti-IL-5 humanized monoclonal antibody	IL-5	Severe eosinophilic asthma	Reduction in asthma exacerbations, blood and sputum eosinophil count. Improved lung function (FEV1). Decreased the use of oral corticosteroids	Approved (Lee et al. 2021; Papi et al. 2018)
4.	Benralizumab	Anti-interleukin-5 receptor α monoclonal antibody	IL-5Rα	Severe eosinophilic asthma	Decreased rates of asthma exacerbation	Approved (Lee et al. 2021)
5.	Dupilumab	Anti-interleukin-4 receptor monoclonal antibody	IL-4R (thus blocks IL-4 and IL-13 pathways)	Severe Asthma	Improved lung function (FEV1) and asthma control. Decreased exacerbation rates Improved MRI ventilation heterogeneity	Approved (Lee et al. 2021)
6.	Lebrikizumab	Anti-human IL-13 monoclonal antibody	IL-13	Uncontrolled asthma with high serum periostin level	Improved lung function (FEV1). Reduced asthma exacerbation rates.	Investigational (Clinical Trial Phase III) (Hanania et al. 2016)

(continued)

Table 4.3 (continued)

Sl. no.	Biologics	Description	Target	Types of asthma	Major effects	Status
7.	Tralokinumab	Anti-human IL-13 monoclonal antibody	IL-13	Severe, uncontrolled asthma	Showed trends of decrease in asthma exacerbation rate	Investigational (Clinical Trial Phase III) (Panettieri Jr. et al. 2018)
8.	Tezepelumab	Human monoclonal antibody that binds to thymic stromal lymphopoietin	TSLP	Severe, Uncontrolled Asthma	Reduction in asthma exacerbations. Improvement in lung function (FEV1), asthma control and quality of life	Investigational (Clinical Trial Phase III) (Menzies-Gow et al. 2021)
9.	Pitrakinra	IL-4 and IL-13 antagonist	IL-4R	Moderate to severe asthma	Reduction of exacerbations and nocturnal awakening	Investigational (Clinical Trial Phase II) (Slager et al. 2012)
10.	Etokimab (ANB020)	Anti-IL-33 antibody	IL-33	Severe eosinophilic asthma	Marked and rapid improvement in lung function (FEV1). Decreased blood eosinophil count	Investigational (Clinical Trial Phase IIa) (Cusack et al. 2021)
11.	Daclizumab	Human monoclonal antibody that binds to α subunit of IL-2 receptor of T cells.	IL-2R	Chronic persistent asthma	Improved lung function and asthma control	Investigational (Clinical Trial Phase II) (Busse et al. 2008)
12.	Eculizumab	Human monoclonal antibody that binds to complement factor C5 and prevents its cleavage	C5	Mild allergic asthma	Lead to improvement in spirometry and reduction in late allergen induced eosinophilia in sputum	Investigational (Clinical Trial Phase II) (Menzella et al. 2015)
13.	Canakinumab	Human anti-IL-1β monoclonal antibody	IL-1β	Mild allergic asthma	Exerts anti-inflammatory effects on late asthmatic response upon allergen inhalation	Investigational (Clinical Trial Phase IIb) (Menzella et al. 2015)

14.	Anrukinzumab	Anti-human IL-13 monoclonal antibody	IL-13	Mild allergic asthma	Marginal improvements noted	Investigational (Clinical Trial Phase II) (Hua et al. 2015)
15.	Itepekimab (REGN3500)	Anti-IL-33 antibody	IL-33	Moderate to severe asthma	Markedly reduced blood eosinophil count. Resulted in better asthma control and improved quality of life	Investigational (Clinical Trial Phase II) (Wechsler et al. 2021)
16.	Astegolimab (MSTT1041A)	Anti-ST2 monoclonal antibody	ST2	Severe asthma	Decreased asthma exacerbation rates	Investigational (Clinical Trial Phase II) (Kelsen et al. 2021)
17.	Quilizumab	Anti-IgE monoclonal antibody	IgE	Allergic asthma	Reduced serum IgE Had no effect on lung function or exacerbations in allergic asthma	Investigational (Clinical Trial Phase II) (Harris et al. 2016)
18.	Anakinra	Recombinant non-glycosylated form of Human IL-1 receptor antagonist	IL-1Ra	Severe, uncontrolled neutrophilic asthma (add on therapy)	Reduction in sputum IL-1β, IL-6, and IL-8. Reduction in neutrophilic airway inflammation	Investigational (Clinical Trial Phase II) (Kyriakopoulos et al. 2021)
19.	Navarixin	CXCR1/2 Antagonist	CXCR2	Neutrophilic asthma (add on therapy)	Considerable reduction in blood and sputum neutrophil, but, no improvement in exacerbation rates	Investigational (Clinical Trial Phase II) (Kyriakopoulos et al. 2021)
20.	Nemiralisib	Inhibitor of phosphoinositide 3-kinase	PI3Kδ	Severe, persistent, uncontrolled asthma	Reduction in IL-5, IL-6, IL-8, and IL-13 levels in induced sputum. Local inhibition of PI3Kδ and reduction of airway inflammation but no marked clinical improvement noted	Investigational (Kyriakopoulos et al. 2021)

(continued)

Table 4.3 (continued)

Sl. no.	Biologics	Description	Target	Types of asthma	Major effects	Status
21.	Duvelisib (IPI-145)	Inhibitor of phosphoinositide 3-kinase	PI3Kδ/γ	Mild asthma	Decrease in mediators of airway inflammation. Clinical improvement in late asthmatic reaction	Investigational (Clinical trial phase IIa) (Kyriakopoulos et al. 2021)
22.	RV-1729	Inhibitor of phosphoinositide 3-kinase (Inhaled)	PI3Kδ/γ	Severe, uncontrolled asthma	Bronchodilation. Reduction in airway inflammation. Reversal of β2-adreno receptor (β2AR) tachyphylaxis	Investigational (Kyriakopoulos et al. 2021)
23.	Ensifentrine (RPL-554)	Dual inhibitor of phosphodiesterase-3 and phosphodiesterase-4	PDE3, PDE4	Severe, uncontrolled neutrophilic asthma (add on therapy)	Bronchodilation. broncho protection. Anti-inflammatory effects	Investigational (Kyriakopoulos et al. 2021)
24.	CHF6001	Phosphodiesterase-4 inhibitor (Inhaled)	PDE4	Asthma (Add on Therapy)	Reduction in late asthmatic response. Inhibition of IFNγ, IL-2, and IL-17	Investigational (Kyriakopoulos et al. 2021)
25.	Imatinib	Tyrosine kinase inhibitor	ABL, c-KIT, PDGF-R	Severe refractory asthma, T2-low asthma	Decrease in airway hyperresponsiveness, mast cell counts, and tryptase production	Investigational (Kyriakopoulos et al. 2021)
26.	Masitinib	Tyrosine kinase inhibitor	c-KIT, PDGF-R, Lck, FAK, FGFR3, CSF1R	Severe, persistent, uncontrolled asthma	Significant reduction in asthma exacerbations	Investigational (Clinical Trial Phase III) (Kyriakopoulos et al. 2021)
27.	AZD7624	MAPK inhibitor (Inhaled)	p38 MAPK	Corticosteroid resistant asthma	Has anti-inflammatory effects	Investigational (Clinical Trial Phase II) (Kyriakopoulos et al. 2021)

modulate immuno-metabolism and regulate the risk of asthma onset (Kozik et al. 2022; Fonseca et al. 2021). However, there exists poor concordance between gut microbial variations in asthma (Abdel-Aziz et al. 2019). Use of both targeted and untargeted metabolomics, analysis of metabolite origins using networks, use of model-based integration of metabolite observation and species abundance, and development of robust tools for multi-omics data integration and statistical analysis can aid in alleviating some of the challenges (Shaffer et al. 2019; Noecker et al. 2016). Currently used biologics mostly entails monoclonal antibodies targeting interleukins, which are costly and effective in a subset of diseases population (Berger et al. 2002). Omics-based studies can provide mechanistic understanding of asthma pathogenesis and help in identification of potential drivers of asthma. This, in turn, shall lead to subsequent development of more precise and targeted therapeutic approaches for effectively alleviating asthma exacerbations.

References

Abdel-Aziz MI, Vijverberg SJH, Neerincx AH, Kraneveld AD, Maitland-van der Zee AH (2019) The crosstalk between microbiome and asthma: exploring associations and challenges. Clin Exp Allergy 49(8):1067–1086

Agache I, Ciobanu C, Agache C, Anghel M (2010) Increased serum IL-17 is an independent risk factor for severe asthma. Respir Med 104(8):1131–1137

Agache I, Eguiluz-Gracia I, Cojanu C, Laculiceanu A, Del Giacco S, Zemelka-Wiacek M, Kosowska A, Akdis CA, Jutel M (2021) Advances and highlights in asthma in 2021. Allergy 76(11):3390–3407

Ash SY, Diaz AA (2017) The role of imaging in the assessment of severe asthma. Curr Opin Pulm Med 23(1):97–102

Barcik W, Boutin RCT, Sokolowska M, Finlay BB (2020) The Role of Lung and Gut Microbiota in the Pathology of Asthma. Immunity 52(2):241–255

Berger M, Shankar V, Vafai A (2002) Therapeutic applications of monoclonal antibodies. Am J Med Sci 324(1):14–30

Bhowmik M, Majumdar S, Dasgupta A, Gupta Bhattacharya S, Saha S (2019) Pilot-scale study of human plasma proteomics identifies ApoE And IL33 as markers in atopic asthma. J Asthma Allergy 12:273–283

Bouchette D, Preuss CV: Zileuton. In: *StatPearls*. Treasure Island (FL); 2022

Bousquet J, Anto JM, Akdis M, Auffray C, Keil T, Momas I, Postma DS, Valenta R, Wickman M, Cambon-Thomsen A et al (2016) Paving the way of systems biology and precision medicine in allergic diseases: the MeDALL success story: Mechanisms of the Development of ALLergy; EU FP7-CP-IP; Project No: 261357; 2010-2015. Allergy 71(11):1513–1525

Brasier AR, Victor S, Ju H, Busse WW, Curran-Everett D, Bleecker E, Castro M, Chung KF, Gaston B, Israel E et al (2010) Predicting intermediate phenotypes in asthma using bronchoalveolar lavage-derived cytokines. Clin Transl Sci 3(4):147–157

Brightling CE, Chanez P, Leigh R, O'Byrne PM, Korn S, She D, May RD, Streicher K, Ranade K, Piper E (2015) Efficacy and safety of tralokinumab in patients with severe uncontrolled asthma: a randomised, double-blind, placebo-controlled, phase 2b trial. Lancet Respir Med 3(9): 692–701

Busse WW, Israel E, Nelson HS, Baker JW, Charous BL, Young DY, Vexler V, Shames RS, Daclizumab Asthma Study G (2008) Daclizumab improves asthma control in patients with moderate to severe persistent asthma: a randomized, controlled trial. Am J Respir Crit Care Med 178(10):1002–1008

Cazzola M, Rogliani P, Calzetta L, Matera MG (2020) Pharmacogenomic response of inhaled corticosteroids for the treatment of asthma: considerations for therapy. Pharmgenomics Pers Med 13:261–271

Chiu CY, Cheng ML, Chiang MH, Wang CJ, Tsai MH, Lin G (2021) Integrated metabolic and microbial analysis reveals host-microbial interactions in IgE-mediated childhood asthma. Sci Rep 11(1):23407

Chung KF (2021) Increasing utility of FeNO as a biomarker of type-2 inflammation in severe asthma. Lancet Respir Med 9(10):1083–1084

Cusack RP, Whetstone CE, Xie Y, Ranjbar M, Gauvreau GM (2021) Regulation of eosinophilia in asthma-new therapeutic approaches for asthma treatment. Cell 10(4)

Das S, Miller M, Broide DH (2017) Chromosome 17q21 genes ORMDL3 and GSDMB in asthma and immune diseases. Adv Immunol 135:1–52

Denner DR, Sangwan N, Becker JB, Hogarth DK, Oldham J, Castillo J, Sperling AI, Solway J, Naureckas ET, Gilbert JA et al (2016) Corticosteroid therapy and airflow obstruction influence the bronchial microbiome, which is distinct from that of bronchoalveolar lavage in asthmatic airways. J Allergy Clin Immunol 137(5):1398–1405 e1393

DeVries A, Vercelli D (2016) Epigenetic mechanisms in asthma. Ann Am Thorac Soc 13(Suppl 1): S48–S50

Fajt ML, Gelhaus SL, Freeman B, Uvalle CE, Trudeau JB, Holguin F, Wenzel SE (2013) Prostaglandin D(2) pathway upregulation: relation to asthma severity, control, and TH2 inflammation. J Allergy Clin Immunol 131(6):1504–1512

Fonseca W, Malinczak CA, Fujimura K, Li D, McCauley K, Li J, Best SKK, Zhu D, Rasky AJ, Johnson CC et al (2021) Maternal gut microbiome regulates immunity to RSV infection in offspring. J Exp Med 218(11)

Forno E, Wang T, Yan Q, Brehm J, Acosta-Perez E, Colon-Semidey A, Alvarez M, Boutaoui N, Cloutier MM, Alcorn JF et al (2017) A multiomics approach to identify genes associated with childhood asthma risk and morbidity. Am J Respir Cell Mol Biol 57(4):439–447

Forno E, Wang T, Qi C, Yan Q, Xu CJ, Boutaoui N, Han YY, Weeks DE, Jiang Y, Rosser F et al (2019) DNA methylation in nasal epithelium, atopy, and atopic asthma in children: a genome-wide study. Lancet Respir Med 7(4):336–346

Gautam Y, Johansson E, Mersha TB (2022) Multi-omics profiling approach to asthma: an evolving paradigm. J Pers Med 12(1)

Golebski K, Kabesch M, Melen E, Potocnik U, van Drunen CM, Reinarts S, Maitland-van der Zee AH, SJH V, Consortium P (2020) Childhood asthma in the new omics era: challenges and perspectives. Curr Opin Allergy Clin Immunol 20(2):155–161

Goleva E, Jackson LP, Harris JK, Robertson CE, Sutherland ER, Hall CF, Good JT Jr, Gelfand EW, Martin RJ, Leung DY (2013) The effects of airway microbiome on corticosteroid responsiveness in asthma. Am J Respir Crit Care Med 188(10):1193–1201

Gomez JL (2019) Epigenetics in asthma. Curr Allergy Asthma Rep 19(12):56

Hagan JB, Laidlaw TM, Divekar R, O'Brien EK, Kita H, Volcheck GW, Hagan CR, Lal D, Teaford HG 3rd, Erwin PJ et al (2017) Urinary leukotriene E4 to determine aspirin intolerance in asthma: a systematic review and meta-analysis. J Allergy Clin Immunol Pract 5(4):990–997 e991

Hanania NA, Korenblat P, Chapman KR, Bateman ED, Kopecky P, Paggiaro P, Yokoyama A, Olsson J, Gray S, Holweg CT et al (2016) Efficacy and safety of lebrikizumab in patients with uncontrolled asthma (LAVOLTA I and LAVOLTA II): replicate, phase 3, randomised, double-blind, placebo-controlled trials. Lancet Respir Med 4(10):781–796

Harris JM, Maciuca R, Bradley MS, Cabanski CR, Scheerens H, Lim J, Cai F, Kishnani M, Liao XC, Samineni D et al (2016) A randomized trial of the efficacy and safety of quilizumab in adults with inadequately controlled allergic asthma. Respir Res 17:29

He Z, Feng J, Xia J, Wu Q, Yang H, Ma Q (2020) Frequency of signs and symptoms in persons with asthma. Respir Care 65(2):252–264

Hekking PP, Loza MJ, Pavlidis S, De Meulder B, Lefaudeux D, Baribaud F, Auffray C, Wagener AH, Brinkman P, Lutter R et al (2017) Transcriptomic gene signatures associated with persistent airflow limitation in patients with severe asthma. Eur Respir J 50(3)

Hernandez-Pacheco N, Pino-Yanes M, Flores C (2019) Genomic predictors of asthma phenotypes and treatment response. Front Pediatr 7:6

Hua F, Ribbing J, Reinisch W, Cataldi F, Martin S (2015) A pharmacokinetic comparison of anrukinzumab, an anti- IL-13 monoclonal antibody, among healthy volunteers, asthma and ulcerative colitis patients. Br J Clin Pharmacol 80(1):101–109

Huang Y, Zhang S, Fang X, Qin L, Fan Y, Ding D, Liu X, Xie M (2018) Plasma miR-199a-5p is increased in neutrophilic phenotype asthma patients and negatively correlated with pulmonary function. PLoS One 13(3):e0193502

Hufnagl K, Pali-Scholl I, Roth-Walter F, Jensen-Jarolim E (2020) Dysbiosis of the gut and lung microbiome has a role in asthma. Semin Immunopathol 42(1):75–93

Jia G, Erickson RW, Choy DF, Mosesova S, Wu LC, Solberg OD, Shikotra A, Carter R, Audusseau S, Hamid Q et al (2012) Periostin is a systemic biomarker of eosinophilic airway inflammation in asthmatic patients. J Allergy Clin Immunol 130(3):647–654 e610

Johansson E, Martin LJ, He H, Chen X, Weirauch MT, Kroner JW, Khurana Hershey GK, Biagini JM (2021) Second-hand smoke and NFE2L2 genotype interaction increases paediatric asthma risk and severity. Clin Exp Allergy 51(6):801–810

Kanazawa J, Kitazawa H, Masuko H, Yatagai Y, Sakamoto T, Kaneko Y, Iijima H, Naito T, Saito T, Noguchi E et al (2019) A cis-eQTL allele regulating reduced expression of CHI3L1 is associated with late-onset adult asthma in Japanese cohorts. BMC Med Genet 20(1):58

Kaplan A (2021) The myth of mild: severe exacerbations in mild asthma: an underappreciated, but preventable problem. Adv Ther 38(3):1369–1381

Kelsen SG, Agache IO, Soong W, Israel E, Chupp GL, Cheung DS, Theess W, Yang X, Staton TL, Choy DF et al (2021) Astegolimab (anti-ST2) efficacy and safety in adults with severe asthma: a randomized clinical trial. J Allergy Clin Immunol 148(3):790–798

Kim H, Ellis AK, Fischer D, Noseworthy M, Olivenstein R, Chapman KR, Lee J (2017a) Asthma biomarkers in the age of biologics. Allergy Asthma Clin Immunol 13:48

Kim LH, Chang H, Namgoong S, Kim JO, Cheong HS, Lee SG, Park JS, Baek AR, Koo SM, Choi IS et al (2017b) Genetic variants of the gasdermin B gene associated with the development of aspirin-exacerbated respiratory diseases. Allergy Asthma Proc 38(1):4–12

Kim SH, Jung HW, Kim M, Moon JY, Ban GY, Kim SJ, Yoo HJ, Park HS (2020) Ceramide/ sphingosine-1-phosphate imbalance is associated with distinct inflammatory phenotypes of uncontrolled asthma. Allergy 75(8):1991–2004

Korematsu S, Yamamoto K, Nagakura T, Miyahara H, Okazaki N, Akiyoshi K, Maeda T, Suenobu S, Izumi T (2010) The indication and effectiveness of low-dose erythromycin therapy in pediatric patients with bronchial asthma. Pediatr Allergy Immunol 21(3):489–492

Kozik AJ, Holguin F, Segal LN, Chatila TA, Dixon AE, Gern JE, Lozupone C, Lukacs N, Lumeng C, Molyneaux PL et al (2022) Microbiome, metabolism, and immunoregulation of asthma: an american thoracic society and national institute of allergy and infectious diseases workshop report. Am J Respir Cell Mol Biol 67(2):155–163

Kyriakopoulos C, Gogali A, Bartziokas K, Kostikas K (2021) Identification and treatment of T2-low asthma in the era of biologics. ERJ Open Res 7(2)

Lajunen TK, Jaakkola JJ, Jaakkola MS (2016) Interleukin 6 SNP rs1800797 associates with the risk of adult-onset asthma. Genes Immun 17(3):193–198

Lama M, Chatterjee M, Chaudhuri TK (2013) Total serum immunoglobulin e in children with asthma. Indian J Clin Biochem 28(2):197–200

Lee Y, Quoc QL, Park HS (2021) Biomarkers for severe asthma: lessons from longitudinal cohort studies. Allergy Asthma Immunol Res 13(3):375–389

Lefaudeux D, De Meulder B, Loza MJ, Peffer N, Rowe A, Baribaud F, Bansal AT, Lutter R, Sousa AR, Corfield J et al (2017) U-BIOPRED clinical adult asthma clusters linked to a subset of sputum omics. J Allergy Clin Immunol 139(6):1797–1807

Li HF, Yan LP, Wang K, Li XT, Liu HX, Tan W (2019) Association between ADAM33 polymorphisms and asthma risk: a systematic review and meta-analysis. Respir Res 20(1):38

Liu J, Liu Y, Wang D, He M, Diao L, Liu Z, Li Y, Tang L, He F, Li D et al: AllerGAtlas 1.0: a human allergy-related genes database. Database (Oxford) 2018, 2018

Loureiro CC, Duarte IF, Gomes J, Carrola J, Barros AS, Gil AM, Bousquet J, Bom AT, Rocha SM: Urinary metabolomic changes as a predictive biomarker of asthma exacerbation. J Allergy Clin Immunol 2014, 133(1):261–263 e261–265

Loverdos K, Bellos G, Kokolatou L, Vasileiadis I, Giamarellos E, Pecchiari M, Koulouris N, Koutsoukou A, Rovina N (2019) Lung microbiome in asthma: current perspectives. J Clin Med 8(11)

Maes T, Cobos FA, Schleich F, Sorbello V, Henket M, De Preter K, Bracke KR, Conickx G, Mesnil C, Vandesompele J et al (2016) Asthma inflammatory phenotypes show differential microRNA expression in sputum. J Allergy Clin Immunol 137(5):1433–1446

Maitre L, de Bont J, Casas M, Robinson O, Aasvang GM, Agier L, Andrusaityte S, Ballester F, Basagana X, Borras E et al (2018) Human Early Life Exposome (HELIX) study: a European population-based exposome cohort. BMJ Open 8(9):e021311

Majumdar S, Ghosh A, Saha S (2018) Modulating Interleukins and their Receptors Interactions with small chemicals using in silico approach for asthma. Curr Top Med Chem 18(13): 1123–1134

Majumdar S, Bhattacharjee S, Jana T, Saha S (2021) DAAB-V2: Updated database of allergy and asthma biomarkers. Allergy 76(12):3829–3832

Mauer Y, Taliercio RM (2020) Managing adult asthma: The 2019 GINA guidelines. Cleve Clin J Med 87(9):569–575

Mauri P, Riccio AM, Rossi R, Di Silvestre D, Benazzi L, De Ferrari L, Dal Negro RW, Holgate ST, Canonica GW (2014) Proteomics of bronchial biopsies: galectin-3 as a predictive biomarker of airway remodelling modulation in omalizumab-treated severe asthma patients. Immunol Lett 162(1 Pt A):2–10

McGeachie MJ, Wu AC, Chang HH, Lima JJ, Peters SP, Tantisira KG (2013) Predicting inhaled corticosteroid response in asthma with two associated SNPs. Pharmacogenomics J 13(4): 306–311

Menzella F, Lusuardi M, Galeone C, Zucchi L (2015) Tailored therapy for severe asthma. Multidiscip Respir Med 10(1):1

Menzies-Gow A, Corren J, Bourdin A, Chupp G, Israel E, Wechsler ME, Brightling CE, Griffiths JM, Hellqvist A, Bowen K et al (2021) Tezepelumab in adults and adolescents with severe, uncontrolled asthma. N Engl J Med 384(19):1800–1809

Miyata J, Arita M (2015) Role of omega-3 fatty acids and their metabolites in asthma and allergic diseases. Allergol Int 64(1):27–34

Modena BD, Bleecker ER, Busse WW, Erzurum SC, Gaston BM, Jarjour NN, Meyers DA, Milosevic J, Tedrow JR, Wu W et al (2017) Gene expression correlated with severe asthma characteristics reveals heterogeneous mechanisms of severe disease. Am J Respir Crit Care Med 195(11):1449–1463

Murphy RC, Pavord ID, Alam R, Altman MC (2021) Management strategies to reduce exacerbations in non-T2 asthma. J Allergy Clin Immunol Pract 9(7):2588–2597

Neveu V, Moussy A, Rouaix H, Wedekind R, Pon A, Knox C, Wishart DS, Scalbert A (2017) Exposome-Explorer: a manually-curated database on biomarkers of exposure to dietary and environmental factors. Nucleic Acids Res 45(D1):D979–D984

Noecker C, Eng A, Srinivasan S, Theriot CM, Young VB, Jansson JK, Fredricks DN, Borenstein E (2016) Metabolic model-based integration of microbiome taxonomic and metabolomic profiles elucidates mechanistic links between ecological and metabolic variation. mSystems 1(1)

North ML, Brook JR, Lee EY, Omana V, Daniel NM, Steacy LM, Evans GJ, Diamond ML, Ellis AK (2017) The Kingston Allergy Birth Cohort: Exploring parentally reported respiratory outcomes through the lens of the exposome. Ann Allergy Asthma Immunol 118(4):465–473

Ntontsi P, Photiades A, Zervas E, Xanthou G, Samitas K (2021) Genetics and epigenetics in asthma. Int J Mol Sci 22(5)

Panettieri RA Jr, Sjobring U, Peterffy A, Wessman P, Bowen K, Piper E, Colice G, Brightling CE (2018) Tralokinumab for severe, uncontrolled asthma (STRATOS 1 and STRATOS 2): two randomised, double-blind, placebo-controlled, phase 3 clinical trials. Lancet Respir Med 6(7): 511–525

Papi A, Brightling C, Pedersen SE, Reddel HK (2018) Asthma. Lancet 391(10122):783–800

Papi A, Blasi F, Canonica GW, Morandi L, Richeldi L, Rossi A (2020) Treatment strategies for asthma: reshaping the concept of asthma management. Allergy Asthma Clin Immunol 16:75

Perera F, Tang WY, Herbstman J, Tang D, Levin L, Miller R, Ho SM (2009) Relation of DNA methylation of 5'-CpG island of ACSL3 to transplacental exposure to airborne polycyclic aromatic hydrocarbons and childhood asthma. PLoS One 4(2):e4488

Peters U, Dixon AE, Forno E (2018) Obesity and asthma. J Allergy Clin Immunol 141(4): 1169–1179

Qian L, Mehrabi Nasab E, Athari SM, Athari SS (2022) Mitochondria signaling pathways in allergic asthma. J Investig Med 70(4):863–882

Quirt J, Hildebrand KJ, Mazza J, Noya F, Kim H (2018) Asthma. Allergy Asthma Clin Immunol 14 (Suppl 2):50

Quoc QL, Choi Y, Thi Bich TC, Yang EM, Shin YS, Park HS (2021) S100A9 in adult asthmatic patients: a biomarker for neutrophilic asthma. Exp Mol Med 53(7):1170–1179

Raita Y, Camargo CA Jr, Bochkov YA, Celedon JC, Gern JE, Mansbach JM, Rhee EP, Freishtat RJ, Hasegawa K (2021) Integrated-omics endotyping of infants with rhinovirus bronchiolitis and risk of childhood asthma. J Allergy Clin Immunol 147(6):2108–2117

Rastogi D, Suzuki M, Greally JM (2013) Differential epigenome-wide DNA methylation patterns in childhood obesity-associated asthma. Sci Rep 3:2164

Ricciardolo FLM, Sprio AE, Baroso A, Gallo F, Riccardi E, Bertolini F, Carriero V, Arrigo E, Ciprandi G (2021) Characterization of T2-Low and T2-high asthma phenotypes in real-life. Biomedicine 9(11)

Roffel MP, Boudewijn IM, van Nijnatten JLL, Faiz A, Vermeulen CJ, van Oosterhout AJ, Affleck K, Timens W, Bracke KR, Maes T et al (2022) Identification of asthma-associated microRNAs in bronchial biopsies. Eur Respir J 59(3)

Rogliani P, Cavalli F, Chetta A, Cazzola M, Calzetta L (2022) Potential drawbacks of ICS/LABA/LAMA triple fixed-dose combination therapy in the treatment of asthma: a quantitative synthesis of safety Profile. J Asthma Allergy 15:565–577

Schofield JPR, Burg D, Nicholas B, Strazzeri F, Brandsma J, Staykova D, Folisi C, Bansal AT, Xian Y, Guo Y et al (2019) Stratification of asthma phenotypes by airway proteomic signatures. J Allergy Clin Immunol 144(1):70–82

Shaffer M, Thurimella K, Quinn K, Doenges K, Zhang X, Bokatzian S, Reisdorph N, Lozupone CA (2019) AMON: annotation of metabolite origins via networks to integrate microbiome and metabolome data. BMC Bioinformatics 20(1):614

Sharma S, Hashmi MF, Chakraborty RK (2022) Asthma medications. In: StatPearls, Treasure Island, FL

Sim S, Choi Y, Park HS (2021) Potential metabolic biomarkers in adult asthmatics. Meta 11(7)

Simpson JL, Phipps S, Baines KJ, Oreo KM, Gunawardhana L, Gibson PG (2014) Elevated expression of the NLRP3 inflammasome in neutrophilic asthma. Eur Respir J 43(4):1067–1076

Simpson JL, Daly J, Baines KJ, Yang IA, Upham JW, Reynolds PN, Hodge S, James AL, Hugenholtz P, Willner D et al (2016) Airway dysbiosis: Haemophilus influenzae and Tropheryma in poorly controlled asthma. Eur Respir J 47(3):792–800

Simpson AJ, Hekking PP, Shaw DE, Fleming LJ, Roberts G, Riley JH, Bates S, Sousa AR, Bansal AT, Pandis I et al (2019) Treatable traits in the European U-BIOPRED adult asthma cohorts. Allergy 74(2):406–411

Singh P, Sharma A, Jha R, Arora S, Ahmad R, Rahmani AH, Almatroodi SA, Dohare R, Syed MA (2020) Transcriptomic analysis delineates potential signature genes and miRNAs associated with the pathogenesis of asthma. Sci Rep 10(1):13354

Sircar G, Saha B, Jana T, Dasgupta A, Gupta Bhattacharya S, Saha S (2015) DAAB: a manually curated database of allergy and asthma biomarkers. Clin Exp Allergy 45(7):1259–1261

Slager RE, Otulana BA, Hawkins GA, Yen YP, Peters SP, Wenzel SE, Meyers DA, Bleecker ER (2012) IL-4 receptor polymorphisms predict reduction in asthma exacerbations during response to an anti-IL-4 receptor alpha antagonist. J Allergy Clin Immunol 130(2):516–522 e514

Sood A, Petersen H, Blanchette CM, Meek P, Picchi MA, Belinsky SA, Tesfaigzi Y (2012) Methylated genes in sputum among older smokers with asthma. Chest 142(2):425–431

Southworth T, Van Geest M, Singh D (2021) Type-2 airway inflammation in mild asthma patients with high blood eosinophils and high fractional exhaled nitric oxide. Clin Transl Sci 14(4): 1259–1264

Spear ML, Hu D, Pino-Yanes M, Huntsman S, Eng C, Levin AM, Ortega VE, White MJ, McGarry ME, Thakur N et al (2019) A genome-wide association and admixture mapping study of bronchodilator drug response in African Americans with asthma. Pharmacogenomics J 19(3): 249–259

Specjalski K, Jassem E (2019) MicroRNAs: potential biomarkers and targets of therapy in allergic diseases? Arch Immunol Ther Exp 67(4):213–223

Stout JW, Visness CM, Enright P, Lamm C, Shapiro G, Gan VN, Adams GK 3rd, Mitchell HE (2006) Classification of asthma severity in children: the contribution of pulmonary function testing. Arch Pediatr Adolesc Med 160(8):844–850

Takahashi K, Pavlidis S, Ng Kee Kwong F, Hoda U, Rossios C, Sun K, Loza M, Baribaud F, Chanez P, Fowler SJ et al (2018) Sputum proteomics and airway cell transcripts of current and ex-smokers with severe asthma in U-BIOPRED: an exploratory analysis. Eur Respir J 51(5)

Takahashi K, Meguro K, Kawashima H, Kashiwakuma D, Kagami SI, Ohta S, Ono J, Izuhara K, Iwamoto I (2019) Serum periostin levels serve as a biomarker for both eosinophilic airway inflammation and fixed airflow limitation in well-controlled asthmatics. J Asthma 56(3): 236–243

Tiotiu A (2018) Biomarkers in asthma: state of the art. Asthma Res Pract 4:10

Tu X, Donovan C, Kim RY, Wark PAB, Horvat JC, Hansbro PM (2021) Asthma-COPD overlap: current understanding and the utility of experimental models. Eur Respir Rev 30(159)

Wark PA, Bucchieri F, Johnston SL, Gibson PG, Hamilton L, Mimica J, Zummo G, Holgate ST, Attia J, Thakkinstian A et al (2007) IFN-gamma-induced protein 10 is a novel biomarker of rhinovirus-induced asthma exacerbations. J Allergy Clin Immunol 120(3):586–593

Wechsler ME, Ruddy MK, Pavord ID, Israel E, Rabe KF, Ford LB, Maspero JF, Abdulai RM, Hu CC, Martincova R et al (2021) Efficacy and safety of itepekimab in patients with moderate-to-severe asthma. N Engl J Med 385(18):1656–1668

Wenzel SE (2012) Asthma phenotypes: the evolution from clinical to molecular approaches. Nat Med 18(5):716–725

Wild CP (2012) The exposome: from concept to utility. Int J Epidemiol 41(1):24–32

Wong EH, Porter JD, Edwards MR, Johnston SL (2014) The role of macrolides in asthma: current evidence and future directions. Lancet Respir Med 2(8):657–670

Xu P, Wang L, Chen D, Feng M, Lu Y, Chen R, Qiu C, Li J (2020) The application of proteomics in the diagnosis and treatment of bronchial asthma. Ann Transl Med 8(4):132

Xu H, Tong L, Gao P, Hu Y, Wang H, Chen Z, Fang L (2021) Combination of ipratropium bromide and salbutamol in children and adolescents with asthma: a meta-analysis. PLoS One 16(2): e0237620

Yang IV, Pedersen BS, Liu A, O'Connor GT, Teach SJ, Kattan M, Misiak RT, Gruchalla R, Steinbach SF, Szefler SJ et al (2015) DNA methylation and childhood asthma in the inner city. J Allergy Clin Immunol 136(1):69–80

Yawn BP (2009) Differential assessment and management of asthma vs chronic obstructive pulmonary disease. Medscape J Med 11(1):20

Ye Q, Liao A, D'Urzo A (2018) FEV1 reversibility for asthma diagnosis: a critical evaluation. Expert Rev Respir Med 12(4):265–267

Zietkowski Z, Skiepko R, Tomasiak MM, Bodzenta-Lukaszyk A (2008) Endothelin-1 in exhaled breath condensate of stable and unstable asthma patients. Respir Med 102(3):470–474

Bronchiectasis

5

Abstract

Bronchiectasis is an obstructive airway disease characterized by persistent bronchial dilation, defective mucociliary clearance, and recurrent microbial infections. Medical imaging and radiomics find an enormous application in disease diagnosis. However, co-existence of other diseases and symptomatic overlap often pose challenge to timely diagnosis. New age omics tools have been recently explored to address such problems. This chapter mentions different probable molecular and metabolic biomarkers of bronchiectasis. Altered airway microbiome and microbial interactome have also been studied for bronchiectasis patients. Clearance of mucus from airways and prevention of recurrent infection lie central to disease control. Different therapeutic approaches followed presently for controlling bronchiectasis have been highlighted. Finally, research areas that can be explored to improve our knowledge for upgrading diagnosis and refining treatment of bronchiectasis patients have been highlighted.

Keywords

Dilated bronchial airways · Haemoptysis · Neutrophil elastase · *Pseudomonas aeruginosa* · Biofilms · Bronchiectasis Radiologically Indexed CT Score (BRICS) · Muco-active agents

5.1 Introduction

Bronchiectasis is a progressive respiratory disorder characterized by persistent, irreversible dilation of the bronchial airways, chronic airway inflammation, impaired mucociliary clearance, and bacterial colonization (Aliberti et al. 2016). Chronic cough, mucopurulent sputum production, and recurrent bacterial infections are commonly observed in patients having bronchiectasis. Other less common

S. Saha et al., *Pulmonomics: Omics Approaches for Understanding Pulmonary Diseases*, https://doi.org/10.1007/978-981-99-3505-5_5

symptoms include dyspnoea, haemoptysis, pleuritic chest pain, wheezing, nail clubbing, and cyanosis (Cohen and Shteinberg 2022). With disease progression, the incidences of recurrent exacerbations increase leading to destruction of the bronchial tree (Polverino et al. 2017). Bronchiectasis was once a neglected disease; however, it gained much interest in the light of revelation of its increasing prevalence in the past decades. It is the third commonest chronic airway inflammatory disorder after asthma and COPD (Cohen and Shteinberg 2022). Bronchiectasis can affect individuals of all age groups and have been linked to deterioration in the quality of life and significant mortality rates, including paediatric mortality (Tino 2018).

5.2 Pathobiology

Several etiological factors, like genetic predisposition, respiratory infections, autoimmune disorders, immune hyper-activation, and immune deficiencies contribute to remodelling, widening, and subsequent dysfunctioning of bronchial airways in bronchiectasis. The cause of bronchiectasis may even be idiopathic (Boyton and Altmann 2016). Initial insults to the respiratory tract in the form of genetic defects (like cystic fibrosis, primary ciliary dyskinesia, Alpha1-antitrypsin deficiency, Mounier-Kuhn syndrome), immune disorders (like combined variable immunoglobulin deficiency, TAP deficiency syndrome, allergic bronchopulmonary aspergillosis, hypersensitivity pneumonitis), autoimmune disorders (like rheumatoid arthritis, Sjogren's syndrome, ankylosing spondylitis, systemic lupus erythematosus, and systemic sclerosis), viral infections (measles, respiratory syncytial viral infection), bacterial infections (*Mycobacterium tuberculosis*, *Bordetella pertussis*, *Haemophilus influenzae*, *Pseudomonas aeruginosa*, *Staphylococcus aureus*, and non-tubercular mycobacterial infections), respiratory diseases (like COPD, asthma, obliterative bronchiolitis, diffuse pan bronchiolitis), or other conditions (like mercury poisoning, yellow nail syndrome) have been commonly implicated for the subsequent onset of bronchiectasis (https://www.ncbi.nlm.nih.gov/books/NBK430810/; O'Donnell 2022; Goyal and Chang 2022). Foreign body aspiration and obstruction also remain an uncommon and often unsuspected cause of bronchiectasis (Palasamudram Shekar et al. 2018). These initial insults lead to inflammation, remodelling, and dilation of bronchial airways. Cartilage destruction, fibrosis, and infiltration of inflammatory immune cells in bronchial airways are commonly noted in bronchiectasis. Hyperplasia and hyperactivity of the mucous glands along with impairment of mucociliary clearance result in accumulation of mucous and enhance further susceptibility to airway infections. Neutrophils and neutrophil elastase (NE) have been implicated to play a central role in airway inflammation, reduced lung function, and increased instances of exacerbations. Elevated levels of chemoattractants like leukotriene B4 (LTB4), CXCL-8, IL-1β, and TNF-α lead to active recruitment of neutrophils in the airways (https://www.ncbi.nlm.nih.gov/books/NBK430810/; Chalmers and Hill 2013). Dysregulation of immune response characterized by impairment of opsonisation and phagocytosis by neutrophils, ineffective clearance of apoptotic cells, and inadequate T cell invasive response are all associated with the pathogenesis of bronchiectasis. Th17 cells have also been suspected to play a vital role in bronchiectasis (https://www.ncbi.nlm.nih.gov/books/NBK430810/; O'Donnell 2022; Flume et al. 2018). Formation and subsequent expansion of neutrophil extracellular traps (NETs) involved in immobilization of microbes have been identified as a measure of disease activity (Keir et al. 2021). Sputum neutrophil elastase (NE) has been linked with bronchiectasis severity index (BSI), dyspnoea score, severe lung function deterioration, and more radiological complications (Chalmers et al. 2017). NE levels were also found to be elevated during disease exacerbations (O'Donnell 2022). Other important criteria like host–microbial interaction, microbial diversity, and blood eosinophil counts are also being assessed to obtain a better understanding of their roles in acute bronchiectasis exacerbations (Richardson et al. 2019; Tunney et al. 2013; Shoemark et al. 2022). Formation of biofilms has also been documented in bronchiectasis. Biofilms were observed in broncho alveolar lavage (BAL) of children with non-cystic fibrosis bronchiectasis, even in the absence of clinically significant infection (Marsh et al. 2015). In another study, upper airway biofilms associated with squamous epithelial cells were observed more in BAL of children with bronchiectasis than in children with protracted bacterial bronchitis. Lower airway biofilms not related to squamous epithelial cells were also observed in BAL of children with bronchiectasis, however, the prevalence of such biofilms was lower in bronchiectasis patients as compared to protracted bacterial bronchitis patients (Marsh et al. 2022). *Pseudomonas aeruginosa* has been primarily implicated in biofilm formation in non-cystic fibrosis bronchiectasis (nCFB) patients as well as cystic fibrosis (CF) patients (Fernandez-Barat et al. 2021; Woo et al. 2016). Although in both nCFB and CF, the overall adaptive strategy of *Pseudomonas aeruginosa* remains the same, the underlying aetiology is different in nCFB and CF (Woo et al. 2016). The genome plasticity of *Pseudomonas aeruginosa* accounts for its diversification and increased survival via generation of different phenotypes associated with iron sequestration and variations in polysaccharides associated with biofilm formation (Woo et al. 2016; Hilliam et al. 2017). Formation of biofilm is one of the major strategies of subverting the immune surveillance and ensuring bacterial persistence in the lung microenvironment in non-cystic fibrosis bronchiectasis (Chalmers and Hill 2013). Biofilm formation also ensures protection from environmental stress and also leads to emergence of antibiotic resistance (Fernandez-Barat et al. 2021). Understanding the role of *Pseudomonas aeruginosa* and biofilm formation in bronchiectasis may improve diagnostic accuracy and therapeutic strategies for controlling the disease exacerbations.

5.3 Clinical Features and Diagnosis

Bronchiectasis is a heterogeneous clinical condition. Bronchiectasis may sometimes complicate other obstructive airway diseases and may also co-exist as a secondary event to different other diseases like cystic fibrosis, alpha-1 antitrypsin deficiency, primary ciliary dyskinesia, inflammatory bowel disease, Mounier–Kuhn syndrome, and immunodeficiency syndromes (O'Donnell 2022). Increased awareness of bronchiectasis is vital for timely diagnosis and accurate management of the disease. Overt features of cough, recurrent mucopurulent expectoration or haemoptysis with or without systemic symptoms are often noted. One must integrate the family history of bronchiectasis, history of childhood infection, history suggestive of autoimmune inflammatory disorders, and smoking history alongside the history of symptoms and frequency of exacerbations (episodic worsening of symptoms). Nail clubbing is frequent in purulent bronchiectasis patients and the auscultation may reveal none, or wheeze or both wheeze and crepitations (Cohen and Shteinberg 2022).

Primarily imaging techniques like chest X-ray, CT, and HRCT scan of chest are commonly used for bronchiectasis diagnosis (https://www.ncbi.nlm.nih.gov/books/NBK430810/). Presence of parallel linear densities, ring shadows, and tram track opacities in chest radiographs indicates thickening and dilation of bronchial walls. Patchy densities and reduced volume due to mucoid impaction of the bronchial tree may also be observed in certain cases. Chest CT scan has a typical pattern for bronchiectasis denoted by dilated bronchial airways and bronchial arteries, tram track appearance, signet ring appearance, varicose bronchial dilation, mosaic lung attenuation, and presence of cystic spaces. Bronchial dilation in bronchiectasis is primarily denoted by the bronchus to the adjacent pulmonary artery ratio or broncho arterial ratio (BAR) greater than 1, absence of bronchial tapering and increased visibility of airways within 1 cm region of pleural surface. Lobar collapse designated by fungal ball or mycetoma formation is observed in specific cases where the patients harbour *Aspergillus fumigatus* colonization and infection in the dilated airways. Multi-detector computed tomography (MDCT) is often chosen over HRCT scanning as it provides high resolution images of thin sections of 1 mm (https://www.ncbi.nlm.nih.gov/books/NBK430810/; Lee et al. 2004). Reduced lung function in bronchiectasis can be monitored using spirometry. Airflow obstruction denoted by reduced FEV1, reduced FVC, and air trapping indicated by increased residual volume (RV) are generally observed. Reduction in D_{LCO} is also observed in some patients. Variation in lung function patterns in spirometry and D_{LCO} across patients are generally not used for bronchiectasis diagnosis, rather used for determining disease severity (Cohen and Shteinberg 2022). Determination of the etiological cause of bronchiectasis holds importance in guiding the course of treatment. Presence of co-existing diseases and congenital diseases is also important (O'Donnell 2022). St. George's Respiratory Questionnaire is also used widely to evaluate exacerbation frequencies and impact of exacerbations on quality of life in bronchiectasis patients (Tino 2018). Arterial blood gas estimation may be needed to document respiratory failure in the disease especially when the arterial oxygen saturation appears towards the lower side in an otherwise ambulatory status (Hadda et al. 2018).

5.4 Biomarkers of Bronchiectasis

Presence of symptoms and imaging primarily lies at the centre of disease diagnosis. Lack of specific molecular biomarkers and in-depth understanding of bronchiectasis pathogenesis often delays diagnosis. Omics approaches serve as an excellent tool for biomarker determination for different pulmonary diseases. Different omics approaches like genomics, transcriptomics, proteomics, and metabolomics have found application in bronchiectasis research. Besides, the airway microbiome of bronchiectasis patients has been studied extensively using culture-based methods and metagenomic techniques. Some of the molecular, metabolic, and microbial signatures of bronchiectasis as observed by omics studies have been tabulated in Table 5.1.

5.5 Genomics

Several single nucleotide polymorphisms and genetic variations have been associated with the risk and severity of bronchiectasis. GC1f isoform of the GC gene encoding for Vitamin D binding protein (VDBP) has been linked with bronchiectasis severity. The GC1s isoform, on the other hand, is associated with milder form of bronchiectasis and lower comorbidity score (Oriano et al. 2021). Variants in DNAH5, CFTR, and epithelial sodium channel have been linked with bronchiectasis disease severity (Guan et al. 2018). Polymorphisms in TAP1, TAP2, and mutations in SLC26A9 have been linked to the onset of idiopathic bronchiectasis (Dogru et al. 2007; Bakouh et al. 2013). Other polymorphisms and variations in FUT2, SERPINA1, and MMP-1 have also been implicated in bronchiectasis (Taylor et al. 2017; Papatheodorou et al. 2010; Hsieh et al. 2013). These are mentioned in detail in Table 5.1.

5.6 Transcriptomics

Transcriptomic studies conducted with sputum exosomal samples from bronchiectasis patients highlighted correlation between expression profiles of microRNAs with bronchiectasis severity index (BSI), frequency of exacerbations, and *Pseudomonas aeruginosa* colonization. Increased miR-223-3p and decreased miR-92b-5p were associated with *Pseudomonas aeruginosa* colonization in bronchiectasis patients (Huang et al. 2019). Overexpression of TLR2, TLR4, CD14, IL8, and IL1β were also observed in the sputum of bronchiectasis patients compared to that in healthy subjects (Simpson et al. 2007). Other studies highlighted overexpression of MMP-8 in bronchiectasis patients and identified Lymphotoxin A as a probable biomarker candidate for bronchiectasis patients with chronic periodontitis (Gupta et al. 2020; Prikk et al. 2001). These transcriptomic signatures have also been enlisted in Table 5.1.

Table 5.1 Different biomarkers of bronchiectasis

Omics approach	Biomarkers	Biological sample	Comments
Genomics	FUT2	Sputum	FUT2 genotype (homozygous secretors) was significantly associated with higher exacerbation rates, lower lung function, and increased frequency of *P. aeruginosa* dominated infection in non-CF bronchiectasis patients (Taylor et al. 2017)
	GC gene encoding vitamin D binding protein	Peripheral blood	The GC1f isoform (rs7041/rs4588 A/G) was associated with increased severity in bronchiectasis patients (Oriano et al. 2021)
	SERPINA1		SERPINA1 p.V213A polymorphism was found to be associated with risk of developing disseminated bronchiectasis (Papatheodorou et al. 2010)
	DNAH5, CFTR, epithelial sodium channel		Presence of biallelic DNAH5 variants, or biallelic CFTR variants along with an epithelial sodium channel variant was associated with greater disease severity in bronchiectasis patients (Guan et al. 2018)
	TAP1		In TAP1-333 polymorphism analysis, Ile/Val genotype was increased while Ile/Ile genotype was decreased in children with idiopathic bronchiectasis. In TAP1-637 polymorphism analysis, Gly/Gly and Asp/Asp genotypes were reduced and Asp/Gly frequency was increased in children with idiopathic

(continued)

Table 5.1 (continued)

Omics approach	Biomarkers	Biological sample	Comments
			bronchiectasis (Dogru et al. 2007)
	TAP2		Ile/Ile genotype was reduced, and Ile/Val genotype was increased in TAP2-379 polymorphism analysis, while Thr/Thr Genotype was reduced, Ala/Ala and Thr/Ala genotypes were increased in TAP2-665 polymorphism analysis in children with idiopathic bronchiectasis (Dogru et al. 2007)
	MMP-1	Peripheral blood monocytes (PBMs)	MMP-1 (−1607G) allele is more frequent in bronchiectasis patients. Presence of at least one MMP-1 (−1607G) allele was associated with increased incidences of hospital admission (Hsieh et al. 2013)
	SLC26A9	DNA	Two missense mutations in SLC26A9, namely p. Val486Ile and p. Arg575Trp were observed in patients with diffuse idiopathic bronchiectasis (Bakouh et al. 2013)
	CHP2	Genome	SNP rs109592 in the CHP2 locus was found to be associated with the nodular bronchiectasis subtype of pulmonary *Mycobacterium avium* complex (MAC) patients in Japanese population (Namkoong et al. 2021)
Transcriptomics	miR-92b-5p, miR-223-3p	Sputum exosome	Reduced expression of miR-92b-5p and increased expression of miR-223-3p were related to *Pseudomonas aeruginosa* colonization

(continued)

Table 5.1 (continued)

Omics approach	Biomarkers	Biological sample	Comments
			in bronchiectasis patients. Bronchiectasis severity index (BSI) correlated with miR-223-3p expression level (Huang et al. 2019)
	TLR2, TLR4, CD14, IL8, IL1β	Sputum	TLR2, TLR4, CD14, IL8, and IL1β were elevated in bronchiectasis patients (Simpson et al. 2007).
	Lymphotoxin A (LTA)	Gingival tissues	LTA may serve as a potential genetic biomarker for bronchiectasis patients with chronic periodontitis (Gupta et al. 2020)
	MMP-8	Lung tissues	MMP-8 mRNA expression was elevated in bronchiectasis patients as compared to healthy subjects. MMP-8 expression was noted in neutrophils, glandular cells, bronchial ciliated epithelial cells, and monocytes/macrophages infiltrating the bronchial epithelial area of bronchiectatic lungs (Prikk et al. 2001)
Proteomics	Neutrophil elastase (NE)	Sputum	Sputum NE serves as a biomarker of disease activity and severity in bronchiectasis patients (Brusselle and Van Braeckel 2017)
	MMP-1, MMP-3, MMP-7, MMP-8, MMP-9, TIMP2, TIMP4		MMP-1, MMP-3, MMP-7, MMP-8, MMP-9, TIMP2, and TIMP4 were significantly elevated in sputum from bronchiectasis patients as compared to that in healthy volunteers (Taylor et al. 2015)
	MMP-2, MMP-8		MMP-2 levels and MMP-8 activity were

(continued)

Table 5.1 (continued)

Omics approach	Biomarkers	Biological sample	Comments
			higher in bronchiectasis patients having *Haemophilus influenzae*-dominated airway infections as compared to bronchiectasis patients with *Pseudomonas aeruginosa* dominated airway infections (Taylor et al. 2015)
	MPO, AZU1, ELANE		Expression of MPO, AZU1, and ELANE were increased in a subset of bronchiectasis patients with acute exacerbations (Huang et al. 2020)
	PIGR, IgHA1, CST3		PIGR, IgHA1, and CST3 expressions were increased in a subset of bronchiectasis patients (Huang et al. 2020)
	IL-8, TNF-α		Sputum IL-8 and TNF-α levels were significantly correlated with HRCT severity score in bronchiectasis patients (Guran et al. 2007)
	OX2G, HSP60	Blood plasma	OX2G and HSP60 expression levels were correlated with bronchiectasis as well as initial and longitudinal Brody bronchiectasis score (BBS) in cystic fibrosis (DeBoer et al. 2017)
	C5a, MMP-10, and MIP-1a		C5a, MMP-10, and MIP-1a were associated with bronchiectasis in children with cystic fibrosis (DeBoer et al. 2017)
	MBL	Serum	Deficiency of MBL in serum is associated with increased disease severity, increased rate of exacerbations, and frequent hospitalization

(continued)

Table 5.1 (continued)

Omics approach	Biomarkers	Biological sample	Comments
			in patients with non-CF bronchiectasis (Chalmers et al. 2013a)
	MMP-1, TGF-β1		Pro-MMP-1, active MMP-1, and TGF-β1 levels in serum were increased in bronchiectasis patients having 1G/1G and 1G/2G genotype of MMP-1 as compared to 2G/2G genotype or normal subjects (Hsieh et al. 2013)
	E-selectin, ICAM-1, VCAM-1		E-selectin, ICAM-1, and VCAM-1 were significantly increased in bronchiectasis patients as compared to healthy subjects (Zheng et al. 2000)
	CRP, fibrinogen		Serum CRP and fibrinogen levels were higher in clinically stable bronchiectasis patients with colonization as compared to non-colonized patients (Ergan Arsava and Coplu 2011)
	CRP, SAA, IL-6		CRP, SAA, and IL6 were significantly elevated in bronchiectasis patients during exacerbations as compared to stable state. These three proteins can serve as predictive biomarkers for exacerbation in bronchiectasis patients (Kapur et al. 2012)
	SOD, CAT	Blood plasma, serum	Superoxide dismutase (SOD) activity was decreased, and catalase (CAT) activity was significantly increased in bronchiectasis patients as compared to healthy

(continued)

Table 5.1 (continued)

Omics approach	Biomarkers	Biological sample	Comments
			controls (Olveira et al. 2013)
	Elastase, MPO, IL-6, IL-8, TNF-α	Broncho alveolar lavage (BAL)	Expression of elastase, MPO, IL-6, IL-8, and TNF-α were significantly elevated in bronchiectasis patients as compared to healthy controls (Angrill et al. 2001)
	MMP-8	Broncho alveolar lavage fluid (BALF)	Proteolytic activation of pro MMP-8 was associated with moderate and severe forms of bronchiectasis (Sepper et al. 1995)
	MMP-8	Broncho alveolar lavage fluid (BALF), lung tissues	MMP-8 was increased and activated in BALF of bronchiectasis patients as compared to healthy subjects. MMP-8 expression was also elevated in lung tissues of bronchiectasis patients (Prikk et al. 2001)
	MMP-8, MMP-9	Endobronchial biopsies (EBBs)	MMP-8 and MMP-9 were overexpressed in bronchiectasis patients (Zheng et al. 2002)
Metabolomics	25-hydroxyvitamin-D	Serum	25-hydroxyvitamin-D deficiency was associated with increased bacterial colonization and disease severity in bronchiectasis patients (Chalmers et al. 2013b)
	Methionine (met), methionine sulfoxide (MetO)	Broncho alveolar lavage fluid (BALF)	MetO (methionine sulfoxide) and % OxMet (percentage of MetO relative to the sum of met and MetO) are found to be correlates of airway neutrophils and bronchiectasis in children with cystic fibrosis (Chandler et al. 2018)
	Methanol, ethanol, acetone/acetoin, 2-propanol, propionate,	Exhaled breath condensate (EBC)	Methanol, ethanol, acetone/acetoin, 2-propanol, and

(continued)

Table 5.1 (continued)

Omics approach	Biomarkers	Biological sample	Comments
	formate, acetate, lactate, saturated fatty acids (SFAs)		propionate were elevated, while formate, acetate, lactate, and saturated fatty acids were reduced in non-cystic fibrosis (nCF) and non-primary ciliary dyskinesia (nPCD) bronchiectasis patients as compared to healthy controls (Paris et al. 2020)
	Methanol, ethanol, 2-propanol, formate, acetate, lactate, saturated fatty acids (SFAs)		Methanol, ethanol, 2-propanol, and lactate were elevated, whereas formate, acetate, and saturated fatty acids were decreased in bronchiectasis patients having primary ciliary dyskinesia (PCD) as compared to healthy subjects (Paris et al. 2020)
	Methanol, acetone/acetoin, ethanol, lactate		Methanol and acetone/acetoin were increased in non-cystic fibrosis (nCF) and non-primary ciliary dyskinesia (nPCD) bronchiectasis patients as compared to bronchiectasis patients having primary ciliary dyskinesia (PCD) Ethanol and lactate were elevated in EBC of bronchiectasis patients having primary ciliary dyskinesia (PCD) as compared to non-cystic fibrosis (nCF) and non-primary ciliary dyskinesia (nPCD) bronchiectasis patients. These metabolites may thus aid in distinguishing between nCF/nPCD bronchiectasis and PCD-associated

(continued)

Table 5.1 (continued)

Omics approach	Biomarkers	Biological sample	Comments
			bronchiectasis (Paris et al. 2020)
	Hydrogen peroxide (H_2O_2)		Exhaled H_2O_2 was elevated in patients with bronchiectasis as compared to healthy subjects (Loukides et al. 2002)
	Superoxide anion, hydrogen peroxide	White blood cells (WBCs)	Intracellular superoxide anion and hydrogen peroxide are significantly increased in bronchiectasis patients compared to the healthy controls (Olveira et al. 2013)
	Thiobarbituric acid reactive substances, 8-isoprostanes	Blood plasma, serum	Thiobarbituric acid reactive substances (TBARs) and 8-isoprostanes are significantly increased in bronchiectasis patients compared to the healthy controls (Olveira et al. 2013)
Metagenomics	*Pasteurellaceae*, *Streptococcaceae*, *Pseudomonadaceae*, *Prevotellaceae*, *Veillonellaceae*, *Actinomycetaceae*	Sputum	Lower airways of patients with non-cystic fibrosis bronchiectasis exhibited a dominance of *Pseudomonadaceae*, *Pasteurellaceae*, and *Streptococcaceae* Significant presence of other anaerobic genera like *Prevotellaceae*, *Actinomycetaceae*, and *Veillonellaceae* (Purcell et al. 2014)
	Pseudomonas aeruginosa, *Haemophilus influenzae*		Presence of *Pseudomonas aeruginosa* was associated with significant reduction in FEV1 (%, pred) and low community diversity Presence of *Pseudomonas aeruginosa* has also been

(continued)

Table 5.1 (continued)

Omics approach	Biomarkers	Biological sample	Comments
			linked to bronchiectasis severity Presence of *Haemophilus influenzae* affects the airway microbial community structure *Pseudomonas aeruginosa* and *Haemophilus influenzae* did not co-exist (Purcell et al. 2014)
	Pseudomonas aeruginosa, *Haemophilus influenzae*, *Moraxella catarrhalis*		*Pseudomonas aeruginosa*, *Haemophilus influenzae* and *Moraxella catarrhalis* were detected in sputum samples from bronchiectasis patients *Haemophilus influenzae* and *Pseudomonas aeruginosa* could shape the microbial community of the airways of bronchiectasis patients (Cox et al. 2017)
	Aspergillus spp., *Cryptococcus* spp., *Clavispora* spp.		Prevalence of *Aspergillus* spp., *Cryptococcus* spp., and *Clavispora* spp. were noted in bronchiectasis patients. *Aspergillus fumigatus* and *Aspergillus terreus* were more dominant and associated with high exacerbation rates in bronchiectasis patients (Mac Aogain et al. 2018)
	Pseudomonas aeruginosa, *Veillonella* spp.		Predominance of *Pseudomonas aeruginosa*, followed by *Veillonella* spp. served as a strong predictor of exacerbation frequency in bronchiectasis patients (Rogers et al. 2014)
	Pseudomonas aeruginosa, *Haemophilus influenzae*, *Streptococcus*	Sputum, Broncho alveolar lavage (BAL)	*Pseudomonas aeruginosa*, *Haemophilus influenzae*, and *Streptococcus*

(continued)

Table 5.1 (continued)

Omics approach	Biomarkers	Biological sample	Comments
	pneumoniae, *Veillonella* spp., *Prevotella* spp., *Neisseria* spp.		*pneumoniae* were identified as characteristic core bacterial population, whereas *Veillonella* spp., *Prevotella* spp. and *Neisseria* spp. were identified as satellite bacterial population in bronchiectasis patients (Rogers et al. 2013)
	Human T-lymphotropic virus type 1 (HTLV-1) subtype C	Serum	HTLV-1 subtype C pro-viral loads were associated with bronchiectasis in indigenous Australian adults at risk of recurrent lower airway tract infections (Einsiedel et al. 2014)
	Coronavirus, rhinovirus, influenza A/B virus	Nasopharyngeal swabs, sputum	Coronavirus, rhinovirus, and influenza A/B virus were found in bronchiectasis patients with exacerbations and were associated with increased expression of serum IL-6 and TNF-α, sputum IL-1β, and TNF-α (Gao et al. 2015)
	Human rhinovirus (HRV-A)	Nasopharyngeal aspirates	HRV-A was most frequently detected in children with bronchiectasis during exacerbations (Kapur et al. 2014) Virus positive exacerbations were associated with increased hospital admissions (Kapur et al. 2014; Budden et al. 2019)

This table enlists various important biomarkers of bronchiectasis identified using different omics approaches

5.7 Proteomics

Several differentially expressed proteins associated with bronchiectasis have been identified by various high- and low-throughput proteomic studies. Different matrix metalloproteinases (MMPs), namely MMP-1, MMP-2, MMP-3, MMP-7, MMP-8, and MMP-9 were found to be overexpressed in bronchiectasis patients (Taylor et al. 2015; Zheng et al. 2002). Of these, MMP-2 and MMP-8 were higher in bronchiectasis patients with *Haemophilus influenzae* infection than in bronchiectasis patients with *Pseudomonas aeruginosa* infections (Taylor et al. 2015). IL-8 and TNF-α levels were also increased in bronchiectasis patients. These proteins have been found to play a crucial role in recruiting neutrophils to the airways and have also been correlated with HRCT severity score (Guran et al. 2007; Angrill et al. 2001). Besides, increased expression of CRP, IL-6, SAA and decreased expression of MBL have been linked with exacerbations in bronchiectasis patients (Chalmers et al. 2013a; Kapur et al. 2012). Other proteins that are differentially expressed in bronchiectasis have been listed in Table 5.1.

5.8 Metabolomics

NMR-based metabolomic study of exhaled breath condensate (EBC) revealed methanol, ethanol, acetone/acetoin, 2-propanol, and propionate to be increased in non-cystic fibrosis (nCF) and non-primary ciliary dyskinesia (nPCD) bronchiectasis patients as compared to healthy controls. Formate, acetate, lactate, and saturated fatty acids were found to be reduced. Besides, methanol and acetone/acetoin levels in nCF/nPCD bronchiectasis patients were more elevated than that in bronchiectasis patients with primary ciliary dyskinesia (PCD). Therefore, these identified metabolites can aid in distinguishing nCF/nPCD bronchiectasis from other forms of bronchiectasis (Paris et al. 2020). Levels of hydrogen peroxide in EBC were found to be increased in bronchiectasis patients as compared to healthy volunteers (Loukides et al. 2002). Besides, superoxide ions, thiobarbituric acid reactive substances and 8-isoprostanes were also increased in bronchiectasis patients (Olveira et al. 2013). The level of methionine sulfoxide and the percentage of methionine sulfoxide relative to the sum of methionine and methionine sulfoxide were found to correlate with bronchiectasis in cystic fibrosis patients (Chandler et al. 2018). These probable metabolic biomarkers of bronchiectasis have also been mentioned in Table 5.1.

5.9 Metagenomics

Pseudomonas aeruginosa, *Haemophilus influenzae*, and *Streptococcus pneumoniae* have been identified as core bacterial population in bronchiectasis patients (Rogers et al. 2013). *Pseudomonas aeruginosa* and *Haemophilus influenzae* have also been documented to modulate the airway microbial flora in bronchiectasis (Purcell et al.

2014). Genotyping of *Pseudomonas aeruginosa* highlighted the diverse nature of clinical isolates of *Pseudomonas aeruginosa* that can colonize the airways of bronchiectasis patients (Eusebio et al. 2015; Woo et al. 2018). Besides, the predominance of *Pseudomonas aeruginosa*, followed by that of *Veillonella* spp. was found to provide robust prediction of exacerbation rates in bronchiectasis patients (Rogers et al. 2014). Integrative microbiome data analysis indicated that both the abundance and the interaction network of *Pseudomonas* spp. are associated with exacerbations in bronchiectasis. It was also highlighted that inclusion of microbial interaction data can improve the performance of bronchiectasis exacerbation prediction models (Mac Aogain et al. 2021). Apart from these core bacterial species, several other bacteria, fungal species, and viruses have been associated with bronchiectasis and its exacerbations as shown in Table 5.1 (Mac Aogain et al. 2018; Einsiedel et al. 2014; Gao et al. 2015; Kapur et al. 2014; Budden et al. 2019).

5.10 Bioinformatics

There are specific bronchiectasis databases for patient registry which contain clinical and CT scans data of the patients. The EMBARC (European Multicentric Bronchiectasis Audit and Research Collaboration) registry is a pan-European observational study of bronchiectasis patients, and this platform is extended in Asia and Australia (Chalmers et al. 2016). EMBARC and Respiratory Research Network of India Registry offer the nature of the Indian patients' characteristics (Dhar et al. 2019). The USA has its own bronchiectasis research registry (Aksamit et al. 2017). The KMBARC (Korean Multicentric Bronchiectasis Audit and Research Collaboration) is another patient registry database in Korea (Lee et al. 2020). Yang et al. performed multivariable logistic analysis with bronchiectasis in Korea and observed that subjects with bronchiectasis were more likely to have other comorbidities like asthma, pulmonary tuberculosis, osteoporosis, and depression as compared to other subjects (Yang et al. 2020). Kim et al. used KMBARC registry data and found that the bronchiectasis patient with COPD had a significantly decreased quality of life (Kim et al. 2021). In conclusion, these studies surmise that bronchiectasis is associated with other lung diseases with decreased quality of life.

5.11 Medical Imaging or Radiomics

Radiology plays an important role in the diagnosis and prognosis of bronchiectasis. Chest HRCT scan of bronchiectasis findings include bronchial diameter to be greater than that of the adjacent pulmonary artery and reduction of normal tapering of terminal bronchioles as they move towards the lung periphery (Milliron et al. 2015). There are CT scoring systems available to assess the disease severity in bronchiectasis. In 1991, Bhalla et al. first introduced a CT scoring system (named as Bhalla score) that considers the structural abnormalities seen in the lungs of cystic fibrosis patients (Bhalla et al. 1991). Subsequently, the Reiff score was developed using the CT parameters like extent of lobular distribution, bronchial dilation, and bronchial wall thickening (Reiff et al. 1995). A simplified and modified Reiff score used the number of lung lobes and the severity of bronchial dilatation and scored it from 0 to 18 (Mandal et al. 2013). The BRICS (Bronchiectasis Radiologically Indexed CT Score) was based on multivariable analysis of the Bhalla score and had shown to predict disease association markers and predict clinical parameters of bronchiectasis severity (Bedi et al. 2018). The HRCT scoring systems in bronchiectasis have evolved over time to assess the severity of the disease and also shown to correlate with other clinical parameters of the disease.

5.12 Multi-Omics and Data Integration

Bronchiectasis is a heterogeneous disease where clinical, chest radiological features along with microbiological milieu in the lung are associated with endotypes and phenotypes of the disease (Jose and Loebinger 2021). Different types of microorganisms like bacteria, fungi, and viruses along with their genes contribute to bronchiectasis exacerbations. So, an integrated multi-biome analysis from bronchiectasis patient's sputum samples was performed using weighted similarity network fusion (WSFN) method (Mac Aogain et al. 2021). This study reported that bronchiectasis patients with higher risk of exacerbations have less microbial diversity, and less complex microbial co-occurrence networks. It was inferred that antibiotics probably target specific interaction networks rather than individual microbes. In conclusion, multi-omics techniques from several sources can be used to generate data to identify phenotypes and endotypes of bronchiectasis patients and dissect the role of microbial interaction networks.

5.13 Present Therapeutic Strategies

The present treatment strategies are an effort to manage the disease and the complications. Mucus clearance and infection management often appear to be most important and in advanced state, the respiratory failure turns out to be a difficult issue. The scope of prevention has been increasing over years. However, bronchiectasis still continues to remain largely incurable. The common medications used in the treatment of bronchiectasis have been enlisted in Table 5.2.

Apart from the pharmacological muco-active agents, Airway Clearance Techniques (ACT) are also practised to improve tracheobronchial clearance. Postural drainage, though aids in mucus clearance, it may increase the risk of gastroesophageal reflux disease (GERD). Physical activity, deep breathing followed by forceful expiration, and percussion of chest wall with cupped hands may aid in improving mucus clearance (https://www.ncbi.nlm.nih.gov/books/NBK430810/; Goyal and Chang 2022; Goyal et al. 2016). Vitamin D supplementation is often recommended in bronchiectasis patients, since, disease severity is related to Vit D deficiency (Boyton and Altmann 2016; Chalmers et al. 2013b). Vaccination against

Table 5.2 Medications used in the treatment of bronchiectasis

Sl. no.	Types of medication	Drug names	Mechanism of action	Comments
1.	Antibiotics	Aminoglycosides (like gentamycin, tobramycin), fluoroquinolones (like ciprofloxacin)	Reduces bacterial load. Breaks the vicious cycle of infection and inflammation. May exert anti-inflammatory and immunomodulatory effects	Inhaled antibiotics are used as maintenance therapy in bronchiectasis Patients with *Pseudomonas* infection (Goyal and Chang 2022)
		Macrolides		Well tolerated in bronchiectasis patients (King 2007) Interferes with mobility and quorum sensing of *Pseudomonas* aeruginosa (Giamarellos-Bourboulis 2008)
		Anti-pseudomonal antibiotics		May be administered intravenously (https://www.ncbi.nlm.nih.gov/books/NBK430810/) Inhaled anti-pseudomonal antibiotics often yields promising results (Antoniu 2018)
		Polymyxins (like colistin)		Nebulized colistin used for bronchiectasis patients with *Pseudomonas* infection (https://www.ncbi.nlm.nih.gov/books/NBK430810/)
		Beta lactam antibiotic (like amoxicillin)		Used in bronchiectasis patients having *Haemophilus influenzae* colonization (https://www.ncbi.nlm.nih.gov/books/NBK430810/)
2.	Mucus clearing agents	Hyperosmolar (HO) agents (like	Change physical properties of mucus,	Inhaled dry powder mannitol is well

(continued)

Table 5.2 (continued)

Sl. no.	Types of medication	Drug names	Mechanism of action	Comments
		hypertonic saline, mannitol)	reduce entanglements formed by mucin polymers by disrupting hydrogen bonds, increase airway hydration, and promote mucus clearance	tolerated and provides airway clearance in adult bronchiectasis patients. Hypertonic saline is tolerated in some patients, but may be discontinued adverse effects like cough, throat irritation and dyspnoea (Goyal and Chang 2022; Nair and Ilowite 2012; Daviskas and Anderson 2006; Maiz Carro and Martinez-Garcia 2019)
		Mucokinetics (like tricyclic nucleotides, ambroxol)	Enhance mucociliary efficiency and promote mucus clearance	Ambroxol hydrochloride along with roxithromycin improved dyspnoea, global CT score, and bronchiectasis score (Nair and Ilowite 2012; Rubin 2007; Liu et al. 2012)
		Mucolytics (like N-acetylcysteine (NAC), inhaled deoxyribonuclease I (DNase))	Break down disulphide linkage in mucus, cause degradation of polymers in mucus, and reduce the viscosity of mucus	Recombinant human DNase was harmful for bronchiectasis patients (Goyal and Chang 2022; Nair and Ilowite 2012)
		Expectorant (like guaifenesin)	Aids in coughing up mucus	Used in bronchiectasis patients (Nair and Ilowite 2012)
3.	Bronchodilators	Long-acting beta agonists and short-acting beta agonists	May aid in mucus clearance	Beneficial for bronchiectasis patients having co-existent asthma (Goyal et al. 2016)
4.	Corticosteroids	Inhaled corticosteroids (ICS), oral corticosteroids (OCS)	May reduce inflammation	ICS reduced airway inflammation and sputum volume but had local adverse effects (King 2007)

(continued)

Table 5.2 (continued)

Sl. no.	Types of medication	Drug names	Mechanism of action	Comments
				OCS may increase risk of infection by *Pseudomonas aeruginosa* (Goyal et al. 2016)

influenza (Hib vaccine), tuberculosis (BCG vaccine), measles, pertussis, and pneumococcal disease is highly recommended in bronchiectasis patients to gain protective immunity against these diseases (https://www.ncbi.nlm.nih.gov/books/NBK430810/; Goyal et al. 2016).

Pulmonary rehabilitation is also found to improve pulmonary function, exercise capacity, and quality of life in bronchiectasis patients (Zanini et al. 2015). Surgical intervention is less common in bronchiectasis patients. Surgery is explored in patients with complications or at an advanced disease stage. Surgery or bronchial artery embolization is conducted in bronchiectasis patients with massive haemoptysis. Lung transplantation is done for severe bronchiectasis in cystic fibrosis patients (https://www.ncbi.nlm.nih.gov/books/NBK430810/; Goyal and Chang 2022).

Several novel treatments are also being evaluated for their efficacy in bronchiectasis. Pidotimod, a synthetic dipeptide molecule with immunomodulatory properties have resulted in decrease in exacerbation in bronchiectasis patients in an open-label adult study (D'Amato et al. 2017). Brensocatib, a reversible inhibitor of dipeptidyl peptidase 1 (DPP-1) has shown promising results in Phase II study and has presently entered Phase III trial. This oral drug brought about reduction in neutrophil serine protease activity in bronchiectasis patients (O'Donnell 2022; Chalmers et al. 2020). Quercetin–chitosan nanoparticle complex was found to inhibit biofilm formation by *Pseudomonas aeruginosa* and quorum sensing. Since, *Pseudomonas aeruginosa* is the primary bacterial pathogen associated with bronchiectasis, use of this nanoparticle complex in bronchiectasis therapeutic regime may reduce over dependence on antibiotics (Tran and Hadinoto 2021).

5.14 Future Perspectives

Research interest in bronchiectasis has been rekindled due to the increased prevalence of this disease. The ever-evolving multi-omics knowledge has helped in unveiling several pathogenic determinants of bronchiectasis. It has improved our understanding of neutrophilic inflammation and microbial involvement in bronchiectasis exacerbations. However, there still exists an enormous knowledge gap in the field of bronchiectasis, which if unravelled can aid in improving disease endotyping and designing precision medicine. Determination of high-risk groups for bronchiectasis and identification of risks factors are of prime importance as modifications of such risk factors may help in delaying the progression of disease. Mechanistic understanding of the phenomena leading to bronchiectasis exacerbations may aid in distinguishing between frequent and infrequent exacerbators. Besides, omics tools should be more extensively used to enable efficient disease endotyping based on identified biomarkers. Association and distinction of bronchiectasis from other neutrophilic inflammatory disorders need to be evaluated to enable differential diagnosis (Goyal et al. 2016). Deeper insight into host–microbial interaction, inter-community microbial interaction, biofilm formation, antibiotic resistance, and quorum sensing is of prime importance in bronchiectasis (Aliberti et al. 2016). Systems biology, multi-omics data integration, and network analyses should be used for the identification of local biomarkers, systemic biomarkers, and treatable traits of bronchiectasis (Amaro et al. 2022). These research areas shall help in moving a step closer towards designing precision medicine for bronchiectasis.

References

Aksamit TR, O'Donnell AE, Barker A, Olivier KN, Winthrop KL, Daniels MLA, Johnson M, Eden E, Griffith D, Knowles M et al (2017) Adult patients with bronchiectasis: a first look at the US bronchiectasis research registry. Chest 151(5):982–992

Aliberti S, Masefield S, Polverino E, De Soyza A, Loebinger MR, Menendez R, Ringshausen FC, Vendrell M, Powell P, Chalmers JD et al (2016) Research priorities in bronchiectasis: a consensus statement from the EMBARC clinical research collaboration. Eur Respir J 48(3): 632–647

Amaro R, Perea L, Sibila O (2022) Future directions in bronchiectasis research. Clin Chest Med 43(1):179–187

Angrill J, Agusti C, De Celis R, Filella X, Rano A, Elena M, De La Bellacasa JP, Xaubet A, Torres A (2001) Bronchial inflammation and colonization in patients with clinically stable bronchiectasis. Am J Respir Crit Care Med 164(9):1628–1632

Antoniu SA (2018) Investigational inhaled therapies for non-CF bronchiectasis. Expert Opin Investig Drugs 27(2):139–146

Bakouh N, Bienvenu T, Thomas A, Ehrenfeld J, Liote H, Roussel D, Duquesnoy P, Farman N, Viel M, Cherif-Zahar B et al (2013) Characterization of SLC26A9 in patients with CF-like lung disease. Hum Mutat 34(10):1404–1414

Bedi P, Chalmers JD, Goeminne PC, Mai C, Saravanamuthu P, Velu PP, Cartlidge MK, Loebinger MR, Jacob J, Kamal F et al (2018) The BRICS (bronchiectasis radiologically indexed CT score): a Multicenter study score for use in idiopathic and Postinfective bronchiectasis. Chest 153(5): 1177–1186

Bhalla M, Turcios N, Aponte V, Jenkins M, Leitman BS, McCauley DI, Naidich DP (1991) Cystic fibrosis: scoring system with thin-section CT. Radiology 179(3):783–788

Boyton RJ, Altmann DM (2016) Bronchiectasis: current concepts in pathogenesis, immunology, and microbiology. Annu Rev Pathol 11:523–554

Brusselle GG, Van Braeckel E (2017) Sputum neutrophil elastase as a biomarker for disease activity in bronchiectasis. Am J Respir Crit Care Med 195(10):1289–1291

Budden KF, Shukla SD, Rehman SF, Bowerman KL, Keely S, Hugenholtz P, Armstrong-James DPH, Adcock IM, Chotirmall SH, Chung KF et al (2019) Functional effects of the microbiota in chronic respiratory disease. Lancet Respir Med 7(10):907–920

Chalmers JD, Hill AT (2013) Mechanisms of immune dysfunction and bacterial persistence in non-cystic fibrosis bronchiectasis. Mol Immunol 55(1):27–34

Chalmers JD, McHugh BJ, Doherty C, Smith MP, Govan JR, Kilpatrick DC, Hill AT (2013a) Mannose-binding lectin deficiency and disease severity in non-cystic fibrosis bronchiectasis: a prospective study. Lancet Respir Med 1(3):224–232

Chalmers JD, McHugh BJ, Docherty C, Govan JR, Hill AT (2013b) Vitamin-D deficiency is associated with chronic bacterial colonisation and disease severity in bronchiectasis. Thorax 68(1):39–47

Chalmers JD, Aliberti S, Polverino E, Vendrell M, Crichton M, Loebinger M, Dimakou K, Clifton I, van der Eerden M, Rohde G et al (2016) The EMBARC European bronchiectasis registry: protocol for an international observational study. ERJ Open Res 2(1)

Chalmers JD, Moffitt KL, Suarez-Cuartin G, Sibila O, Finch S, Furrie E, Dicker A, Wrobel K, Elborn JS, Walker B et al (2017) Neutrophil elastase activity is associated with exacerbations and lung function decline in bronchiectasis. Am J Respir Crit Care Med 195(10):1384–1393

Chalmers JD, Haworth CS, Metersky ML, Loebinger MR, Blasi F, Sibila O, O'Donnell AE, Sullivan EJ, Mange KC, Fernandez C et al (2020) Phase 2 trial of the DPP-1 inhibitor Brensocatib in bronchiectasis. N Engl J Med 383(22):2127–2137

Chandler JD, Margaroli C, Horati H, Kilgore MB, Veltman M, Liu HK, Taurone AJ, Peng L, Guglani L, Uppal K et al (2018) Myeloperoxidase oxidation of methionine associates with early cystic fibrosis lung disease. Eur Respir J 52(4)

Cohen R, Shteinberg M (2022) Diagnosis and evaluation of bronchiectasis. Clin Chest Med 43(1): 7–22

Cox MJ, Turek EM, Hennessy C, Mirza GK, James PL, Coleman M, Jones A, Wilson R, Bilton D, Cookson WO et al (2017) Longitudinal assessment of sputum microbiome by sequencing of the 16S rRNA gene in non-cystic fibrosis bronchiectasis patients. PLoS One 12(2):e0170622

D'Amato M, Simioli F, Martino M, Sorrentino N, Porzio M, Stanziola AA, Molino A, Mormile M (2017) Open label case-control study to assess Pidotimod efficacy in non CF bronchiectasis disease: a pilot study. Eur Respir J 50

Daviskas E, Anderson SD (2006) Hyperosmolar agents and clearance of mucus in the diseased airway. J Aerosol Med 19(1):100–109

DeBoer EM, Kroehl ME, Wagner BD, Accurso FJ, Harris JK, Lynch DA, Sagel SD, Deterding RR (2017) Proteomic profiling identifies novel circulating markers associated with bronchiectasis in cystic fibrosis. Proteomics Clin Appl:11(9-10)

Dhar R, Singh S, Talwar D, Mohan M, Tripathi SK, Swarnakar R, Trivedi S, Rajagopala S, D'Souza G, Padmanabhan A et al (2019) Bronchiectasis in India: results from the European multicentre bronchiectasis audit and research collaboration (EMBARC) and respiratory research network of India registry. Lancet Glob Health 7(9):e1269–e1279

Dogru D, Ozbas Gerceker F, Yalcin E, Cobanoglu N, Pekcan S, Ozcelik U, Kiper N, Ozguc M (2007) The role of TAP1 and TAP2 gene polymorphism in idiopathic bronchiectasis in children. Pediatr Pulmonol 42(3):237–241

Einsiedel L, Cassar O, Goeman E, Spelman T, Au V, Hatami S, Joseph S, Gessain A (2014) Higher human T-lymphotropic virus type 1 subtype C proviral loads are associated with bronchiectasis in indigenous Australians: results of a case-control study. Open Forum Infect Dis 1(1):ofu023

Ergan Arsava B, Coplu L (2011) Does airway colonization cause systemic inflammation in bronchiectasis? Tuberk Toraks 59(4):340–347

Eusebio N, Amorim AA, Gamboa F, Araujo R (2015) Molecular identification and genotyping of Pseudomonas aeruginosa isolated from cystic fibrosis and non-cystic fibrosis patients with bronchiectasis. Pathog Dis 73(2):1–7

Fernandez-Barat L, Alcaraz-Serrano V, Amaro R, Torres A (2021) Pseudomonas aeruginosa in bronchiectasis. Semin Respir Crit Care Med 42(4):587–594

Flume PA, Chalmers JD, Olivier KN (2018) Advances in bronchiectasis: endotyping, genetics, microbiome, and disease heterogeneity. Lancet 392(10150):880–890

Gao YH, Guan WJ, Xu G, Lin ZY, Tang Y, Lin ZM, Gao Y, Li HM, Zhong NS, Zhang GJ et al (2015) The role of viral infection in pulmonary exacerbations of bronchiectasis in adults: a prospective study. Chest 147(6):1635–1643

Giamarellos-Bourboulis EJ (2008) Macrolides beyond the conventional antimicrobials: a class of potent immunomodulators. Int J Antimicrob Agents 31(1):12–20
Goyal V, Chang AB (2022) Bronchiectasis in childhood. Clin Chest Med 43(1):71–88
Goyal V, Grimwood K, Marchant J, Masters IB, Chang AB (2016) Pediatric bronchiectasis: no longer an orphan disease. Pediatr Pulmonol 51(5):450–469
Guan WJ, Li JC, Liu F, Zhou J, Liu YP, Ling C, Gao YH, Li HM, Yuan JJ, Huang Y et al (2018) Next-generation sequencing for identifying genetic mutations in adults with bronchiectasis. J Thorac Dis 10(5):2618–2630
Gupta A, Singh N, Kumar A, Verma UP, Verma AK, Shyam H, Lal N, Kant S, Kumari A (2020) Differential expression of inflammatory responsive genes between chronic periodontitis and periodontally affected bronchiectasis patients. Mol Biol Res Commun 9(4):169–172
Guran T, Ersu R, Karadag B, Akpinar IN, Demirel GY, Hekim N, Dagli E (2007) Association between inflammatory markers in induced sputum and clinical characteristics in children with non-cystic fibrosis bronchiectasis. Pediatr Pulmonol 42(4):362–369
Hadda V, Chawla G, Tiwari P, Madan K, Khan MA, Mohan A, Khilnani GC, Guleria R (2018) Noninvasive ventilation for acute respiratory failure due to noncystic fibrosis bronchiectasis. Indian J Crit Care Med 22(5):326–331
Hilliam Y, Moore MP, Lamont IL, Bilton D, Haworth CS, Foweraker J, Walshaw MJ, Williams D, Fothergill JL, De Soyza A et al (2017) Pseudomonas aeruginosa adaptation and diversification in the non-cystic fibrosis bronchiectasis lung. Eur Respir J 49(4)
Hsieh MH, Chou PC, Chou CL, Ho SC, Joa WC, Chen LF, Sheng TF, Lin HC, Wang TY, Chang PJ et al (2013) Matrix metalloproteinase-1 polymorphism (−1607G) and disease severity in non-cystic fibrosis bronchiectasis in Taiwan. PLoS One 8(6):e66265
Huang Y, Chen CL, Yuan JJ, Li HM, Han XR, Chen RC, Guan WJ, Zhong NS (2019) Sputum Exosomal microRNAs profiling reveals critical pathways modulated by Pseudomonas aeruginosa colonization in bronchiectasis. Int J Chron Obstruct Pulmon Dis 14:2563–2573
Huang J-J, Gao Y, Barton A, Richardson H, Leyah H, Barclay H, Kuzmanova E, Keir H, Dicker A, Chalmers J (2020) Sputum proteomics identifies distinct endotypes of exacerbations in bronchiectasis. Eur Respir J:56
Jose RJ, Loebinger MR (2021) Clinical and radiological phenotypes and Endotypes. Semin Respir Crit Care Med 42(4):549–555
Kapur N, Masters IB, Morris PS, Galligan J, Ware R, Chang AB (2012) Defining pulmonary exacerbation in children with non-cystic fibrosis bronchiectasis. Pediatr Pulmonol 47(1):68–75
Kapur N, Mackay IM, Sloots TP, Masters IB, Chang AB (2014) Respiratory viruses in exacerbations of non-cystic fibrosis bronchiectasis in children. Arch Dis Child 99(8):749–753
Keir HR, Shoemark A, Dicker AJ, Perea L, Pollock J, Giam YH, Suarez-Cuartin G, Crichton ML, Lonergan M, Oriano M et al (2021) Neutrophil extracellular traps, disease severity, and antibiotic response in bronchiectasis: an international, observational, multicohort study. Lancet Respir Med 9(8):873–884
Kim SH, Kim C, Jeong I, Lee SJ, Kim TH, Lee CY, Oh YM, Lee H, Kim Y (2021) Chronic obstructive pulmonary disease is associated with decreased quality of life in bronchiectasis patients: findings from the KMBARC registry. Front Med (Lausanne) 8:722124
King P (2007) Is there a role for inhaled corticosteroids and macrolide therapy in bronchiectasis? Drugs 67(7):965–974
Lee JH, Kim YK, Kwag HJ, Chang JH (2004) Relationships between high-resolution computed tomography, lung function and bacteriology in stable bronchiectasis. J Korean Med Sci 19(1): 62–68
Lee H, Choi H, Sim YS, Park S, Kim WJ, Yoo KH, Lee SJ, Kim TH, Yang B, Jeong I et al (2020) KMBARC registry: protocol for a multicentre observational cohort study on non-cystic fibrosis bronchiectasis in Korea. BMJ Open 10(1):e034090
Liu JF, Zhong XN, He ZY, Zhong DJ, Bai J, Zhang JQ, Zhong W (2012) Impact of treatment with low dose roxithromycin on stable bronchiectasis. Zhonghua Jie He He Hu Xi Za Zhi 35(11): 824–827

Loukides S, Bouros D, Papatheodorou G, Lachanis S, Panagou P, Siafakas NM (2002) Exhaled H (2)O(2) in steady-state bronchiectasis: relationship with cellular composition in induced sputum, spirometry, and extent and severity of disease. Chest 121(1):81–87

Mac Aogain M, Chandrasekaran R, Lim AYH, Low TB, Tan GL, Hassan T, Ong TH, Hui Qi Ng A, Bertrand D, Koh JY et al (2018) Immunological corollary of the pulmonary mycobiome in bronchiectasis: the CAMEB study. Eur Respir J 52(1)

Mac Aogain M, Narayana JK, Tiew PY, Ali N, Yong VFL, Jaggi TK, Lim AYH, Keir HR, Dicker AJ, Thng KX et al (2021) Integrative microbiomics in bronchiectasis exacerbations. Nat Med 27(4):688–699

Maiz Carro L, Martinez-Garcia MA (2019) Nebulized hypertonic saline in noncystic fibrosis bronchiectasis: a comprehensive review. Ther Adv Respir Dis 13:1753466619866102

Mandal P, Patel D, Hill A (2013) A simplified CT scoring system in non cystic fibrosis bronchiectasis. Eur Respir J:42

Marsh RL, Thornton RB, Smith-Vaughan HC, Richmond P, Pizzutto SJ, Chang AB (2015) Detection of biofilm in bronchoalveolar lavage from children with non-cystic fibrosis bronchiectasis. Pediatr Pulmonol 50(3):284–292

Marsh RL, Binks MJ, Smith-Vaughan HC, Janka M, Clark S, Richmond P, Chang AB, Thornton RB (2022) Prevalence and subtyping of biofilms present in bronchoalveolar lavage from children with protracted bacterial bronchitis or non-cystic fibrosis bronchiectasis: a cross-sectional study. Lancet Microbe 3(3):e215–e223

Milliron B, Henry TS, Veeraraghavan S, Little BP (2015) Bronchiectasis: mechanisms and imaging clues of associated common and uncommon diseases. Radiographics 35(4):1011–1030

Nair GB, Ilowite JS (2012) Pharmacologic agents for mucus clearance in bronchiectasis. Clin Chest Med 33(2):363–370

Namkoong H, Omae Y, Asakura T, Ishii M, Suzuki S, Morimoto K, Kawai Y, Emoto K, Oler AJ, Szymanski EP et al (2021) Genome-wide association study in patients with pulmonary Mycobacterium avium complex disease. Eur Respir J 58(2)

O'Donnell AE (2022) Bronchiectasis - a clinical review. N Engl J Med 387(6):533–545

Olveira G, Olveira C, Dorado A, Garcia-Fuentes E, Rubio E, Tinahones F, Soriguer F, Murri M (2013) Cellular and plasma oxidative stress biomarkers are raised in adults with bronchiectasis. Clin Nutr 32(1):112–117

Oriano M, Aliberti S, Guerini FR, Agliardi C, Di Francesco C, Gelmini A, Terranova L, Zanzottera M, Marchisio P, Clerici M et al (2021) The isoform GC1f of the vitamin D binding protein is associated with bronchiectasis severity. Biomedicine 9(11)

Palasamudram Shekar S, Bajarano P, Hadeh A, Rojas E, Gillenwater SR, Savage E, Mehta JP (2018) Case of missing plastic: foreign body bronchiectasis. Cureus 10(7):e2974

Papatheodorou A, Makrythanasis P, Kaliakatsos M, Dimakou A, Orfanidou D, Roussos C, Kanavakis E, Tzetis M: Development of novel microarray methodology for the study of mutations in the SERPINA1 and ADRB2 genes--their association with Obstructive Pulmonary Disease and Disseminated Bronchiectasis in Greek patients. Clin Biochem 2010, 43(1-2):43-50

Paris D, Palomba L, Mirra V, Borrelli M, Corcione A, Santamaria F, Maniscalco M, Motta A (2020) NMR profiling of exhaled breath condensate defines different metabolic phenotypes of non-cystic fibrosis bronchiectasis. Int J Mol Sci 21(22)

Polverino E, Goeminne PC, McDonnell MJ, Aliberti S, Marshall SE, Loebinger MR, Murris M, Canton R, Torres A, Dimakou K et al (2017) European Respiratory Society guidelines for the management of adult bronchiectasis. Eur Respir J 50(3)

Prikk K, Maisi P, Pirila E, Sepper R, Salo T, Wahlgren J, Sorsa T (2001) In vivo collagenase-2 (MMP-8) expression by human bronchial epithelial cells and monocytes/macrophages in bronchiectasis. J Pathol 194(2):232–238

Purcell P, Jary H, Perry A, Perry JD, Stewart CJ, Nelson A, Lanyon C, Smith DL, Cummings SP, De Soyza A (2014) Polymicrobial airway bacterial communities in adult bronchiectasis patients. BMC Microbiol 14:130

Reiff DB, Wells AU, Carr DH, Cole PJ, Hansell DM (1995) CT findings in bronchiectasis: limited value in distinguishing between idiopathic and specific types. AJR Am J Roentgenol 165(2): 261–267

Richardson H, Dicker AJ, Barclay H, Chalmers JD (2019) The microbiome in bronchiectasis. Eur Respir Rev 28(153)

Rogers GB, van der Gast CJ, Cuthbertson L, Thomson SK, Bruce KD, Martin ML, Serisier DJ (2013) Clinical measures of disease in adult non-CF bronchiectasis correlate with airway microbiota composition. Thorax 68(8):731–737

Rogers GB, Zain NM, Bruce KD, Burr LD, Chen AC, Rivett DW, McGuckin MA, Serisier DJ (2014) A novel microbiota stratification system predicts future exacerbations in bronchiectasis. Ann Am Thorac Soc 11(4):496–503

Rubin BK (2007) Mucolytics, expectorants, and mucokinetic medications. Respir Care 52(7): 859–865

Sepper R, Konttinen YT, Ding Y, Takagi M, Sorsa T (1995) Human neutrophil collagenase (MMP-8), identified in bronchiectasis BAL fluid, correlates with severity of disease. Chest 107(6):1641–1647

Shoemark A, Shteinberg M, De Soyza A, Haworth CS, Richardson H, Gao Y, Perea L, Dicker AJ, Goeminne PC, Cant E et al (2022) Characterization of eosinophilic bronchiectasis: a European multicohort study. Am J Respir Crit Care Med 205(8):894–902

Simpson JL, Grissell TV, Douwes J, Scott RJ, Boyle MJ, Gibson PG (2007) Innate immune activation in neutrophilic asthma and bronchiectasis. Thorax 62(3):211–218

Taylor SL, Rogers GB, Chen AC, Burr LD, McGuckin MA, Serisier DJ (2015) Matrix metalloproteinases vary with airway microbiota composition and lung function in non-cystic fibrosis bronchiectasis. Ann Am Thorac Soc 12(5):701–707

Taylor SL, Woodman RJ, Chen AC, Burr LD, Gordon DL, McGuckin MA, Wesselingh S, Rogers GB (2017) FUT2 genotype influences lung function, exacerbation frequency and airway microbiota in non-CF bronchiectasis. Thorax 72(4):304–310

Tino G (2018) Bronchiectasis: phenotyping an orphan disease. Am J Respir Crit Care Med 197(11): 1371–1373

Tran TT, Hadinoto K (2021) A potential quorum-sensing inhibitor for bronchiectasis therapy: quercetin-chitosan nanoparticle complex exhibiting superior inhibition of biofilm formation and swimming motility of Pseudomonas aeruginosa to the native quercetin. Int J Mol Sci 22(4)

Tunney MM, Einarsson GG, Wei L, Drain M, Klem ER, Cardwell C, Ennis M, Boucher RC, Wolfgang MC, Elborn JS (2013) Lung microbiota and bacterial abundance in patients with bronchiectasis when clinically stable and during exacerbation. Am J Respir Crit Care Med 187(10):1118–1126

Woo TE, Duong J, Jervis NM, Rabin HR, Parkins MD, Storey DG (2016) Virulence adaptations of Pseudomonas aeruginosa isolated from patients with non-cystic fibrosis bronchiectasis. Microbiology (Reading) 162(12):2126–2135

Woo TE, Lim R, Surette MG, Waddell B, Bowron JC, Somayaji R, Duong J, Mody CH, Rabin HR, Storey DG et al (2018) Epidemiology and natural history of Pseudomonas aeruginosa airway infections in non-cystic fibrosis bronchiectasis. ERJ Open Res 4(2)

Yang B, Jang HJ, Chung SJ, Yoo SJ, Kim T, Kim SH, Shin YM, Kang HK, Kim JS, Choi H et al (2020) Factors associated with bronchiectasis in Korea: a national database study. Ann Transl Med 8(21):1350

Zanini A, Aiello M, Adamo D, Cherubino F, Zampogna E, Sotgiu G, Chetta A, Spanevello A (2015) Effects of pulmonary rehabilitation in patients with non-cystic fibrosis bronchiectasis: a retrospective analysis of clinical and functional predictors of efficacy. Respiration 89(6): 525–533

Zheng L, Tipoe G, Lam WK, Leung RY, Ho JC, Shum IH, Ooi GC, Ip MS, Tsang KW (2000) Up-regulation of circulating adhesion molecules in bronchiectasis. Eur Respir J 16(4):691–696
Zheng L, Lam WK, Tipoe GL, Shum IH, Yan C, Leung R, Sun J, Ooi GC, Tsang KW (2002) Overexpression of matrix metalloproteinase-8 and -9 in bronchiectatic airways in vivo. Eur Respir J 20(1):170–176

Restrictive Pulmonary Diseases

6

Abstract

Restrictive pulmonary diseases refers to "a group of lung disorders associated with restricted lung expansion and reduced lung volume". These pulmonary diseases can ensue from intrinsic parenchymal pathologies or by extrinsic causes and contribute to significant morbidity and mortality. Pulmonary function tests, 6-min walk test, and HRCT scans of thorax are extensively used for diagnostic purposes. Immunosuppressants and antifibrotic agents are commonly used for therapy. The complexity and heterogeneity of restrictive lung disorders make the differential diagnosis and efficient disease management complicated. Other challenges encountered include varying response to current therapeutics and adverse effects related to medications. Disease phenotyping and biomarker identification using advanced omics tools are, therefore, crucial and revolutionizing for the diagnosis and management of restrictive respiratory disorders.

Keywords

Restrictive pulmonary diseases · Interstitial lung diseases (ILDs) · Total lung capacity (TLC) · Tissue scarring · Fibrotic lesions · Honeycomb appearance · HRCT scans

6.1 Introduction

Restrictive pulmonary diseases represent a heterogeneous group of parenchymal, pleural, and extra-pulmonary respiratory disorders characterized by decreased lung distensibility leading to reduced lung volume, reduced lung capacities, compromised lung expansion, and lung compliance (Robinson 2016). Restrictive pulmonary diseases are detected by clinical suspicion and subsequent evaluation including spirometry, measurement of diffusion capacity, chest X-ray, and HRCT chest. The

S. Saha et al., *Pulmonomics: Omics Approaches for Understanding Pulmonary Diseases*, https://doi.org/10.1007/978-981-99-3505-5_6

later remains the main mode of diagnosis most of the time. The global prevalence of restrictive respiratory diseases is much less as compared to the prevalence of obstructive respiratory diseases (https://www.ncbi.nlm.nih.gov/books/NBK560880/). Nevertheless, restrictive pulmonary diseases have been significantly associated with increasing morbidity and mortality rates. In 2019, restrictive pulmonary disorders ranked 30th amongst the major causes of death across the world (Jeganathan and Sathananthan 2022).

Both the pulmonary parenchymal and extra-parenchymal restrictions imply reduced vital capacity (VC) with preservation or increase in forced expiratory airflow in respect to the vital capacity [forced expiratory volume in the first second/forced vital capacity, e.g. FEV1/FVC] (Aggarwal and Agarwal 2007). This pulmonary restrictive pattern in spirometry is accompanied by gradual reduction of the diffusion capacity [measured for carbon monoxide; (D_{LCO})] and the ventilatory capacity which are relatively preserved in extra-pulmonary causes of restriction (https://www.ncbi.nlm.nih.gov/books/NBK560880/).

The parenchymal restrictive pulmonary diseases are mainly constituted by diffuse parenchymal lung disease (DPLD), also known as interstitial lung diseases. DPLD, an essentially non-infective and nonmalignant problem of diffuse distribution of the lungs can develop from over 300 aetiologies. The pathology is also variable and it bears either potentially fibrotic or established fibrotic elements. The knowledge especially regarding the aetiology of a DPLD is complex and fast evolving as the conditions evolve from a wide array of heterogeneous factors. Therefore, the etiological evaluation is often exhaustive (Kulshrestha et al. 2020). The treatment has been evolving and is based on some general principles, the etiological factors involved, the physiological status, the complications developed (especially respiratory failure), and the concomitant comorbidities. The appreciation of treatment response is often difficult, so also the job of monitoring. The emotional and the financial issues complicate the scenario. With better understanding of the pathogenesis, the scope of biomarker-research has expanded widely but effective and clinically applicable biomarkers beyond radiological interpretations are limited. Here, in the chapter, we will provide an overview of them.

6.2 Clinical Features

Restricted pulmonary diseases are largely clinically characterized by reduced capacity of exercise, shortness of breath especially with exertion and cough that often increases with the severity of the disease (Laveneziana 2010; Das et al. 2017). The progression is marked with decline in lung function and respiratory failure both in acute and chronic cases of restrictive respiratory diseases (Laveneziana 2010). The hypoxemia (reduced arterial oxygen content) present initially on exercise progresses even to the resting state as the disease progresses. There are other clinical clues that often help to suspect the condition. The complications of the disease are also often apparent clinically. Interstitial lung diseases (ILDs) are often exclusively characterized by the "3C features"—coughing, clubbing of nails, and coarse crackle sound in chest auscultation (Wallis and Spinks 2015).

6.3 Aetiology and Pathogenesis

The restricted diseases of lungs may result from intrinsic as well as extrinsic causes (as shown in Fig. 6.1). The intrinsic cause of restrictive lung diseases includes alterations and damages in distal lung parenchyma. Intrinsic restrictive lung diseases are characterized by inflammatory changes or dysregulated fibrogenesis or both (https://www.ncbi.nlm.nih.gov/books/NBK560880/). The process usually develops deep distally in broncho-epithelial domains mostly in the respiratory part of the lungs. Sometimes, it is initiated at distal bronchioles (hypersensitivity pneumonitis) and sometimes it ensues from alveolar epithelial damage with disturbed healing (https://www.ncbi.nlm.nih.gov/books/NBK560880/; Dalphin and Didier 2013). DPLD is associated with deposition of collagen in the interstitium and replacement of normal lung tissues with scar-tissues (fibrosis) that occurs and increases with disease progression (https://www.ncbi.nlm.nih.gov/books/NBK560880/). It results in reticular appearance, features of traction (traction bronchiectasis), or even honeycomb-like appearance and they are well picked up in HRCT chest (Raghu et al. 2018). Peripheral bronchial structures are also involved in such parenchymal damage.

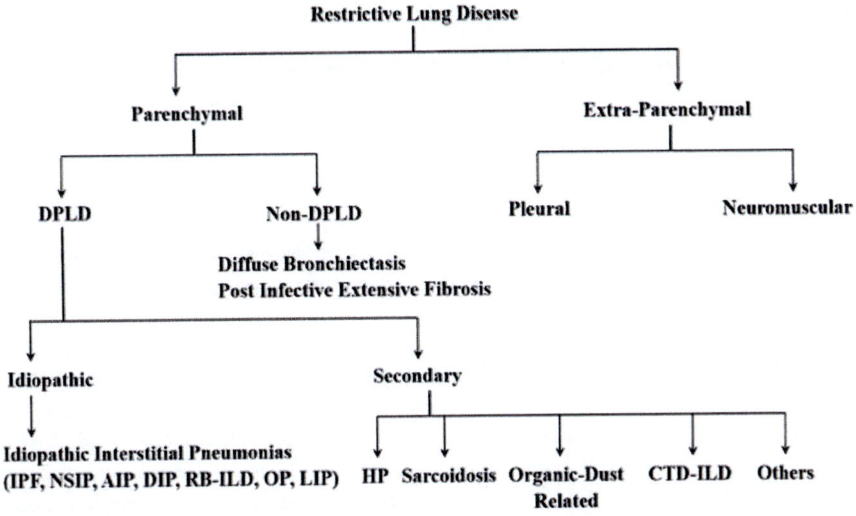

Fig. 6.1 The overview of classification of restrictive pulmonary problems. (*IPF* idiopathic pulmonary fibrosis, *NSIP* non-specific interstitial pneumonia, *AIP* acute interstitial pneumonia, *RB-ILD* respiratory bronchiolitis interstitial pneumonia, *OP* organizing pneumonia, *LIP* lymphocytic interstitial pneumonia, *HP* hypersensitivity pneumonitis, *CTD-ILD*: connective tissue disease ILD)

Restrictive pulmonary diseases may also occur due to extrinsic or extra-parenchymal reasons leading to impediments on inspiration (as shown in Fig. 6.1). The common extrinsic causes of restrictive pulmonary disorders include:

i. Pleural thickening or diseases of the pleura
ii. Restrictions in chest wall movements from chest wall problems
iii. Fusion of thoracic vertebrae or costovertebral joints and deformations in rib-cage or spine as kyphoscoliosis
iv. Limitation in the functioning of the neuromuscular apparatus leading to inadequate ventilation despite normal lung parenchyma (https://www.ncbi.nlm.nih.gov/books/NBK560880/; West 2013).

Certain symptoms are specifically associated with restrictive diseases triggered by extrinsic causes. These include increased body mass index (BMI), history or presence of neuromuscular diseases, and deviation of spinal curvature (https://www.ncbi.nlm.nih.gov/books/NBK560880/).

In general, elderly and obese individuals, smokers, and females are at a high risk of developing restrictive respiratory diseases (Esposito et al. 2015; Kurth and Hnizdo 2015). Individuals exposed to occupational or environmental hazards like coal dust, asbestos, and other hazardous dust specks are highly susceptible to tissue scarring and alterations associated with restrictive lung diseases (Mannino et al. 2012).

The primary symptom of restrictive pulmonary diseases is the insidious progression of exertional dyspnoea.

6.4 Different Types of Restrictive Pulmonary Diseases

The classification scheme of restrictive lung diseases elaborated in Fig. 6.1 largely captures the different intrinsic or parenchymal and extrinsic or extra-parenchymal entities.

Interstitial lung diseases (ILDs) triggered by intrapulmonary or intrinsic restriction broadly comprises of the following diseased conditions—Idiopathic Pulmonary Fibrosis (IPF), Sarcoidosis, Hypersensitivity Pneumonitis (HP), Interstitial Pneumonia with Autoimmune Features (IPAF), Non-Specific Interstitial Pneumonia (NSIP), Pneumoconiosis, Connective Tissue Disease associated ILD (CTD-ILD), drug induced ILDs, occupational ILDs (like asbestosis), and unclassifiable ILD (uILD) (Cottin 2019). These different ILDs exhibit significant overlap in symptoms and disease behaviour as depicted by Fig. 6.2 (Wells et al. 2018; Fernandes et al. 2019).

Presently, experts have decided to classify the DPLDs as per the pathophysiology relating to development of predominant and accelerated fibrosis (progressive fibrosis) signifying a relatively rapid progressive course with reduction in functional capacity of lungs and poor prognosis. IPF (idiopathic pulmonary fibrosis) has been the classical prototype of such progressive fibrotic phenotype of disease but DPLD from other aetiologies can also turn to a similar state. They are grouped as PF-ILD

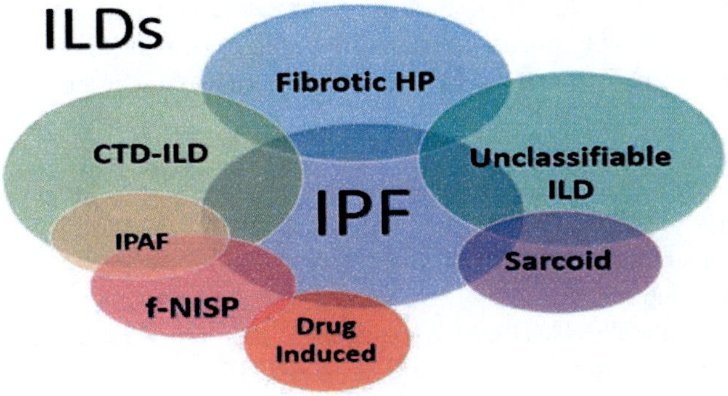

Fig. 6.2 Venn diagram depicting the overlap of disease behaviour between the different interstitial lung diseases (ILDs)—idiopathic pulmonary fibrosis (IPF) and other interstitial lung diseases (ILDs)

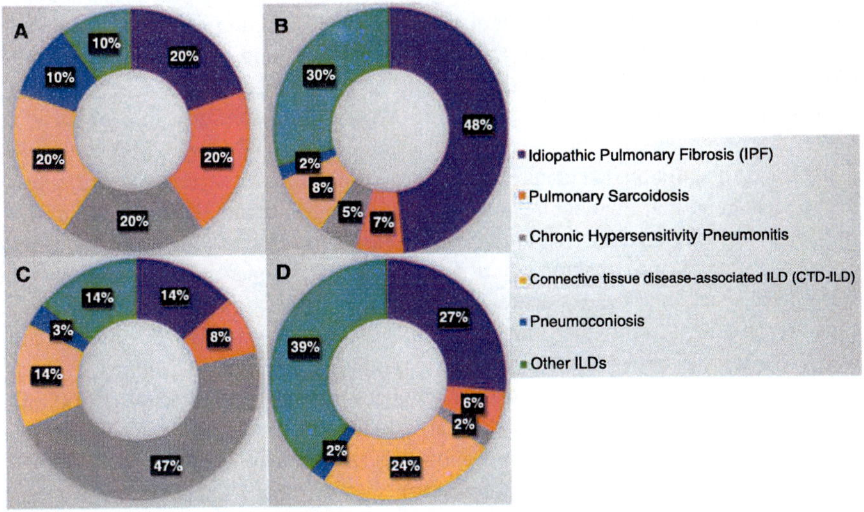

Fig. 6.3 The relative distribution of the different interstitial lung diseases (ILDs) in (**a**) United States, (**b**) Europe, (**c**) India, and (**d**) China. Information on the relative distribution of ILDs have been derived from (Lederer and Martinez 2018; Guenther et al. 2018; Singh et al. 2017; Ban et al. 2018)

(progressive fibrotic-interstitial lung disease). PF-ILD should be treated with antifibrotic medications as nintedanib or pirfenidone (Albera et al. 2021).

However, the relative prevalence of the different types of ILDs varies across the different countries in the world. Such distribution across the USA, Europe, India, and China has been represented in Fig. 6.3 (Lederer and Martinez 2018; Guenther et al. 2018; Singh et al. 2017; Ban et al. 2018). In general, across the globe, the most common and predominant forms of ILDs include Idiopathic Pulmonary Fibrosis

(IPF) and Sarcoidosis (Cottin et al. 2018; Dhooria et al. 2018). In India, however, chronic hypersensitivity pneumonitis dominates the scenario (Singh et al. 2017).

The extrinsic or extra-parenchymal restrictive diseases can be broadly divided into three groups as follows:

i. Pleural conditions including pleural effusions, pleural thickening, and pleural scarring
ii. Conditions related to chest wall like kyphosis, scoliosis, pectus excavatum, and pectus carinatum
iii. Neuromuscular diseases like phrenic neuropathies, amyotrophic lateral sclerosis, and muscular dystrophy (Kurth and Hnizdo 2015)

6.5 Diagnosis of Restrictive Pulmonary Diseases

Pulmonary function tests (like spirometry, D_{LCO}, and lung volume measurements), 6-min walk test, and imaging techniques (like HRCT scanning and MRI) are widely used for the diagnosis of tissue scarring and restrictive lung diseases (Cottin 2019). On primary identification of restriction, further group-specific and disease-specific investigations are planned and experts from multiple disciplines join together to make a consensus on the possible disease-diagnosis through MDDs (multi-disciplinary discussions) (Furini et al. 2019). MDD is found to improve the quality of diagnosis and help management significantly (Biglia et al. 2019; Meyer 2016).

Tissue fibrosis in restrictive lung diseases result in reduced lung compliance and decreased inspiratory capacity, often represented by a typical restrictive pattern in spirometry, characterized by decreased total lung capacity (TLC) and forced vital capacity (FVC). As per the American Thoracic Society (ATS), the presence of restrictive lung disorder is diagnosed by a predicted TLC value less than 80%, as determined in spirometry (https://www.ncbi.nlm.nih.gov/books/NBK560880/; Aggarwal and Agarwal 2007; Brack et al. 2002). The disease severity can also be stratified based on the extent of reduction in TLC (as listed in Table 6.1) (https://www.ncbi.nlm.nih.gov/books/NBK560880/). ATS/ ERS has also provided guidelines for categorization of restrictive pulmonary diseases based on vital capacity (VC) in 1991 and FEV1% in 2005 (Aggarwal and Agarwal 2007).

Table 6.1 Classification of restrictive pulmonary diseases based on disease severity

Severity of restrictive pulmonary diseases	Total lung capacity (TLC) [Predicted] (%)
Mild restrictive pulmonary disease	70–80
Moderate restrictive pulmonary disease	60–70
Moderately severe restrictive pulmonary disease	50–60
Severe restrictive pulmonary disease	<50

This table enlists the different degrees of severity of restrictive lung diseases determined in terms of predicted total lung capacity (TLC)

Table 6.2 Changes in important lung function parameters in restrictive pulmonary diseases

Sl. no.	Important lung function parameters	Trends in restrictive pulmonary diseases
1.	Forced vital capacity (FVC)	Reduced
2.	Forced expiratory volume in one second (FEV1)	Slightly reduced or normal
3.	FEV1/FVC	Normal or increased
4.	Diffusing capacity of lung for carbon monoxide	Decreased (for intrinsic restrictive pulmonary diseases)
5.	Total lung capacity (TLC)	Decreased (<80%)
6.	Functional residual capacity (FRC)	Reduced
7.	Residual volume (RV)	Decreased

This table contains the general trends of lung function parameters in restrictive respiratory diseases as noted from pulmonary function tests (PFTs)

Decreased functional residual capacity (FRC) and residual volume (RV) are observed in restrictive lung disease, and spirometry plays a good role (https://www.ncbi.nlm.nih.gov/books/NBK493170/). Patients with restrictive disorders often fail to exhale for a prolonged period to empty the lungs to the residual volume. This leads to decreased Forced Vital Capacity (FVC) and Forced Expiratory Volume in one second (FEV1) values in spirometric analysis. Since, the decrease in FVC appears to be more pronounced as compared to FEV1, the FEV1/FVC ratio either remains unaltered or gets increased (Keddissi et al. 2013).

Progressive increase and thickening of fibrotic tissues also interfere with gaseous exchange between lung tissues and blood, particularly in intrinsically triggered restrictive pulmonary diseases. Therefore, the diffusing capacity of lungs for carbon monoxide (D_{LCO}) is exclusively reduced in case of intrinsic pulmonary restriction. Extrinsic pulmonary restriction, on the other hand, is characterized by restricted spirometric pattern and unaltered D_{LCO} (https://www.ncbi.nlm.nih.gov/books/NBK556149/; Johnson and Theurer 2014).

The standard trend of fluctuations in lung parameters for restrictive respiratory diseases has been summarized in Table 6.2.

Functional exercise capacity in patients with restrictive pulmonary diseases is determined by the 6-minute walk test (6MWT). Exercise capacity, as detected by 6MWT, is impaired in restrictive pulmonary diseases like idiopathic pulmonary fibrosis (IPF) due to altered ventilatory mechanics and circulatory complications. Parameters like desaturation and 6-minute walk distance (6MWD) obtained in the 6MWT serve as indicators of mortality in patients with restrictive lung diseases, particularly IPF (Lancaster 2018).

High-resolution computed tomography (HRCT) scanning serves as an imaging diagnostic tool for restrictive pulmonary diseases. It enables visualization of the texture and extent of tissue scarring, fibrotic lesions, and the characteristic honeycomb appearance. The presence of traction bronchiectasis and honeycombing reflects a predominant fibrotic pathology. Thus, HRCT scan together with spirometry, D_{LCO} and 6-minute walk test allow diagnosis and monitoring of progressive deterioration in patients with restrictive respiratory disorders (Cottin 2019).

6.6 Treatment of Restrictive Pulmonary Diseases

Therapy of restrictive lung diseases depends on the underlying aetiology. Immunosuppressants like steroids, methotrexate, mycophenolate mofetil, azathioprine, and cyclophosphamide are commonly used to reduce inflammation and suppress immune responses in different restrictive lung disorders (Yasuoka 2015; Coker 2009). Different anti-fibrotic drugs like pirfenidone and nintedanib are also used to delay fibrotic disease progression, especially in IPF (Robinson 2016). Oxygen therapy is often used as a supportive care for improving shortness of breath and functional or exercise capacity in patients with ILDs (Bell et al. 2017). Management of comorbidities and pulmonary rehabilitation is employed to achieve better control of symptoms of restrictive lung diseases. Severe impairment in lung function can be restored using surgical corrections in pleural pathologies (Johari et al. 2016). Lung transplantation can be explored as a well-tolerated and effective last resort therapy for chronic respiratory failure in patients with ILD (Sulica et al. 2001).

6.7 Manifold Challenges Related to Restrictive Pulmonary Diseases

Current diagnostic tools have enabled improved diagnosis of restrictive respiratory diseases, however, often differential diagnosis of ILDs becomes complicated and demands invasive diagnosis using lung biopsies (Tolle 2022). Lack of adequate awareness and late referrals also poses hindrance to efficient diagnosis and disease management (Distler et al. 2020). Stage of initiating treatment also largely depends on effective disease diagnosis. Besides, restrictive respiratory diseases like ILDs have variable clinical course and variable response to current therapeutics. Some patients respond to the standard therapeutics at an early stage but show rapid deterioration at a later stage. Besides, the prolonged usage of immunosuppressant drugs is associated with local and systemic adverse effects. The adverse effects of antifibrotics also matter. The safety and efficacy of using combinatorial therapy are also potential causes of concern (Tolle 2022; Valenzuela et al. 2020). The present therapeutic strategies are mostly preventive or supportive; they can slow down disease progression but hardly cure an ILD. However, there is a huge dearth of effective curative treatment for most of the restrictive pulmonary diseases. High mortality rate is associated with restrictive lung diseases. Thus, there is a pressing need to explore deep into disease pathobiology for efficient identification of the entities, deciding their drug targets and designing of potential therapeutics.

6.8 Conclusion

Understanding the complexity and heterogeneity of restrictive pulmonary disorders at molecular level holds the key to overcoming the challenges of present-day diagnostics and therapeutic strategies. Such knowledge of the underlying molecular pathways shall also enable efficient improvisation or innovations in the field of novel diagnostic tools for better disease phenotyping and the discovery of newer therapeutics (Wheelock et al. 2013). Use of modernized, high-throughput omics approaches, followed by integrative analysis shall unravel molecular mechanisms and important biomarkers involved in driving the pathophysiology of different restrictive pulmonary diseases (Kan et al. 2017; Zheng et al. 2022). Biomarkers thus identified can aid in resolving the unmet needs of efficient disease management by paving the way for novel and targeted therapeutic strategies with minimal side effects. The ongoing active research in biomarker development may lead to better and easier class, group, and disease-specific diagnosis, treatment, and monitoring of restrictive lung diseases in future. We have dealt with some important members of the interstitial lung disease as idiopathic pulmonary fibrosis and sarcoidosis separately in the book.

References

Aggarwal AN, Agarwal R (2007) The new ATS/ERS guidelines for assessing the spirometric severity of restrictive lung disease differ from previous standards. Respirology 12(5):759–762

Albera C, Verri G, Sciarrone F, Sitia E, Mangiapia M, Solidoro P (2021) Progressive fibrosing interstitial lung diseases: a current perspective. Biomedicine 9(9)

Ban C, Yan W, Xie B, Zhu M, Liu Y, Zhang S, Ye Q, Ren Y, Jiang D, Geng J et al (2018) Spectrum of interstitial lung disease in China from 2000 to 2012. Eur Respir J 52(3)

Bell EC, Cox NS, Goh N, Glaspole I, Westall GP, Watson A, Holland AE (2017) Oxygen therapy for interstitial lung disease: a systematic review. Eur Respir Rev 26(143)

Biglia C, Ghaye B, Reychler G, Koenig S, Yildiz H, Lacroix V, Tamirou F, Hoton D, Pieters T, Froidure A (2019) Multidisciplinary management of interstitial lung diseases: a real-life study. Sarcoidosis Vasc Diffuse Lung Dis 36(2):108–115

Brack T, Jubran A, Tobin MJ (2002) Dyspnea and decreased variability of breathing in patients with restrictive lung disease. Am J Respir Crit Care Med 165(9):1260–1264

Coker RK (2009) Management strategies for pulmonary sarcoidosis. Ther Clin Risk Manag 5(3): 575–584

Cottin V (2019) Treatment of progressive fibrosing interstitial lung diseases: a milestone in the management of interstitial lung diseases. Eur Respir Rev 28(153)

Cottin V, Hirani NA, Hotchkin DL, Nambiar AM, Ogura T, Otaola M, Skowasch D, Park JS, Poonyagariyagorn HK, Wuyts W et al (2018) Presentation, diagnosis and clinical course of the spectrum of progressive-fibrosing interstitial lung diseases. Eur Respir Rev 27(150)

Dalphin JC, Didier A (2013) Environmental causes of the distal airways disease. Hypersensitivity pneumonitis and rare causes. Rev Mal Respir 30(8):669–681

Das V, Desai U, Joshi JM (2017) Clinical profile of interstitial lung disease at a tertiary care centre, India. Pneumonologie 30(1)

Dhooria S, Agarwal R, Sehgal IS, Prasad KT, Garg M, Bal A, Aggarwal AN, Behera D (2018) Spectrum of interstitial lung diseases at a tertiary center in a developing country: a study of 803 subjects. PLoS One 13(2):e0191938

Distler O, Assassi S, Cottin V, Cutolo M, Danoff SK, Denton CP, Distler JHW, Hoffmann-Vold AM, Johnson SR, Muller Ladner U et al (2020) Predictors of progression in systemic sclerosis patients with interstitial lung disease. Eur Respir J 55(5)

Esposito DB, Lanes S, Donneyong M, Holick CN, Lasky JA, Lederer D, Nathan SD, O'Quinn S, Parker J, Tran TN (2015) Idiopathic pulmonary fibrosis in United States automated claims. incidence, prevalence, and algorithm validation. Am J Respir Crit Care Med 192(10): 1200–1207

Fernandes L, Nasser M, Ahmad K, Cottin V (2019) Interstitial pneumonia with autoimmune features (IPAF). Front Med (Lausanne) 6:209

Furini F, Carnevale A, Casoni GL, Guerrini G, Cavagna L, Govoni M, Scire CA (2019) The role of the multidisciplinary evaluation of interstitial lung diseases: systematic literature review of the current evidence and future perspectives. Front Med (Lausanne) 6:246

Guenther A, Krauss E, Tello S, Wagner J, Paul B, Kuhn S, Maurer O, Heinemann S, Costabel U, Barbero MAN et al (2018) The European IPF registry (eurIPFreg): baseline characteristics and survival of patients with idiopathic pulmonary fibrosis. Respir Res 19(1):141

Jeganathan N, Sathananthan M (2022) The prevalence and burden of interstitial lung diseases in the USA. ERJ Open Res 8(1)

Johari J, Sharifudin MA, Ab Rahman A, Omar AS, Abdullah AT, Nor S, Lam WC, Yusof MI (2016) Relationship between pulmonary function and degree of spinal deformity, location of apical vertebrae and age among adolescent idiopathic scoliosis patients. Singap Med J 57(1): 33–38

Johnson JD, Theurer WM (2014) A stepwise approach to the interpretation of pulmonary function tests. Am Fam Physician 89(5):359–366

Kan M, Shumyatcher M, Himes BE (2017) Using omics approaches to understand pulmonary diseases. Respir Res 18(1):149

Keddissi JI, Elya MK, Farooq SU, Youness HA, Jones KR, Awab A, Kinasewitz GT (2013) Bronchial responsiveness in patients with restrictive spirometry. Biomed Res Int 2013:498205

Kulshrestha R, Dhanda H, Pandey A, Singh A, Kumar R (2020) Immunopathogenesis and therapeutic potential of macrophage influx in diffuse parenchymal lung diseases. Expert Rev Respir Med 14(9):917–928

Kurth L, Hnizdo E (2015) Change in prevalence of restrictive lung impairment in the U.-S. population and associated risk factors: the National Health and Nutrition Examination Survey (NHANES) 1988–1994 and 2007–2010. Multidiscip Respir Med 10(1):7

Lancaster LH (2018) Utility of the six-minute walk test in patients with idiopathic pulmonary fibrosis. Multidiscip Respir Med 13:45

Laveneziana P (2010) Qualitative aspects of exertional dyspnea in patients with restrictive lung disease. Multidiscip Respir Med 5(3):211–215

Lederer DJ, Martinez FJ (2018) Idiopathic pulmonary fibrosis. N Engl J Med 378(19):1811–1823

Mannino DM, MA MB, Tan W, Kocabas A, Anto J, Vollmer WM, Buist AS, Group BCR (2012) Restricted spirometry in the Burden of Lung Disease Study. Int J Tuberc Lung Dis 16(10): 1405–1411

Meyer KC (2016) Multidisciplinary discussions and interstitial lung disease diagnosis: how useful is a meeting of the minds? Lancet Respir Med 4(7):529–531

Raghu G, Remy-Jardin M, Myers JL, Richeldi L, Ryerson CJ, Lederer DJ, Behr J, Cottin V, Danoff SK, Morell F et al (2018) Diagnosis of Idiopathic Pulmonary Fibrosis. An Official ATS/ERS/JRS/ALAT Clinical Practice Guideline. Am J Respir Crit Care Med 198(5):e44–e68

Robinson HC (2016) Respiratory conditions update: restrictive lung disease. FP Essent 448:29–34

Singh S, Collins BF, Sharma BB, Joshi JM, Talwar D, Katiyar S, Singh N, Ho L, Samaria JK, Bhattacharya P et al (2017) Interstitial lung disease in India. Results of a prospective registry. Am J Respir Crit Care Med 195(6):801–813

Sulica R, Teirstein A, Padilla ML (2001) Lung transplantation in interstitial lung disease. Curr Opin Pulm Med 7(5):314–322

Tolle LB (2022) Challenges in the diagnosis and management of patients with fibrosing interstitial lung disease. Case Rep Pulmonol 2022:9942432

Valenzuela C, Torrisi SE, Kahn N, Quaresma M, Stowasser S, Kreuter M (2020) Ongoing challenges in pulmonary fibrosis and insights from the nintedanib clinical programme. Respir Res 21(1):7

Wallis A, Spinks K (2015) The diagnosis and management of interstitial lung diseases. BMJ 350: h2072

Wells AU, Brown KK, Flaherty KR, Kolb M, Thannickal VJ (2018) Group IPFCW: what's in a name? That which we call IPF, by any other name would act the same. Eur Respir J 51(5)

West JB. Pulmonary pathophysiology: the essentials. 2013

Wheelock CE, Goss VM, Balgoma D, Nicholas B, Brandsma J, Skipp PJ, Snowden S, Burg D, D'Amico A, Horvath I et al (2013) Application of 'omics technologies to biomarker discovery in inflammatory lung diseases. Eur Respir J 42(3):802–825

Yasuoka H (2015) Recent treatments of interstitial lung disease with systemic sclerosis. Clin Med Insights Circ Respir Pulm Med 9(Suppl 1):97–110

Zheng P, Sun S, Wang J, Cheng ZJ, Lei KC, Xue M, Zhang T, Huang H, Zhang XD, Sun B (2022) Integrative omics analysis identifies biomarkers of idiopathic pulmonary fibrosis. Cell Mol Life Sci 79(1):66

7 Idiopathic Pulmonary Fibrosis

Abstract

The unknown aetiology of irreversible fibrotic scarring of lung tissues in Idiopathic Pulmonary Fibrosis (IPF) has resulted in the lack of a gold standard for disease diagnosis. Presently, medical imaging like HRCT scanning remains the best choice for disease diagnosis. However, recently several molecular, metabolic, and microbiome signatures associated with IPF have been identified using leading edge omics approaches. These probable biomarker candidates can be explored for improved, non-invasive disease diagnosis. The current treatment regime of IPF primarily relies on anti-fibrotic agents which slows down progression of fibrosis, but fails to cure the disease completely. Different biologics and small molecule modulators are being investigated clinically for their safety and efficacy in IPF patients. However, several research areas need to be explored to bridge knowledge gaps and find solution to unanswered questions related to IPF for better diagnosis and management of the disease.

Keywords

Fibrosis · Extracellular matrix (ECM) · Honeycomb appearance · HRCT scans · MUC5B · Nintedanib · Pirfenidone

7.1 Introduction

Idiopathic Pulmonary Fibrosis (IPF) is a specific variety of relatively rapid chronic progressive IIP (idiopathic interstitial pneumonia) characterized by UIP (usual interstitial pneumonia) pattern in HRCT chest and histology. An irreversible scarring and fibrotic remodelling of the lung parenchyma characterizes the disease (Wollin et al. 2015; van Geffen et al. 2021). This incurable disease is associated with poor prognosis and poor survival (median survival time is about 3–5 years) (Yu et al.

S. Saha et al., *Pulmonomics: Omics Approaches for Understanding Pulmonary Diseases*, https://doi.org/10.1007/978-981-99-3505-5_7

2018; Funke and Geiser 2015). Chronic shortness of breath, non-productive cough, bibasilar inspiratory crackles, exercise intolerance, progressive decline in lung function, and gradual deterioration in quality of life are commonly observed in IPF patients (Kolb et al. 2017; Fraser and Hoyles 2016; Rindlisbacher et al. 2018). The available therapeutic options are limited, and they are only capable of slowing down the progression of this fibrotic disease. This lack of effective therapy has added to the disease burden and mortality (Martinez et al. 2017). In the recent past, there has been a steady rise in the prevalence of IPF all over the world along with increased incidences of hospitalization and deaths (Lederer and Martinez 2018).

7.2 Pathophysiology

Excessive deposition of collagen rich extracellular matrix upon chronic or acute injury leads to lung fibrosis (Selvarajah et al. 2021). This ensues from either disturbed tissue healing (dysregulated fibro-genesis) or secondary fibrosis from inflammation. Genetic predisposition together with exposure to environmental factors (like smoke, pollutants, toxins, radiations, and infectious agents) or other diseased conditions (like gastro oesophageal reflux, autoimmune reactions), connective tissue diseases (like rheumatoid arthritis, scleroderma, dermatomyositis, polymyositis, and systemic lupus erythematosus) lead to fibrotic remodelling of the lung parenchyma (van Geffen et al. 2021). Tissue fibrosis can affect different organs (as shown in Fig. 7.1) and lead to gradual loss of function of those organs (Zeisberg and Kalluri 2013; Sun et al. 2022; Zahr et al. 2016; Wang et al. 2022; Lenti and Di Sabatino 2019). Fibrosis of lung tissues is termed as pulmonary fibrosis. Idiopathic pulmonary fibrosis (IPF), the form of lung fibrosis without an identifiable cause is possibly the most intriguing and rapidly progressing form of fibrotic disorder (Selvarajah et al. 2021).

Although IPF has an unknown aetiology, the complex interplay of genetic susceptibility and environmental triggers is thought to initiate the pathogenesis of IPF via an activation phase (as shown schematically in Fig. 7.2). Altered gene expression and dysregulation of various pathways in response to acute or chronic injuries of alveolar epithelium, eventually lead to fibrotic remodelling of lung tissues (Maher et al. 2007). Fibroblasts form an important cellular component of the fibrotic tissues. These fibroblasts differentiate into contractile myofibroblasts with huge potential of producing extracellular matrix (ECM). Accumulation of ECM rich in collagen, hyaluronan, laminin, elastin, fibronectin, and other glycoproteins leads to thickening of alveolar walls and compromises with the gaseous exchange capacity of the alveoli. Several immune cells, namely mesenchymal stromal cells, myeloid-derived suppressor cells, dendritic cells, macrophages, regulatory B cells (Bregs), and T cells (Tregs), have also been implicated in mediating the transition of inflammation to fibrotic proliferation in IPF (van Geffen et al. 2021). The fibrotic proliferation ensues from "fibroblastic foci" formed in the interstitium of the lung parenchyma that displays laying of the collagen fibre in addition to ECM. These foci are the characteristic pathological correlates of IPF and they consist of discrete

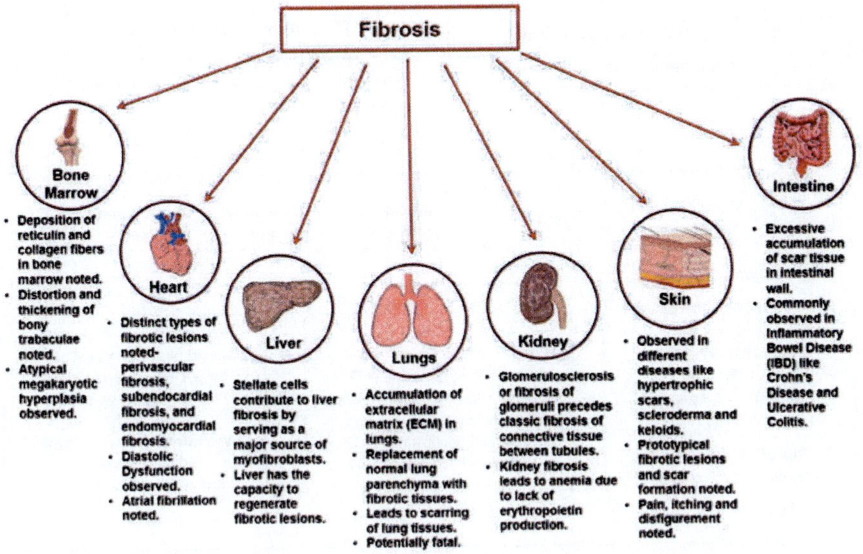

Fig. 7.1 Different organs affected in fibrosis. Apart from lungs, fibrosis may occur in different organs including heart, liver, intestine, kidney, bone marrow, and skin. Information on fibrosis of different organs have been derived from (Zeisberg and Kalluri 2013; Sun et al. 2022; Zahr et al. 2016; Wang et al. 2022)

clusters of fibroblasts and myofibroblasts. They, alongside the epithelial-mesenchymal transition, destroyed alveolar septae and disturbed blood flow characterize IPF (Yamaguchi et al. 2017). Such fibrogenic activities lead to irreversible scarring and thickening of lung parenchyma (Khan et al. 2021).

Imaging techniques like chest radiograph, computed tomography (CT), and high-resolution computed tomography (HRCT) are used for the diagnosis of IPF. However, presently there is no gold standard for effective diagnosis of IPF. Of these imaging techniques, volumetric scanning of the chest in HRCT is of prime importance in assessing usual interstitial pneumonia (UIP) patterns. Characteristic UIP pattern includes honeycomb appearance with or without traction bronchiectasis or traction bronchiolectasis. Presence of a typical UIP pattern without any identified cause is a diagnostic indicator of IPF. Presence of probable UIP pattern (denoted by lower lobe or peripheral bilateral reticulation with traction bronchiectasis without honeycomb appearance) has also been observed in many IPF patients (Raghu et al. 2018). Invasive lung biopsy is recommended by ATS/ERS/JRS/ALAT guidelines in patients with a probable UIP pattern on HRCT scan (Raghu et al. 2018). Video-assisted thoracoscopic surgery is often explored as an alternative option to surgical lung biopsy (SLB) for IPF diagnosis (Lederer and Martinez 2018; Raghu et al. 2018; Richeldi et al. 2018). The complications associated with these invasive diagnostic procedures for IPF are often bothersome and delay disease detection. Therefore,

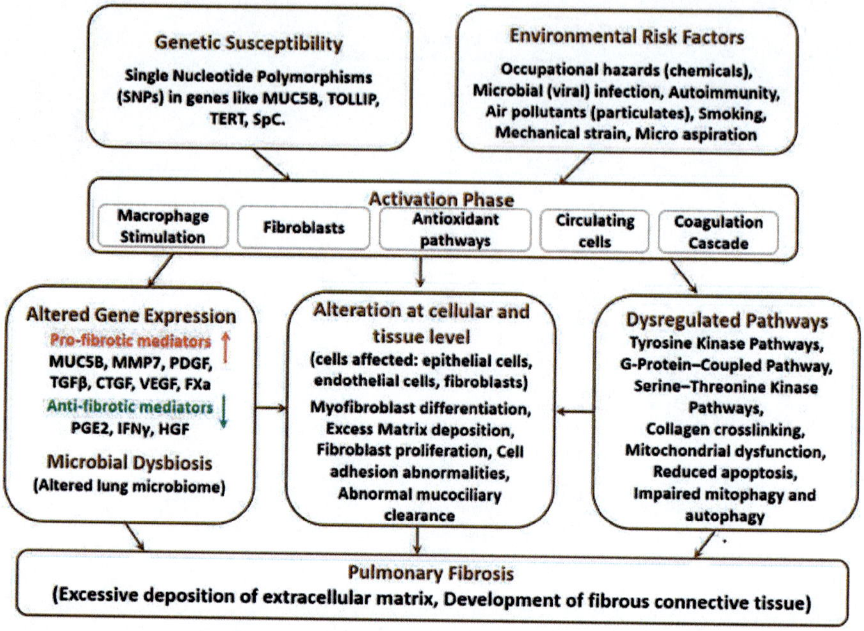

Fig. 7.2 Schematic representation of the disease pathogenesis of idiopathic pulmonary fibrosis (IPF)

improvements of existing techniques and screening of novel biomarkers are largely required to enable rapid and accurate diagnosis of IPF.

7.3 Biomarkers of IPF

The presence of challenges in the existent diagnostic methods, the variable course of IPF progression, and the lack of proper surrogate endpoints for disease outcomes in IPF patients highlights the dire need for reproducible biomarkers of IPF (Zhang and Kaminski 2012). Besides, biomarkers can aid in timely diagnosis of IPF patients alongside improved characterization and phenotyping of IPF. Different modern omics approaches have enabled screening of various biological samples for identification of novel and potential diagnostic, prognostic, and disease susceptibility biomarkers of IPF. The contribution of each of the omics approaches in dissecting and identifying novel biomarker candidates involved in IPF pathogenesis has been discussed below.

7.4 Genomics

Genome wide association study (GWAS) identified SNPs in MUC5B (rs35705950), SPPL2C (rs17690703), and TOLLIP (rs111521887, rs5743894, rs5743890) to be linked with increased susceptibility to IPF. Minor allele of rs5743890 of TOLLIP was associated with increased risk of mortality in IPF patients (Noth et al. 2013). The T allele of rs35705950 in MUC5B gene was associated with prolonged survival in IPF patients undergoing anti-fibrotic therapy (Biondini et al. 2021). SNP in TERT intronic region (rs 2736100) was found to be significantly associated with IPF (Mushiroda et al. 2008). Variants in TERC, TERT, SFTPC, SFTPA2, DKC1, PARN, TINF2, NAF1, and ABCA3 have also been associated with IPF (Kropski and Blackwell 2019). Important SNPs and genetic variants related to IPF have been documented in Table 7.1.

7.5 Epigenomics

Epigenetic alterations like DNA methylation and histone modification have been extensively studied and strongly associated with the development of IPF (Yang and Schwartz 2015). Defective histone modifications (like acetylation) have been associated with the repression of anti-fibrotic gene expressions (like IP-10, cyclooxygenase-2, and caveolin-1) in IPF (Martinez et al. 2017; Coward et al. 2010; Coward et al. 2009). DNA hypermethylation was found to mediate reduced E-prostanoid 2 (EP2) receptor expression and thus block the anti-fibrotic effects of prostaglandin E2 (PGE2) in IPF patients (Huang et al. 2010). DNA hypermethylation and reduced expression of Thy1 have been noted in IPF patients (Sanders et al. 2008). Epigenomic study of DNA methylation profile of IPF lungs and subsequent comparative analysis with that of lung cancer patients and healthy subjects have highlighted IPF patients to possess an intermediate methylation profile between than of lung cancer and control. Four hundred and two differentially methylated CpG islands were found in both IPF and lung cancer patients. The same study identified hypomethylation of long interspersed nuclear element 1 (LINE-1) retrotransposons to be absent in IPF lungs as opposed to that in lung cancer (Rabinovich et al. 2012). Some of the other epigenetic modifications associated with IPF have been enlisted in Table 7.1.

7.6 Transcriptomics

Extensive transcriptomic studies using BALF and lung tissue specimens from IPF patients have highlighted enrichment of pathways including the WNT/β-catenin pathway, p53/hypoxia pathway, PDGF pathway, extracellular matrix pathway, lung development pathway, IGF1 pathway, cell adhesion pathway, TGF-β pathway, mitochondrial function, and HIF-1-alpha pathway (Vukmirovic and Kaminski 2018; Newton and Herzog 2021). Several differentially expressed molecules, namely

Table 7.1 Different biomarkers of idiopathic pulmonary fibrosis (IPF)

Omics approach	Biomarkers	Biological sample	Comments
Genomics	TOLLIP	Peripheral blood leukocytes	TOLLIP SNP rs5743890 is related to worse survival in IPF patients (Bonella et al. 2021)
	TLR3	Lung fibroblasts	L412F polymorphism (rs3775291) of TLR3 is associated with rapid decline in lung function and greater mortality in IPF patients (O'Dwyer et al. 2013)
	MUC5B	Peripheral blood leukocytes, lungs	MUC5B promoter SNP rs35705950 is a strong risk factor for familial IPF and sporadic IPF. IPF patients with SNP rs35705950 have better survival as compared to patients without the said SNP (Bonella et al. 2021; Yang et al. 2015)
	SFTPC, TERT, TERC, RTEL1, DKC1	Blood, lung tissues	Rare variants in SFTPC, TERT, TERC, RTEL1, and DKC1 are associated with predisposition to familial pulmonary fibrosis (Cogan et al. 2015)
	AKAP13		SNP rs62025270 near the gene AKAP13 is associated with susceptibility to IPF (Allen et al. 2017)
	TOLLIP, IL-4	Blood	SNPs of TOLLIP (rs111521887) and IL-4 (rs2070874) may serve as indicators of disease course modifying factors in IPF (Sterclova et al. 2021)
	IL-8	Lung biopsy, broncho alveolar lavage (BAL)	SNP in the promoter of IL-8 (rs4073T > A) may increase the risk of IPF via upregulation of IL-8 (Ahn et al. 2011)
Transcriptomics	Matrix metallopeptidase 7 (MMP7)	Serum, blood plasma, broncho	MMP7 is significantly increased in IPF patients

(continued)

Table 7.1 (continued)

Omics approach	Biomarkers	Biological sample	Comments
		alveolar lavage (BAL), lungs	(Vukmirovic and Kaminski 2018)
	miR-302c, miR-423, miR-210, miR-376c, miR-185	Lung Biopsies	miR-302c, miR-423, miR-210, miR-376c, and miR-185 are overexpressed in rapidly progressing IPF patients (Oak et al. 2011)
	C-X-C motif chemokine 13 (CXCL13)	Blood plasma, lungs	CXCL13 is significantly increased in IPF patients. It can serve as a prognostic biomarker in IPF patients (Vuga et al. 2014)
	Chemokine (C-C motif) ligand 8 (CCL8)	Blood plasma, broncho alveolar lavage (BAL), lungs	CCL8 was significantly elevated in IPF patients. CCL8 may aid in differential diagnosis and prediction of survival in IPF patients (Lee et al. 2017)
	Osteopontin (SPP1)	Broncho alveolar lavage (BAL), lungs	SPP1 is increased in IPF patients (Pardo et al. 2005)
	Cartilage oligomeric matrix protein (COMP)	Lungs	COMP is increased in IPF patients (Vuga et al. 2013)
	CXCL8, ICAM1, LIF, PTGS2, IL7R, CSF2, TGFBR2, SDC3, L1CAM	Mesenchymal stromal cell (MSC)-like cells from lung biopsies	Increased expression of these genes was noted in IPF patients (Boesch et al. 2020)
	C-X-C motif chemokine 14 (CXCL14)	Lung biopsies	CXCL14 is significantly increased in IPF patients. It can serve as systemic biomarker for IPF patients with increased sonic hedgehog pathway activity and can monitor therapeutic efficacy of hedgehog antagonist therapy in IPF patients (Jia et al. 2017)
	BTNL9, PLLP	Lung tissues	Expression of BTNL9 and PLLP was reduced in lung tissues of IPF patients (Zheng et al. 2022)
	GPX8, VCAM1, ARRB1, SGTA, GABARAPL1		GPX8 and VCAM1 were increased, while ARRB1,

(continued)

Table 7.1 (continued)

Omics approach	Biomarkers	Biological sample	Comments
			SGTA, and GABARAPL1 were decreased in IPF patients (Qian et al. 2021)
	COL1A1, COL1A2, COL3A1, COL5A2, COL6A3, COL14A1, COL18A1		Upregulation of COL1A1, COL1A2, COL3A1, COL5A2, COL6A3, COL14A1, and COL18A1 was observed in IPF patients (Tian et al. 2019; Selman et al. 2006; Konishi et al. 2009)
	miR-21, miR-155		miR-21 and miR-155 were significantly increased in IPF patients (Oak et al. 2011; Yamada et al. 2013)
	miR-29		miR-29 was significantly reduced in IPF and can serve as a potential therapeutic target for IPF (Cushing et al. 2011)
	let-7d		let-7d was significantly reduced in IPF patients (Pandit et al. 2010)
	MUC5B		MUC5B expression was significantly increased in IPF lungs (Seibold et al. 2011)
Epigenomics	CLDN5, ZNF467, TP53INP1, DDAH1	Lung tissues	CLDN5, ZNF467, TP53INP1, and DDAH1 genes were hypermethylated in IPF and were associated with reduced mRNA expression in IPF patients (Sanders et al. 2012)
	CASZ1, FOXC1, MXD4, ZDHHC4		Differential methylation at sites near the transcription factors CASZ1, FOXC1, MXD4, and ZDHHC4 may be associated with IPF pathogenesis (Yang et al. 2014)
	p14(ARF)	Pulmonary fibroblasts	Hypermethylation and silencing of p14(ARF) may be linked to apoptosis resistance in lung

(continued)

Table 7.1 (continued)

Omics approach	Biomarkers	Biological sample	Comments
			fibroblasts from IPF patients (Cisneros et al. 2012)
	Fas		Reduced Fas expression mediated by histone acetylation and methylation leads to apoptosis resistance in lung fibroblasts from IPF patients (Huang et al. 2013)
	miR-17–92 cluster	Lung tissues, pulmonary fibroblasts	Hypermethylation of the miR-17–92 cluster reduces expression of these miRNAs and in turn promotes expression of target genes in IPF patients (Dakhlallah et al. 2013)
Proteomics	Matrix metallopeptidase 10 (MMP10)	Serum, broncho alveolar lavage (BAL), lungs	MMP10 is significantly increased in IPF patients. MMP10 may serve as a biomarker for monitoring severity and prognosis in IPF patients (Sokai et al. 2015)
	MMP7, MMP1	Peripheral blood, serum	MMP7 was significantly increased in IPF patients as compared to healthy controls. MMP1 was significantly increased in IPF patients as compared to sarcoidosis and non-IPF UIP (Morais et al. 2015) These proteins may serve as potential blood biomarkers of IPF. MMP7 expression may further aid in monitoring disease progression (Rosas et al. 2008)
	Periplakin (PPL)	Serum	Circulating autoantibodies directed towards periplakin (PPL) were noted in the serum of IPF patients, and the presence of circulating anti-PPL

(continued)

Table 7.1 (continued)

Omics approach	Biomarkers	Biological sample	Comments
			autoantibodies was associated with disease severity (Taille et al. 2011)
	Periostin		Serum periostin levels are inversely associated with lung functions in IPF patients (Okamoto et al. 2011)
	MIG		Serum MIG was elevated in IPF patients compared to sarcoidosis patients and healthy controls (Antoniou et al. 2006)
	KNG1, AMBP, AHSG, SERPINA1, CRP		KNG1, AMBP, and AHSG were significantly downregulated in IPF patients, while SERPINA1 and CRP were significantly upregulated in IPF patients. KNG1, AMBP, AHSG, SERPINA1, and CRP may serve as specific IPF biomarkers (Niu et al. 2017)
	ICAM5, PIGR, SPON1, ROBO2	Blood plasma	ICAM5, PIGR, SPON1, and ROBO2 expressions were higher in severe IPF. These proteins can serve as indicators for disease severity (Todd et al. 2019)
	ACTG1, FN1, SERPINC1, ECM1, PPBP		ACTG1, FN1, SERPINC1, and ECM1 were reduced in IPF plasma samples, while PPBP was increased in IPF plasma samples (Moodley et al. 2019)
	FCN2, LGMN, CTSS, sVEGFR2, ICOS, TRY3		Increased expression of FCN2, LGMN, CTSS, sVEGFR2 and decreased expression of ICOS, TRY3 are associated with better survival in IPF patients (Khan et al. 2021)
	SFTPD, MMP1, MMP9, TIMP3, CCL18, CCL5,		SFTPD, MMP1, MMP9, TIMP3, CCL18, CCL5,

(continued)

Table 7.1 (continued)

Omics approach	Biomarkers	Biological sample	Comments
	C4A, C4B, CCL22, PGAM1, GPD1, THBS1, VWF, CCL17, BPI, OLR1, CAPG, SPARC, FN1, PF4, PPBP, ICAM5, CXCL13, PGD, C1R, HIST1H1C, GDF15, S100A9, FN1, PDGFB, ANXA6, VAV1, APP, HK2, S100A6, HNRNPA2B1, PDGFA		C4A, C4B, CCL22, PGAM1, GPD1, THBS1, VWF, CCL17, BPI, OLR1, CAPG, SPARC, FN1, PF4, PPBP, ICAM5, CXCL13, PGD, C1R, HIST1H1C, GDF15, S100A9, FN1, PDGFB, ANXA6, VAV1, APP, HK2, S100A6, HNRNPA2B1, and PDGFA were significantly elevated in IPF patients (Todd et al. 2019)
	CCL28, CCL25, TLR4, UNC5D, TNFSF12, PRL, CCL4L1, MAP2K4, PDGFB, AKT2, ICAM5, CA6, SPON1, CXCL11, OLR1		CCL28, CCL25, TLR4, UNC5D, TNFSF12, PRL, CCL4L1, MAP2K4, PDGFB, AKT2, ICAM5, CA6, SPON1, CXCL11, and OLR1 were increased in blood plasma of IPF patients (Sivakumar et al. 2021)
	MMP3, CKB, CKM, ADSL, SHH, CA6, AGER, HSPB1, TFF1, PRKCA, PRKACA		MMP3, CKB, CKM, ADSL, SHH, CA6, AGER, HSPB1, TFF1, PRKCA, and PRKACA were significantly reduced in IPF patients (Todd et al. 2019)
	GFAP, SAA1, BCAN, HIST3H2A, H2AFZ, NPPB, PRSS2, IGHE, CHI3L1, REG1A, PLA2G2A, IGFBP2, EPO, CKM, CTSD, TIMP1, ANGPT2, ACY1, LDHB		GFAP, SAA1, BCAN, HIST3H2A, H2AFZ, NPPB, PRSS2, IGHE, CHI3L1, REG1A, PLA2G2A, IGFBP2, EPO, CKM, CTSD, TIMP1, ANGPT2, ACY1, and LDHB were reduced in blood plasma of IPF patients (Sivakumar et al. 2021)
	CCL11, CCL22, CXCL12, TPSB2, MDK, KLK7, CCL21, SERPINA5, FRZB, GFRA1, VTN, CCL5, CCL3L1, CXCL9, RSPO3, SORCS2,	Blood Plasma, Lung Tissues	CCL11, CCL22, CXCL12, TPSB2, MDK, KLK7, CCL21, SERPINA5, FRZB, GFRA1, VTN, CCL5, CCL3L1, CXCL9, RSPO3, SORCS2,

(continued)

Table 7.1 (continued)

Omics approach	Biomarkers	Biological sample	Comments
	CCL17, CTSV, GZMA, PDE5A, CTSF, CHL1, MSTN, SPARC, A2M		CCL17, CTSV, GZMA, PDE5A, CTSF, CHL1, MSTN, SPARC, and A2M were increased in both plasma and lung tissues of IPF patients (Sivakumar et al. 2021)
	IL1RL1, IL1R2, S100A12, IL18R1, CD177, RETN, TNNT2, FSTL3, HIST2H2BE		IL1RL1, IL1R2, S100A12, IL18R1, CD177, RETN, TNNT2, FSTL3, and HIST2H2BE were significantly reduced in both plasma and lung tissues of IPF patients as compared to healthy subjects (Sivakumar et al. 2021)
	S100A9 or Calgranulin B	Broncho Alveolar Lavage Fluid (BALF)	S100A9 is significantly elevated in BALF of IPF patients and can serve as a biomarker for discriminating IPF from other fibrotic diseases. It can also serve as an indicator of disease severity (Bargagli et al. 2008; Hara et al. 2012)
	Migration Inhibitory Factor (MIF)		Expression of MIF was significantly elevated in BAL of IPF patients (Bargagli et al. 2009)
	ApoA-I		ApoA-I protein was significantly reduced in BALF of IPF patients (Foster et al. 2015)
	CXCL7, CCL18, CCL24, OPN (osteopontin)		CXCL7, CCL18, CCL24, and OPN were increased in BALF of IPF patients (Foster et al. 2015)
	BTNL9, PLLP	Lung Tissues	Expression of BTNL9 and PLLP was significantly reduced in lung tissues of IPF patients (Zheng et al. 2022)
	COL1A1, HSP90AA1, CTSB, SCGB1A1, AGR2, LGALS7, DMBT1, MME, CAV1		COL1A1, HSP90AA1, CTSB, SCGB1A1, AGR2, LGALS7, and DMBT1 were increased in

(continued)

Table 7.1 (continued)

Omics approach	Biomarkers	Biological sample	Comments
			IPF patients Expression of MME and CAV1 was reduced in IPF patients (Tian et al. 2019)
	HBA, HBB		Reduction in haemoglobin (Hb) α and β monomers was noted in IPF patients (Ishikawa et al. 2010)
Metabolomics	Lysophosphatidylcholine (LysoPC)	Serum	LysoPC levels were significantly elevated, and LysoPC may serve as a potential serum biomarker for IPF patients (Rindlisbacher et al. 2018)
	3-hydroxydecanoyl carnitine		3-hydroxydecanoyl carnitine was significantly elevated in serum from IPF patients compared to healthy subjects (Rindlisbacher et al. 2018)
	3-Deoxyvitamin D3	Blood plasma	3-Deoxyvitamin D3 was elevated in IPF patients (Yan et al. 2017)
	Stigmasteryl Ester (16:1), (E,E)-3,7,11-trimethyl-2,6,10-dodecatrienyl dodecanoate		Stigmasteryl ester (16:1) and (E,E)-3,7,11-trimethyl-2,6,10-dodecatrienyl dodecanoate were elevated in IPF patients (Yan et al. 2017)
	Diacylglycerols DG (O-16:0/18:1) DG (18:0/18:2/0:0)[iso2] Triacylglycerol TG (18:4/20:3/22:0) [iso6]		Both diacylglycerols [DG (O-16:0/18:1) DG (18:0/18:2/0:0)[iso2]] and triacylglycerol [TG (18:4/20:3/22:0) [iso6]] were elevated in IPF patients (Yan et al. 2017)
	Proline, allysine, 4-Hydroxyproline, Valine, Leucine/isoleucine, Alanine	Exhaled breath	Proline, allysine, 4-hydroxyproline, valine, leucine/isoleucine, and alanine were significantly elevated in exhaled breath from IPF patients (Gaugg et al. 2019).
	Glucose, adenosine-tri-phosphate (ATP)	Lung tissues	Reduction of glucose and ATP levels was noted in IPF lung tissues (Kang et al. 2016)

(continued)

Table 7.1 (continued)

Omics approach	Biomarkers	Biological sample	Comments
	Inosine, hypoxanthine, glucose-6-phosphate, dihydroxyacetone phosphate, lactic acid, glycine, cysteine, glutamic acid, proline		Inosine, hypoxanthine, glucose-6-phosphate, dihydroxyacetone phosphate, lactic acid, glycine, cysteine, glutamic acid, and proline were elevated in lung tissues of IPF patients. These differentially expressed metabolites could serve as potential biomarkers for IPF (Kang et al. 2016)
	Sphinganine, sphingosine, sphingosine-1-phosphate (S1P), sphingomyelin		Sphinganine, sphingosine, sphingosine-1-phosphate (S1P), and sphingomyelin were reduced in IPF patients (Zhao et al. 2017)
	Fumarate, aspartate, succinyl carnitine, fructose 1,6- bisphosphate (fructose 1,6-BP), phosphoenolpyruvate		Fumarate, aspartate, succinyl carnitine, fructose 1,6-BP, and phosphoenolpyruvate were reduced in IPF patients (Zhao et al. 2017)
	Proline, creatine, putrescine, spermidine, 4-hydroxyproline, glycine, fructose, carnitine		Proline, creatine, putrescine, spermidine, 4-hydroxyproline, glycine, fructose, and carnitine were increased in IPF patients (Zhao et al. 2017)
	Caproate, caprylate, myristate, palmitoleate, acylcarnitines, hexanoylcarnitine, octanoylcarnitine, palmitoylcarnitine, succinylcarnitine		Long and medium chain fatty acids caproate, caprylate, myristate, and palmitoleate were significantly increased in lung tissues from IPF patients. Medium chain acylcarnitines, hexanoylcarnitine, octanoylcarnitine, palmitoylcarnitine, and succinylcarnitine were significantly decreased in IPF lungs (Zhao et al. 2017)
	Glutamate, glutamine		Glutamate and glutamine levels were increased in

(continued)

Table 7.1 (continued)

Omics approach	Biomarkers	Biological sample	Comments
			IPF lungs (Zhao et al. 2017)
	Haem, biliverdin, bilirubin		Haem and biliverdin levels were decreased in IPF lungs, while bilirubin was elevated in IPF lung tissues (Zhao et al. 2017)
Metagenomics	*Prevotella* spp. *Veillonella* spp. *Escherichia* spp.	Broncho alveolar lavage (BAL)	*Prevotella* spp., *Veillonella* spp., and *Escherichia* spp. were more predominant in IPF (Han et al. 2014)
	Streptococcus spp. *Staphylococcus* spp.		Presence of *Streptococcus* spp. or *Staphylococcus* spp. was associated with decline in lung function, disease progression and mortality in IPF patients (Han et al. 2014)
	Firmicutes, Proteobacteria, Bacteroidetes, Streptococcaceae, Veillonellaceae	Broncho alveolar lavage fluid (BALF)	Predominance of Firmicutes, Proteobacteria, and Bacteroidetes phyla was noted in IPF patients. Relative abundance of Firmicutes, Streptococcaceae, and Veillonellaceae was associated with FVC, distance in 6-min walk test and serum level of surfactant D protein in IPF patients (Takahashi et al. 2018)
	Streptococcus spp. *Pseudobutyrivibrio* spp. *Anaerorhabdus* spp.		β-diversity was reduced in IPF patients. Increased abundance of *Streptococcus* spp., *Pseudobutyrivibrio* spp., and *Anaerorhabdus* spp. were noted in IPF patients (Tong et al. 2019)
	Veillonella spp. *Neisseria* spp. *Streptococcus* spp. *Haemophilus* spp.		Increased abundance of *Veillonella* spp., *Neisseria* spp., *Streptococcus* spp., and *Haemophilus* spp. were noted in IPF patients (Amati et al. 2022)

(continued)

Table 7.1 (continued)

Omics approach	Biomarkers	Biological sample	Comments
	Lactobacillus spp. *Paracoccus* spp. *Akkermansia* spp.	Lung tissues	Increased abundance of *Lactobacillus* spp., *Paracoccus* spp., and *Akkermansia* spp. were noted in IPF patients (Yoon et al. 2021)
	Caulobacter spp. *Azonexus* spp. *Undibacterium* spp.		Reduced population of *Caulobacter* spp., *Azonexus* spp., and *Undibacterium* spp. were noted in IPF patients (Yoon et al. 2021)
	Streptococcus spp. *Sphingomonas* spp. *Clostridium* spp.		Increased relative abundance of *Streptococcus* spp., *Sphingomonas* spp., and *Clostridium* spp. were independently related to mortality in IPF patients (Yoon et al. 2021)
	Azonexus spp. *Pelomonas* spp.		*Azonexus* spp. and *Pelomonas* spp. were associated with IPF diagnosis in multivariable logistic analysis adjusted by gender and age (Yoon et al. 2021)
	Curvibacter spp. *Thioprofundum* spp. *Anoxybacillus* spp. *Enterococcus* spp.		*Curvibacter* spp. and *Thioprofundum* spp. were correlated positively to FVC in IPF patients *Anoxybacillus* spp. and *Enterococcus* spp. were inversely correlated to FVC in IPF patients (Yoon et al. 2021).
	Aquabacterium spp. *Peptoniphilus* spp. *Fusobacterium* spp. *Phycicoccus* spp.		Relative abundance of *Aquabacterium* spp. and *Peptoniphilus* spp. was correlated positively with the distance covered in 6-min walk test in IPF patients Relative abundance of *Fusobacterium* spp. and *Phycicoccus* spp. were inversely correlated with the distance covered in 6-min walk test in IPF patients (Yoon et al. 2021)

This table enlists various important biomarkers of IPF identified using different omics approaches

MMP7, OPN, WISP1, IGBP3, IGBP5, PTPN11, FKBP10, RXFP1, and PINK1 have been identified as potential regulators of fibrosis (Pardo et al. 2005; Zuo et al. 2002; Konigshoff et al. 2009; Veraldi and Feghali-Bostwick 2012; Tzouvelekis et al. 2017; Staab-Weijnitz et al. 2015; Tan et al. 2016; Bueno et al. 2015). Different dysregulated miRNAs like miR-21, let-d, and miR-29 have also been linked to pulmonary fibrosis (Yu et al. 2018). Transcriptomic analyses of PBMCs have identified 52 gene signatures associated with severe prognosis in IPF patients (Herazo-Maya et al. 2013). Several blood transcriptomic signatures like AMP, OLFM4, CTSG, DEFA3, CEACAM6, HLTF, GABBR1, PACSIN1, and IGHM have also been identified as potential candidates for novel diagnostic biomarkers of IPF (Yang et al. 2012). Other IPF specific transcriptomic signatures have been mentioned in Table 7.1.

7.7 Proteomics

Global and targeted proteomic studies identified different proteins that could serve as potential diagnostic or prognostic biomarkers of IPF. Krebs von den Lungen-6 (KL-6), surfactant proteins (SP-A and SP-D), matrix metalloproteinases (MMP-1, MMP-7, MMP-8, MMP-9, and MMP-10), tissue inhibitors of MMP-1 (TIMP-1), and mediators of inflammation (like CD28 and CCL18) have been identified as potential peripheral blood based diagnostic biomarkers of IPF (Sokai et al. 2015; Rosas et al. 2008; Newton and Herzog 2021; van den Blink et al. 2010; Stainer et al. 2021). Many of these serum proteins have also been evaluated for their role as prognostic biomarkers of IPF. Serum KL-6 was found to be elevated in patients with acute exacerbations as compared to stable IPF patients (Stainer et al. 2021). MMP-7, MMP-12, MMP-13, and TIMP-4 protein levels were found to enable classification of IPF patients based on disease severity (Todd et al. 2020). Serum SP-A, SP-D, and YKL-40 levels were associated with increased mortality in IPF patients (Stainer et al. 2021; Inoue et al. 2020). Besides, serum SP-A, SP-D, and MMP-7 were also screened as potential therapeutic biomarkers of IPF for monitoring disease progression upon anti-fibrotic therapy (Stainer et al. 2021). Other potential biomarker candidates of IPF obtained from blood and other biological samples like BALF and lung tissues have been listed in Table 7.1.

7.8 Metabolomics

High-throughput metabolomic studies have highlighted several metabolic pathways, whose disruption may be associated with IPF pathogenesis (Newton and Herzog 2021). Studies conducted using NMR spectroscopy and mass spectroscopy have revealed changes in metabolic pathways like glycolysis, citric acid cycle, fatty acid metabolism, ornithine aminotransferase pathways, adenosine triphosphate degradation, and glutathione biosynthesis to be associated with fibrotic changes in IPF (Yu et al. 2018; Kang et al. 2016; Zhao et al. 2017). Serum lysophosphatidylcholine

(LysoPC), the precursor of lysophosphatidic acid (LPA) was identified as a potential biomarker for IPF (Rindlisbacher et al. 2018). Sphingosine 1-phosphate (S1P) was also found to play a crucial role in IPF pathogenesis. Besides, arginine metabolites, namely putrescine, creatine, 4-hydroxy proline, and spermidine were found to be increased in IPF lungs (Roque and Romero 2021). Alveolar nitric oxide levels were found to be significantly elevated in IPF patients as compared to healthy controls. Levels of hydrogen peroxide and 8-isoprostane in exhaled breath condensate of IPF patients were also different from healthy subjects (Hayton et al. 2019). Other metabolites that are differentially expressed in IPF have been tabulated in Table 7.1.

7.9 Metagenomics

Lung microbial dysbiosis have been extensively observed in IPF patients. Alterations in the lung microbiome have also been implicated in IPF pathogenesis and progression. Increased bacterial load in broncho alveolar lavage fluid (BALF) was identified as a probable biomarker for rapidly progressive IPF (Chioma et al. 2021). Metagenomic studies using 16SrRNA sequencing have highlighted the presence of pathogenic strains of *Staphylococcus* spp. and *Streptococcus* spp. in IPF lungs and BALF. Presence of these microbes was strongly associated with decline in FVC, reduced D_{LCO}, increased instances of acute exacerbations and rapid disease progression (Han et al. 2014; Chioma et al. 2021). *Streptococcus* spp., *Pseudobutyrivibrio* spp., and *Anaerorhabdus* spp. were found to be enriched in IPF lungs as compared to that in healthy controls. Besides, translocation of 32.84% of the lung microbiome from the oral cavity was also noted in the said study (Tong et al. 2019). Other changes in lung microbiome in IPF patients have been documented in Table 7.1. Untangling these altered host–microbiome interactions in IPF may aid in better control of disease outcomes.

7.10 Medical Imaging or Radiomics

Medical imaging of the chest has a critical role in IPF evaluation and diagnosis of suspected IPF patients (Mohning et al. 2019). Along with clinical features, imaging features of IPF give an accurate diagnosis of the disease. Radiographic patterns of IPF include usual interstitial pneumonia (UIP), subpleural and basal predominant, and honeycombing with or without peripheral traction bronchiectasis (Raghu et al. 2018). High resolution computed tomography (HRCT) and hyperpolarized (HP) 129Xe magnetic resonance imaging (MRI) are widely used for assessing the IPF disease progression (Mammarappallil et al. 2019). Non-invasive nature of imaging diagnosis makes it a popular choice for the physician. However, to analyse imaging data an expert in radiology is required. Artificial intelligence has revolutionized the medical imaging field, and a software based on deep learning of HRCT images can accurately quantify parenchymal lesions and airway volumes of IPF patients (Handa et al. 2022). Fibrosis-Net, another tool, can predict pulmonary fibrosis progression from chest CT images (Wong et al. 2021). Thus, combining HRCT and MRI scans of IPF patients with artificial intelligence can accurately diagnose heterogeneity of IPF patients and improve the treatment regimen.

7.11 Bioinformatics

Experimental expression data, mainly transcriptomics and proteomics, have been generated on the lung tissue samples of IPF patients and on bleomycin induced animal models to understand the disease pathogenesis. IPF gained more attention after the worldwide SARS-CoV-2 outbreak, where COVID-19-associated lung fibrosis was reported. In the last two years (2020–2022), four databases were reported focusing on pulmonary fibrosis (PulmonDB, ILDGDB, Fibromine, and FibROAD) (Sun et al. 2022; Villasenor-Altamirano et al. 2020; Li et al. 2020; Fanidis et al. 2021). PulmonDB is focussed on only transcriptomics datasets, whereas ILDGDB and fibromine contain both transcriptomics and proteomics datasets and provide mining tools to extract the specific gene/protein sets. FibROAD provides information about 17 different organs affected by fibrosis including lung, liver, skin, and heart. It also provides information about several OMICs datasets like RNA-seq, single-cell RNA-seq, miRNA-seq, ChIP-seq, ATAC-seq MeDIP-seq, and MBD-seq. Another resource, the IPF Cell Atlas provides information about single-cell RNA sequencing of immune cells like alveolar macrophages and epithelial cells (Neumark et al. 2020). Radiology is an important tool for diagnosis of IPF patients. Medical imaging of HRCT chest scan data of IPF patients is compiled in the MedGIFT database (Depeursinge et al. 2012). These databases provide valuable resources of multi-omics datasets associated with IPF and also provide visualization tools to mine and comprehend the complex datasets.

7.12 Multi-Omics and Data Integration

Integration of multi-omics datasets associated with IPF was used primarily to identify potential biomarkers, to understand the molecular patterns or regulatory factors, and to identify the molecular subgroups (Zheng et al. 2022; Konigsberg et al. 2021; McDonough et al. 2019; Zhang et al. 2021). An integrative approach named Data Integration Analysis for Biomarker discovery using Latent cOmponents (DIABLO) was used for identifying key molecular signatures of IPF from multi-omics assays including coding and noncoding transcriptomes, DNA methylomes, and proteomes (Konigsberg et al. 2021; Singh et al. 2019). A weighted gene co-expression analysis (WGCNA) and KEGG enrichment modules were used from 220 IPF patients' gene expression profiles to identify three major disease subtypes (Zhang et al. 2021). Integration of multi-omics datasets from IPF patients and animal models was analysed to identify key genes and regulatory pathways.

7.13 Current Applications

Treatment recommendations for IPF were mostly limited to anti-inflammatory and immunosuppressive agents until the recent past (Sgalla et al. 2020). Recent scientific advances identified two novel anti-fibrotic drugs—nintedanib and pirfenidone. In 2015, ATS/ERS/JRS/ALAT guidelines recommended the use of nintedanib and pirfenidone, along with antacids for IPF patients (Raghu et al. 2015). These drugs significantly slow down the progression of fibrosis and improve lung function in IPF patients (Raghu 2017). The detailed mode of action, effect, and adverse reactions of these drugs in IPF patients have been tabulated in Table 7.2.

Although nintedanib and pirfenidone slow down the disease progression, these anti-fibrotic medications fail to prevent deterioration in IPF patients. Therefore, several antibodies, biologics, small molecule inhibitors, and drugs are still being investigated for identifying effective and potential therapies of IPF. The different drug candidates and biologics under clinical trial for IPF have been listed in Table 7.3.

Multipotent mesenchymal stem cells (MSCs) have also been evaluated in clinical trials for IPF (Serrano-Mollar 2018). MSCs were safe and well tolerated in IPF patients without further worsening of fibrotic symptoms (Tzouvelekis et al. 2013; Chambers et al. 2014; Glassberg et al. 2017). Oxygen therapy is recommended as a palliative care for improving oxygen levels in IPF patients having significantly decreased oxygen saturation (Fraser and Hoyles 2016). Pulmonary rehabilitation (PR) has also been effective in improving the quality of life and capability of doing physical activity in IPF patients (Jackson et al. 2014). Lung transplantation has been associated with increased survival rate in IPF patients, but is largely limited by the scarcity of donor organs (van Geffen et al. 2021; Kumar et al. 2018). Devising an efficient cure for IPF is challenging and relies to a large extent on discovery of novel therapeutic targets using omics approaches.

7.14 Future Perspectives

Increased mortality of IPF has brought the disease under limelight. Use of omics approach has highlighted differentially expressed biomolecules and metabolites. However, the etio-pathogenesis of IPF remains largely unknown (Glass et al. 2022). The role of different respiratory viruses and host receptors in the onset and progression of IPF needs to be studied in detail (Moghoofei et al. 2022). Animal models of fibrosis can further enrich our mechanistic understanding of IPF disease biology. Although, mouse models hold promise but the development of animal model of IPF is highly challenging (Blackwell et al. 2014). Integrative multi-omics studies using samples from IPF patients should therefore be conducted with a view of obtaining an insight into disease aetiology, identifying biomarkers, and therapeutic targets. These therapeutic targets can be further used to design new drugs for clinical investigation and subsequent use in IPF patients. Improvement in disease diagnosis can also be immensely beneficial in mediating timely therapeutic

Table 7.2 Drugs commonly used to treat IPF

Sl. no.	Drug	Description	Mode of action	Status	Effects in IPF patients	Side effects
1.	Nintedanib	Intracellular triple tyrosine kinase inhibitor	Binds competitively to receptors of FGF, PDGF, and VEGF, thereby blocking downstream signalling events and fibrogenic pathways. May exert additional effect by inhibiting Src tyrosine kinases, namely Lyn, Lck, and Flt-3	FDA approved (2014)	Effectively slows down decline in forced vital capacity (FVC) in moderately advanced IPF patients Improves lung function and reduces instances of acute exacerbations in IPF patients	Mostly well tolerated in patients for prolonged period. May exert gastrointestinal side effects like diarrhoea and nausea (Wollin et al. 2015; van Geffen et al. 2021; Kolb et al. 2017; Fraser and Hoyles 2016)
2.	Pirfenidone	Orally bioavailable pyridine derivative	Serves as an anti-fibrotic agent, anti-inflammatory agent, and anti-oxidant. Inhibits the TGF-β pathway (exact mechanism of action is unknown). Reduces key fibrotic cytokines like TGF-β, FGF, and IL-1β	FDA Approved (2014)	Effectively slows down decline in forced vital capacity (FVC) in moderately advanced IPF patients Brings about decrease in proliferation and differentiation of fibroblasts Leads to reduction in lung collagen content and fibrosis score Long-term therapy reduces mortality associated with IPF	Continuous therapy is mostly well tolerated in IPF patients. May exert gastrointestinal side effects (like nausea, diarrhoea and dyspepsia) and photo-toxicity (like skin rashes) (van Geffen et al. 2021; Kolb et al. 2017; Fraser and Hoyles 2016; Swigris and Fairclough 2010)

Table 7.3 Drugs and biologics that are under clinical investigation for use in IPF

Sl. no.	Drugs/ biologics	Description	Target	Major effects	Status
1.	Pamrevlumab (FG-3019)	Humanized monoclonal antibody against connective tissue growth factor (CTGF)	CTGF	Improved lung function and quality of life in IPF patients Significantly decreased the progression of pulmonary fibrosis	Investigational (Clinical Trial Phase III) (Sgalla et al. 2020; Richeldi et al. 2019)
2.	GLPG1690 (Ziritaxestat)	Selective autotaxin (ATX) inhibitor	ATX	Lead to significant improvement in lung function Study was terminated as risk benefit profile was not supportive for further study	Investigational (Clinical Trial Phase III) (Richeldi et al. 2019)
3.	PRM-151	Recombinant human pentraxin-2 (PTX-2) protein	–	Inhibits the differentiation of monocytes into macrophages and prevents TGF-β production. Significantly reduced the rate of decline in forced vital capacity (FVC)	Investigational (Clinical Trial Phase III) (Richeldi et al. 2019)
4.	PBI-4050	3-pentylbenzeneacetic acid sodium salt, orally active small molecule compound	GPR40, GPR84	Decreases the expression of pro-fibrotic and pro-inflammatory markers, thereby exerting anti-fibrotic effect Safe and well tolerated	Investigational (Clinical Trial Phase II) (Sgalla et al. 2020; Richeldi et al. 2019; Khalil et al. 2019)
5.	IW001	A type V collagen (Col(V)) oral immunotherapeutic agent	Antibodies to auto-antigen Col (V)	Safe and well tolerated Stabilizes FVC and lung function in Col(V)—reactive IPF patients	Investigational (Clinical Trial Phase I) (Richeldi et al. 2019; Wilkes et al. 2015)
6.	Belumosudil (KD025) (SLx-2119)	Orally available ROCK2 (Rho associated coiled-coil containing protein kinase 2) inhibitor	ROCK2	Effectively controls decline in lung function	Investigational (Clinical Trial Phase II) (Richeldi et al. 2019)

7.	GB0139 (TD139)	Small molecule inhibitor of galectin-3 (Gal-3)	Gal-3	Reduced expression of plasma biomarkers associated with the progression of IPF	Investigational (Clinical Trial Phase II) (Hirani et al. 2021)
8.	BMS-986020	High affinity lysophosphatidic acid receptor 1 (LPA1) antagonist	LPA1	Significantly slowed the rate of decline in lung function, but was associated with increase in hepatic enzymes and occurrence of cholecystitis (therefore, study was terminated)	Investigational (Clinical Trial Phase II) (Palmer et al. 2018)
9.	MN-001 (Tipelukast)	Orally bioavailable small molecule with anti-inflammatory and anti-fibrotic activities	LTR, PDE3/4, 5-LO	Studies are ongoing	Investigational (Clinical Trial Phase II) (Ptasinski et al. 2021)
10.	CC-90001	Selective c-Jun N-terminal kinase (JNK) inhibitor	JNK	Associated with lack of decline in FVC in IPF patients	Investigational (Clinical Trial Phase II) (Popmihajlov et al. 2022)
11.	Treprostinil	A prostacyclin vasodilator	PTGIR, PPARD, P2RY12	Leads to improvement in FVC in IPF patients	Investigational (Clinical Trial Phase III) (Nathan et al. 2022)
12.	Rituximab	Anti-CD20 monoclonal antibody	CD20	Results pending	Investigational (Clinical Trial Phase II) (Sgalla et al. 2020)
13.	LTI-03	A 7 amino acid long caveolin-1-derived peptide	–	Safe and well tolerated in healthy subjects	Investigational (Clinical Trial Phase Ia) (Ptasinski et al. 2021)
14.	TRK-250	siRNA-based oligonucleotide	TGF-β1	Studies are ongoing	Investigational (Clinical Trial Phase Ia) (Ptasinski et al. 2021)
15.	HEC68498	Selective inhibitor of class 1 isozymes of phosphoinositide 3-kinase (PI3K) and mammalian target of rapamycin (mTOR)	PI3K, mTOR	Studies are ongoing	Investigational (Clinical Trial Phase I) (Ptasinski et al. 2021)
16.	Omipalisib (GSK2126458)	Small molecule pyridyl sulphonamide inhibitor of phosphatidylinositol 3-kinase (PI3K) (has potential anti-neoplastic activity)	PI3K	Well tolerated in IPF patients Decreases aberrant glucose signalling in fibrotic portions of IPF lungs	Investigational (Clinical Trial Phase I) (Ptasinski et al. 2021)

(continued)

Table 7.3 (continued)

Sl. no.	Drugs/ biologics	Description	Target	Major effects	Status
17.	BBT-877	Autotaxin inhibitor	NPP2 or ENPP2	Safe and well tolerated in healthy subjects	Investigational (Clinical Trial Phase I) (Ptasinski et al. 2021)
18.	Nalbuphine	Opioid receptor agonist	Opioid receptor	Studies are ongoing	Investigational (Clinical Trial Phase II) (Ptasinski et al. 2021)
19.	GLPG1205	Selective G-protein coupled receptor 84 (GPR84) antagonist	GPR84	Caused smaller declines in FVC and total lung volume in IPF patients	Investigational (Clinical Trial Phase II) (Ptasinski et al. 2021)
20.	PLN-74809	Small molecule integrin antagonist	Integrin (αVβ1, αVβ6)	Studies initiated	Investigational (Clinical Trial Phase II) (Ptasinski et al. 2021)
21.	GDC-3280	Small molecule Growth factor antagonist based on pirfenidone's phenyl pyridone scaffold	–	Well tolerated in healthy subjects	Investigational (Clinical Trial Phase I) (Ptasinski et al. 2021)
22.	ND-L02-s0201 (BMS-986263)	Anti-HSP47 siRNA lipid nanoparticle	HSP47	Recruitment not yet started	Investigational (Clinical Trial Phase II) (Ptasinski et al. 2021)
23.	BI 1015550	Oral, phosphodiesterase 4B (PDE4B) inhibitor with anti-inflammatory and anti-fibrotic activities	PDE4B	Studies are ongoing	Investigational (Clinical Trial Phase II) (Ptasinski et al. 2021)

invention. Hyperpolarized Xenon MRI has been newly introduced and needs to be explored extensively to revolutionize the visualization and evaluation of fibrotic lesions in IPF patients. Further clinical improvisation can make this novel MRI technique highly beneficial for IPF diagnosis and make it capable for monitoring disease prognosis and therapeutic response (Mammarappallil et al. 2019). Besides, computational tools, machine learning techniques, and artificial intelligence can be widely used in IPF research for merging clinical parameters with imaging data, for designing multi-parametric mathematical and predictive models for IPF disease progression. In a nutshell, these future research avenues shall pave the way for better diagnosis, improved control of disease parameters and may aid in bringing about reduction in IPF associated mortality rate.

References

Ahn MH, Park BL, Lee SH, Park SW, Park JS, Kim DJ, Jang AS, Park JS, Shin HK, Uh ST et al (2011) A promoter SNP rs4073T>A in the common allele of the interleukin 8 gene is associated with the development of idiopathic pulmonary fibrosis via the IL-8 protein enhancing mode. Respir Res 12:73

Allen RJ, Porte J, Braybrooke R, Flores C, Fingerlin TE, Oldham JM, Guillen-Guio B, Ma SF, Okamoto T, John AE et al (2017) Genetic variants associated with susceptibility to idiopathic pulmonary fibrosis in people of European ancestry: a genome-wide association study. Lancet Respir Med 5(11):869–880

Amati F, Stainer A, Mantero M, Gramegna A, Simonetta E, Suigo G, Voza A, Nambiar AM, Cariboni U, Oldham J et al (2022) Lung microbiome in idiopathic pulmonary fibrosis and other interstitial lung diseases. Int J Mol Sci 23(2)

Antoniou KM, Tzouvelekis A, Alexandrakis MG, Sfiridaki K, Tsiligianni I, Rachiotis G, Tzanakis N, Bouros D, Milic-Emili J, Siafakas NM (2006) Different angiogenic activity in pulmonary sarcoidosis and idiopathic pulmonary fibrosis. Chest 130(4):982–988

Bargagli E, Olivieri C, Prasse A, Bianchi N, Magi B, Cianti R, Bini L, Rottoli P (2008) Calgranulin B (S100A9) levels in bronchoalveolar lavage fluid of patients with interstitial lung diseases. Inflammation 31(5):351–354

Bargagli E, Olivieri C, Nikiforakis N, Cintorino M, Magi B, Perari MG, Vagaggini C, Spina D, Prasse A, Rottoli P (2009) Analysis of macrophage migration inhibitory factor (MIF) in patients with idiopathic pulmonary fibrosis. Respir Physiol Neurobiol 167(3):261–267

Biondini D, Cocconcelli E, Bernardinello N, Lorenzoni G, Rigobello C, Lococo S, Castelli G, Baraldo S, Cosio MG, Gregori D et al (2021) Prognostic role of MUC5B rs35705950 genotype in patients with idiopathic pulmonary fibrosis (IPF) on antifibrotic treatment. Respir Res 22(1): 98

Blackwell TS, Tager AM, Borok Z, Moore BB, Schwartz DA, Anstrom KJ, Bar-Joseph Z, Bitterman P, Blackburn MR, Bradford W et al (2014) Future directions in idiopathic pulmonary fibrosis research. An NHLBI workshop report. Am J Respir Crit Care Med 189(2):214–222

Boesch M, Baty F, Brutsche MH, Tamm M, Roux J, Knudsen L, Gazdhar A, Geiser T, Khan P, Hostettler KE (2020) Transcriptomic profiling reveals disease-specific characteristics of epithelial cells in idiopathic pulmonary fibrosis. Respir Res 21(1):165

Bonella F, Campo I, Zorzetto M, Boerner E, Ohshimo S, Theegarten D, Taube C, Costabel U (2021) Potential clinical utility of MUC5B und TOLLIP single nucleotide polymorphisms (SNPs) in the management of patients with IPF. Orphanet J Rare Dis 16(1):111

Bueno M, Lai YC, Romero Y, Brands J, St Croix CM, Kamga C, Corey C, Herazo-Maya JD, Sembrat J, Lee JS et al (2015) PINK1 deficiency impairs mitochondrial homeostasis and promotes lung fibrosis. J Clin Invest 125(2):521–538

Chambers DC, Enever D, Ilic N, Sparks L, Whitelaw K, Ayres J, Yerkovich ST, Khalil D, Atkinson KM, Hopkins PM (2014) A phase 1b study of placenta-derived mesenchymal stromal cells in patients with idiopathic pulmonary fibrosis. Respirology 19(7):1013–1018

Chioma OS, Hesse LE, Chapman A, Drake WP (2021) Role of the microbiome in interstitial lung diseases. Front Med (Lausanne) 8:595522

Cisneros J, Hagood J, Checa M, Ortiz-Quintero B, Negreros M, Herrera I, Ramos C, Pardo A, Selman M (2012) Hypermethylation-mediated silencing of p14(ARF) in fibroblasts from idiopathic pulmonary fibrosis. Am J Physiol Lung Cell Mol Physiol 303(4):L295–L303

Cogan JD, Kropski JA, Zhao M, Mitchell DB, Rives L, Markin C, Garnett ET, Montgomery KH, Mason WR, McKean DF et al (2015) Rare variants in RTEL1 are associated with familial interstitial pneumonia. Am J Respir Crit Care Med 191(6):646–655

Coward WR, Watts K, Feghali-Bostwick CA, Knox A, Pang L (2009) Defective histone acetylation is responsible for the diminished expression of cyclooxygenase 2 in idiopathic pulmonary fibrosis. Mol Cell Biol 29(15):4325–4339

Coward WR, Watts K, Feghali-Bostwick CA, Jenkins G, Pang L (2010) Repression of IP-10 by interactions between histone deacetylation and hypermethylation in idiopathic pulmonary fibrosis. Mol Cell Biol 30(12):2874–2886

Cushing L, Kuang PP, Qian J, Shao F, Wu J, Little F, Thannickal VJ, Cardoso WV, Lu J (2011) miR-29 is a major regulator of genes associated with pulmonary fibrosis. Am J Respir Cell Mol Biol 45(2):287–294

Dakhlallah D, Batte K, Wang Y, Cantemir-Stone CZ, Yan P, Nuovo G, Mikhail A, Hitchcock CL, Wright VP, Nana-Sinkam SP et al (2013) Epigenetic regulation of miR-17–92 contributes to the pathogenesis of pulmonary fibrosis. Am J Respir Crit Care Med 187(4):397–405

Depeursinge A, Vargas A, Platon A, Geissbuhler A, Poletti PA, Muller H (2012) Building a reference multimedia database for interstitial lung diseases. Comput Med Imaging Graph 36(3):227–238

Fanidis D, Moulos P, Aidinis V (2021) Fibromine is a multi-omics database and mining tool for target discovery in pulmonary fibrosis. Sci Rep 11(1):21712

Foster MW, Morrison LD, Todd JL, Snyder LD, Thompson JW, Soderblom EJ, Plonk K, Weinhold KJ, Townsend R, Minnich A et al (2015) Quantitative proteomics of bronchoalveolar lavage fluid in idiopathic pulmonary fibrosis. J Proteome Res 14(2):1238–1249

Fraser E, Hoyles RK (2016) Therapeutic advances in idiopathic pulmonary fibrosis. Clin Med (Lond) 16(1):42–51

Funke M, Geiser T (2015) Idiopathic pulmonary fibrosis: the turning point is now! Swiss Med Wkly 145:w14139

Gaugg MT, Engler A, Bregy L, Nussbaumer-Ochsner Y, Eiffert L, Bruderer T, Zenobi R, Sinues P, Kohler M (2019) Molecular breath analysis supports altered amino acid metabolism in idiopathic pulmonary fibrosis. Respirology 24(5):437–444

Glass DS, Grossfeld D, Renna HA, Agarwala P, Spiegler P, DeLeon J, Reiss AB (2022) Idiopathic pulmonary fibrosis: current and future treatment. Clin Respir J 16(2):84–96

Glassberg MK, Minkiewicz J, Toonkel RL, Simonet ES, Rubio GA, DiFede D, Shafazand S, Khan A, Pujol MV, LaRussa VF et al (2017) Allogeneic human mesenchymal stem cells in patients with idiopathic pulmonary fibrosis via intravenous delivery (AETHER): a phase I safety clinical trial. Chest 151(5):971–981

Han MK, Zhou Y, Murray S, Tayob N, Noth I, Lama VN, Moore BB, White ES, Flaherty KR, Huffnagle GB et al (2014) Lung microbiome and disease progression in idiopathic pulmonary fibrosis: an analysis of the COMET study. Lancet Respir Med 2(7):548–556

Handa T, Tanizawa K, Oguma T, Uozumi R, Watanabe K, Tanabe N, Niwamoto T, Shima H, Mori R, Nobashi TW et al (2022) Novel artificial intelligence-based technology for chest computed tomography analysis of idiopathic pulmonary fibrosis. Ann Am Thorac Soc 19(3): 399–406

Hara A, Sakamoto N, Ishimatsu Y, Kakugawa T, Nakashima S, Hara S, Adachi M, Fujita H, Mukae H, Kohno S (2012) S100A9 in BALF is a candidate biomarker of idiopathic pulmonary fibrosis. Respir Med 106(4):571–580

Hayton C, Terrington D, Wilson AM, Chaudhuri N, Leonard C, Fowler SJ (2019) Breath biomarkers in idiopathic pulmonary fibrosis: a systematic review. Respir Res 20(1):7

Herazo-Maya JD, Noth I, Duncan SR, Kim S, Ma SF, Tseng GC, Feingold E, Juan-Guardela BM, Richards TJ, Lussier Y et al (2013) Peripheral blood mononuclear cell gene expression profiles predict poor outcome in idiopathic pulmonary fibrosis. Sci Transl Med 5(205):205ra136

Hirani N, MacKinnon AC, Nicol L, Ford P, Schambye H, Pedersen A, Nilsson UJ, Leffler H, Sethi T, Tantawi S et al (2021) Target inhibition of galectin-3 by inhaled TD139 in patients with idiopathic pulmonary fibrosis. Eur Respir J 57(5)

Huang SK, Fisher AS, Scruggs AM, White ES, Hogaboam CM, Richardson BC, Peters-Golden M (2010) Hypermethylation of PTGER2 confers prostaglandin E2 resistance in fibrotic fibroblasts from humans and mice. Am J Pathol 177(5):2245–2255

Huang SK, Scruggs AM, Donaghy J, Horowitz JC, Zaslona Z, Przybranowski S, White ES, Peters-Golden M (2013) Histone modifications are responsible for decreased Fas expression and apoptosis resistance in fibrotic lung fibroblasts. Cell Death Dis 4:e621

Inoue Y, Kaner RJ, Guiot J, Maher TM, Tomassetti S, Moiseev S, Kuwana M, Brown KK (2020) Diagnostic and prognostic biomarkers for chronic fibrosing interstitial lung diseases with a progressive phenotype. Chest 158(2):646–659

Ishikawa N, Ohlmeier S, Salmenkivi K, Myllarniemi M, Rahman I, Mazur W, Kinnula VL (2010) Hemoglobin alpha and beta are ubiquitous in the human lung, decline in idiopathic pulmonary fibrosis but not in COPD. Respir Res 11:123

Jackson RM, Gomez-Marin OW, Ramos CF, Sol CM, Cohen MI, Gaunaurd IA, Cahalin LP, Cardenas DD (2014) Exercise limitation in IPF patients: a randomized trial of pulmonary rehabilitation. Lung 192(3):367–376

Jia G, Chandriani S, Abbas AR, DePianto DJ, N'Diaye EN, Yaylaoglu MB, Moore HM, Peng I, DeVoss J, Collard HR et al (2017) CXCL14 is a candidate biomarker for Hedgehog signalling in idiopathic pulmonary fibrosis. Thorax 72(9):780–787

Kang YP, Lee SB, Lee JM, Kim HM, Hong JY, Lee WJ, Choi CW, Shin HK, Kim DJ, Koh ES et al (2016) Metabolic Profiling Regarding Pathogenesis of Idiopathic Pulmonary Fibrosis. J Proteome Res 15(5):1717–1724

Khalil N, Manganas H, Ryerson CJ, Shapera S, Cantin AM, Hernandez P, Turcotte EE, Parker JM, Moran JE, Albert GR et al (2019) Phase 2 clinical trial of PBI-4050 in patients with idiopathic pulmonary fibrosis. Eur Respir J 53(3)

Khan T, Dasgupta S, Ghosh N, Chaudhury K (2021) Proteomics in idiopathic pulmonary fibrosis: the quest for biomarkers. Mol Omics 17(1):43–58

Kolb M, Bonella F, Wollin L (2017) Therapeutic targets in idiopathic pulmonary fibrosis. Respir Med 131:49–57

Konigsberg IR, Borie R, Walts AD, Cardwell J, Rojas M, Metzger F, Hauck SM, Fingerlin TE, Yang IV, Schwartz DA (2021) Molecular signatures of idiopathic pulmonary fibrosis. Am J Respir Cell Mol Biol 65(4):430–441

Konigshoff M, Kramer M, Balsara N, Wilhelm J, Amarie OV, Jahn A, Rose F, Fink L, Seeger W, Schaefer L et al (2009) WNT1-inducible signaling protein-1 mediates pulmonary fibrosis in mice and is upregulated in humans with idiopathic pulmonary fibrosis. J Clin Invest 119(4): 772–787

Konishi K, Gibson KF, Lindell KO, Richards TJ, Zhang Y, Dhir R, Bisceglia M, Gilbert S, Yousem SA, Song JW et al (2009) Gene expression profiles of acute exacerbations of idiopathic pulmonary fibrosis. Am J Respir Crit Care Med 180(2):167–175

Kropski JA, Blackwell TS (2019) Progress in understanding and treating idiopathic pulmonary fibrosis. Annu Rev Med 70:211–224

Kumar A, Kapnadak SG, Girgis RE, Raghu G (2018) Lung transplantation in idiopathic pulmonary fibrosis. Expert Rev Respir Med 12(5):375–385

Lederer DJ, Martinez FJ (2018) Idiopathic Pulmonary Fibrosis. N Engl J Med 378(19):1811–1823
Lee JU, Cheong HS, Shim EY, Bae DJ, Chang HS, Uh ST, Kim YH, Park JS, Lee B, Shin HD et al (2017) Gene profile of fibroblasts identify relation of CCL8 with idiopathic pulmonary fibrosis. Respir Res 18(1):3
Lenti MV, Di Sabatino A (2019) Intestinal fibrosis. Mol Asp Med 65:100–109
Li Y, Wu G, Shang Y, Qi Y, Wang X, Ning S, Chen H (2020) ILDGDB: a manually curated database of genomics, transcriptomics, proteomics and drug information for interstitial lung diseases. BMC Pulm Med 20(1):323
Maher TM, Wells AU, Laurent GJ (2007) Idiopathic pulmonary fibrosis: multiple causes and multiple mechanisms? Eur Respir J 30(5):835–839
Mammarappallil JG, Rankine L, Wild JM, Driehuys B (2019) New developments in imaging idiopathic pulmonary fibrosis with hyperpolarized xenon magnetic resonance imaging. J Thorac Imaging 34(2):136–150
Martinez FJ, Collard HR, Pardo A, Raghu G, Richeldi L, Selman M, Swigris JJ, Taniguchi H, Wells AU (2017) Idiopathic pulmonary fibrosis. Nat Rev Dis Primers 3:17074
McDonough JE, Kaminski N, Thienpont B, Hogg JC, Vanaudenaerde BM, Wuyts WA (2019) Gene correlation network analysis to identify regulatory factors in idiopathic pulmonary fibrosis. Thorax 74(2):132–140
Moghoofei M, Mostafaei S, Kondori N, Armstrong ME, Babaei F (2022) Bacterial and viral coinfection in idiopathic pulmonary fibrosis patients: the prevalence and possible role in disease progression. BMC Pulm Med 22(1):60
Mohning MP, Richards JC, Huie TJ (2019) Idiopathic pulmonary fibrosis: the radiologist's role in making the diagnosis. Br J Radiol 92(1099):20181003
Moodley YP, Corte TJ, Oliver BG, Glaspole IN, Livk A, Ito J, Peters K, Lipscombe R, Casey T, Tan DBA (2019) Analysis by proteomics reveals unique circulatory proteins in idiopathic pulmonary fibrosis. Respirology 24(11):1111–1114
Morais A, Beltrao M, Sokhatska O, Costa D, Melo N, Mota P, Marques A, Delgado L (2015) Serum metalloproteinases 1 and 7 in the diagnosis of idiopathic pulmonary fibrosis and other interstitial pneumonias. Respir Med 109(8):1063–1068
Mushiroda T, Wattanapokayakit S, Takahashi A, Nukiwa T, Kudoh S, Ogura T, Taniguchi H, Kubo M, Kamatani N, Nakamura Y et al (2008) A genome-wide association study identifies an association of a common variant in TERT with susceptibility to idiopathic pulmonary fibrosis. J Med Genet 45(10):654–656
Nathan SD, Behr J, Cottin V, Lancaster L, Smith P, Deng CQ, Pearce N, Bell H, Peterson L, Flaherty KR (2022) Study design and rationale for the TETON phase 3, randomised, controlled clinical trials of inhaled treprostinil in the treatment of idiopathic pulmonary fibrosis. BMJ Open Respir Res 9(1)
Neumark N, Cosme C Jr, Rose KA, Kaminski N (2020) The idiopathic pulmonary fibrosis cell atlas. Am J Physiol Lung Cell Mol Physiol 319(6):L887–L893
Newton CA, Herzog EL (2021) Molecular markers and the promise of precision medicine for interstitial lung disease. Clin Chest Med 42(2):357–364
Niu R, Liu Y, Zhang Y, Zhang Y, Wang H, Wang Y, Wang W, Li X (2017) iTRAQ-based proteomics reveals novel biomarkers for idiopathic pulmonary fibrosis. PLoS One 12(1): e0170741
Noth I, Zhang Y, Ma SF, Flores C, Barber M, Huang Y, Broderick SM, Wade MS, Hysi P, Scuirba J et al (2013) Genetic variants associated with idiopathic pulmonary fibrosis susceptibility and mortality: a genome-wide association study. Lancet Respir Med 1(4):309–317
Oak SR, Murray L, Herath A, Sleeman M, Anderson I, Joshi AD, Coelho AL, Flaherty KR, Toews GB, Knight D et al (2011) A micro RNA processing defect in rapidly progressing idiopathic pulmonary fibrosis. PLoS One 6(6):e21253
O'Dwyer DN, Armstrong ME, Trujillo G, Cooke G, Keane MP, Fallon PG, Simpson AJ, Millar AB, McGrath EE, Whyte MK et al (2013) The Toll-like receptor 3 L412F polymorphism and

disease progression in idiopathic pulmonary fibrosis. Am J Respir Crit Care Med 188(12): 1442–1450

Okamoto M, Hoshino T, Kitasato Y, Sakazaki Y, Kawayama T, Fujimoto K, Ohshima K, Shiraishi H, Uchida M, Ono J et al (2011) Periostin, a matrix protein, is a novel biomarker for idiopathic interstitial pneumonias. Eur Respir J 37(5):1119–1127

Palmer SM, Snyder L, Todd JL, Soule B, Christian R, Anstrom K, Luo Y, Gagnon R, Rosen G (2018) Randomized, double-blind, placebo-controlled, phase 2 trial of BMS-986020, a lysophosphatidic acid receptor antagonist for the treatment of idiopathic pulmonary fibrosis. Chest 154(5):1061–1069

Pandit KV, Corcoran D, Yousef H, Yarlagadda M, Tzouvelekis A, Gibson KF, Konishi K, Yousem SA, Singh M, Handley D et al (2010) Inhibition and role of let-7d in idiopathic pulmonary fibrosis. Am J Respir Crit Care Med 182(2):220–229

Pardo A, Gibson K, Cisneros J, Richards TJ, Yang Y, Becerril C, Yousem S, Herrera I, Ruiz V, Selman M et al (2005) Up-regulation and profibrotic role of osteopontin in human idiopathic pulmonary fibrosis. PLoS Med 2(9):e251

Popmihajlov Z, Sutherland DJ, Horan GS, Ghosh A, Lynch DA, Noble PW, Richeldi L, Reiss TF, Greenberg S (2022) CC-90001, a c-Jun N-terminal kinase (JNK) inhibitor, in patients with pulmonary fibrosis: design of a phase 2, randomised, placebo-controlled trial. BMJ Open Respir Res 9(1)

Ptasinski VA, Stegmayr J, Belvisi MG, Wagner DE, Murray LA (2021) Targeting alveolar repair in idiopathic pulmonary fibrosis. Am J Respir Cell Mol Biol 65(4):347–365

Qian W, Cai X, Qian Q, Zhang X (2021) Identification and validation of potential biomarkers and pathways for idiopathic pulmonary fibrosis by comprehensive bioinformatics analysis. Biomed Res Int 2021:5545312

Rabinovich EI, Kapetanaki MG, Steinfeld I, Gibson KF, Pandit KV, Yu G, Yakhini Z, Kaminski N (2012) Global methylation patterns in idiopathic pulmonary fibrosis. PLoS One 7(4):e33770

Raghu G (2017) Idiopathic pulmonary fibrosis: lessons from clinical trials over the past 25 years. Eur Respir J 50(4)

Raghu G, Rochwerg B, Zhang Y, Garcia CA, Azuma A, Behr J, Brozek JL, Collard HR, Cunningham W, Homma S et al (2015) An Official ATS/ERS/JRS/ALAT clinical practice guideline: treatment of idiopathic pulmonary fibrosis. An update of the 2011 clinical practice guideline. Am J Respir Crit Care Med 192(2):e3–e19

Raghu G, Remy-Jardin M, Myers JL, Richeldi L, Ryerson CJ, Lederer DJ, Behr J, Cottin V, Danoff SK, Morell F et al (2018) Diagnosis of idiopathic pulmonary fibrosis. An Official ATS/ERS/JRS/ALAT clinical practice guideline. Am J Respir Crit Care Med 198(5):e44–e68

Richeldi L, Wilson KC, Raghu G (2018) Diagnosing idiopathic pulmonary fibrosis in 2018: bridging recommendations made by experts serving different societies. Eur Respir J 52(3)

Richeldi L, Baldi F, Pasciuto G, Macagno F, Panico L (2019) Current and future idiopathic pulmonary fibrosis therapy. Am J Med Sci 357(5):370–373

Rindlisbacher B, Schmid C, Geiser T, Bovet C, Funke-Chambour M (2018) Serum metabolic profiling identified a distinct metabolic signature in patients with idiopathic pulmonary fibrosis – a potential biomarker role for LysoPC. Respir Res 19(1):7

Roque W, Romero F (2021) Cellular metabolomics of pulmonary fibrosis, from amino acids to lipids. Am J Physiol Cell Physiol 320(5):C689–C695

Rosas IO, Richards TJ, Konishi K, Zhang Y, Gibson K, Lokshin AE, Lindell KO, Cisneros J, Macdonald SD, Pardo A et al (2008) MMP1 and MMP7 as potential peripheral blood biomarkers in idiopathic pulmonary fibrosis. PLoS Med 5(4):e93

Sanders YY, Pardo A, Selman M, Nuovo GJ, Tollefsbol TO, Siegal GP, Hagood JS (2008) Thy-1 promoter hypermethylation: a novel epigenetic pathogenic mechanism in pulmonary fibrosis. Am J Respir Cell Mol Biol 39(5):610–618

Sanders YY, Ambalavanan N, Halloran B, Zhang X, Liu H, Crossman DK, Bray M, Zhang K, Thannickal VJ, Hagood JS (2012) Altered DNA methylation profile in idiopathic pulmonary fibrosis. Am J Respir Crit Care Med 186(6):525–535

Seibold MA, Wise AL, Speer MC, Steele MP, Brown KK, Loyd JE, Fingerlin TE, Zhang W, Gudmundsson G, Groshong SD et al (2011) A common MUC5B promoter polymorphism and pulmonary fibrosis. N Engl J Med 364(16):1503–1512

Selman M, Pardo A, Barrera L, Estrada A, Watson SR, Wilson K, Aziz N, Kaminski N, Zlotnik A (2006) Gene expression profiles distinguish idiopathic pulmonary fibrosis from hypersensitivity pneumonitis. Am J Respir Crit Care Med 173(2):188–198

Selvarajah B, Azuelos I, Anastasiou D, Chambers RC (2021) Fibrometabolism-An emerging therapeutic frontier in pulmonary fibrosis. Sci Signal 14(697)

Serrano-Mollar A (2018) Cell therapy in idiopathic pulmonary fibrosis(dagger). Med Sci (Basel) 6(3)

Sgalla G, Flore M, Siciliano M, Richeldi L (2020) Antibody-based therapies for idiopathic pulmonary fibrosis. Expert Opin Biol Ther 20(7):779–786

Singh A, Shannon CP, Gautier B, Rohart F, Vacher M, Tebbutt SJ, Le Cao KA (2019) DIABLO: an integrative approach for identifying key molecular drivers from multi-omics assays. Bioinformatics 35(17):3055–3062

Sivakumar P, Ammar R, Thompson JR, Luo Y, Streltsov D, Porteous M, McCoubrey C, Cantu E 3rd, Beers MF, Jarai G et al (2021) Integrated plasma proteomics and lung transcriptomics reveal novel biomarkers in idiopathic pulmonary fibrosis. Respir Res 22(1):273

Sokai A, Handa T, Tanizawa K, Oga T, Uno K, Tsuruyama T, Kubo T, Ikezoe K, Nakatsuka Y, Tanimura K et al (2015) Matrix metalloproteinase-10: a novel biomarker for idiopathic pulmonary fibrosis. Respir Res 16:120

Staab-Weijnitz CA, Fernandez IE, Knuppel L, Maul J, Heinzelmann K, Juan-Guardela BM, Hennen E, Preissler G, Winter H, Neurohr C et al (2015) FK506-binding protein 10, a potential novel drug target for idiopathic pulmonary fibrosis. Am J Respir Crit Care Med 192(4):455–467

Stainer A, Faverio P, Busnelli S, Catalano M, Della Zoppa M, Marruchella A, Pesci A, Luppi F (2021) Molecular biomarkers in idiopathic pulmonary fibrosis: state of the art and future directions. Int J Mol Sci 22(12)

Sterclova M, Kishore A, Sikorova K, Skibova J, Petrek M, Vasakova M (2021) Effect of genotype on the disease course in idiopathic pulmonary fibrosis despite antifibrotic treatment. Biomed Rep 15(5):87

Sun YZ, Hu YF, Zhang Y, Wei SY, Yang BL, Xu YP, Rong ZL, Wang D, Yang B (2022) FibROAD: a manually curated resource for multi-omics level evidence integration of fibrosis research. Database (Oxford) (2022)

Swigris J, Fairclough D (2010) Pirfenidone in idiopathic pulmonary fibrosis. Eur Respir J 36(3): 695–696. author reply 696-698

Taille C, Grootenboer-Mignot S, Boursier C, Michel L, Debray MP, Fagart J, Barrientos L, Mailleux A, Cigna N, Tubach F et al (2011) Identification of periplakin as a new target for autoreactivity in idiopathic pulmonary fibrosis. Am J Respir Crit Care Med 183(6):759–766

Takahashi Y, Saito A, Chiba H, Kuronuma K, Ikeda K, Kobayashi T, Ariki S, Takahashi M, Sasaki Y, Takahashi H (2018) Impaired diversity of the lung microbiome predicts progression of idiopathic pulmonary fibrosis. Respir Res 19(1):34

Tan J, Tedrow JR, Dutta JA, Juan-Guardela B, Nouraie M, Chu Y, Trejo Bittar H, Ramani K, Biswas PS, Veraldi KL et al (2016) Expression of RXFP1 is decreased in idiopathic pulmonary fibrosis. Implications for relaxin-based therapies. Am J Respir Crit Care Med 194(11): 1392–1402

Tian Y, Li H, Gao Y, Liu C, Qiu T, Wu H, Cao M, Zhang Y, Ding H, Chen J et al (2019) Quantitative proteomic characterization of lung tissue in idiopathic pulmonary fibrosis. Clin Proteomics 16:6

Todd JL, Neely ML, Overton R, Durham K, Gulati M, Huang H, Roman J, Newby LK, Flaherty KR, Vinisko R et al (2019) Peripheral blood proteomic profiling of idiopathic pulmonary fibrosis biomarkers in the multicentre IPF-PRO Registry. Respir Res 20(1):227

Todd JL, Vinisko R, Liu Y, Neely ML, Overton R, Flaherty KR, Noth I, Newby LK, Lasky JA, Olman MA et al (2020) Circulating matrix metalloproteinases and tissue metalloproteinase inhibitors in patients with idiopathic pulmonary fibrosis in the multicenter IPF-PRO Registry cohort. BMC Pulm Med 20(1):64

Tong X, Su F, Xu X, Xu H, Yang T, Xu Q, Dai H, Huang K, Zou L, Zhang W et al (2019) Alterations to the lung microbiome in idiopathic pulmonary fibrosis patients. Front Cell Infect Microbiol 9:149

Tzouvelekis A, Paspaliaris V, Koliakos G, Ntolios P, Bouros E, Oikonomou A, Zissimopoulos A, Boussios N, Dardzinski B, Gritzalis D et al (2013) A prospective, non-randomized, no placebo-controlled, phase Ib clinical trial to study the safety of the adipose derived stromal cells-stromal vascular fraction in idiopathic pulmonary fibrosis. J Transl Med 11:171

Tzouvelekis A, Yu G, Lino Cardenas CL, Herazo-Maya JD, Wang R, Woolard T, Zhang Y, Sakamoto K, Lee H, Yi JS et al (2017) SH2 domain-containing phosphatase-2 is a novel antifibrotic regulator in pulmonary fibrosis. Am J Respir Crit Care Med 195(4):500–514

van den Blink B, Wijsenbeek MS, Hoogsteden HC (2010) Serum biomarkers in idiopathic pulmonary fibrosis. Pulm Pharmacol Ther 23(6):515–520

van Geffen C, Deissler A, Quante M, Renz H, Hartl D, Kolahian S (2021) Regulatory Immune Cells in Idiopathic Pulmonary Fibrosis: Friends or Foes? Front Immunol 12:663203

Veraldi KL, Feghali-Bostwick CA (2012) Insulin-like growth factor binding proteins-3 and -5: central mediators of fibrosis and promising new therapeutic targets. Open Rheumatol J 6:140–145

Villasenor-Altamirano AB, Moretto M, Maldonado M, Zayas-Del Moral A, Munguia-Reyes A, Romero Y, Garcia-Sotelo JS, Aguilar LA, Aldana-Assad O, Engelen K et al (2020) PulmonDB: a curated lung disease gene expression database. Sci Rep 10(1):514

Vuga LJ, Milosevic J, Pandit K, Ben-Yehudah A, Chu Y, Richards T, Sciurba J, Myerburg M, Zhang Y, Parwani AV et al (2013) Cartilage oligomeric matrix protein in idiopathic pulmonary fibrosis. PLoS One 8(12):e83120

Vuga LJ, Tedrow JR, Pandit KV, Tan J, Kass DJ, Xue J, Chandra D, Leader JK, Gibson KF, Kaminski N et al (2014) C-X-C motif chemokine 13 (CXCL13) is a prognostic biomarker of idiopathic pulmonary fibrosis. Am J Respir Crit Care Med 189(8):966–974

Vukmirovic M, Kaminski N (2018) Impact of transcriptomics on our understanding of pulmonary fibrosis. Front Med (Lausanne) 5:87

Wang Y, Huang B, Jin T, Ocansey DKW, Jiang J, Mao F (2022) Intestinal fibrosis in inflammatory bowel disease and the prospects of mesenchymal stem cell therapy. Front Immunol 13:835005

Wilkes DS, Chew T, Flaherty KR, Frye S, Gibson KF, Kaminski N, Klemsz MJ, Lange W, Noth I, Rothhaar K (2015) Oral immunotherapy with type V collagen in idiopathic pulmonary fibrosis. Eur Respir J 45(5):1393–1402

Wollin L, Wex E, Pautsch A, Schnapp G, Hostettler KE, Stowasser S, Kolb M (2015) Mode of action of nintedanib in the treatment of idiopathic pulmonary fibrosis. Eur Respir J 45(5): 1434–1445

Wong A, Lu J, Dorfman A, McInnis P, Famouri M, Manary D, Lee JRH, Lynch M (2021) Fibrosis-net: a tailored deep convolutional neural network design for prediction of pulmonary fibrosis progression from chest CT images. Front Artif Intell 4:764047

Yamada M, Kubo H, Ota C, Takahashi T, Tando Y, Suzuki T, Fujino N, Makiguchi T, Takagi K, Suzuki T et al (2013) The increase of microRNA-21 during lung fibrosis and its contribution to epithelial-mesenchymal transition in pulmonary epithelial cells. Respir Res 14:95

Yamaguchi M, Hirai S, Tanaka Y, Sumi T, Miyajima M, Mishina T, Yamada G, Otsuka M, Hasegawa T, Kojima T et al (2017) Fibroblastic foci, covered with alveolar epithelia exhibiting epithelial-mesenchymal transition, destroy alveolar septa by disrupting blood flow in idiopathic pulmonary fibrosis. Lab Investig 97(3):232–242

Yan F, Wen Z, Wang R, Luo W, Du Y, Wang W, Chen X (2017) Identification of the lipid biomarkers from plasma in idiopathic pulmonary fibrosis by lipidomics. BMC Pulm Med 17(1): 174

Yang IV, Schwartz DA (2015) Epigenetics of idiopathic pulmonary fibrosis. Transl Res 165(1): 48–60

Yang IV, Luna LG, Cotter J, Talbert J, Leach SM, Kidd R, Turner J, Kummer N, Kervitsky D, Brown KK et al (2012) The peripheral blood transcriptome identifies the presence and extent of disease in idiopathic pulmonary fibrosis. PLoS One 7(6):e37708

Yang IV, Pedersen BS, Rabinovich E, Hennessy CE, Davidson EJ, Murphy E, Guardela BJ, Tedrow JR, Zhang Y, Singh MK et al (2014) Relationship of DNA methylation and gene expression in idiopathic pulmonary fibrosis. Am J Respir Crit Care Med 190(11):1263–1272

Yang IV, Fingerlin TE, Evans CM, Schwarz MI, Schwartz DA (2015) MUC5B and Idiopathic Pulmonary Fibrosis. Ann Am Thorac Soc 12(Suppl 2):S193–S199

Yoon HY, Moon SJ, Song JW (2021) Lung tissue microbiome is associated with clinical outcomes of idiopathic pulmonary fibrosis. Front Med (Lausanne) 8:744523

Yu G, Ibarra GH, Kaminski N (2018) Fibrosis: Lessons from OMICS analyses of the human lung. Matrix Biol 68-69:422–434

Zahr AA, Salama ME, Carreau N, Tremblay D, Verstovsek S, Mesa R, Hoffman R, Mascarenhas J (2016) Bone marrow fibrosis in myelofibrosis: pathogenesis, prognosis and targeted strategies. Haematologica 101(6):660–671

rjZeisberg M, Kalluri R: Cellular mechanisms of tissue fibrosis. 1. Common and organ-specific mechanisms associated with tissue fibrosis. Am J Physiol Cell Physiol 2013, 304(3):C216–C225

Zhang Y, Kaminski N (2012) Biomarkers in idiopathic pulmonary fibrosis. Curr Opin Pulm Med 18(5):441–446

Zhang N, Guo Y, Wu C, Jiang B, Wang Y (2021) Identification of the molecular subgroups in idiopathic pulmonary fibrosis by gene expression profiles. Comput Math Methods Med 2021: 7922594

Zhao YD, Yin L, Archer S, Lu C, Zhao G, Yao Y, Wu L, Hsin M, Waddell TK, Keshavjee S et al (2017) Metabolic heterogeneity of idiopathic pulmonary fibrosis: a metabolomic study. BMJ Open Respir Res 4(1):e000183

Zheng P, Sun S, Wang J, Cheng ZJ, Lei KC, Xue M, Zhang T, Huang H, Zhang XD, Sun B (2022) Integrative omics analysis identifies biomarkers of idiopathic pulmonary fibrosis. Cell Mol Life Sci 79(1):66

Zuo F, Kaminski N, Eugui E, Allard J, Yakhini Z, Ben-Dor A, Lollini L, Morris D, Kim Y, DeLustro B et al (2002) Gene expression analysis reveals matrilysin as a key regulator of pulmonary fibrosis in mice and humans. Proc Natl Acad Sci U S A 99(9):6292–6297

Sarcoidosis

8

Abstract

The complex, heterogeneous clinical presentation of sarcoidosis leads to misdiagnosis and delayed initialization of treatment. Recent advancement in omics approaches has boosted biomarker research and improved our understanding of sarcoidosis pathogenesis. Different biomarkers identified by different omics approaches have been discussed. Integration of multi-omics datasets and development of mathematical models have further aided in identification of novel therapeutic targets. Immunomodulators and corticosteroids are commonly used to hinder the progression of granulomatous inflammation to life-threatening forms of the disease. Biologics and potential drugs candidates under clinical trial for use in sarcoidosis have also been documented. In the end, glimpses of the future directions of research in the field of sarcoidosis have been described.

Keywords

Non-caseating granulomas · Multisystem disorder · Swollen lymph nodes · Radiographic examination · Immunomodulators

8.1 Introduction

Sarcoidosis is a complex multisystem disorder characterized by granulomatous inflammation. The prevalence and distribution of sarcoidosis are largely influenced by ethnicity and the geographical location (Arkema and Cozier 2020). It affects the lungs more commonly (in ~90% of the cases) than other organs (Moor et al. 2020). The "other" organs and systems affected are the lymphatic system and reticuloendothelial system (lymph nodes, liver, spleen), eyes, skin, joints, brain and peripheral nerves, heart, and kidneys (Jain et al. 2020; https://www.ncbi.nlm.nih.gov/books/NBK430687/). Sarcoidosis is associated with heterogeneous clinical presentations in

S. Saha et al., *Pulmonomics: Omics Approaches for Understanding Pulmonary Diseases*, https://doi.org/10.1007/978-981-99-3505-5_8

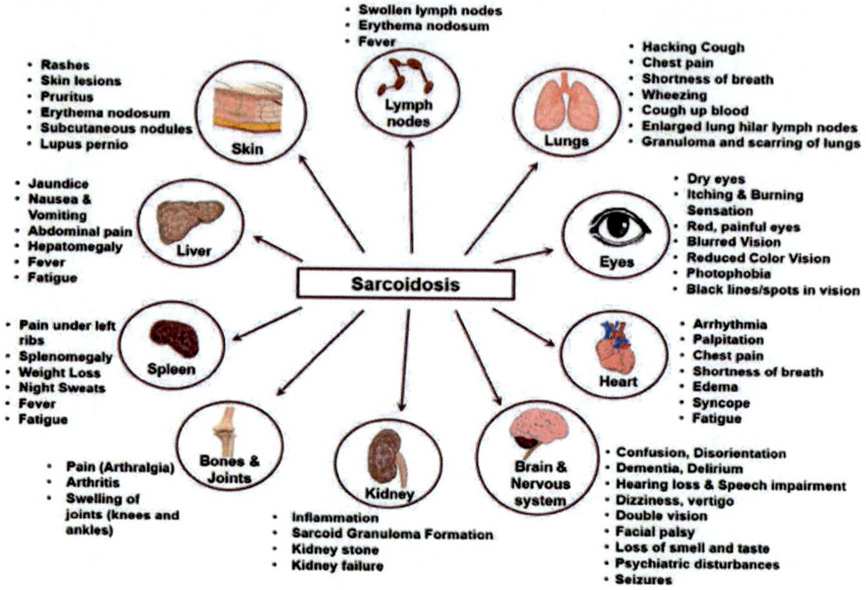

Fig. 8.1 Different organs affected in sarcoidosis and associated symptoms. Although, over 80–90% of sarcoidosis cases involve lungs, extra-pulmonary sarcoidosis may occur in different organs including heart, liver, spleen, eyes, lymph nodes, bones, skin, and nervous system

almost every system involvement. The major symptoms of pulmonary sarcoidosis include persistent dry cough, shortness of breath, chest pain, and fatigue (Ungprasert et al. 2019). Patients with extra-pulmonary sarcoidosis may exhibit a wide plethora of symptoms depending upon the organs affected. Details of the symptoms of both pulmonary and extra-pulmonary sarcoidosis have been summarized in Fig. 8.1. A benign clinical course is generally associated with most cases of sarcoidosis. Presence of comorbidities and complications like pulmonary hypertension influences the symptomatology and the outcome of the disease. The cardiac, renal, or neurologic complications are associated with increased morbidity and mortality (Millward et al. 2021; Drent et al. 2021). Average mortality of sarcoidosis is around 1–5% (Moor et al. 2020). However, there has been substantial increase in sarcoidosis burden and mortality rate in the past decades (Arkema and Cozier 2018; Gerke 2020).

8.2 Etiopathology

The exact aetiology of pulmonary sarcoidosis is yet to be discovered, but a complex interplay of genetic predisposition and environmental triggers have been implicated to drive the onset of sarcoidosis. The formation and the presence of non-caseating granulomas and chronic inflammation are noted in the lungs (Valentonyte et al. 2005). The formation of such epithelioid cell granulomas is driven by dysregulated immune system and is associated with the infiltration of immune cells (namely CD4+ T cells, macrophages, giant cells, CD8+ T cells, and B cells) in the surrounding region (Moor et al. 2020; Guerrero et al. 2020; Broos et al. 2013).

8.3 Clinical Features

The clinical presentation of the disease may vary from being asymptomatic to subclinical inflammation that would ultimately progress towards Lofgren syndrome characterized by an acute form of inflammation (Valentonyte et al. 2005). In most cases, the symptoms resolve spontaneously without any medical intervention. However, in certain cases, such prolonged granulomatous inflammation may progressively lead to fibrotic changes in lungs and decline in lung function, thereby paving way for a severe form of life-threatening disease, characterized by chronic respiratory failure (https://www.ncbi.nlm.nih.gov/books/NBK430687/; Valentonyte et al. 2005). Different organ and system involvement can give rise to a myriad of presentations and is elaborated in Fig. 8.1.

Diagnosis of sarcoidosis is challenging due to the ability of the disease to masquerade other conditions. The diagnosis relies on effectively excluding the chance of occurrence of other granulomatous diseases like tuberculosis, coccidiomycosis, and histoplasmosis (Rosen 2007; Cooper and Suau 2022). Tuberculin skin test is often conducted to exclude co-presence of active or latent tuberculosis. Other non-specific serological tests like serum alkaline phosphatase and angiotensin-converting enzyme (ACE) concentration are also important (https://www.ncbi.nlm.nih.gov/books/NBK430687/). An isolated increase in serum alkaline phosphatase is indicative of granulomatous hepatic involvement in a proper clinical context. SACE has been important in diagnosis but the sensitivity is low (Wang et al. 2022). Imaging tests are of great value in sarcoidosis diagnosis. Commonly used imaging techniques include chest X-ray, computed tomography (CT) scan, high-resolution CT (HRCT) scans, fluorine-18-fluorodeoxyglucose-positron emission tomography (FDG-PET), and single-photon emission computed tomography (SPECT). Radiographic examination of lungs allows classification of pulmonary sarcoidosis into four major stages of disease progression (Ungprasert et al. 2019). The radiographic signatures associated with each of the four stages of pulmonary sarcoidosis have been highlighted in Fig. 8.2. Pulmonary function tests (PFTs) are also performed to assess the restrictive or obstructive pattern in advanced sarcoidosis patients. Approximately, 10% of the advanced cases exhibit an obstructive pattern in PFTs (https://www.ncbi.nlm.nih.gov/books/NBK430687/). The diffusing capacity for carbon monoxide (D_{LCO}) test reveals a reduction in D_{LCO} in patients with pulmonary sarcoidosis. The 6-minute walk distance is reduced in majority of sarcoidosis patients (Baughman et al. 2007). The diagnosis is confirmed by histopathological detection of non-caseating granuloma with exclusion of other possible causes for so as tuberculosis and fungal infections. Trans-bronchial lung biopsy is done commonly; often endobronchial biopsy is also confirmatory (Shorr et al. 2001). The mediastinal lymph nodes are sampled by ultrasound guided

Stages of Sarcoidosis	Radiographic Features
Stage I	**Mediastinal and bilateral hilar adenopathy present. No pulmonary infiltrates observed.**
Stage II	**Mediastinal and bilateral hilar adenopathy present. Pulmonary infiltrates and reticular opacities observed.**
Stage III	**Pulmonary infiltrates and reticular opacities observed. Shrinking hilar adenopathy noted.**
Stage IV	**Reticular opacities and pulmonary fibrosis with volume loss observed. No adenopathy noted.**

Fig. 8.2 Classification of pulmonary sarcoidosis based on radiographic tests. Radiographic observations have led to the classification of pulmonary sarcoidosis into four stages

transbronchial needle aspiration (EBUS-TBNA) or with help of mediastinoscopic biopsies (https://www.ncbi.nlm.nih.gov/books/NBK430687/). Biopsy from other organs is often useful as skin, liver, or bone marrow biopsy.

The different radiographic stages and the varied tempo of clinical presentation of sarcoidosis enable phenotyping of sarcoidosis. Four different phenotypic categories have been recognized and linked with therapeutic recommendations. These are (i) asymptomatic (needs no treatment), (ii) acute (self-limited disease with a duration less than 2 years; treated with corticosteroids), (iii) chronic (treated with antimetabolites and second-line drugs), and (iv) advanced (to be treated with biologics along with standard treatment) (Baughman and Grutters 2015).

8.4 Biomarkers of Pulmonary Sarcoidosis

The complexities and complications involved in the diagnosis of sarcoidosis via invasive techniques demand the identification of specific, non-invasive, reliable, and highly reproducible biomarkers of sarcoidosis that would ease and enable rapid detection of the disease. Identification of biomarkers depicting disease progression and response to therapeutics is of vital importance in the management of sarcoidosis. Since long, there have been no ideal biomarker for sarcoidosis (Kraaijvanger et al. 2020). The advent and advancement in omics have favoured the quest of biomarkers for sarcoidosis. Some of the probable biomarker candidates for sarcoidosis identified using different omics approaches have been documented in Table 8.1.

8.5 Genomics

Genome-wide association studies (GWAS) is extensively used as the gold standard for determining single nucleotide polymorphisms (SNPs) and other genetic variants associated with increased susceptibility to sarcoidosis. Variations in the MHC loci (including HLA-DRB1, HLA-DQB1, HLA-DRA, and HLA-DPB1) have been

Table 8.1 Different biomarkers of pulmonary sarcoidosis

Omics approach	Biomarkers	Biological samples	Comments
Genomics	NOTCH4	Genomic DNA	NOTCH4 SNP rs715299 was identified as a sarcoidosis-associated locus (Adrianto et al. 2012)
	BTNL2	Blood	Greater frequency of the A allele of BTNL2 SNP rs2076530 was observed in patients with sarcoidosis (Li et al. 2006)
	HLA-DRA/BTNL2		SNP in the HLA-DRA/ BTNL2 region (rs3135365, rs3177928, rs6937545, rs5007259) were associated with sarcoidosis (Wolin et al. 2017)
	ANXA11	DNA	rs1049550 of ANXA11 was related to sarcoidosis susceptibility. Two SNPs (rs61860052 and rs4377299) were found to be associated with sarcoidosis susceptibility in African Americans (Levin et al. 2013)
	IRF5		Haplotype comprising of two functional SNPs of IRF5 (rs10954213A and rs2280714A) are associated with susceptibility to sarcoidosis in Japanese population (Tanizawa et al. 2013)
Transcriptomics	TLR2, RIPK2, CLEC7A, NLRP3, MAP2K, CYBB, ATP6AP1, LAMP2, SERPINA1	Monocytes	TLR2, RIPK2, CLEC7A, NLRP3, MAP2K, CYBB, ATP6AP1, LAMP2, and SERPINA1 were upregulated in sarcoidosis patients (Talreja et al. 2017)
	LAPTM4B, RPSA, RPL10A, PSME3		LAPTM4B, RPSA, RPL10A, and PSME3 were downregulated in

(continued)

Table 8.1 (continued)

Omics approach	Biomarkers	Biological samples	Comments
			sarcoidosis patients (Talreja et al. 2017)
	PECAM1, ITGAM, ITGB1, CXCR4, ATG12, HIF1A, CCR1, IER3, mTOR	Blood (CD14 monocytes)	PECAM1, ITGAM, ITGB1, CXCR4, ATG12, HIF1A, CCR1, IER3, and mTOR were increased in sarcoidosis patients (Garman et al. 2020b)
	STAT1, IL10RA, PTEN		STAT1, IL10RA, and PTEN were downregulated in sarcoidosis patients (Garman et al. 2020b)
	HBEGF, SAP30	Peripheral Blood	HBEGF and SAP30 were increased in uncomplicated and complicated sarcoidosis patients (Zhou et al. 2012)
	SESN3, NOG, ZNF671, CRIP1, KLRB1, MEI1, ZNF614, CX3CR1, APOBEC3D, ZNF540, KIAA1147, RBM12B, FKBP1A, SERTAD1, FITM2, TSHZ2		SESN3, NOG, ZNF671, CRIP1, KLRB1, MEI1, ZNF614, CX3CR1, APOBEC3D, ZNF540, KIAA1147, RBM12B, FKBP1A, SERTAD1, FITM2, and TSHZ2 were decreased in uncomplicated and complicated sarcoidosis patients (Zhou et al. 2012)
	CXCL10, GBP5, STAT1, AIM2, ICAM1, JAK2	Peripheral blood, orbital tissues	CXCL10, GBP5, STAT1, AIM2, ICAM1, and JAK2 were increased in sarcoidosis (Schupp et al. 2017)
	miR-146a, miR-150	Broncho alveolar lavage (BAL)	miR-146a and miR-150 are elevated in sarcoidosis patients and are inversely related to lung function (Arger et al. 2020)
	IFN-γ, CCL5, CXCL9		IFN-γ, CCL5, and CXCL9 were increased in sarcoidosis (Schupp et al. 2017)

(continued)

Table 8.1 (continued)

Omics approach	Biomarkers	Biological samples	Comments
	Cathelicidin antimicrobial peptide (CAMP)		CAMP levels were reduced in severe stages of sarcoidosis (Schupp et al. 2017)
	MBTPS1, CBLB, ATP10A, EFHA2, STAT4, ITGA6, ZFYVE9, TMEM263, URI1, SATB1, IL6ST, FIGNL1, GALNT12, ERCC6L2, SORCS3, MTERFD2	Peripheral blood mononuclear cells (PBMCs)	MBTPS1, CBLB, ATP10A, EFHA2, STAT4, ITGA6, ZFYVE9, TMEM263, URI1, SATB1, IL6ST, FIGNL1, GALNT12, ERCC6L2, SORCS3, and MTERFD2 were decreased with disease severity in sarcoidosis patients (Zhou et al. 2017)
	miR-196a		miR-196a was downregulated in uncomplicated sarcoidosis patients, and further decreased in complicated sarcoidosis patients (Zhou et al. 2017)
	miR-23a, miR-23b, miR-30c, miR-93, miR-143, miR-185, miR-223		miR-23a, miR-23b, miR-30c, miR-93, miR-143, miR-185, and miR-223 were elevated in uncomplicated sarcoidosis patients and further increased in complicated sarcoidosis patients (Zhou et al. 2017)
	miR-16, miR-20a	Peripheral blood mononuclear cells (PBMCs), broncho alveolar lavage (BAL)	miR-16 and miR-20a levels were significantly increased in sarcoidosis patients having lung volume restriction (Kiszalkiewicz et al. 2016)
	miR-27b, miR-221, miR-192		miR-27b, miR-221, and miR-192 levels were higher in Stage I sarcoidosis patients as compared to Stage II-IV sarcoidosis patients (Kiszalkiewicz et al. 2016)

(continued)

Table 8.1 (continued)

Omics approach	Biomarkers	Biological samples	Comments
	miR-33, miR-34		miR-33 and miR-34 are increased in sarcoidosis (Arger et al. 2020)
	MMP12, ADAMDEC1	Broncho alveolar lavage (BAL), lung tissues	MMP12 and ADAMDEC1 expression levels were positively correlated to severity of sarcoidosis (Arger et al. 2020)
	NOTCH4, STAB1	Granuloma from lungs and lymph nodes	NOTCH4 was downregulated in pulmonary sarcoidosis. Reduced expression of STAB1 was noted in lymph nodes in sarcoidosis (Casanova et al. 2020)
	ADAMST1, NPR1, CXCL2, FABP4		ADAMST1, NPR1, CXCL2, and FABP4 were exclusively expressed in sarcoidosis granuloma derived from lungs and lymph nodes (Casanova et al. 2020)
Epigenomics	HLA-DPB2, CXCL7, CCL16, MSMB, KIR3DL2, HMGA1, ZDHHC2, PDXK, HLA-DPB1, HLA-DQA2	Lungs	HLA-DPB2, CXCL7, CCL16, MSMB, KIR3DL2, HMGA1, ZDHHC2, PDXK, HLA-DPB1, and HLA-DQA2 were differentially methylated and expressed in progressive sarcoidosis patients as compared to remitting sarcoidosis patients (Yang et al. 2019)
Proteomics	ACE	Serum	Increased serum ACE level serves as a diagnosing biomarker for sarcoidosis (Kahkouee et al. 2016)
	Lysozyme		Increased serum concentration of lysozyme is noted on the onset of the disease; therefore, lysozyme can serve as a prognostic

(continued)

Table 8.1 (continued)

Omics approach	Biomarkers	Biological samples	Comments
			biomarker of sarcoidosis (Kraaijvanger et al. 2020; Bergantini et al. 2019)
	IL-2R		IL-2R levels are elevated in sarcoidosis patients. Serum levels of IL-2R along with other imaging studies can aid in sarcoidosis diagnosis (Kraaijvanger et al. 2020)
	Chitotriosidase (CTO)		Increased levels are noted in progressive sarcoidosis. The level decreases upon treatment with immunosuppressant drugs (Kraaijvanger et al. 2020; Bergantini et al. 2019)
	Serum amyloid A (SAA)		SAA is elevated in sarcoidosis patients, and SAA level correlates with decline in lung function (Kraaijvanger et al. 2020)
	Neopterin		Neopterin levels are increased in sarcoidosis (Kraaijvanger et al. 2020)
	YKL-40		YKL-40 levels are elevated in patients with active sarcoidosis as compared to those with inactive sarcoidosis (Kraaijvanger et al. 2020)
	sCD163		sCD163 levels are increased in sarcoidosis (Kraaijvanger et al. 2020)
	C-C motif chemokine ligand 18 (CCL18)		CCL18 levels are increased in patients with active sarcoidosis (Kraaijvanger et al. 2020)
	ICAM1, LEP	Serum, granulomas	ICAM1 and LEP were increased in sarcoidosis.

(continued)

Table 8.1 (continued)

Omics approach	Biomarkers	Biological samples	Comments
			These proteins may aid in differential diagnosis of sputum-negative tuberculosis and sarcoidosis (Du et al. 2015)
	ARPC3, HLA-DQA1, FABP3, LAMC1, GPDA, EFHD2, 1B15, CD5L, LCN1, SFPQ	Broncho alveolar lavage fluid (BALF)	ARPC3, HLA-DQA1, FABP3, LAMC1, GPDA, EFHD2, 1B15, CD5L, LCN1, and SFPQ were significantly decreased in progressive sarcoidosis patients as compared to non-progressive sarcoidosis patients (Bhargava et al. 2020)
	GATA5, CMBL, PDXK, HBB, AXA81, HBA, ANXA3, GSTM3, GSTA1, LEG1H		GATA5, CMBL, PDXK, HBB, AXA81, HBA, ANXA3, GSTM3, GSTA1, and LEG1H levels were significantly higher in progressive sarcoidosis patients as compared to non-progressive sarcoidosis patients (Bhargava et al. 2020)
	AK1BA, PDXK, HBD, RPS2, CALM1, UCHL1, PGS2, CHIT1, CAH3, SBSN		AK1BA, PDXK, HBD, RPS2, CALM1, UCHL1, PGS2, CHIT1, CAH3, and SBSN were significantly higher in sarcoidosis patients (Bhargava et al. 2020)
	RXYLT1, PEPC, CS, KLK1, CYT5, CAV1, HMGA1, CYBR1, PLLP, ACTG1		RXYLT1, PEPC, CS, KLK1, CYT5, CAV1, HMGA1, CYBR1, PLLP, and ACTG1 were significantly reduced in sarcoidosis patients (Bhargava et al. 2020)
	IP-10, MIG		IP-10 and MIG were upregulated in BALF of sarcoidosis patients as compared to IPF patients and healthy controls (Giusti 2020)
	ALB, PCDHA2, AAT		ALB and PCDHA2 levels were increased in

(continued)

Table 8.1 (continued)

Omics approach	Biomarkers	Biological samples	Comments
			sarcoidosis, whereas AAT was decreased in sarcoidosis (Kriegova et al. 2006)
	EPHB1, RAB11A, VCP, GSN, GSS, ZFP1, MyD88, TOP2B, DPP9, PLD, AP2B1, DGKB1, ORP1, DYNC1LI1, ALDHA1, GILT	Alveolar macrophages	EPHB1, RAB11A, VCP, GSN, GSS, ZFP1, MyD88, TOP2B, DPP9, PLD, AP2B1, DGKB1, ORP1, DYNC1LI1, ALDHA1, and GILT were significantly increased in sarcoidosis patients (Silva et al. 2013)
	VIM, RHOA, ATP1A1, HSP70, AK5, GOLIM4		VIM, RHOA, ATP1A1, HSP70, AK5, and GOLIM4 were significantly reduced in sarcoidosis patients (Silva et al. 2013)
	MUC5A, MIPT3, PDCD4, FCGBP	Broncho alveolar lavage (BAL) cells	MUC5A, MIPT3, PDCD4, and FCGBP were significantly increased in sarcoidosis patients (Bhargava et al. 2020)
	ANXA3, CD177, P85A, E9PMC5, ITB2, PERM, C163A		ANXA3, CD177, P85A, E9PMC5, ITB2, PERM, and C163A were significantly decreased in sarcoidosis patients (Bhargava et al. 2020)
	Vitamin D-binding protein(VDBP)	Broncho alveolar lavage fluid (BALF) Exosomes	VDBP was higher in BALF exosomes from sarcoidosis patients. Thus, it can serve as exosome derived biomarker for sarcoidosis (Martinez-Bravo et al. 2017)
	Krebs von den Lungen-6 (KL-6)	Serum, broncho alveolar lavage fluid (BALF)	KL-6 levels are elevated in sarcoidosis patients, highest level of KL-6 is noted in stage IV pulmonary sarcoidosis. KL-6 may aid in disease monitoring in sarcoidosis patients

(continued)

Table 8.1 (continued)

Omics approach	Biomarkers	Biological samples	Comments
			(Kraaijvanger et al. 2020)
	CXCL9, CXCL10, CXCL11		CXCL9, CXCL10, and CXCL11 are elevated in sarcoidosis patients. CXCL10 and CXCL11 levels are associated with decline in lung function. CXCL9 and CXCL11 levels are positively correlated with the total number of organs affected in sarcoidosis (Kraaijvanger et al. 2020)
Metabolomics	Methanol, butyrate, lactate, acetate, N-butyrate	Saliva	Methanol and butyrate were decreased, while lactate, acetate, and N-butyrate were increased in sarcoidosis patients (Duchemann et al. 2016)
	Glutamine, isoleucine, succinate	Serum	Glutamine, isoleucine, and succinate levels were significantly decreased in sarcoidosis patients (Geamanu et al. 2016).
	3-hydroxybutyrate, acetoacetate, carnitine, cysteine, homocysteine, pyruvate, trimethylamine N-oxide		These metabolites were significantly increased in sarcoidosis patients compared to healthy subjects (Geamanu et al. 2016)
	Lactate, glutamate, pyruvate, taurine, sucrose, hypotaurine, O-phosphocholine, ethanolamine phosphate, adenosine-5 monophosphate, LL-2,6-diamino heptanedioate, N-methyl-D-aspartic acid	Blood plasma	These metabolites aid in distinguishing between veterans and civilians with sarcoidosis These metabolites are significantly increased in veterans with sarcoidosis compared to civilians with sarcoidosis (Banoei et al. 2019)
	Glutamine, acetone, creatinine, L-valine, L-tryptophan, L-cysteic acid, L-methionine,		These metabolites are significantly decreased in veterans with sarcoidosis compared to

(continued)

Table 8.1 (continued)

Omics approach	Biomarkers	Biological samples	Comments
	propionate, N-acetyl glycine, L-arginine, L-phenylalanine, 5-dihydroorotate, 2-amino-2-methyl propionate, α-hydroxy isobutyric acid, 4-imidazole acetic acid, mevalolactone, SN-glycerol 3-phosphate		civilians with sarcoidosis. These metabolites thus aid in distinguishing between veterans and civilians with sarcoidosis (Banoei et al. 2019)
Metagenomics	*Atopobium* spp. *Fusobacterium* spp.	Broncho alveolar lavage (BAL)	*Atopobium* spp. and *Fusobacterium* spp. were noted in sarcoidosis patients and were identified as sarcoidosis-associated bacteria (Inaoka et al. 2019)
	Cutibacterium acnes		*Cutibacterium acnes* is found in BAL of most of sarcoidosis patients (Inaoka et al. 2019).

classically associated with sarcoidosis (Garman et al. 2020a; Calender et al. 2020). SNPs of IL23R (rs7517847, rs11465804, and rs11209026) have also been linked with increased susceptibility to sarcoidosis (Kim et al. 2011). Polymorphism in SH2B3 (rs3184504) have been related to greater risk of developing sarcoidosis, whereas SNPs in RAB23 (rs1040461 and rs61860052) have been correlated with increased chance of developing sarcoidosis-associated uveitis (Devalliere and Charreau 2011; Davoudi et al. 2018). Other SNPs associated with sarcoidosis have been tabulated in Table 8.1.

8.6 Epigenomics

Epigenetic modifications like histone modifications and DNA methylation have emerged as important regulatory mechanisms associated with aberrant gene expression in different pulmonary diseases (Konigsberg et al. 2021). The presence of epigenetic changes has been proposed to be involved in the pathogenesis of sarcoidosis. However, till date only one epigenomics study has been conducted for sarcoidosis (Garman et al. 2020a). Differential methylation patterns and altered expression of several genes like CXCL7, HLA-DPB2, CCL16, MSMB, KIR3DL2, HMGA1, ZDHHC2, PDXK, HLA-DPB1, and HLA-DQA2 were observed exclusively in progressive sarcoidosis patients (also shown in Table 8.1) (Yang et al. 2019). This study provides an evidence of epigenetic changes leading to sarcoidosis and also demonstrates the necessity of further epigenomics studies for sarcoidosis.

8.7 Transcriptomics

Several high-throughput omics studies have been conducted using different biological samples like blood, monocytes, broncho alveolar lavage (BAL), and lung tissues to unravel transcriptomic signatures associated with sarcoidosis. Microarray-based longitudinal study of blood samples demonstrated increased expression of pattern recognition receptors and genes involved in interferon signalling in sarcoidosis patients. Downregulation of genes of the T cell receptor (TCR) signalling pathway was also observed in sarcoidosis patients compared to healthy subjects (Su et al. 2014). Similarity in blood transcriptomic profile has been noted in tuberculosis and sarcoidosis (Maertzdorf et al. 2012; Koth et al. 2011; Chai et al. 2020). However, difference in the degree of transcriptional activity and transcriptional response to therapeutics were observed in sarcoidosis and tuberculosis (Bloom et al. 2013; Casanova et al. 2020). Besides these, other RNA sequencing and transcriptomic analyses identified novel genes and pathways associated with sarcoidosis (Talreja et al. 2017; Zhou et al. 2017). Several altered gene and miRNA expressions have also been correlated with disease severity and response to common treatments. Some of these differentially expressed genes and miRNAs associated with sarcoidosis have been enlisted in Table 8.1.

8.8 Proteomics

Powerful and sensitive high-throughput proteomic technologies enable identification of important protein markers of sarcoidosis from various biological samples like blood, macrophages, broncho alveolar lavage fluid (BALF), exosomes, granulomas, and lung tissues (Guerrero et al. 2020). Proteomics analysis of extracellular vesicles from serum identified increased CD14 and LBP as candidate biomarkers for sarcoidosis (Futami et al. 2022). Several serum and BALF proteins like CXCL5, CXCL9, CXCL10, MCP-1, MCP-3, MCP-4, TRAIL, TNF, OSM, CDCP1, and SCF have also been identified that enable efficient distinction between sarcoidosis and idiopathic pulmonary fibrosis (IPF) (Majewski et al. 2021). Leptin and ICAM-1 protein expression profiles in serum enable distinction between sarcoidosis and sputum-negative tuberculosis (Du et al. 2015). Besides these, serum levels of proteins like chitotriosidase, KL-6, ACE, YKL40, neopterin, and lysozyme can be explored as diagnostic or prognostic biomarkers of sarcoidosis (Kraaijvanger et al. 2020; Bergantini et al. 2019). Other differentially expressed protein signatures of pulmonary sarcoidosis have been tabulated in Table 8.1.

8.9 Metabolomics

Different techniques like proton nuclear magnetic resonance spectroscopy (1H NMR), inductively coupled plasma mass spectrometry (ICP-MS) and hydrophilic interaction liquid chromatography mass spectrometry (HILIC-MS) have been used to identify important metabolic alterations associated with sarcoidosis. Palmitoylcarnitine and *p*-coumaroylagmatine were identified as blood plasma derived potential diagnostic metabolic biomarkers of fibrosing pulmonary sarcoidosis (Mirsaeidi et al. 2016). Apart from blood plasma samples, serum and salivary samples have also been studied to identify changes in metabolome in sarcoidosis patients. The metabolic signatures of sarcoidosis identified in these studies have been documented in Table 8.1.

8.10 Metagenomics

Microbial dysbiosis along the gut–lung axis have also been associated with granulomatous inflammation in pulmonary sarcoidosis. Presently, little knowledge is available regarding the role of microbiome in the pathogenesis of sarcoidosis (Crouser et al. 2017). However, several metagenomics studies have been conducted for dissecting the microbial composition and lineage in sarcoidosis. 16S rRNA sequencing of broncho alveolar lavage fluid (BALF) revealed an increased abundance of *Atopobium* spp. and *Fusobacterium* spp. in sarcoidosis patients as compared to healthy controls (Chioma et al. 2021). Analyses of BALF exhibited similar microbial profile in both pulmonary sarcoidosis patients and rheumatoid arthritis patients. Increased abundance of *Prevotella* spp., *Treponema* spp., *Porphyromonas* spp. and reduced predominance of *Burkholderia* spp., *Actinomyces* spp., were noted in both the diseases as compared to that in healthy subjects (Scher et al. 2016). A few microbial signatures of sarcoidosis have been listed in Table 8.1. Pyro sequencing-based study of lower airway microbiome in patients with idiopathic interstitial pneumonia (IIP), non-IIP, and sarcoidosis revealed the presence of Prevotellaceae, Acidaminococcaceae, and Streptococcaceae in majority of the patients, although no significant difference in lung microbiome was noted as compared to healthy subjects (Garzoni et al. 2013). An observational study on pulmonary microbiota failed to determine distinctive microbial signatures of sarcoidosis as compared to other interstitial lung diseases (Becker et al. 2019). Another metagenomics study involving the use of lymph node biopsies, Kveim reagent samples, BALF, and fresh granulomatous spleen samples highlighted enrichment of certain microbial species in single sarcoidosis sample types, but failed to determine similar enrichments in multiple sarcoidosis sample types (Clarke et al. 2018). Further studies are required for effective identification of altered microbial signatures involved in the pathogenesis and onset of sarcoidosis.

8.11 Bioinformatics

Databases based on epidemiological studies provide critical information of the risk factors and distribution of sarcoidosis. Few national epidemiological databases specifically for sarcoidosis are available for Swedish, Taiwanese, and USA residents (Arkema et al. 2016; Wu et al. 2017; Baughman et al. 2016). Overall, it was observed that the incidence and prevalence of sarcoidosis were found to be highest in Nordic countries and African Americans. The molecular data focusing on genetic factors on Sarcoidosis are available in genome-wide association studies (GWAS) and whole exome sequencing (WES) databases (Dehghan 2018; Jaganathan et al. 2020).

8.12 Medical Imaging or Radiomics

Radiology plays an essential role in the diagnosis and evaluation of sarcoidosis patients as evident from Fig. 8.2. Medical imaging techniques are involved in the assessment of thoracic sarcoidosis using chest X-ray and high-resolution CT (HRCT) imaging, and positron emission-computed tomography (PET-CT) (Keijsers and Grutters 2020). FDG-PET/CT can be used especially when conventional tests are inconclusive (Keijsers and Grutters 2020). It can detect treatable active disease especially in cardiac and fibrotic pulmonary sarcoidosis (Keijsers and Grutters 2020). The extra-thoracic involvement has been evaluated using HRCT, magnetic resonance imaging (MRI), and positron emission-computed tomography (PET-CT) (Larici et al. 2018). Cardiac MRI remains the choice of investigation for cardiac involvement in sarcoidosis (Komada et al. 2016). Bilateral hilar lymphadenopathy (BHL) with peri-lymphatic micronodular imaging pattern is observed in 50–80% of sarcoidosis cases (Nunes et al. 2012). At a later stage of sarcoidosis, lymph nodes usually become less or non-apparent and calcification is seen in more than 20% of cases on chest radiography (Israel et al. 1981). Chest HRCT is a preferred choice for sarcoidosis than X-ray in difficult cases and in suspected pulmonary complications including development of pulmonary hypertension (Tana et al. 2020; Duong and Bonham 2018).

8.13 Multi-Omics and Data Integration

Multi-omics data analyses on sarcoidosis were performed to identify novel candidate genes, to identify key regulators, and to identify molecular phenotypes. A study based on positional integration analysis of 38 heterogeneous datasets from genomic, transcriptomic, proteomic, and phenomic identifies potential candidate genes like NELFE, EGFL7, AGPAT2, NRC3, and NEU1 (Hocevar et al. 2018). In another study, sarcoidosis protein–protein interaction network was developed using differentially expressed genes and eventually six hub proteins (ICOS, CTLA4, FLT3LG, CD33, GPR29, and ITGA4) were identified, which can be potential biomarkers or main targets for sarcoidosis treatment (Tazyeen et al. 2022). A national multicentre research study was also performed on Genomic Research in Sarcoidosis (GRADS Sarc), funded by the National Heart, Lung, and Blood Institute (NHLBI) with an aim to integrate clinical, transcriptomics, and lung microbiome data to identify novel molecular phenotypes and personalized approaches for sarcoidosis (Moller et al. 2015). Hao et al. developed a mathematical model of sarcoidosis based on several parameters like macrophages, Th1, Treg, Th17 cells activation along with other cytokines that predicts therapeutic targets and possible treatment (Hao et al. 2014). Taken together, both national level large scale multi-omics studies like GRADS and small-scale individual studies were performed to understand this complex disease using multi-omics approaches to find novel biomarkers and target genes.

8.14 Current Applications

The asymptomatic nature of sarcoidosis and the lack of awareness often result in delayed disease diagnosis and initiation of treatment. This makes treatment of sarcoidosis challenging. The standard therapeutics used for pulmonary sarcoidosis are primarily aimed at restricting the granulomatous inflammation and preventing the progression towards a life-threatening disease (Li et al. 2021). Oral corticosteroids are used as the first-line therapy for sarcoidosis. Although, glucocorticoids lead to short-term improvement in lung function, they are associated with serious side effects upon prolonged usage in a dose-dependent manner (Drent et al. 2021). Patients unresponsive to oral corticosteroids are generally subjected to immunomodulators like methotrexate, azathioprine, leflunomide, mycophenolate mofetil, cyclophosphamide, and hydroxychloroquine. Of these corticosteroid-sparing medications, methotrexate is the most commonly used second-line therapy for sarcoidosis (Moor et al. 2020). The impact of these drugs in sarcoidosis patients along with the associated adverse effects has been tabulated in Table 8.2.

Antidepressant medications and dietary supplementation are often to reduce symptoms in sarcoidosis patients, who do not respond to corticosteroids or immunosuppressants. Lung transplantation is rarely performed in patients with pulmonary sarcoidosis, but leads to improved post-surgical survival rate (Drent et al. 2021). Besides, several biologics and potential drug candidates are also being evaluated for their safety and efficacy in sarcoidosis patients as enlisted in Table 8.3.

The development of therapeutics for sarcoidosis is challenging. The discovery of biomarkers and novel drug targets for sarcoidosis using different omics tools can only aid in circumventing the challenges and complexities associated with efficient blocking of fibrotic pathways and granulomatous inflammation in sarcoidosis.

Table 8.2 Drugs commonly used to treat Pulmonary Sarcoidosis

Sl. no.	Drug	Description	Effect in sarcoidosis patients	Side effects
1.	Systemic glucocorticoid (prednisolone, methylprednisolone)	Exerts anti-inflammatory effects	Used as the first-line treatment for sarcoidosis. Suppresses cytokine production responsible for persistent granuloma formation May lead to improvement in symptoms in short term. Cannot cure the disease. Cannot alter the outcome of the disease	Muscle weakness, fluid retention, osteonecrosis, osteoporosis, weight gain, mood swings, insomnia, hypertension, cataract, glaucoma, hyperglycaemia, gastritis, opportunistic infections (Gerke 2020; Li et al. 2021; Melani et al. 2021)
2.	Methotrexate	An antimetabolite that competitively inhibits dihydrofolate reductase (DHFR), an enzyme involved in tetrahydrofolate synthesis. Inhibits adenosine deaminase	Can reduce hormonal secretion Improves lung function upon prolonged usage (6 months) Exerts steroid-sparing effect	Fatigue, dyspepsia, alopecia, mucositis, hypersensitivity pneumonia, and opportunistic infections. In extreme cases, it may lead to liver and bone marrow toxicity (Moor et al. 2020; Gerke 2020; Baughman and Grutters 2015; Li et al. 2021; Melani et al. 2021)
3.	Azathioprine	Prodrug of 6-mercaptopurine. Inhibits purine synthesis. It also causes immunosuppression	Can reduce hormonal secretion Improves lung function and diffusing capacity (D_{LCO}). Tapers use of steroids	Fatigue, myalgia, dyspepsia, alopecia, blurred vision, mucositis, hypersensitivity reactions, jaundice, and opportunistic infections (Gerke 2020; Baughman and Grutters 2015; Li et al. 2021; Melani et al. 2021)

(continued)

Table 8.2 (continued)

Sl. no.	Drug	Description	Effect in sarcoidosis patients	Side effects
4.	Leflunomide	An immunomodulatory drug. Inhibits mitochondrial enzyme dihydrofolate dehydrogenase (DHODH), which is involved in the de novo synthesis of uridine monophosphate (rUMP), needed for RNA and DNA synthesis. Blocks the expansion of activated T cells	Used as a combination in second-line treatment for recurrent sarcoidosis patients. Significant but slight improvement in FVC noted for pulmonary sarcoidosis patients. Good response noted in ~83% of cases with extra-pulmonary sarcoidosis	Fatigue, skin rashes, dyspepsia, alopecia, diarrhoea, mucositis, hypersensitivity pneumonia, and opportunistic infections. Severe side effects include peripheral neuropathy and systemic hypertension (Gerke 2020; Baughman and Grutters 2015; Li et al. 2021; Melani et al. 2021)
5.	Mycophenolate mofetil	Prodrug of mycophenolic acid (MPA). Inhibits inosine-5′-monophosphate dehydrogenase. Suppresses cell-mediated immune responses and antibody formation	Exerts steroid-sparing effect. Has effect on lung function in sarcoidosis patients, especially those who are intolerant to methotrexate. Improves clinical conditions in sarcoidosis patients	Nausea, vomiting, diarrhoea, anaemia, and hyperglycaemia. Severe adverse effects include cytomegalovirus viraemia, hepatitis, other opportunistic infections, and leukopenia (Gerke 2020; Baughman and Grutters 2015; Melani et al. 2021; Hamzeh et al. 2014)
6.	Cyclophosphamide	An alkylating agent which exerts cytotoxic effect by cross-linking of strands of DNA and RNA and by inhibiting protein synthesis	Cytotoxic to both resting and dividing lymphocytes. Suppresses over active immune system and reduces reactive immune cells in sarcoidosis. Reduces inflammation	Fatigue, dyspepsia, alopecia, mucositis, hypersensitivity reactions, and opportunistic infections Extreme side effects include haemorrhagic cystitis, liver

(continued)

Table 8.2 (continued)

Sl. no.	Drug	Description	Effect in sarcoidosis patients	Side effects
				toxicity, bone marrow toxicity, and cytopenia (Gerke 2020; Melani et al. 2021)
7.	Hydroxychloroquine	Inhibits acidification of lysosomal enzymes needed for antigen presentation and TLR2 activation. Inhibits conversion of 25-(OH) vitamin D to 1, 25-(OH) vitamin D, thus blocking hypercalciuria and hypercalcaemia in sarcoidosis. Has immunomodulatory effects	May interfere with antigen presentation. Inhibits TLR signalling and T cell activation. Decreases release of inflammatory cytokines by B cells and T cells Particularly beneficial in cutaneous disease	Skin rashes, skin pigmentation, and gastrointestinal intolerance. Severe adverse effects include retinopathy, neuromyopathy, and cardiomyopathy (Gerke 2020; Li et al. 2021; Melani et al. 2021)

8.15 Future Perspectives

High end omics approaches have improved our existing knowledge of the complexity of non-caseating granulomatous inflammation in sarcoidosis. Nonetheless, our understanding still remains inadequate and incomplete. Future research priorities for sarcoidosis shall include refinement of imaging tools (like CT, PET, and MRI for better diagnosis) and continued quest of biomarkers using omics tools (Kraaijvanger et al. 2020). Omics approaches should be exhaustively used to identify biomarkers of sarcoidosis that can enable rapid differential diagnosis of the disease from other granulomatous diseases like tuberculosis. Identified biomarkers can be explored further for robust, precise subclassification of sarcoidosis (Kaiser et al. 2019). Mechanistic differences between various phenotypes and sub-phenotypes of sarcoidosis need to be elucidated. Development of animal model is of prime importance in improving our mechanistic understanding of the disease pathobiology (Crouser et al. 2017). Besides, multivariate mathematical models based on symptom scores, clinical parameters, cytokine profiles, imaging scores, and cell types can be constructed to enable robust prediction of disease onset, progression, and therapeutic response in clinical settings (Hao et al. 2014). Single cell omics data can also be employed to generate mathematical models using machine learning techniques to enable better prediction of sarcoidosis (Crouser et al. 2017). These new futuristic

Table 8.3 Drugs and biologics that are under clinical investigation for use in Pulmonary Sarcoidosis

Sl. no.	Biologics	Description	Target	Major effects in pulmonary sarcoidosis	Major effects in extra-pulmonary sarcoidosis	Status
1.	Infliximab	Chimeric monoclonal antibody targeting TNF	TNF	Significant improvement in forced vital capacity (FVC)	Causes improvement in refractory neurologic sarcoidosis and osteo-sarcoidosis Resolved cutaneous nodules in case of cutaneous sarcoidosis Long-term therapy beneficial for multi-organ sarcoidosis Did not exert any effect in articular sarcoidosis	Investigational Clinical Trial (Phase III) (Moor et al. 2020; Drent et al. 2021; Gerke 2020; Baughman and Grutters 2015; Melani et al. 2021; Saketkoo and Baughman 2016; Ogbue et al. 2020)
2.	Adalimumab	Anti TNFα monoclonal antibody	TNFα	Was effective in patients with refractory pulmonary sarcoidosis, at high dose and after long-term therapy	Suppressed myocardial infarction in cardiac sarcoidosis, when used in combination with methotrexate Improved lesion area and quality of life in cutaneous sarcoidosis Reduced inflammation in patients with sarcoid uveitis Did not exert any effect in articular sarcoidosis	Investigational Clinical Trial (Phase II) (Drent et al. 2021; Gerke 2020; Baughman and Grutters 2015; Melani et al. 2021; Saketkoo and Baughman 2016; Ogbue et al. 2020)
3.	Rituximab	Humanized monoclonal antibody against CD20	CD20	Improvement in FVC and 6-min walk test in patients with refractory pulmonary sarcoidosis.	Improvements noted in patients with ocular sarcoidosis	Investigational Clinical Trial (Phase II) (Drent et al. 2021; Gerke 2020; Baughman and Grutters

(continued)

Table 8.3 (continued)

Sl. no.	Biologics	Description	Target	Major effects in pulmonary sarcoidosis	Major effects in extra-pulmonary sarcoidosis	Status
						2015; Melani et al. 2021; Saketkoo and Baughman 2016)
4.	Canakinumab (ACZ885)	Humanized monoclonal antibody against IL-1β	IL-1β	Being investigated for pulmonary sarcoidosis	–	Investigational Clinical Trial (Phase II) (Melani et al. 2021)
5.	Abatacept	Fusion protein (Fc region of IgG1 fused to CTLA-4 extracellular domain)	CTLA-4	Being investigated for pulmonary sarcoidosis	–	Investigational Clinical Trial (Phase II) (Gerke 2020; Melani et al. 2021; Saketkoo and Baughman 2016)
6.	Anakinra	IL-1 receptor antagonist	IL-1 activity	–	Being investigated for cardiac sarcoidosis.	Investigational Clinical Trial (Phase II) (Melani et al. 2021)
7.	Atorvastatin	Statin medication	HMG-CoA reductase, IFN-γ	Reduction in risk for flare ups in mild to moderate pulmonary sarcoidosis	–	Investigational Clinical Trial (Phase II) (Melani et al. 2021)
8.	Apremilast	Inhibitor of phosphodiesterase 4 (PDE4)	PDE4	–	Effective in some cutaneous sarcoidosis patients with persistent lesions	Investigational Clinical Trial (Phase III) (Gerke 2020; Baughman and Grutters 2015; Melani et al. 2021)
9.	Pentoxifylline	Phosphodiesterase 4 (PDE4) inhibitor	PDE4	Exerted steroid-sparing effects and slightly reduced instances of flares in pulmonary sarcoidosis	–	Investigational Clinical Trial (Phase II) (Gerke 2020; Baughman and Grutters 2015; Melani et al. 2021)

10.	Roflumilast	Phosphodiesterase 4 (PDE4) inhibitor	PDE4	Reduced flares in chronic fibrotic pulmonary sarcoidosis	–	Investigational Clinical Trial (Phase IV) (Gerke 2020; Melani et al. 2021)
11.	Tofacitinib	Small molecule inhibitor of Janus Kinase (JAK)	JAK	Improvement of respiratory function and reduction in corticosteroid usage was noted in patients with pulmonary sarcoidosis	Improvements noted in case of skin sarcoidosis	Investigational Clinical Trial (Phase I) (Melani et al. 2021)
12.	Vasoactive intestinal peptide (VIP)	28 amino acid peptide belonging to glucagon/secretin superfamily	–	Safe and well tolerated in sarcoidosis patients Exerts immuno-regulatory effect in patients	–	Investigational Clinical Trial (Phase II) (Prasse et al. 2010)

research perspectives may elucidate the elusive etiopathologies of sarcoidosis and thus, aid in improving disease management.

References

Adrianto I, Lin CP, Hale JJ, Levin AM, Datta I, Parker R, Adler A, Kelly JA, Kaufman KM, Lessard CJ et al (2012) Genome-wide association study of African and European Americans implicates multiple shared and ethnic specific loci in sarcoidosis susceptibility. PLoS One 7(8): e43907

Arger NK, O'Connor B, Koth LL (2020) Molecular profiling in sarcoidosis. Curr Opin Pulm Med 26(5):562–567

Arkema EV, Cozier YC (2018) Epidemiology of sarcoidosis: current findings and future directions. Ther Adv Chronic Dis 9(11):227–240

Arkema EV, Cozier YC (2020) Sarcoidosis epidemiology: recent estimates of incidence, prevalence and risk factors. Curr Opin Pulm Med 26(5):527–534

Arkema EV, Grunewald J, Kullberg S, Eklund A, Askling J (2016) Sarcoidosis incidence and prevalence: a nationwide register-based assessment in Sweden. Eur Respir J 48(6):1690–1699

Banoei MM, Iupe I, Bazaz RD, Campos M, Vogel HJ, Winston BW, Mirsaeidi M (2019) Metabolomic and metallomic profile differences between veterans and civilians with pulmonary sarcoidosis. Sci Rep 9(1):19584

Baughman RP, Grutters JC (2015) New treatment strategies for pulmonary sarcoidosis: antimetabolites, biological drugs, and other treatment approaches. Lancet Respir Med 3(10): 813–822

Baughman RP, Sparkman BK, Lower EE (2007) Six-minute walk test and health status assessment in sarcoidosis. Chest 132(1):207–213

Baughman RP, Field S, Costabel U, Crystal RG, Culver DA, Drent M, Judson MA, Wolff G: Sarcoidosis in America. (2016) Analysis based on health care use. Ann Am Thorac Soc 13(8): 1244–1252

Becker A, Vella G, Galata V, Rentz K, Beisswenger C, Herr C, Walter J, Tierling S, Slevogt H, Keller A et al (2019) The composition of the pulmonary microbiota in sarcoidosis – an observational study. Respir Res 20(1):46

Bergantini L, Bianchi F, Cameli P, Mazzei MA, Fui A, Sestini P, Rottoli P, Bargagli E (2019) Prognostic biomarkers of sarcoidosis: a comparative study of serum chitotriosidase, ACE, lysozyme, and KL-6. Dis Markers 2019:8565423

Bhargava M, Viken KJ, Barkes B, Griffin TJ, Gillespie M, Jagtap PD, Sajulga R, Peterson EJ, Dincer HE, Li L et al (2020) Novel protein pathways in development and progression of pulmonary sarcoidosis. Sci Rep 10(1):13282

Bloom CI, Graham CM, Berry MP, Rozakeas F, Redford PS, Wang Y, Xu Z, Wilkinson KA, Wilkinson RJ, Kendrick Y et al (2013) Transcriptional blood signatures distinguish pulmonary tuberculosis, pulmonary sarcoidosis, pneumonias and lung cancers. PLoS One 8(8):e70630

Broos CE, van Nimwegen M, Hoogsteden HC, Hendriks RW, Kool M, van den Blink B (2013) Granuloma formation in pulmonary sarcoidosis. Front Immunol 4:437

Calender A, Weichhart T, Valeyre D, Pacheco Y (2020) Current insights in genetics of sarcoidosis: functional and clinical impacts. J Clin Med 9(8)

Casanova NG, Gonzalez-Garay ML, Sun B, Bime C, Sun X, Knox KS, Crouser ED, Sammani N, Gonzales T, Natt B et al (2020) Differential transcriptomics in sarcoidosis lung and lymph node granulomas with comparisons to pathogen-specific granulomas. Respir Res 21(1):321

Chai Q, Lu Z, Liu Z, Zhong Y, Zhang F, Qiu C, Li B, Wang J, Zhang L, Pang Y et al (2020) Lung gene expression signatures suggest pathogenic links and molecular markers for pulmonary tuberculosis, adenocarcinoma and sarcoidosis. Commun Biol 3(1):604

Chioma OS, Hesse LE, Chapman A, Drake WP (2021) Role of the microbiome in interstitial lung diseases. Front Med (Lausanne) 8:595522

Clarke EL, Lauder AP, Hofstaedter CE, Hwang Y, Fitzgerald AS, Imai I, Biernat W, Rekawiecki B, Majewska H, Dubaniewicz A et al (2018) Microbial Lineages in sarcoidosis. A metagenomic analysis tailored for low-microbial content samples. Am J Respir Crit Care Med 197(2): 225–234

Cooper D, Suau S (2022) Sarcoidosis. Emerg Med Clin North Am 40(1):149–157

Crouser ED, Fingerlin TE, Yang IV, Maier LA, Nana-Sinkam P, Collman RG, Kaminski N (2017) Application of "Omics" and systems biology to sarcoidosis research. Ann Am Thorac Soc 14 (Suppl_6):S445–S451

Davoudi S, Chang VS, Navarro-Gomez D, Stanwyck LK, Sevgi DD, Papavasileiou E, Ren A, Uchiyama E, Sullivan L, Lobo AM et al (2018) Association of genetic variants in RAB23 and ANXA11 with uveitis in sarcoidosis. Mol Vis 24:59–74

Dehghan A (2018) Genome-wide association studies. Methods Mol Biol 1793:37–49

Devalliere J, Charreau B (2011) The adaptor Lnk (SH2B3): an emerging regulator in vascular cells and a link between immune and inflammatory signaling. Biochem Pharmacol 82(10): 1391–1402

Drent M, Crouser ED, Grunewald J (2021) Challenges of sarcoidosis and its management. N Engl J Med 385(11):1018–1032

Du SS, Zhao MM, Zhang Y, Zhang P, Hu Y, Wang LS, Zhou Y, Li QH, Li Y, Du YK et al (2015) Screening for differentially expressed proteins relevant to the differential diagnosis of sarcoidosis and tuberculosis. PLoS One 10(9):e0132466

Duchemann B, Triba MN, Guez D, Rzeznik M, Savarin P, Nunes H, Valeyre D, Bernaudin JF, Le Moyec L (2016) Nuclear magnetic resonance spectroscopic analysis of salivary metabolome in sarcoidosis. Sarcoidosis Vasc Diffuse Lung Dis 33(1):10–16

Duong H, Bonham CA (2018) Sarcoidosis-associated pulmonary hypertension: pathophysiology, diagnosis, and treatment. Clin Pulm Med 25(2):52–60

Futami Y, Takeda Y, Koba T, Narumi R, Nojima Y, Ito M, Nakayama M, Ishida M, Yoshimura H, Naito Y et al (2022) Identification of CD14 and lipopolysaccharide-binding protein as novel biomarkers for sarcoidosis using proteomics of serum extracellular vesicles. Int Immunol 34(6): 327–340

Garman L, Montgomery CG, Rivera NV (2020a) Recent advances in sarcoidosis genomics: epigenetics, gene expression, and gene by environment (G x E) interaction studies. Curr Opin Pulm Med 26(5):544–553

Garman L, Pelikan RC, Rasmussen A, Lareau CA, Savoy KA, Deshmukh US, Bagavant H, Levin AM, Daouk S, Drake WP et al (2020b) Single cell transcriptomics implicate novel monocyte and T cell immune dysregulation in sarcoidosis. Front Immunol 11:567342

Garzoni C, Brugger SD, Qi W, Wasmer S, Cusini A, Dumont P, Gorgievski-Hrisoho M, Muhlemann K, von Garnier C, Hilty M (2013) Microbial communities in the respiratory tract of patients with interstitial lung disease. Thorax 68(12):1150–1156

Geamanu A, Gupta SV, Bauerfeld C, Samavati L (2016) Metabolomics connects aberrant bioenergetic, transmethylation, and gut microbiota in sarcoidosis. Metabolomics 12

Gerke AK (2020) Treatment of sarcoidosis: a multidisciplinary approach. Front Immunol 11: 545413

Giusti C (2020) Sarcoidosis and the alpha chemokine MIG. Clin Ter 171(2):e161–e166

Guerrero CR, Maier LA, Griffin TJ, Higgins L, Najt CP, Perlman DM, Bhargava M (2020) Application of proteomics in sarcoidosis. Am J Respir Cell Mol Biol 63(6):727–738

Hamzeh N, Voelker A, Forssen A, Gottschall EB, Rose C, Mroz P, Maier LA (2014) Efficacy of mycophenolate mofetil in sarcoidosis. Respir Med 108(11):1663–1669

Hao W, Crouser ED, Friedman A (2014) Mathematical model of sarcoidosis. Proc Natl Acad Sci U S A 111(45):16065–16070

Hocevar K, Maver A, Kunej T, Peterlin B (2018) Sarcoidosis related novel candidate genes identified by multi-omics integrative analyses. OMICS 22(5):322–331

Inaoka PT, Shono M, Kamada M, Espinoza JL (2019) Host-microbe interactions in the pathogenesis and clinical course of sarcoidosis. J Biomed Sci 26(1):45
Israel HL, Lenchner G, Steiner RM (1981) Late development of mediastinal calcification in sarcoidosis. Am Rev Respir Dis 124(3):302–305
Jaganathan D, Bohra A, Thudi M, Varshney RK (2020) Fine mapping and gene cloning in the post-NGS era: advances and prospects. Theor Appl Genet 133(5):1791–1810
Jain R, Yadav D, Puranik N, Guleria R, Jin JO (2020) Sarcoidosis: causes, diagnosis, clinical features, and treatments. J Clin Med 9(4)
Kahkouee S, Samadi K, Alai A, Abedini A, Rezaiian L (2016) Serum ACE level in sarcoidosis patients with typical and atypical HRCT manifestation. Pol J Radiol 81:458–461
Kaiser Y, Eklund A, Grunewald J (2019) Moving target: shifting the focus to pulmonary sarcoidosis as an autoimmune spectrum disorder. Eur Respir J 54(1)
Keijsers RGM, Grutters JC (2020) In which patients with sarcoidosis is FDG PET/CT indicated? J Clin Med 9(3)
Kim HS, Choi D, Lim LL, Allada G, Smith JR, Austin CR, Doyle TM, Goodwin KA, Rosenbaum JT, Martin TM (2011) Association of interleukin 23 receptor gene with sarcoidosis. Dis Markers 31(1):17–24
Kiszalkiewicz J, Piotrowski WJ, Pastuszak-Lewandoska D, Gorski P, Antczak A, Gorski W, Domanska-Senderowska D, Migdalska-Sek M, Czarnecka KH, Nawrot E et al (2016) Altered miRNA expression in pulmonary sarcoidosis. BMC Med Genet 17:2
Komada T, Suzuki K, Ishiguchi H, Kawai H, Okumura T, Hirashiki A, Naganawa S (2016) Magnetic resonance imaging of cardiac sarcoidosis: an evaluation of the cardiac segments and layers that exhibit late gadolinium enhancement. Nagoya J Med Sci 78(4):437–446
Konigsberg IR, Maier LA, Yang IV (2021) Epigenetics and sarcoidosis. Eur Respir Rev 30:160
Koth LL, Solberg OD, Peng JC, Bhakta NR, Nguyen CP, Woodruff PG (2011) Sarcoidosis blood transcriptome reflects lung inflammation and overlaps with tuberculosis. Am J Respir Crit Care Med 184(10):1153–1163
Kraaijvanger R, Janssen Bonas M, Vorselaars ADM, Veltkamp M (2020) Biomarkers in the diagnosis and prognosis of sarcoidosis: current use and future prospects. Front Immunol 11:1443
Kriegova E, Melle C, Kolek V, Hutyrova B, Mrazek F, Bleul A, du Bois RM, von Eggeling F, Petrek M (2006) Protein profiles of bronchoalveolar lavage fluid from patients with pulmonary sarcoidosis. Am J Respir Crit Care Med 173(10):1145–1154
Larici AR, Glaudemans AW, Del Ciello A, Slart RH, Calandriello L, Gheysens O (2018) Radiological and nuclear medicine imaging of sarcoidosis. Q J Nucl Med Mol Imaging 62(1):14–33
Levin AM, Iannuzzi MC, Montgomery CG, Trudeau S, Datta I, McKeigue P, Fischer A, Nebel A, Rybicki BA (2013) Association of ANXA11 genetic variation with sarcoidosis in African Americans and European Americans. Genes Immun 14(1):13–18
Li Y, Wollnik B, Pabst S, Lennarz M, Rohmann E, Gillissen A, Vetter H, Grohe C (2006) BTNL2 gene variant and sarcoidosis. Thorax 61(3):273–274
Li Y, Liang Z, Zheng Y, Qiao J, Wang P (2021) Pulmonary sarcoidosis: from clinical features to pathology- narrative review. Ann Palliat Med 10(3):3438–3444
Maertzdorf J, Weiner J 3rd, Mollenkopf HJ, Network TB, Bauer T, Prasse A, Muller-Quernheim J, Kaufmann SH (2012) Common patterns and disease-related signatures in tuberculosis and sarcoidosis. Proc Natl Acad Sci U S A 109(20):7853–7858
Majewski S, Zhou X, Milkowska-Dymanowska J, Bialas AJ, Piotrowski WJ, Malinovschi A (2021) Proteomic profiling of peripheral blood and bronchoalveolar lavage fluid in interstitial lung diseases: an explorative study. ERJ Open Res 7(1)
Martinez-Bravo MJ, Wahlund CJ, Qazi KR, Moulder R, Lukic A, Radmark O, Lahesmaa R, Grunewald J, Eklund A, Gabrielsson S (2017) Pulmonary sarcoidosis is associated with exosomal vitamin D-binding protein and inflammatory molecules. J Allergy Clin Immunol 139(4):1186–1194

Melani AS, Bigliazzi C, Cimmino FA, Bergantini L, Bargagli E (2021) A comprehensive review of sarcoidosis treatment for pulmonologists. Pulm Ther 7(2):325–344

Millward K, Fiddler CA, Thillai M (2021) Update on sarcoidosis guidelines. Curr Opin Pulm Med 27(5):484–489

Mirsaeidi M, Banoei MM, Nienow CK, Abassi T, Hakim A, Schraufnagel D, Winston BW, Sweiss N, Baughman R, Garcia JG et al (2016) Plasma metabolomic profile in fibrosing pulmonary sarcoidosis. Sarcoidosis Vasc Diffuse Lung Dis 33(1):29–38

Moller DR, Koth LL, Maier LA, Morris A, Drake W, Rossman M, Leader JK, Collman RG, Hamzeh N, Sweiss NJ et al (2015) Rationale and design of the genomic research in alpha-1 antitrypsin deficiency and sarcoidosis (GRADS) study. Sarcoidosis protocol. Ann Am Thorac Soc 12(10):1561–1571

Moor CC, Kahlmann V, Culver DA, Wijsenbeek MS (2020) Comprehensive care for patients with sarcoidosis. J Clin Med 9(2)

Nunes H, Uzunhan Y, Gille T, Lamberto C, Valeyre D, Brillet PY (2012) Imaging of sarcoidosis of the airways and lung parenchyma and correlation with lung function. Eur Respir J 40(3): 750–765

Ogbue OD, Malhotra P, Akku R, Jayaprakash TP, Khan S (2020) Biologic therapies in sarcoidosis and uveitis: a review. Cureus 12(7):e9057

Prasse A, Zissel G, Lutzen N, Schupp J, Schmiedlin R, Gonzalez-Rey E, Rensing-Ehl A, Bacher G, Cavalli V, Bevec D et al (2010) Inhaled vasoactive intestinal peptide exerts immunoregulatory effects in sarcoidosis. Am J Respir Crit Care Med 182(4):540–548

Rosen Y (2007) Pathology of sarcoidosis. Semin Respir Crit Care Med 28(1):36–52

Saketkoo LA, Baughman RP (2016) Biologic therapies in the treatment of sarcoidosis. Expert Rev Clin Immunol 12(8):817–825

Scher JU, Joshua V, Artacho A, Abdollahi-Roodsaz S, Ockinger J, Kullberg S, Skold M, Eklund A, Grunewald J, Clemente JC et al (2016) The lung microbiota in early rheumatoid arthritis and autoimmunity. Microbiome 4(1):60

Schupp JC, Vukmirovic M, Kaminski N, Prasse A (2017) Transcriptome profiles in sarcoidosis and their potential role in disease prediction. Curr Opin Pulm Med 23(5):487–492

Shorr AF, Torrington KG, Hnatiuk OW (2001) Endobronchial biopsy for sarcoidosis: a prospective study. Chest 120(1):109–114

Silva E, Souchelnytskyi S, Kasuga K, Eklund A, Grunewald J, Wheelock AM (2013) Quantitative intact proteomics investigations of alveolar macrophages in sarcoidosis. Eur Respir J 41(6): 1331–1339

Su R, Li MM, Bhakta NR, Solberg OD, Darnell EP, Ramstein J, Garudadri S, Ho M, Woodruff PG, Koth LL (2014) Longitudinal analysis of sarcoidosis blood transcriptomic signatures and disease outcomes. Eur Respir J 44(4):985–993

Talreja J, Farshi P, Alazizi A, Luca F, Pique-Regi R, Samavati L (2017) RNA-sequencing Identifies novel pathways in sarcoidosis monocytes. Sci Rep 7(1):2720

Tana C, Donatiello I, Coppola MG, Ricci F, Maccarone MT, Ciarambino T, Cipollone F, Giamberardino MA (2020) CT findings in pulmonary and abdominal sarcoidosis. Implications for diagnosis and classification. J Clin Med 9(9)

Tanizawa K, Handa T, Nagai S, Sato H, Yamada R, Ito I, Kubo T, Ito Y, Watanabe K, Aihara K et al (2013) Interferon regulatory factor 5 polymorphisms in sarcoidosis. Mod Rheumatol 23(6): 1158–1165

Tazyeen S, Ahmed MM, Farooqui A, Alam A, Malik MZ, Saeed M, Ahmad I, Abohashrh M, Singh RKB, Ishrat R (2022) Identification of key regulators in Sarcoidosis through multidimensional systems biological approach. Sci Rep 12(1):1236

Ungprasert P, Ryu JH, Matteson EL (2019) Clinical manifestations, diagnosis, and treatment of sarcoidosis. Mayo Clin Proc Innov Qual Outcomes 3(3):358–375

Valentonyte R, Hampe J, Huse K, Rosenstiel P, Albrecht M, Stenzel A, Nagy M, Gaede KI, Franke A, Haesler R et al (2005) Sarcoidosis is associated with a truncating splice site mutation in BTNL2. Nat Genet 37(4):357–364

Wang W, Ma Y, Zhang Y, Lin J, He J, Nong Y, Zhang X, Jia Y (2022) Diagnostic and staging value of serum angiotensin-converting enzyme in sarcoidosis. Comput Math Methods Med 2022: 4657502

Wolin A, Lahtela EL, Anttila V, Petrek M, Grunewald J, van Moorsel CHM, Eklund A, Grutters JC, Kolek V, Mrazek F et al (2017) SNP variants in major histocompatibility complex are associated with sarcoidosis susceptibility-a joint analysis in four European populations. Front Immunol 8: 422

Wu CH, Chung PI, Wu CY, Chen YT, Chiu YW, Chang YT, Liu HN (2017) Comorbid autoimmune diseases in patients with sarcoidosis: a nationwide case-control study in Taiwan. J Dermatol 44(4):423–430

Yang IV, Konigsberg I, MacPhail K, Li L, Davidson EJ, Mroz PM, Hamzeh N, Gillespie M, Silveira LJ, Fingerlin TE et al (2019) DNA methylation changes in lung immune cells are associated with granulomatous lung disease. Am J Respir Cell Mol Biol 60(1):96–105

Zhou T, Zhang W, Sweiss NJ, Chen ES, Moller DR, Knox KS, Ma SF, Wade MS, Noth I, Machado RF et al (2012) Peripheral blood gene expression as a novel genomic biomarker in complicated sarcoidosis. PLoS One 7(9):e44818

Zhou T, Casanova N, Pouladi N, Wang T, Lussier Y, Knox KS, Garcia JGN (2017) Identification of Jak-STAT signaling involvement in sarcoidosis severity via a novel microRNA-regulated peripheral blood mononuclear cell gene signature. Sci Rep 7(1):4237

9 Pulmonary Vascular Diseases

Abstract

Pulmonary vascular diseases or pulmonary vasculopathies represent diseases of pulmonary vasculature that are associated with altered pulmonary vascular resistance and compromised pulmonary circulation. The etiological spectrum of pulmonary vasculopathy is extremely complex. Symptomatic presentation often makes disease diagnosis difficult. Primarily, imaging techniques including CT scanning, CT angiography, pulmonary angiography, and right heart catheterization are used for disease diagnosis along with a host of investigations to ascertain the aetiology at times. According to the nature of the pathology, different drugs such as anticoagulants, pulmonary vasodilators, diuretics, and inotropes are used for therapy. Manifold challenges are encountered in disease diagnosis and streamlining therapy that often fail to ensure cure or even good relief. The dearth of a proper disease biomarker is often a cause of concern. Modernized omics tools, should therefore, be used to overcome these challenges, by mediating deep phenotyping and discovery of non-invasive biomarkers for pulmonary vascular diseases.

Keywords

Pulmonary vascular diseases (PVD) · Vasculature · Pulmonary vascular resistance (PVR) · Pulmonary hypertension · Pulmonary oedema · Pulmonary embolism · Pulmonary arteriovenous malformations · CT angiography

9.1 Introduction

Pulmonary vascular diseases (PVD) encompass a complex group of disease conditions in which blood vessels within the lungs are affected, leading to impaired pulmonary circulation. Diseases of the pulmonary vascular system, also referred to

S. Saha et al., *Pulmonomics: Omics Approaches for Understanding Pulmonary Diseases*, https://doi.org/10.1007/978-981-99-3505-5_9

as pulmonary vasculopathies may be congenital, secondary, or even idiopathic (https://www.ncbi.nlm.nih.gov/books/NBK553875/; Marini et al. 2018). Pulmonary vasculopathy includes inflammation of the vessels (vasculitis), endothelial dysfunction, hypertrophy of the vascular walls, destruction of the vascularity, increase in the pulmonary arterial pressure (pulmonary hypertension) with or without concomitant left heart disease (Groshong et al. 2008). The increase in pulmonary vascular resistance (PVR) impairs alveolar gaseous exchange and results in abnormalities in arterial blood gas profiles (Melot and Naeije 2011). Due to the progressive nature of PVDs, these diseases can be life threatening with high morbidity and mortality rates (Lau et al. 2011; Khan et al. 2022; Russo et al. 2022; Chamarthy et al. 2018). The prevalence of PVDs is noted to be higher in developing countries as compared to developed countries across the globe (Butrous et al. 2008). However, the overall global burden of PVDs remains underestimated owing to lack of awareness and precise early-stage diagnosis.

9.2 Aetiology and Clinical features

The pulmonary vascular system is primarily affected in pulmonary vascular diseases (PVD). Structurally the pulmonary vascular system is different from systemic vasculature with a huge metabolically active endothelial system (Huertas et al. 2018; Ai et al. 2021). This system is distinct as it acts as a conduit between the right and left heart and functions primarily to shed carbon dioxide and upload oxygen onto haemoglobin at the alveolo-capillary level of transit (https://www.ncbi.nlm.nih.gov/books/NBK539907/). Thus, it maintains the continuous supply of oxygen to the system for metabolism. The system has several peculiarities. It is a low-pressure system with widest capillary network of any other organ system in the body; the pulmonary artery carries the same volume of deoxygenated blood (the cardiac output) from the right heart to lungs alone that is usually meant to indicate the total ejected blood volume from the left heart to the rest of the body (Townsley 2012; https://www.ncbi.nlm.nih.gov/books/NBK525948/; https://www.ncbi.nlm.nih.gov/books/NBK518997/). As the whole circulation passes through the lungs constantly, the endothelium and the vasculature as a whole are subject to exposure and lodgement of several metabolic products and other circulating substances including toxins and infection. While working as filter, the pulmonary circulatory system and especially the capillaries are often burdened with deposition of many such unwarranted substances that can lead to different kinds of pathologies (Huertas et al. 2018; Suresh and Shimoda 2016).

Past history of anomalous pulmonary venous connection, blood clotting disorders, increasing age, and obesity often serve as risk factors for pulmonary vascular diseases (Loomba et al. 2016). Pulmonary vascular diseases may also be triggered by congenital or acquired defects or may be due to some apparently unknown (idiopathic) reasons (Diller and Gatzoulis 2007; Dexter 1979; Firth et al. 2010). Different lung diseases can also compromise the pulmonary vasculature structurally or functionally (Ai et al. 2020; Hopkins and McLoughlin 2002; Chen et al. 2016). This may change the relationship of ventilation (V) with pulmonary perfusion (Q) and such V-Q mismatch can lead to hypoxemia or low oxygen content in arterial blood. Change in pulmonary vascular resistance (PVR) is also observed in case of pulmonary vascular diseases (Melot and Naeije 2011). Pulmonary vascular resistance provides the measure of the resistance encountered during blood flow from pulmonary artery to the left atrium of the heart (https://www.ncbi.nlm.nih.gov/books/NBK554380/). Increase in PVR is noted in case of pulmonary arterial hypertension (PAH), chronic thromboembolic pulmonary hypertension (CTEPH), pulmonary embolism, and host of other causes. Reduction in PVR is observed in patients with pulmonary arteriovenous malformations having hereditary haemorrhagic telangiectasia. An increased physiologic dead space is often noted in pulmonary vascular diseases with raised PVR (Melot and Naeije 2011). Protracted hypoxemia can cause increased vasomotor tone in pulmonary arterioles and cause eventual remodelling (Voelkel and Tuder 2000). Similarly, for a problem in the venous drainage to the left heart, there may be increased back-pressure build up in the venules and capillaries that ultimately jeopardize the normal pulmonary vascular function (Egbe et al. 2018). Pulmonary vasculature also becomes involved as a target in different systemic disease alone or in addition to concomitant lung affections. They can cause inflammation of the pulmonary arterioles and capillaries (pulmonary vasculitis) (Brown 2006). The vascular system can also be affected by malignancies, especially angiosarcoma (Grafino et al. 2016).

Whatever be the nature and cause of the insult, the pulmonary vascular pathologies exhibit a wide range of symptomatic presentations including haemorrhage, pulmonary vascular remodelling causing pulmonary hypertension, embolism, pulmonary oedema, and pulmonary arteriovenous malformations. They often result in secondary parenchymal changes. Physiologically, they produce hypoxemia initially on exertion and subsequently at rest (Melot and Naeije 2011). Clinically it translates to progressive shortness of breath, chest pain, syncope, oedema, and features of right heart failure that ensues when the relatively thin myocardium fails to cope up with increased pulmonary pressure head (Melot and Naeije 2011; Kalva 2018; Tarbox and Swaroop 2013).

9.3 Different Pulmonary Vascular Diseases

Commonly recognized pulmonary vascular diseases are enumerated in Table 9.1 that provides basic information about them.

Pulmonary embolism is often under-diagnosed especially in cases of small emboli or due to the presence of atypical symptoms (Morrone and Morrone 2018). On the other hand, pulmonary oedema is not a disease in itself. However, it may lead to further complications and hence can be life threatening in patients with lung and heart diseases (West 2013). The pathophysiology of pulmonary hypertension is shrouded in mystery. Recently, several omics-based studies have been conducted to improve our understanding of the disease and to identify non-invasive biomarkers for better disease diagnosis (Anwar et al. 2016). Pulmonary arteriovenous

Table 9.1 Commonly recognized pulmonary vascular diseases

Pulmonary vascular diseases	Basic features
Pulmonary embolism	Potentially fatal condition with a high mortality rate (Assouline et al. 2022) Disruption of blood flow into pulmonary artery or its branches occurs due to thrombotic emboli (blood clots) or nonthrombotic emboli (like fat, air, amniotic fluid, tumour cells) (https://www.ncbi.nlm.nih.gov/books/NBK560551/; West 2013; Kahn and de Wit 2022) Pulmonary thrombotic emboli arise from detached portions of deep venous thrombi (Tarbox and Swaroop 2013) Impairs gaseous exchange (Tarbox and Swaroop 2013) May lead to pulmonary infarction (Tarbox and Swaroop 2013)
Pulmonary oedema	Also known as pulmonary congestion Abnormal accumulation of fluid in extravascular tissues and alveolar air spaces in the lungs is observed (West 2013) Can be cardiogenic and non-cardiogenic Leads to reduced alveolar gaseous exchange and eventually causes respiratory failure (https://www.ncbi.nlm.nih.gov/books/NBK557611/) It is a life-threatening disorder (Chioncel et al. 2015)
Pulmonary hypertension	Characterized by elevated pulmonary vascular resistance and narrowing of blood vessels within the pulmonary vasculature (https://www.ncbi.nlm.nih.gov/books/NBK482463/) Patients with pulmonary hypertension have increased pulmonary arterial pressure (West 2013; Galie et al. 2016) Symptoms like angina (chest pain), dyspnoea on exertion, palpitations, fatigue, syncope, and lower extremity oedema are common (McLaughlin et al. 2009)
Pulmonary arteriovenous malformations	Rare condition and mostly congenital (https://www.ncbi.nlm.nih.gov/books/NBK559289/) Usually observed in case of hereditary haemorrhagic telangiectasia (HHT) (Chamarthy et al. 2018; Majumdar and McWilliams 2020) Structurally abnormal and direct capillary free communication between pulmonary arteries and pulmonary veins (Majumdar and McWilliams 2020) Results in an anatomical right to left shunt (Shovlin 2014) Leads to impairment of gaseous exchange and natural filtration (mediated by pulmonary capillary bed) in the lungs (Majumdar and McWilliams 2020; Shovlin 2014) Common symptoms include dyspnoea, finger clubbing, cyanosis, and chest pain (Majumdar and McWilliams 2020)

This table enlists the different pulmonary vascular diseases along with their basic features

malformations being rare are often under-studied. Diagnosis and monitoring of the disease require more attention to overcome the challenges associated with disease detection and therapy (Majumdar and McWilliams 2020).

9.4 Diagnosis of Pulmonary Vascular Diseases

Imaging modalities like chest radiograph, computed tomography (CT) scanning (with or without contrast), and magnetic resonance imaging (MRI) are commonly used for the diagnosis of pulmonary vascular diseases. Doppler echocardiography aids in disease assessment by complimenting the findings from radiological and invasive studies (Roberts and Forfia 2011). It acts as a good screening tool. Transpulmonary pressure gradient and diastolic pressure gradient also find application in disease diagnosis especially for pulmonary hypertension (Naeije et al. 2013). Ventilation-perfusion (V/Q) scan is used to assess airflow and blood flow distribution in the lungs. They are especially useful in diagnosing emboli and clots in patients with pulmonary embolism (https://www.ncbi.nlm.nih.gov/books/NBK564428/). Fluorodeoxyglucose positron emission tomography (FDG-PET) has recently found application in studying large vessel vasculitis (https://www.ncbi.nlm.nih.gov/books/NBK553875/). However, CT scan continues to be the primary tool for suspecting a diagnosis of PVD and studying pulmonary vasculature. Thin section multi-detector CT angiography and magnetic resonance angiography provide high-resolution images of pulmonary vasculature. These techniques aid in disease localization and post-therapeutic assessment (Kalva 2018). Novel state of art techniques like dual scanner and dual energy CT scanners hold promise for improved diagnosis and surveillance of patients with pulmonary vascular diseases (Vlahos et al. 2022). Measurement of pulmonary vascular pressure and pulmonary vascular resistance is solely dependent on invasive imaging tests. Right heart catheterization and pulmonary angiography provide functional and anatomical details of pulmonary circulation in patients with pulmonary vascular diseases (Kalva 2018).

9.5 Treatment of Pulmonary Vascular Diseases

The complex therapeutic strategy for pulmonary vascular diseases includes an initial risk evaluation, preliminary medication and further therapeutic interventions, prolonged follow-ups, and requisite adjustments of initial treatment (Galie et al. 2021). However, till date, there is no complete cure to pulmonary vasculopathies. The therapeutic regime and medications used for pulmonary vascular diseases are largely guided by the underlying causes. Anticoagulants and thrombolytic (or clot-busting) medication are the mainstay therapies for pulmonary embolism (Becattini et al. 2022). Oxygen supplementation and surgical pulmonary embolectomy are also explored in severe cases of acute pulmonary embolism (Belohlavek et al. 2013; Iaccarino et al. 2018). Pulmonary oedema is managed by nitrates, diuretics, morphine, inotropes and with ventilatory support in extreme cases (Purvey and Allen 2017). Calcium channel blockers, endothelin receptor antagonists, phosphodiesterase 5 inhibitors, prostacyclin analogues, prostacyclin receptor agonist, soluble guanylate cyclase stimulators, anticoagulants, diuretics, inotropic agents (like digoxin) are often prescribed to patients with pulmonary hypertension. Besides these, supervised rehabilitation programmes and interventional therapies (like balloon atrial septostomy, Potts shunt, and pulmonary artery denervation) are also performed for pulmonary hypertension patients (Humbert et al. 2022). Pulmonary arteriovenous malformations are classically treated by surgical interventions employing endovascular techniques or transcatheter endovascular treatment (Chamarthy et al. 2018). Pulmonary arteriovenous malformations patients with diffuse lesions are often subjected to lung transplantation (Meek et al. 2011). The same remains an option for other PVDs in advanced stage (Reitz et al. 1982; Mendeloff et al. 2002).

9.6 Challenges and Future Research Avenues for Pulmonary Vascular Diseases

Majority of the studies have focused on understanding the aetiology and pathophysiology of the different pulmonary vascular diseases. However, there is still a dearth of non-invasive biomarkers and phenotypic stratification of pulmonary vascular diseases. Omics and imaging-based deep phenotyping of pulmonary vascular diseases is of prime importance in streamlining treatment (Oldham et al. 2021). Pulmonary vascular diseases may remain asymptomatic or have symptoms that overlap with other respiratory diseases which can make the diagnosis difficult (Vachiery and Gaine 2012; Saygin and Domsic 2019). Most of the present reliable, confirmatory diagnostic strategies depend on invasive imaging or measurement techniques. Besides, poor awareness and ignorance amongst the concerned population often lead to delayed diagnosis of disease and treatment (Mocumbi et al. 2020). Identification of non-invasive biomarkers using multi-omics and biological network-based approaches can aid in overcoming the use of invasive diagnostic tools and mediate timely diagnosis. Advancements in the field of non-invasive imaging modalities can also be explored as an alternative diagnostic approach. Presently, there is no complete cure to pulmonary vascular diseases. Multi-omics studies can aid in identifying crucial biomolecules which may be explored as novel drug targets for pulmonary vascular diseases. Besides, biomarkers identified by omics tools can also aid in guiding and monitoring therapeutic efficacy (McMahon and Bryan 2017; Odler et al. 2018). Interventional therapies and lung transplantation are often considered for therapeutic purposes. However, these surgical interventions are often risky and can lead to several complications. Besides, they require high-level expertise that is often not readily available. Very few selected centres are available for such complicated procedures, especially in developing countries (Hasan et al. 2020). Screening of predictive biomarkers of pulmonary vascular diseases may aid in risk identification and pave the way for preventive medications to slow down disease progression and minimize disease severity at a later stage (Goss et al. 2017). Lastly, the high mortality associated with pulmonary vascular diseases is a growing cause of concern. Therefore, screening and searching diagnostic, predictive, and therapeutic biomarkers are currently of paramount importance in mediating in better disease diagnosis and management.

9.7 Conclusion

Unveiling the pathophysiological mechanisms and aetiologies of pulmonary vascular diseases is of great significance in overcoming the challenges related to pulmonary vasculopathies. New age omics technologies can aid in improving our mechanistic understanding of disease biology alongside efficient characterization of patients having pulmonary vascular diseases (Hemnes 2018). Multi-omics data integration and analyses may serve as a window for endo-phenotyping of individuals at higher risk for developing pulmonary vascular diseases (Leopold and Hemnes 2021). It shall also enable identification of non-invasive molecular markers that could be efficiently used for disease diagnosis, for monitoring disease progression and therapeutic efficacy without invasive interventions (Santos-Gomes et al. 2022). Such omics-based studies shall also pave the way for identification of novel therapeutic targets that could be explored for improving clinical care and treatment regime (Brittain and Chan 2016). Present omics-based studies focusing on pulmonary vasculopathies are in their preliminary stage and are mostly based on a single omics platform. However, coupling of clinical data, imaging data, and integrative omics data should be extensively performed for pulmonary vascular diseases for better diagnosis and control. Omics-based investigation of pulmonary hypertension has been discussed in details in the subsequent chapter in this book.

References

Ai J, Hong W, Wu M, Wei X (2021) Pulmonary vascular system: a vulnerable target for COVID-19. MedComm (2020) 2(4):531–547

Anwar A, Ruffenach G, Mahajan A, Eghbali M, Umar S (2016) Novel biomarkers for pulmonary arterial hypertension. Respir Res 17(1):88

Assouline B, Assouline-Reinmann M, Giraud R, Levy D, Saura O, Bendjelid K, Combes A, Schmidt M (2022) Management of high-risk pulmonary embolism: what is the place of extracorporeal membrane oxygenation? J Clin Med 11(16)

Becattini C, Kokorin VA, Lesniak W, Marin-Leon I, Medrano FJ, Morbidoni L, Marra AM, Biskup E, Riera-Mestre A, Dicker D (2022) Pulmonary embolism – an EFIM guideline critical appraisal and adaptation for practicing clinicians. Eur J Intern Med 96:5–12

Belohlavek J, Dytrych V, Linhart A (2013) Pulmonary embolism, part II: management. Exp Clin Cardiol 18(2):139–147

Brittain EL, Chan SY (2016) Integration of complex data sources to provide biologic insight into pulmonary vascular disease (2015 Grover Conference Series). Pulm Circ 6(3):251–260

Brown KK (2006) Pulmonary vasculitis. Proc Am Thorac Soc 3(1):48–57

Butrous G, Ghofrani HA, Grimminger F (2008) Pulmonary vascular disease in the developing world. Circulation 118(17):1758–1766

Chamarthy MR, Park H, Sutphin P, Kumar G, Lamus D, Saboo S, Anderson M, Kalva SP (2018) Pulmonary arteriovenous malformations: endovascular therapy. Cardiovasc Diagn Ther 8(3): 338–349

Chen Y, Liu C, Lu W, Li M, Hadadi C, Wang EW, Yang K, Lai N, Huang J, Li S et al (2016) Clinical characteristics and risk factors of pulmonary hypertension associated with chronic respiratory diseases: a retrospective study. J Thorac Dis 8(3):350–358

Chioncel O, Collins SP, Ambrosy AP, Gheorghiade M, Filippatos G (2015) Pulmonary oedema-therapeutic targets. Card Fail Rev 1(1):38–45

Dexter L (1979) Pulmonary vascular disease in acquired and congenital heart disease. Arch Intern Med 139(8):922–928

Diller GP, Gatzoulis MA (2007) Pulmonary vascular disease in adults with congenital heart disease. Circulation 115(8):1039–1050

Egbe AC, Reddy YNV, Khan AR, Al-Otaibi M, Akintoye E, Obokata M, Borlaug BA (2018) Venous congestion and pulmonary vascular function in Fontan circulation: implications for prognosis and treatment. Int J Cardiol 271:312–316

Firth AL, Mandel J, Yuan JX (2010) Idiopathic pulmonary arterial hypertension. Dis Model Mech 3(5–6):268–273

Galie N, Humbert M, Vachiery JL, Gibbs S, Lang I, Torbicki A, Simonneau G, Peacock A, Vonk Noordegraaf A, Beghetti M et al (2016) 2015 ESC/ERS Guidelines for the diagnosis and treatment of pulmonary hypertension: The Joint Task Force for the Diagnosis and Treatment of Pulmonary Hypertension of the European Society of Cardiology (ESC) and the European Respiratory Society (ERS): Endorsed by: Association for European Paediatric and Congenital Cardiology (AEPC), International Society for Heart and Lung Transplantation (ISHLT). Eur Heart J 37(1):67–119

Galie N, Dardi F, Manes A, Palazzini M (2021) Progress in the treatment of pulmonary vascular diseases. G Ital Cardiol (Rome) 22(10):781–792

Goss KN, Everett AD, Mourani PM, Baker CD, Abman SH (2017) Addressing the challenges of phenotyping pediatric pulmonary vascular disease. Pulm Circ 7(1):7–19

Grafino M, Alves P, de Almeida MM, Garrido P, Hasmucrai D, Teixeira E, Sotto-Mayor R (2016) Angiosarcoma of the lung. J Bras Pneumol 42(1):68–70

Groshong SD, Tomashefski JF Jr, Cool CD (2008) Pulmonary vascular disease. Dail Hammar's Pulmonary Pathol:1032–1087

Hasan B, Hansmann G, Budts W, Heath A, Hoodbhoy Z, Jing ZC, Koestenberger M, Meinel K, Mocumbi AO, Radchenko GD et al (2020) Challenges and special aspects of pulmonary hypertension in middle- to low-income regions: JACC state-of-the-art review. J Am Coll Cardiol 75(19):2463–2477

Hemnes AR (2018) Using omics to understand and treat pulmonary vascular disease. Front Med (Lausanne) 5:157

Hopkins N, McLoughlin P (2002) The structural basis of pulmonary hypertension in chronic lung disease: remodelling, rarefaction or angiogenesis? J Anat 201(4):335–348

Huertas A, Guignabert C, Barbera JA, Bartsch P, Bhattacharya J, Bhattacharya S, Bonsignore MR, Dewachter L, Dinh-Xuan AT, Dorfmuller P et al (2018) Pulmonary vascular endothelium: the orchestra conductor in respiratory diseases: highlights from basic research to therapy. Eur Respir J 51(4)

Humbert M, Kovacs G, Hoeper MM, Badagliacca R, Berger RMF, Brida M, Carlsen J, Coats AJS, Escribano-Subias P, Ferrari P et al (2022) 2022 ESC/ERS Guidelines for the diagnosis and treatment of pulmonary hypertension. Eur Respir J

Iaccarino A, Frati G, Schirone L, Saade W, Iovine E, D'Abramo M, De Bellis A, Sciarretta S, Greco E (2018) Surgical embolectomy for acute massive pulmonary embolism: state of the art. J Thorac Dis 10(8):5154–5161

Kahn SR, de Wit K (2022) Pulmonary embolism. N Engl J Med 387(1):45–57

Kalva SP (2018) Pulmonary vascular disease: diagnosis and endovascular therapy. Cardiovasc Diagn Ther 8(3):199–200

Khan NA, Alharbi AF, Alshehri AQ, Attieh AI, Farouk HH, Alshammri HH, Alqahtani HA, Alassaf MF, Alrejaye MS, Aljthalin RA et al (2022) Early diagnosis of pulmonary embolism related to clinical presentation and vital signs in the emergency department at King Saud Medical City. Cureus 14(7):e27087

Lau EM, Manes A, Celermajer DS, Galie N (2011) Early detection of pulmonary vascular disease in pulmonary arterial hypertension: time to move forward. Eur Heart J 32(20):2489–2498

Leopold JA, Hemnes AR (2021) Integrative omics to characterize and classify pulmonary vascular disease. Clin Chest Med 42(1):195–205

Loomba RS, Aggarwal S, Arora RR, Anderson R (2016) Bodily isomerism is an independent risk factor for pulmonary hypertension in adults with congenital heart disease. Ther Adv Respir Dis 10(3):194–199

Majumdar S, McWilliams JP (2020) Approach to pulmonary arteriovenous malformations: a comprehensive update. J Clin Med 9(6)

Marini TJ, He K, Hobbs SK, Kaproth-Joslin K (2018) Pictorial review of the pulmonary vasculature: from arteries to veins. Insights Imaging 9(6):971–987

McLaughlin VV, Archer SL, Badesch DB, Barst RJ, Farber HW, Lindner JR, Mathier MA, McGoon MD, Park MH, Rosenson RS et al (2009) ACCF/AHA 2009 expert consensus document on pulmonary hypertension a report of the American College of Cardiology Foundation Task Force on Expert Consensus Documents and the American Heart Association developed in collaboration with the American College of Chest Physicians; American Thoracic Society, Inc.; and the Pulmonary Hypertension Association. J Am Coll Cardiol 53(17): 1573–1619

McMahon TJ, Bryan NS (2017) Biomarkers in pulmonary vascular disease: gauging response to therapy. Am J Cardiol 120(8S):S89–S95

Meek ME, Meek JC, Beheshti MV (2011) Management of pulmonary arteriovenous malformations. Semin Intervent Radiol 28(1):24–31

Melot C, Naeije R (2011) Pulmonary vascular diseases. Compr Physiol 1(2):593–619

Mendeloff EN, Meyers BF, Sundt TM, Guthrie TJ, Sweet SC, de la Morena M, Shapiro S, Balzer DT, Trulock EP, Lynch JP et al (2002) Lung transplantation for pulmonary vascular disease. Ann Thorac Surg 73(1):209–217. discussion 217–209

Mocumbi A, Mastala A, Guambe P, Dzudie A (2020) Pulmonary vascular disease in Africa: Lessons from registries. Glob Cardiol Sci Pract 2020(1):e202002

Morrone D, Morrone V (2018) Acute pulmonary embolism: focus on the clinical picture. Korean Circ J 48(5):365–381

Naeije R, Vachiery JL, Yerly P, Vanderpool R (2013) The transpulmonary pressure gradient for the diagnosis of pulmonary vascular disease. Eur Respir J 41(1):217–223

Odler B, Foris V, Gungl A, Muller V, Hassoun PM, Kwapiszewska G, Olschewski H, Kovacs G (2018) Biomarkers for pulmonary vascular remodeling in systemic sclerosis: a pathophysiological approach. Front Physiol 9:587

Oldham WM, Hemnes AR, Aldred MA, Barnard J, Brittain EL, Chan SY, Cheng F, Cho MH, Desai AA, Garcia JGN et al (2021) NHLBI-CMREF workshop report on pulmonary vascular disease classification: JACC state-of-the-art review. J Am Coll Cardiol 77(16):2040–2052

Purvey M, Allen G (2017) Managing acute pulmonary oedema. Aust Prescr 40(2):59–63

Reitz BA, Wallwork JL, Hunt SA, Pennock JL, Billingham ME, Oyer PE, Stinson EB, Shumway NE (1982) Heart-lung transplantation: successful therapy for patients with pulmonary vascular disease. N Engl J Med 306(10):557–564

Roberts JD, Forfia PR (2011) Diagnosis and assessment of pulmonary vascular disease by Doppler echocardiography. Pulm Circ 1(2):160–181

Russo F, Turco A, Adamo M, D'Ascenzo F, Iannaccone M, De Benedictis M, Cianfanelli L, Branca L, Rolandi A, Ferlini M et al (2022) Percutaneous treatment of acute pulmonary embolism: state of the art. G Ital Cardiol (Rome) 23(7):13–21

Santos-Gomes J, Gandra I, Adao R, Perros F, Bras-Silva C (2022) An overview of circulating pulmonary arterial hypertension biomarkers. Front Cardiovasc Med 9:924873

Saygin D, Domsic RT (2019) Pulmonary arterial hypertension in systemic sclerosis: challenges in diagnosis, screening and treatment. Open Access Rheumatol 11:323–333

Shovlin CL (2014) Pulmonary arteriovenous malformations. Am J Respir Crit Care Med 190(11): 1217–1228

Suresh K, Shimoda LA (2016) Lung circulation. Compr Physiol 6(2):897–943

Tarbox AK, Swaroop M (2013) Pulmonary embolism. Int J Crit Illn Inj Sci 3(1):69–72

Townsley MI (2012) Structure and composition of pulmonary arteries, capillaries, and veins. Compr Physiol 2(1):675–709

Vachiery JL, Gaine S (2012) Challenges in the diagnosis and treatment of pulmonary arterial hypertension. Eur Respir Rev 21(126):313–320

Vlahos I, Jacobsen MC, Godoy MC, Stefanidis K, Layman RR (2022) Dual-energy CT in pulmonary vascular disease. Br J Radiol 95(1129):20210699

Voelkel NF, Tuder RM (2000) Hypoxia-induced pulmonary vascular remodeling: a model for what human disease? J Clin Invest 106(6):733–738

West JB: Pulmonary pathophysiology : the essentials; 2013

Pulmonary Hypertension

10

Abstract

The progressive and devastating disorder of pulmonary hypertension is associated with vascular remodelling and vascular rarefaction. This escalates pulmonary arterial pressures and pulmonary vascular resistance and causes untimely death. Right heart catheterization is currently the gold standard for confirmed disease diagnosis. Recent omics tools have been elaborately utilized to identify non-invasive biomarkers for overcoming the use of present invasive diagnostic tools. Integration of multi-omics data and network analyses have also been conducted for identification of novel drug targets. The current medications used primarily aim to restrict disease progression and severity. This chapter also discusses the different biologics and drugs that are being clinically investigated for pulmonary hypertension. Finally, future research directions for pulmonary hypertension have been described in the end.

Keywords

Pulmonary vasculature · Vascular remodelling · Pulmonary arterial pressure (PAP) · Pulmonary vascular resistance (PVR) · Right ventricular (RV) dysfunction · Right heart catheterization · BMPR2 · NT-proBNP

10.1 Introduction

Pulmonary hypertension is a disease of pulmonary vasculature that is characterized by increased pulmonary arterial pressure (PAP) and pulmonary vascular resistance (PVR). Pulmonary hypertension patients have a resting mean PAP value ≥25 mmHg, as detected by right heart catheterization (Thenappan et al. 2018; Galie et al. 2016). The definition of pulmonary hypertension has been made as mean PAP ≥ 20 mmHg from its previous criterion of pulmonary hypertension

S. Saha et al., *Pulmonomics: Omics Approaches for Understanding Pulmonary Diseases*, https://doi.org/10.1007/978-981-99-3505-5_10

patients have a resting mean value ≥25 mmHg (Humbert et al. 2022a). Patients mostly display symptoms like shortness of breath, fatigue, and angina, which often increase upon exertion (https://www.ncbi.nlm.nih.gov/books/NBK482463/). Vascular remodelling in pulmonary hypertension eventually results in right ventricular hypertrophy (RVH), right-sided cardiac failure, and premature death in pulmonary hypertension patients (Rajagopal et al. 2021). The progressive and fatal disease of pulmonary hypertension can arise from different aetiologies. Although the overall prevalence of the disease is rare, it has gained much attention in recent past owing to the dearth of non-invasive diagnostic tools, absence of proper cure, and high mortality rate.

10.2 Etiopathology

Based on the underlying aetiology, pathogenesis, and hemodynamic characterization, the World Health Organization (WHO) has classified pulmonary hypertension into five major groups as shown in Fig. 10.1 (Galie et al. 2016; McGlinchey and Johnson 2014; Bisserier et al. 2020). Each of these five groups has distinct characteristics as defined in terms of mean pulmonary arterial pressure (mPAP)

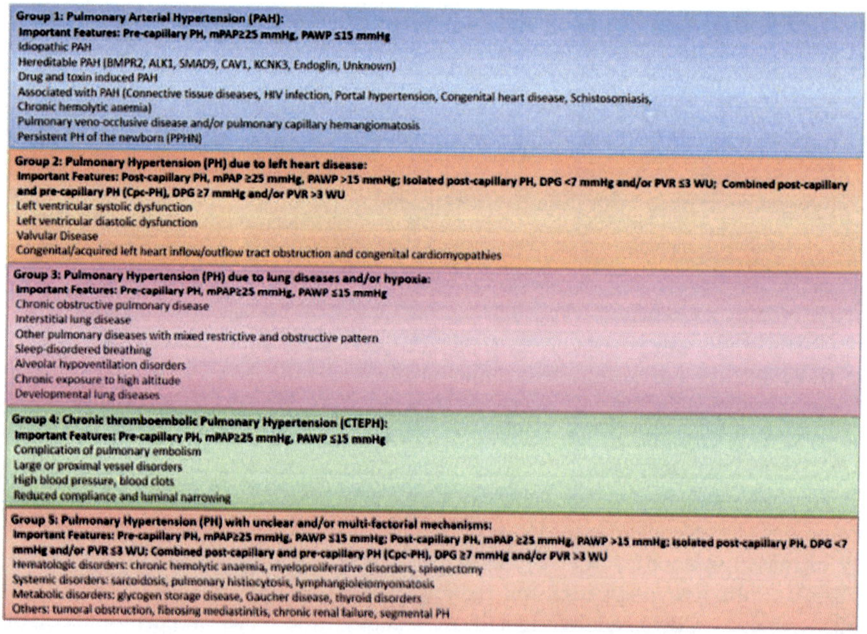

Fig. 10.1 WHO classification of pulmonary hypertension based on aetiology and hemodynamic parameters. Hemodynamic characteristics and aetiologies of each of the five groups have been mentioned. Information on each of these groups have been derived from (Galie et al. 2016; McGlinchey and Johnson 2014; Bisserier et al. 2020)

and pulmonary artery wedge pressure (PAWP). Cardiac output (CO), diastolic pressure gradient (DPG), and pulmonary vascular resistance (PVR) are also taken into consideration. Pulmonary arterial hypertension, pulmonary hypertension due to lung diseases, chronic thromboembolic pulmonary hypertension, and pulmonary hypertension with unclear and/or multifactorial mechanisms all exhibit a mean pulmonary arterial pressure ≥25 mmHg and pulmonary artery wedge pressure ≤15 mmHg. A pulmonary artery wedge pressure >15 mmHg is noted in case of pulmonary hypertension due to heart diseases as well as in pulmonary hypertension with unclear and/or multifactorial mechanisms (Galie et al. 2016).

There are several predisposing risk factors for the development of pulmonary hypertension. Heterozygous mutations in BMPR2 gene have been implicated in more than 70% of pulmonary arterial hypertension (PAH) cases. In several cases of PAH, the underlying cause remains to be unknown or idiopathic.

The different types of PAH are all characterized by vascular remodelling. Vascular remodelling includes pulmonary artery endothelial cell (PAEC) proliferation, pulmonary artery smooth muscle cell (PASMC) proliferation, fibroblast proliferation, cellular migration, medial and adventitial layer hypertrophy, and extracellular matrix (ECM) deposition (Zolty 2021). Formation of vascular lesions in PAH lungs is associated with an increase in pulmonary vascular resistance (PVR) and pulmonary arterial pressure (PAP) (Guignabert et al. 2015; Guignabert et al. 2013). The underlying pathobiology of pulmonary arterial hypertension is complex and involves the complicated interplay of several signalling pathways, inflammatory responses, altered cellular energetics, and resistance to apoptosis as summarized in Fig. 10.2.

10.3 Clinical Features

The common clinical symptoms of pulmonary hypertension are non-specific and generally include shortness of breath, angina, fatigue, weakness, bendopnea, syncope, and oedema. Exercise or physical exertion induced dyspnoea along with fatigue and over-exhaustion is also commonly reported in pulmonary hypertension (PH) patients. Dry cough, exercise induced nausea and vomiting, hoarseness, hemoptysis, and arrhythmias are also noted in PH patients in rare cases. Worsening of the symptoms occurs with the gradual progression of the disease (Hoeper et al. 2017). Right-sided heart failure occurs at a later stage (Frost et al. 2019).

Physical examination of PH reveals signs like left parasternal lift, augmented second heart sound (P2 component), RV third heart sound, right ventricular lift, jugular venous distension, elevated jugular venous pressure, pan systolic murmur of tricuspid regurgitation, pulmonary regurgitation murmurs, and S3 galloping sound (Galie et al. 2016; Frost et al. 2019).

Electrocardiography, transthoracic echocardiography (TTE), chest radiography, pulmonary function tests (PFTs) and arterial blood gases, cardiopulmonary exercise testing (CPET), blood tests and immunological assays, ventilation/perfusion lung scanning, computed tomography (CT) scanning, high-resolution CT (HRCT),

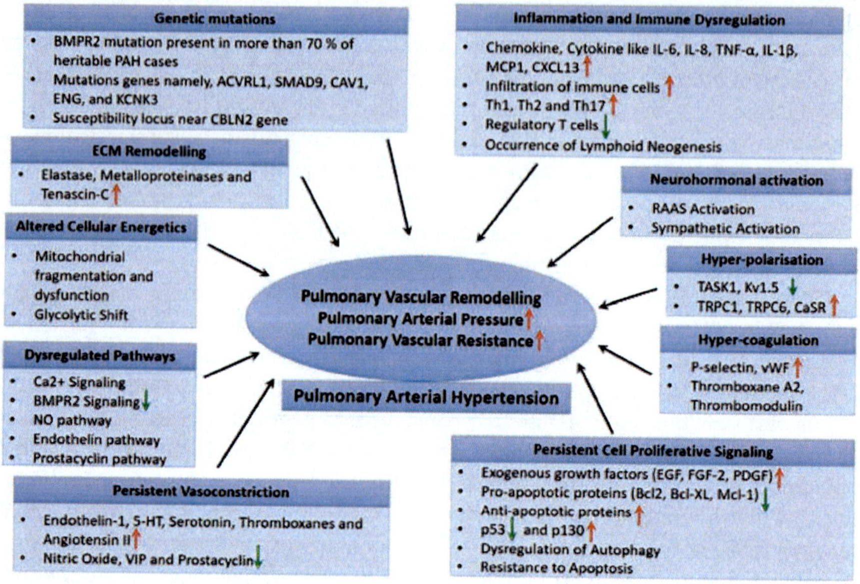

Fig. 10.2 Pathogenesis of pulmonary arterial hypertension. The different mechanisms that leads to pulmonary vascular remodelling, increased pulmonary arterial pressure, and pulmonary vascular resistance in pulmonary arterial hypertension have been derived from (Zolty 2021; Guignabert et al. 2015; Guignabert et al. 2013)

contrast-enhanced CT, pulmonary angiography, cardiac magnetic resonance imaging, and right heart catheterization (RHC) are widely used for the diagnosis of pulmonary hypertension (Galie et al. 2016). Amongst these, transthoracic echocardiography is the most important non-invasive diagnostic tool, whereas the invasive technique of right heart catheterization is the only diagnostic tool that confirms the presence of pulmonary hypertension (Frost et al. 2019). Due to the invasive nature of the confirmatory diagnostic tool, there is an urgent need to screen biomarkers using integrated omics approaches for improving disease diagnosis.

10.4 Biomarkers of Pulmonary Hypertension

Pulmonary hypertension is associated with a significantly high level of morbidity and mortality. Identification of novel, non-invasive biomarkers for the different groups of pulmonary hypertension shall not only aid in disease diagnosis at an early stage but also aid in stratifying disease severity, monitoring disease prognosis, guiding therapy, and studying therapeutic responses. Identified biomarkers may even serve as surrogate endpoints. Different omics strategies like genomics, metagenomics, epigenomics, transcriptomics, proteomics, and metabolomics have paved the way for identification of the key pathways and differentially expressed bio-molecules involved in the pathogenesis in pulmonary hypertension. Improvement in radiomics, imaging modalities, and bioinformatics tools have also used to improve disease diagnosis by boosting our understanding of the disease pathobiology. The different potential biomarker candidates identified by various omics techniques have been summarized in Table 10.1.

10.5 Genomics

Genome-wide association studies and meta-analysis studies have identified several single nucleotide polymorphisms (SNPs) and variations associated with pulmonary hypertension. These genetic determinants of risk, susceptibility, and survival in pulmonary hypertension are of immense importance in conferring protection to high-risk groups. SNPs of BMPR2 (namely rs6435156 and rs1048829), ACVRL1 (rs121909287), and SMAD9 (rs397514716) genes were associated with increased risk of essential hypertension (Chen et al. 2019). BMPR2 mutations are of prime significance as they influence phenotypes in male pulmonary arterial hypertension patients (Liu et al. 2012). Besides, BMPR2 haploinsufficiency has also been linked to primary pulmonary hypertension (Machado et al. 2001). Minor alleles of rs3740297 and rs7923671 of BMP9 gene are associated with increased susceptibility to idiopathic pulmonary arterial hypertension (Guo et al. 2020). Polymorphisms in VEGF (rs833061T > C), SOD2 (rs4880), 5-HTT (L/S polymorphism), and END1 (rs5370) were associated with susceptibility of developing pulmonary arterial hypertension (Zhuo et al. 2017; Xu et al. 2017; Jiao et al. 2019). SNPs of CAV1 gene (namely rs8713 and rs1049337) were linked with increased risk of pulmonary hypertension in COPD patients (Li et al. 2019). Likewise, rs12434438 (AA genotype) of HIF1A and rs755622*C of MIF may be linked with the onset of pulmonary arterial hypertension in systemic sclerosis patients (Takagi et al. 2020; Bossini-Castillo et al. 2017). Some other SNPs associated with pulmonary hypertension have been documented in Table 10.1.

10.6 Epigenomics

Several studies have been conducted to determine epigenetic modifications involved in the pathogenesis of pulmonary hypertension. Most of these studies have been performed in rats and transgenic mice models of pulmonary hypertension. Few studies have been conducted in patients having pulmonary hypertension (Hudson and Farkas 2021). DNA methylation and histone modifications have been implicated in pulmonary hypertension (Pullamsetti et al. 2016). Epigenetic targets have immense potential as prognostic biomarkers and therapeutic targets in pulmonary hypertension. Increased nuclear histone acetyltransferase (HAT) and decreased nuclear histone deacetylase (HDAC) activity have been observed in the lung tissues derived from idiopathic pulmonary arterial hypertension patients (Mumby et al. 2017). Genes involved in lipid transport pathway were found to be differentially

Table 10.1 Different biomarkers of pulmonary hypertension

Omics approach	Biomarkers	Biological samples	Comments
Genomics	SIRT3	Blood	SIRT3 loss of function SNP rs11246020 was found to be associated with idiopathic pulmonary arterial hypertension (Paulin et al. 2014)
	CPS1, NOTCH3		SNPs in CPS1 (rs192759073, rs1047883, and rs2229589) and NOTCH3 (rs1044008) were significantly associated with persistent pulmonary hypertension of the new born (Liu et al. 2019)
	SIRT3, UCP2		SNPs in SIRT3 (rs11246020) and UCP2 (rs659366) were positively correlated with disease severity and outcome in pulmonary arterial hypertension patients (Zhang et al. 2021)
	EPAS1		SNP of EPAS1 (rs17034984) was associated with persistent pulmonary hypertension of the new born (Nakwan et al. 2022)
	COX2		Minor alleles for rs5275 and rs689470 of COX2 gene serve as a significant risk factor for onset of pulmonary arterial hypertension (Durmus et al. 2022)
	ARHGEF18		SNP of ARHGEF18 (rs3745357) may serve as a biomarker of genetic susceptibility to non-idiopathic pulmonary arterial hypertension (Li et al. 2018)
	GNG2		SNP rs11157866 of GNG2 was associated with improved 6 min walk distance and improvement in functional class in pulmonary arterial hypertension patients (Benza et al. 2015)

	NOX3, TBX4		SNPs in NOX3 (rs6557421) and TBX4 (rs3744439) were associated with susceptibility to pulmonary hypertension (Yin et al. 2018)
	COL18A1	Serum	Missense variant of COL18A1 (rs12483377) was related to reduced mortality in pulmonary arterial hypertension patients (Damico et al. 2015)
	EDN1		T allele of rs2070699 of EDN1 was associated with increased risk of neonatal persistent pulmonary hypertension (Mei et al. 2016)
	SOX17, HLA-DPA1/DPB1	Lung tissues, endothelial cells	SNPs in SOX17, namely rs10103692 and rs13266183 were associated with pulmonary arterial hypertension. HLA-DPA1/DPB1 rs2856830 was strongly related to survival in pulmonary arterial hypertension patients (Rhodes et al. 2019)
	CBLN2	Lung Tissues	The G allele of rs2217560 in CBLN2 gene is associated with an increased risk for both idiopathic and familial pulmonary arterial hypertension. Another SNP of CBLN2 (rs9916909) was also associated with idiopathic and familial pulmonary arterial hypertension (Germain et al. 2013)
	TBX4	DNA	Mutations in TBX4 (small patella syndrome) were linked with childhood-onset pulmonary arterial hypertension (Kerstjens-Frederikse et al. 2013)
	NOS3	Blood, Buccal cells	SNP in the NOS3 gene (rs1799983) was associated with trans pulmonary gradient (TPG), mean pulmonary artery pressure (mPAP), and diastolic pulmonary gradient (DPG) in patients

(continued)

Table 10.1 (continued)

Omics approach	Biomarkers	Biological samples	Comments
			having pulmonary hypertension and left heart failure patient (HF-PH) (Duarte et al. 2018)
Epigenomics	DES, ETS2, TBX1/5, TERT, MYH7, ZAP70, STAT5A, COL1A1, TNFSF4, TCAP, OXT, GNG8, GATA2, ID3, LDB3, NFAM1, ACTB, IGFBP1, GPR182, ADCY4, BCL2L11, ARHGDIB, ABCA1, ADIPOQ, APOA4, FGF, TNFSF4, CX3CR1	Pulmonary endothelial cells (PEC)	Promoters of DES, ETS2, TBX1/5, TERT, MYH7, ZAP70, STAT5A, COL1A1, TNFSF4, TCAP, OXT, GNG8, GATA2, ID3, LDB3, NFAM1, ACTB, IGFBP1, GPR182, ADCY4, BCL2L11, ARHGDIB, ABCA1, ADIPOQ, and APOA4 genes were differentially methylated in pulmonary arterial hypertension patients DNA hypermethylation was noted in promoters of FGF, TNFSF4, and CX3CR1 (Hautefort et al. 2017)
	GIGYF1, HLA-DRB1, RPL17, HIC1, OSR2, MN1, ADAMTS19, HLA-DRB5, NOSTRIN, JRK, RB1, CDKN1A, PIK3R1, LMO2, STAT3, LIMS1, AHSA2, CAMKK2, C6orf58, HHLA2, SSH1, CTBP2, ANGPT1, PIK3CA, HRAS	Pulmonary arterial smooth muscle cells (PASMCs)	GIGYF1, HLA-DRB1, RPL17, HIC1, OSR2, MN1, ADAMTS19, HLA-DRB5, and NOSTRIN genes were hypermethylated in chronic thromboembolic pulmonary hypertension patients JRK, RB1, CDKN1A, PIK3R1, LMO2, STAT3, LIMS1, AHSA2, CAMKK2, C6orf58, SSH1, HHLA2, CTBP2, ANGPT1, PIK3CA, and HRAS genes were hypomethylated in chronic thromboembolic pulmonary hypertension (CTEPH) patients (Wang et al. 2018)
	BMPR2	Peripheral blood mononuclear cells (PBMCs)	BMPR2 promoter is hyper methylated in patients suffering from heritable pulmonary arterial hypertension (Liu et al. 2017)
	BMPR2	Blood	Hypermethylation of BMPR2 was observed in valvular heart disease patients having pulmonary

			artery hypertension (VHD-PAH) (Li et al. 2021a)
	PRKCA, LIF, PRKAG2, SPRED2, FGFR2, SKP2, HCN2, Protamine 1		PRKCA, LIF, PRKAG2, SPRED2, FGFR2, SKP2, HCN2, and Protamine 1 were differentially methylated in patients having rheumatic heart disease and secondary pulmonary arterial hypertension (RHD-PAH) (Zheng et al. 2017)
	SOD3	Pulmonary arterial smooth muscle cells (PASMCs), lung tissues	Histone deacetylation is associated with reduced SOD3 expression in idiopathic pulmonary arterial hypertension patients (Nozik-Grayck et al. 2016)
	SOCS3, GNAS, ITGAL, NCOR2, NFIC	Circulating CD4+ T cells	SOCS3, GNAS, ITGAL, NCOR2, and NFIC were differentially regulated in pulmonary arterial hypertension patients (Benincasa et al. 2022)
Transcriptomics	SMAD6, Granzyme A	Lung	SMAD6 and Granzyme A are decreased in patients with pulmonary arterial hypertension (Sasagawa et al. 2016)
	TNF, CSF3		TNF and CSF3 signalling play a crucial role in pulmonary arterial hypertension (Stearman et al. 2019; Hemnes et al. 2020)
	CCDC80, AGR2		CCDC80 and AGR2 are increased in patients with pulmonary arterial hypertension. Increased CCDC80 may be associated with the pathogenesis of pulmonary arterial hypertension (Sasagawa et al. 2016)
	p70S6K	Blood	p70S6K expression was associated with pulmonary arterial hypertension (Elinoff et al. 2020)

(continued)

Table 10.1 (continued)

Omics approach	Biomarkers	Biological samples	Comments
	HIF1-α, KLF10, TRPC1, Sestrin1, SMAD5		HIF1-α, KLF10, TRPC1, Sestrin1, and SMAD5 are differentially regulated in pulmonary arterial hypertension (Hemnes et al. 2020; Rhodes et al. 2020)
	miR-140-5p		miR-140-5p was reduced in pulmonary arterial hypertension patients as compared to healthy controls (Rothman et al. 2016)
	miR23a, miR-130, miR-191, miR-204, miR-145, miR-27a, miR-328, miR-1-2, miR-199, and miR-744		Different microRNAs, namely miR23a, miR-130, miR-191, miR-204, miR-145, miR-27a, miR-328, miR-1-2, miR-199, and miR-744 may serve as potential biological markers of idiopathic pulmonary arterial hypertension (Sarrion et al. 2015)
	AMD1		AMD1 was reduced in pulmonary arterial hypertension patients (Rhodes et al. 2020)
	miR-4632	Serum	miR-4632 was reduced in patients with pulmonary arterial hypertension (Hemnes et al. 2020).
	PDGF-β	Blood vessel	PDGF-β was elevated in pulmonary arterial hypertension patients (Saygin et al. 2020)
	MMP-9, VEGF	Peripheral blood mono nuclear cells (PBMCs)	MMP-9 and VEGF expression was higher in mild pulmonary arterial hypertension patients as compared to severe pulmonary arterial hypertension patients (Grigoryev et al. 2008)
	ADM, IL7R, ZFP36, GLUL, JUND, BCL6, EREG, CXCL2, MMP25		ADM, IL7R, ZFP36, GLUL, JUND, BCL6, EREG, CXCL2, and MMP25 were associated

			with pulmonary arterial hypertension (Grigoryev et al. 2008)
	TNF-α, GM-CSF		TNF-α and GM-CSF were significantly enriched in patients having pulmonary arterial hypertension (Hemnes et al. 2020; Elinoff et al. 2020)
Proteomics	NT-proBNP	Blood	NT-proBNP level greater than 1400 ng/L is associated with increased mortality in patients with pulmonary arterial hypertension (Galie et al. 2016; Hemnes et al. 2020) NT-proBNP may serve as a crucial indicator of successful disease control in pulmonary arterial hypertension patients (Galie et al. 2016)
	TRAIL, CCL5, CCL7, CCL4, MIF, IL-12, IL-17, IL-10, IL-7, VEGF, IL-8, IL-4, PDGF-β, IL-6, CCL11		Cytokine profiling enabled clustering of pulmonary arterial hypertension patients based on cytokine expression. Cluster 1 was characterized by upregulation of TRAIL, CCL5, CCL7, CCL4, and MIF. Cluster 3 was characterized by upregulation of IL-12, IL-17, IL-10, IL-7, and VEGF. Cluster 4 was characterized by upregulation of IL-8, IL-4, PDGF-β, IL-6, and CCL11 (Sweatt et al. 2019)
	C4a, vitronectin, alpha-1-antitrypsin		C4a, vitronectin, and alpha-1-antitrypsin may serve as diagnostic indicators of idiopathic pulmonary arterial hypertension (Qin et al. 2021)
	MDK, FSTL3		MDK and FSTL3 may serve as diagnostic indicators of systemic sclerosis associated pulmonary arterial hypertension (SSc-PAH) (Qin et al. 2021)
	DBH, ADIPO, ANPEP		DBH, ADIPO, and ANPEP may serve as prognostic markers of pulmonary arterial

(continued)

Table 10.1 (continued)

Omics approach	Biomarkers	Biological samples	Comments
			hypertension in congenital heart disease patients (Qin et al. 2021)
	Endoglin		Endoglin expression was significantly elevated in Group I pulmonary arterial hypertension patients (Malhotra et al. 2013)
	NT-proBNP, Ghrelin	Plasma	Plasma NT-proBNP and ghrelin levels were significantly elevated in idiopathic pulmonary arterial hypertension patients as compared to healthy subjects (Yang et al. 2013)
	Adrenomedulin, CD40L, Von Willebrand factor (vWF) and its antigen (vWF:Ag)		Adrenomedulin, CD40L, vWF, and vWF:Ag were significantly elevated in pulmonary arterial hypertension patients as compared to healthy subjects (Santos-Gomes et al. 2022)
	TIMP-1		Elevated TIMP-1 level may serve as a circulating biomarker for diagnosing pulmonary hypertension in COPD patients (He et al. 2021)
	HDL-4, HDL-4-Apo A-2		Decreased plasma levels of HDL-4 (small HDL particles transporting fibrinolytic proteins) are related to poor outcomes in both idiopathic and heritable pulmonary arterial hypertension patients. HDL-4-Apo A-2 was significantly associated with survival (Harbaum et al. 2019)
	IGFBP6, IGF2R, IGFBP7, IGFBP2, pRKAA1, ANGPTL4, OLR1, APOA1, CD36		IGFBP6, IGF2R, IGFBP7, IGFBP2, pRKAA1, ANGPTL4, OLR1, APOA1, and CD36 were differentially expressed in pulmonary arterial hypertension patients (Hemnes et al. 2020; Hemnes et al. 2019)

	Cardiac troponin T (cTnT)	Serum	Cardiac troponin T (cTnT) expression in the serum may serve as an indicator of poor prognosis in patients with chronic pre-capillary pulmonary hypertension (Torbicki et al. 2003)
	MCP1	Plasma, Serum	MCP1 expression was increased in idiopathic pulmonary arterial hypertension (IPAH) patients and connective tissue disease-associated pulmonary arterial hypertension (CTD-PAH) patients (Santos-Gomes et al. 2022)
	Endostatin		Endostatin expression was elevated in idiopathic pulmonary arterial hypertension (IPAH) patients, connective tissue disease-associated pulmonary arterial hypertension (CTD-PAH) patients, and in pulmonary arterial hypertension patients having complications of congenital heart disease (CHD-PAH (Santos-Gomes et al. 2022)
	BMPR2	Lung tissues	BMPR2 expression was reduced in patients with primary pulmonary hypertension (Atkinson et al. 2002)
	OPG	Serum, lung tissues	Significant elevation of OPG was observed in idiopathic pulmonary arterial hypertension patients (McGlinchey and Johnson 2014; Lawrie et al. 2008)
	CLIC4, Arf6	Pulmonary artery endothelial cells (PAECs)	Novel pathway including increased levels of CLIC4 and Arf6 as an effector of CLIC4 was found to be involved in pulmonary arterial hypertension (Harbaum et al. 2021; Abdul-Salam et al. 2019)
	TCTP	Blood outgrowth endothelial cells, Lungs	TCTP is over expressed in both idiopathic and hereditary pulmonary arterial hypertension patients. TCTP mediates endothelial pro-survival

(continued)

Table 10.1 (continued)

Omics approach	Biomarkers	Biological samples	Comments
			and pulmonary vascular remodelling (Lavoie et al. 2014)
	VEGF	Arterial Vascular Endothelial cells	VEGF expression was increased in idiopathic pulmonary arterial hypertension (IPAH) patients and in pulmonary arterial hypertension patients having complications of congenital heart disease (CHD-PAH) (Santos-Gomes et al. 2022)
	NDUFA9, UQCRC2, UQCRC1, ATP5L, ATP5B, IDH2, OGDH, SLC25A4, ECH1, GLO1, FBP2, ATP2A1, PEBP1, MYH1, MYH2, MYH7, MYLPF, APOA1, PDLIM3	Muscle cells	NDUFA9, UQCRC2, UQCRC1, ATP5L, ATP5B, IDH2, OGDH, SLC25A4, and ECH1 were downregulated in pulmonary arterial hypertension patients as compared to healthy subjects GLO1, FBP2, ATP2A1, PEBP1, MYH1, MYH2, MYH7, MYLPF, APOA1, and PDLIM3 were upregulated in pulmonary arterial hypertension patients as compared to healthy subjects (Malenfant et al. 2015).
Metabolomics	Glutamate, tryptophan, N2,N2-dimethylguanosine, N1-methylinosine, fatty acid acylcarnitines, Malate, Fumarate, Polyamines, Phosphatidylcholines, Steroids, Sphingomyelins	Plasma	Amino acids (like glutamate, tryptophan), tRNA-specific modified nucleosides (N2,N2-dimethylguanosine, N1-methylinosine), fatty acid acylcarnitines, TCA cycle intermediates (malate, fumarate), and polyamine metabolites were increased in patients with pulmonary arterial hypertension as compared to healthy controls Phosphatidylcholines, steroids, and sphingomyelins were decreased in patients with

			pulmonary arterial hypertension as compared to healthy controls (Rhodes et al. 2017a)
	Long chain acylcarnitines, phosphatidylcholines		Long chain acylcarnitines and phosphatidylcholine levels were increased in plasma of pulmonary arterial hypertension patients (Hemnes et al. 2019)
	Arginine, ornithine, citrulline, dimethyl arginine		Arginine, ornithine, citrulline, symmetric and asymmetric dimethyl arginine are related to right ventricular-pulmonary vascular (RV-PV) dysfunction. The ratio of arginine to the sum of ornithine and citrulline has been identified as a potential biomarker in pulmonary hypertension patients with sickle cell disease and heart failure (Lewis et al. 2016)
	Uric acid, allantoin, xanthine, xanthosine, inosine		Purine metabolites like uric acid, allantoin, xanthine, xanthosine, and inosine were closely related to right ventricular-pulmonary vascular (RV-PV) dysfunction. Uric acid was associated with pulmonary hypertension (Lewis et al. 2016)
	Serotonin		Serotonin expression is elevated in plasma of idiopathic pulmonary arterial hypertension patients (Santos-Gomes et al. 2022)
	IDO-TMs		Indoleamine 2, 3-dioxygenase tryptophan metabolites (IDO-TMs) were increased in pulmonary arterial hypertension patients and correlated strongly with right ventricular-pulmonary vascular (RV-PV) dysfunction (Lewis et al. 2016)
	Homocysteine		Homocysteine levels are elevated in pulmonary arterial hypertension patients compared to healthy subjects. Plasma homocysteine levels are

(continued)

Table 10.1 (continued)

Omics approach	Biomarkers	Biological samples	Comments
			also increased in pulmonary arterial hypertension patients with congenital heart disease (Santos-Gomes et al. 2022; Sanli et al. 2012)
	D-Dimer		D-dimer level was positively correlated with pulmonary artery pressure and inversely related to oxygen saturation and 6-min walk distance. Plasma D-dimer level may play a vital role in evaluating primary pulmonary hypertension patients and identifying individuals at greater risk for developing severe disease (Shitrit et al. 2002a) Elevated D-dimer levels in primary pulmonary hypertension patients may serve as indicator of endogenous fibrinolysis in disease pathogenesis (Shitrit et al. 2002b)
	Uric Acid	Serum	Serum uric acid level was significantly elevated in primary pulmonary hypertension patients. Serum uric acid levels were found to be associated with cardiac index, mean right atrial pressures, disease severity and mortality rate in primary pulmonary hypertension (Nagaya et al. 1999; Hoeper et al. 1999) Level of uric acid was increased in patients with severe pulmonary arterial hypertension (PAH) and was related to mortality. Thus uric acid could be explored as a prognostic biomarker of clinical severity for PAH (Bendayan et al. 2003; Savale et al. 2021)

	Asymmetric dimethyl arginine (ADMA)	Serum, plasma	ADMA levels are increased in patients with pulmonary arterial hypertension (PAH), idiopathic pulmonary arterial hypertension (IPAH), chronic thromboembolic pulmonary hypertension (CTEPH), and pulmonary arterial hypertension patients with congenital heart disease (CHD-PAH). In IPAH patients, ADMA level is related to mean pulmonary arterial pressure (mPAP), right atrial pressure (RAP), pulmonary vascular resistance (PVR) index, mixed venous oxygen saturation (SvO2), cardiac index, and survival rate. ADMA levels were positively associated with RAP and negatively with cardiac output, cardiac index, SvO2, and survival rate in CHD-PAH patients. In CTEPH, ADMA levels showed correlation with PVR, SvO2, mPAP, RAP, cardiac output, and cardiac index (Sanli et al. 2012; Zhang et al. 2015; Kielstein et al. 2005; Giannakoulas et al. 2014; Gorenflo et al. 2001; Skoro-Sajer et al. 2007)
	Kynurenine		Kynurenine levels were significantly elevated in pulmonary arterial hypertension patients and were strongly correlated to mean pulmonary arterial pressure (mPAP) and adverse clinical course (Rhodes et al. 2017a; Nagy et al. 2017; Jasiewicz et al. 2016)
	cGMP (cyclic guanosine monophosphate)	Plasma, urine	Plasma cGMP levels were associated with disease severity in pulmonary arterial hypertension patients (Ghofrani et al. 2002) Urinary cGMP levels were elevated in primary pulmonary hypertension patients and were indicative of severe haemodynamic impairment

(continued)

Table 10.1 (continued)

Omics approach	Biomarkers	Biological samples	Comments
			(Bogdan et al. 1998) cGMP serves as an important therapeutic target in pulmonary arterial hypertension (Santos-Gomes et al. 2022)
	15-F2t-isoprostane		Plasma concentrations of 15-F2t-isoprostane were significantly elevated and related to high mortality in pulmonary arterial hypertension patients (Zhang et al. 2014). 15-F2t-isoprostane levels were elevated in urines of primary pulmonary hypertension patients (Robbins et al. 2005) Urinary 15-F2t-isoprostane levels are independently associated with mortality in pulmonary arterial hypertension patients. Urinary 15-F2t-isoprostane level may serve as prognostic biomarker of pulmonary arterial hypertension (Cracowski et al. 2012; Dromparis and Michelakis 2012)
	12-hydroxyeicosanoids (12-HETE), 15-hydroxyeicosanoids (15-HETE), 5-hydroxyeicosanoids (5-HETE), 9-hydroxyoctadecadienoic acid (9-HODE), 13-hydroxyoctadecadienoic acid (13-HODE), 5-oxo-eicosatetraenoic acid	Plasma, lung tissues	Plasma 12-hydroxyeicosanoids and 15-hydroxyeicosanoids serve as independent predictors of survival in pulmonary arterial hypertension patients (Al-Naamani et al. 2016) Eicosanoids (like 9-HODE, 13-HODE, 5-HETE, 12-HETE, and 15-HETE) were elevated in plasma of both idiopathic and associated pulmonary arterial hypertension patients (Ross et al. 2015) 12-HETE, 15-HETE, and 5-oxo-eicosatetraenoic

			acid were increased in lung tissues of pulmonary hypertension patients (Bowers et al. 2004) In pulmonary hypertension, elevated levels of 15-HETE lead to hypoxia-mediated pulmonary vascular remodelling, fibrosis, and angiogenesis (Santos-Gomes et al. 2022; Ma et al. 2011)
	Methyl isobutyl ketone, 2,2,4,4-Tetramethyltetrahydrofuran, aniline, *p*-menthone, benzothiazole, *m*-cymene or *o*-cymene, N-ethyl-benzamine, 2-menthene, 1,6-dioxacyclododecane-7,12-dione, 2-Ethyl-3-hydroxyhexyl 2-methylpropanoate	Exhaled breath condensate (EBC)	Peaks for ten metabolites (namely methyl isobutyl ketone, aniline, *p*-menthone, 2,2,4,4-tetramethyltetrahydrofuran, benzothiazole, *m*-cymene or *o*-cymene, N-ethyl-benzamine, 2-menthene, 1,6-dioxacyclododecane-7,12-dione and 2-Ethyl-3-hydroxyhexyl 2-methylpropanoate) have been noted exclusively in idiopathic pulmonary arterial hypertension (IPAH) patients (Mansoor et al. 2014).
	Nitric oxide (NO), NO products, glutathione	Lungs, broncho alveolar lavage fluid (BALF)	Nitric oxide (NO) was decreased in primary pulmonary hypertension (PPH) patients NO products were decreased, and glutathione was increased in BALF of primary pulmonary hypertension (PPH) patients as compared to healthy controls (Kaneko et al. 1998)
	Nitric oxide (NO), sphingosine-1-phosphate (S1P) metabolites, heme metabolites, arginine	Lung tissues	Increased nitric oxide (NO), sphingosine-1-phosphate (S1P) metabolites, heme metabolites, and decreased arginine levels were noted in advanced stage pulmonary arterial hypertension patients (Zhao et al. 2015)
	Glucose, fructose, sorbitol, fructose-6-phosphate, phosphoenolpyruvate (PEP), fructose 1,6-bisphosphate, 3-phosphoglycerate		Levels of glucose, fructose, sorbitol, and fructose-6-phosphate were elevated, whereas fructose 1,6-bisphosphate, phosphoenolpyruvate (PEP) and 3-phosphoglycerate were reduced in

(continued)

Table 10.1 (continued)

Omics approach	Biomarkers	Biological samples	Comments
			pulmonary arterial hypertension patients (Zhao et al. 2014)
	Citrate, *cis*-aconitate, succinate, succinyl carnitine		Increased levels of citrate, cis-aconitate, succinate, and succinyl carnitine were observed in pulmonary arterial hypertension patients (Zhao et al. 2014)
	Tetradecanedioate, hexadecanedioate, octadecanedioate, adrenate, caprylate, caproate, palmitoleate, myristate		Dicarboxylic fatty acids (like tetradecanedioate, hexadecanedioate, and octadecanedioate), adrenate, and long/medium-chain free fatty acid products (like caprylate, caproate, palmitoleate, and myristate) were significantly accumulated in lung tissues obtained from pulmonary arterial hypertension patients (Zhao et al. 2014)
Metagenomics	*Lachnospiraceae bacterium GAM79, Anaerostipes rhamnosivorans, Amedibacterium intestinale, Ruminococcus bicirculans, Ruminococcus albus*	Stool	*Lachnospiraceae bacterium GAM79* was significantly decreased in stools of patients with pulmonary arterial hypertension Increased abundance of *Anaerostipes rhamnosivorans* and reduced prevalence of *Amedibacterium intestinale, Ruminococcus bicirculans*, and *Ruminococcus albus* were associated with severity of pulmonary arterial hypertension (Jose et al. 2022)
	Firmicutes (*Clostridium, Staphylococcus, Blautia, Streptococcus, Roseburia, Rumonococcus*), *Bacteroidetes* (*Prevotella*), *Actinobacteria* (*Rothia, Bifidobacterium, Collinsella, Coriobacteriales*), *Proteobacteria*		Increased abundance of *Firmicutes* (like *Clostridium, Staphylococcus, Blautia, Streptococcus, Roseburia*, and *Rumonococcus*), *Bacteroidetes* (like *Prevotella*), *Actinobacteria* (like *Rothia, Bifidobacterium, Collinsella*, and *Coriobacteriales*), and *Proteobacteria* (like

	(*Citrobacter, Desulfovibrio, Enterobacter, Escherichia, Klebsiella, Pseudomonas*)		*Citrobacter, Desulfovibrio, Enterobacter, Escherichia, Klebsiella,* and *Pseudomonas*) were noted in faecal samples from patients with pulmonary arterial hypertension (Kim et al. 2020)
	Firmicutes (*Coprococcus, Butyrivibrio, Clostridia, Lachnospiraceae, Eubacterium, Akkermansia, Lactococcus, Subdoligranulum*), *Bacteroidetes* (*Bacteroides*), *Proteobacteria* (*Parasutterella*)		Decreased prevalence of *Firmicutes* (like *Coprococcus, Butyrivibrio, Clostridia, Lachnospiraceae, Eubacterium, Akkermansia, Lactococcus,* and *Subdoligranulum*), *Bacteroidetes* (like *Bacteroides*) and *Proteobacteria* (like *Parasutterella*) were observed in faecal samples from patients with pulmonary arterial hypertension (Kim et al. 2020)
	Firmicutes, Proteobacteria, Fusobacteria, Chloroflexi, Bacteroidetes, Saccharibacteria, SR1_Absconditabacteria	Pharyngeal swabs	Abundance of *Firmicutes* (like *Streptococcaceae* and *Streptococcus*), *Proteobacteria* (like *Lautropia* and *Ralstonia*), *Fusobacteria* (like *Leptotrichiaceae* and *Leptotrichia*), and *Chloroflexi* were increased in pulmonary hypertension patients Prevalence of *Firmicutes* (like *Carnobacteriaceae* and *Granulicatella*), *Bacteroidetes* (like *Prevotella, Porphyromonadaceae, Flavobacteriaceae, Alloprevotella* and *Capnocytophage*), *Actinobacteria* (like *Rothia*), *Proteobacteria* (like *Haemophilus*), *Saccharibacteria,* and *SR1_Absconditabacteria* were decreased in pulmonary hypertension patients (Zhang et al. 2020)

This table enlists various important biomarkers of pulmonary hypertension identified using different omics approaches methylated in pulmonary arterial hypertension patients (Hautefort et al. 2017). Different other epigenetic signatures identified in pulmonary hypertension patients have been tabulated in Table 10.1.

10.7 Transcriptomics

Several high-throughput transcriptomic studies have been conducted for pulmonary hypertension. Genes involved in different cellular functions, namely cell proliferation, inflammation, immunity, and extracellular matrix turnover were found to be dysregulated in pulmonary hypertension patients (Harbaum et al. 2021). SLC4A4, PPARGC1A, DEPDC1, PCDH18, ZNF208, IL6R, CENPN, ST18, NOX4, NKAIN2, ZNF423, and ZNF117 were overexpressed, whereas CAB39, PDE7A, TMPO, CEP85L, HPGD, NAA15, SEC23A, ODF2L, MBNL1, and TNKS2 were decreased in idiopathic pulmonary arterial hypertension patients as compared to healthy volunteers (Sarrion et al. 2015). miRNA-124 was overexpressed in blood outgrowth endothelial cells obtained from pulmonary arterial hypertension patients as compared to healthy subjects (Harbaum et al. 2021). Reduced plasma miRNA-150 level was found to be related to poor survivability in pulmonary arterial hypertension patients (Rhodes et al. 2013). A list of different other genes and miRNAs that are differentially expressed in pulmonary hypertension have been provided in Table 10.1.

10.8 Proteomics

High-throughput proteomic studies have been extensively conducted using various biological samples from pulmonary hypertension patients. Plasma GDF-15 level was found to be elevated in naïve pulmonary arterial hypertension (PAH) patients. Plasma GDF-15 level greater than 1200 ng/L was associated with increased mortality in PAH (Anwar et al. 2016). Serum level of different proteins, namely PTX3, IP-10, endothelin-1, CysC, Copeptin, sST2, sVEGFR1, CRP, and Gal-3 were elevated in PAH patients. These proteins shall be explored further as novel serum biomarkers of PAH (McGlinchey and Johnson 2014; Santos-Gomes et al. 2022). Elevated serum PIM1 expression in PAH was related to higher mortality risk (McGlinchey and Johnson 2014). Proteomics and phosphoproteomic-based studies have highlighted different dysregulated proteins and altered phosphorylation status in PAH patients. Proteins like EIF2A, EIF3C, EIF4B, EIF4G3, EIF5B, HSPA5, and RPL3 were increased significantly, whereas AKT1, MAPK1, CALM3, GNA11, MGST1, NOS3, PRKACA, and SOD1 were decreased significantly in PAH patients. EIF5B, TGOLN2, PABPN1, and ZC3H4 were differentially phosphorylated in PAH patients (Xu et al. 2019). Nine proteins, namely ST2, TIMP-1, TIMP-2, IGFBP-1, plasminogen, ApoE, EPO, complement factor H, and factor D could identify PAH patients with high mortality rate (Rhodes et al. 2017b). Proteomic analyses have also highlighted vital pathways like glycolysis, fatty acid β oxidation, HIF signalling, EIF2 signalling, TGF-β signalling, mTOR signalling, chemokine signalling, nitric oxide signalling, RhoA signalling, and cAMP-PKA signalling to be associated with PAH (Qin et al. 2021). Other proteomic signatures of pulmonary hypertension have been documented in Table 10.1.

10.9 Metabolomics

Profiling of metabolites in pulmonary hypertension patients using modern metabolomics tools has improved our understanding of the contribution of circulatory peptides and metabolites in the onset of different forms of pulmonary hypertension (Lewis 2014). Heterogeneous global metabolomic profile is observed in pulmonary hypertension patients as compared to healthy subjects. Uric acid, asymmetric dimethyl arginine (ADMA), cyclic guanosine monophosphate (cGMP), indoleamine 2, 3-dioxygenase tryptophan metabolites (IDO-TMs), and kynurenine were identified as important metabolic signatures of pulmonary hypertension (Santos-Gomes et al. 2022). Aminogram-based study of plasma amino acid levels revealed most of the studied amino acids to be over expressed in pulmonary hypertension patients. Only arginine and tryptophan were reduced in pulmonary hypertension patients, whereas histidine and tyrosine levels were comparable between the patients and the controls. Fisher ratio (branched-chain amino acids/ aromatic amino acids) was found to decrease with an increase in disease severity in pulmonary hypertension (Yanagisawa et al. 2015). Other metabolic signatures related to pulmonary hypertension have been listed in Table 10.1. A combination of targeted proteomics, non-targeted metabolomics, and integrative bioinformatics identified several metabolic pathways that are dysregulated in patients with pulmonary arterial hypertension. Increased oxidative stress, altered tri-carboxylic acid (TCA) cycle flux, dysfunctional arginine pathway, glutamate metabolism, and nitric oxide pathway were among the commonly dysregulated pathways (Xu et al. 2019). Defects in fatty acid metabolism have also been documented in pulmonary arterial hypertension patients (Brittain et al. 2016).

10.10 Metagenomics

Dysbiosis of gut and airway microbiome has been documented in patients suffering from pulmonary hypertension. Gut microbial dysbiosis and change in gut microbial metabolite profiles have been implicated in mediating the pathogenesis of pulmonary arterial hypertension by modulating the host inflammatory immune response via microbial metabolites (Wu et al. 2022). Besides, right heart failure in pulmonary arterial hypertension patients has been linked with increased intestinal permeability as well as higher circulating levels of soluble CD14 and gut bacterial endotoxin (Thenappan et al. 2019). Increased serum LPS levels noted in untreated pulmonary arterial hypertension patients as compared to treated ones, suggested that increased tissue permeability and increased bacterial translocation were associated with severe impairment of cardiopulmonary functioning in pulmonary arterial hypertension patients (Ranchoux et al. 2017). The changes in gut microbial profile in pulmonary hypertension patients have been summarized in Table 10.1. Alterations in airway microbiome have also been documented in pulmonary hypertension patients (as shown in Table 10.1) (Zhang et al. 2020). Changes in both airway and gut microbiome highlight the involvement of the gut–lung axis in disease pathogenesis (Ranchoux et al. 2017). Exploring the crosstalk of the airway and gut microbiome in pulmonary hypertension patients using integrative omics and systems biology approaches can open up new avenues of using microbiota as novel therapeutic targets (Huang et al. 2022).

10.11 Bioinformatics

There is a patient registry database for PH patients named the Registry to Evaluate Early and Long-term PH Disease Management (REVEAL) in the USA, and the PH patients were categorized based on World Health Organization (WHO) functional class as mentioned in Fig. 10.1 (McGoon et al. 2008; McGoon and Miller 2012). It is the richest source of data on PH patients compiled from 1990 to the present and has been used for developing algorithms to accurately identify and validate patients for administrative claims (Papani et al. 2018; Mathai et al. 2019). The algorithms have also been developed from REVEAL data for PH-specific therapy and diagnostic procedures (Gillmeyer et al. 2019). In the future, functional and early survival outcomes in the general PH population and PH subsets could be predicted with higher accuracy using machine learning tools from the REVEAL datasets.

10.12 Medical Imaging or Radiomics

Non-invasive medical imaging is the most popular choice for evaluating PH patients, and several imaging tools are utilized in the evaluation of PH, such as chest X-ray (CXR), computed tomography (CT), magnetic resonance imaging (MRI), and positron emission tomography (PET) (Ascha et al. 2017; Aluja Jaramillo et al. 2018; Altschul et al. 2019; Remy-Jardin et al. 2021). CXR can detect cardiomegaly due to right atrial (RA) and right ventricle (RV) enlargement and pruning of peripheral blood vessels. However, CT is more popular in the diagnosis of PH patients for its fast scanning, and good spatial and temporal resolution (Altschul et al. 2019). MRI can determine right ventricular function in patients with PH like right ventricular ejection fraction (Remy-Jardin et al. 2021). PET is advised to distinguish or associate group 4 PH patients to pulmonary artery sarcoma patients (Altschul et al. 2019). In summary, in the absence of any biomarker in practice, medical imaging is a key player in detecting and classifying PH patients.

10.13 Multi-Omics and Data Integration

Integrated data analyses were performed with differentially expressed genes (DEGs) from the transcriptional data of blood and lung tissues of pulmonary hypertension (PH) patients and healthy controls (Tang et al. 2022; Yang et al. 2022). The DEGs were annotated with gene ontology terms and further used for construction of Protein-Protein Interaction Networks (PPINs) and pathway enrichment analysis. The identified hub genes from the blood sample in the network analysis were enriched at extracellular exosome, haemoglobin complex, and oxygen transporter activity (Tang et al. 2022). In another study with the whole blood of PH patients, several DEGs were implicated in pathways involved in angiogenesis, EGF receptor signalling pathway, and VEGF signalling pathway (Ezenwa et al. 2020). A study using lung tissue of PH patients identified MAPK6 as a potential biomarker with an AUC of 100% and in another study, CDC5L and DDX46 were identified from lung tissue as marker genes and therapeutic targets for PH using integrated bioinformatics analysis (Yang et al. 2022; Li et al. 2021b). Ten hub genes (HSP90AA1, CDC5L, MDM2, LRRK2, CFTR, IQGAP1, CAND1, TOP2A, DDX21, and HIF1A) were identified from PPIN analysis from two microarray datasets of pulmonary arterial hypertension (PAH) patients (Wei et al. 2022). Several integrated studies were performed to identify potential biomarkers or target genes in PH patients focusing mainly on transcriptomics studies.

10.14 Present Therapeutics

Currently, there is no specific drug for the treatment and cure of pulmonary hypertension. The present medications primarily act as antagonists to endothelin pathway, agonists, and analogues of prostacyclin pathway and targeting nitric oxide pathway (Zhang et al. 2022). Endothelin receptor antagonists, calcium channel blockers, phosphodiesterase 5 inhibitors, prostacyclin analogues, prostacyclin receptor agonist, soluble guanylate cyclase stimulators (sGCS), anticoagulants, diuretics, and inotropes are commonly used for the treatment of pulmonary hypertension (Humbert et al. 2022b; https://rarediseases.org/rare-diseases/pulmonary-arterial-hypertension/). The details of these medicines have been mentioned in Table 10.2. These drugs mostly improve exercise tolerance and quality of life. In fewer cases, these drugs may prove to be beneficial controlling worsening of the disease and reducing mortality (Zhang et al. 2022).

Oxygen therapy is also given to pulmonary hypertension patients to improve capacity to tolerate exercise (Humbert et al. 2022b). Modern surgical interventions are increasing used in patients with pulmonary hypertension. Surgical pulmonary endarterectomy (PEA), lung transplantation, and pulmonary artery balloon dilatation are performed to provide relief to pulmonary hypertension patients (Zhang et al. 2022).

Recently, omics studies have unveiled some novel therapeutic targets. Several novel drug candidates and biologics have entered preclinical and clinical trials for

Table 10.2 Drugs for the treatment of pulmonary arterial hypertension (PAH)

Sl. no.	Category of drugs	Drug names	Function
1.	Endothelin receptor antagonists	Bosentan, ambrisentan, macitentan	Functions by reducing endothelin level in blood, thereby preventing vasoconstriction of blood vessels in lungs It results in reduction of blood pressure in pulmonary arteries (Raja and Raja 2011; Varela-Chinchilla et al. 2022)
2.	Phosphodiesterase 5 inhibitors	Sildenafil, tadalafil	Leads to relaxation of blood vessels by blocking phosphodiesterase 5 enzyme. Increases blood flow to lungs (Raja and Raja 2011; Varela-Chinchilla et al. 2022)
3.	Prostacyclin analogues	Epoprostenol, treprostinil (intravenous administration), iloprost (nebulized)	Functions as potent vasodilators and increases blood flow (Raja and Raja 2011)
4.	Calcium channel blockers	Nifedipine, diltiazem, nicardipine, amlodipine	Serves as vasodilators and primarily dilates blood vessels of heart, lungs, fingers, and toes (Galie et al. 2016)
5.	Soluble guanylate cyclase stimulators (sGCS)	Riociguat	Stimulates the nitric oxide-sGC-cGMP pathway. Activation of this pathway and stimulation of soluble guanylate cyclase (sGC) lead to vasodilation (Raja and Raja 2011; Varela-Chinchilla et al. 2022)
6.	Prostacyclin receptor agonists	Selexipag	Leads to vasodilation (Galie et al. 2016)
7.	Anticoagulant medicines (blood thinners)	Warfarin	Prevents formation of blood clots and hence prevents blocking of pulmonary arteries by blood clots (Lai et al. 2022).
8.	Diuretics	Furosemide, bumetiaide, spironolactone	Prevents water and salt retention from the system (Galie et al. 2016; Varela-Chinchilla et al. 2022)
9.	Inotropic agents (cardiac glycosides)	Digoxin	Strengthens heart muscle contractions and aids in more efficient pumping by heart (Varela-Chinchilla et al. 2022)

This table enlists the different medications widely used for the treatment of pulmonary arterial hypertension

evaluating their safety and effectiveness in pulmonary hypertension patients. A list of few drug candidates and biologics that are clinically investigated for pulmonary hypertension have been provided in Table 10.3.

Several other probable drug candidates like tamoxifen, trimetazidine, olaparib, metformin, nab-rapamycin, dehydroepiandrosterone, angiotensin-converting

Table 10.3 Biologics and other drugs tested for the treatment of pulmonary hypertension

Sl. no.	Name	Description	Comments	Clinical investigation stage
1.	Sotatercept	Fusion protein comprising of the extracellular domain of human activin receptor type IIA that is fused to the Fc domain of human IgG1	Reduced pulmonary vascular resistance in pulmonary arterial hypertension patients receiving background therapy (Zolty 2021; Sommer et al. 2021; Komrokji et al. 2018)	Phase III
2.	Imatinib	Tyrosine kinase inhibitor	Exhibited prolonged efficiency in patients with severe pulmonary arterial hypertension in Phase II trial Although effective in pulmonary arterial hypertension patients in Phase III trial, it had severe adverse effects. So it was discontinued (Zolty 2021; Raja and Raja 2011; Sommer et al. 2021)	Phase III
3.	Selonsertib	Inhibitor of apoptosis signal-regulating kinase 1 (ASK1)	Safe and well tolerated in patients with pulmonary arterial hypertension, but did not lead to significant clinical improvement (Zolty 2021; Sommer et al. 2021)	Phase II
4.	Aviptadil	Vasoactive intestinal peptide (VIP)	Improved oxygenation in patients with pulmonary hypertension. Exerts modest, short lived pulmonary vasodilating effect. Reduced the workload of the right ventricle (Raja and Raja 2011; Leuchte et al. 2008)	Phase II
5.	Levosimendan	Calcium sensitizer having inotropic properties	Significantly improved pulmonary capillary wedge pressure (PCWP) during supine exercise and significantly decreased pulmonary arterial pressure in pulmonary hypertension patients with preserved	Phase II

(continued)

Table 10.3 (continued)

Sl. no.	Name	Description	Comments	Clinical investigation stage
			ejection fraction (Burkhoff et al. 2021)	
6.	Rituximab	Anti-CD20 antibody	Brings about change in pulmonary vascular resistance (PVR) in patients with systemic sclerosis-associated pulmonary arterial hypertension (Zolty 2021; Sommer et al. 2021)	Phase II
7.	FK506 (Tacrolimus)	Macrolide having immunosuppressive properties Inhibits calcineurin	Well tolerated at low levels. Increased BMPR2 expression in subsets of patients having pulmonary arterial hypertension	Phase II
8.	Racecadotril	Inhibitor of neprilysin	Increased plasma level of ANP and cGMP; and reduced pulmonary vascular resistance in patients with pulmonary arterial hypertension (Sommer et al. 2021; Hobbs et al. 2019)	Phase II
9.	Anakinra	IL-1 receptor antagonist (IL-1ra)	Reduced inflammation, improved ventilatory efficiency and exercise capacity in patients with pulmonary arterial hypertension	Phase IB/II
10.	Fulvestrant	Oestrogen receptor antagonist	Increased stroke volume, increased 6-minute walk distance, and decreased 16α-hydroxyestradiol (16OHE2) and circulating haematopoietic progenitor cells (HPCs) in pulmonary arterial hypertension patients (Zolty 2021; Sommer et al. 2021)	Phase II
11.	Tocilizumab	Anti-IL-6 antibody	Did not exhibit any significant change in pulmonary vascular resistance in pulmonary	Phase II

(continued)

Table 10.3 (continued)

Sl. no.	Name	Description	Comments	Clinical investigation stage
			arterial hypertension (Zolty 2021; Sommer et al. 2021)	
12.	Anastrozole	Aromatase Inhibitor	Safe and well tolerated. Reduced 17β-oestradiol levels and improved 6-minute-walk distance in pulmonary arterial hypertension patients (Zolty 2021; Sommer et al. 2021; Sitbon et al. 2019)	Phase II
13.	Apabetalone	Small molecule inhibitor of bromo-domain containing protein 4 (BRD4)	Reduced pulmonary vascular resistance (PVR), improved cardiac output (CO) and stroke volume (SV) in patients with pulmonary arterial hypertension (Zolty 2021; Sommer et al. 2021; Hsieh et al. 2022)	Phase I
14.	Ranolazine	Selective late-sodium current inhibitor	Improved right ventricular (RV) function in patients having pre-capillary pulmonary hypertension (Sommer et al. 2021; Sitbon et al. 2019).	Phase IV
15.	Bardoxolone methyl	Semi-synthetic triterpenoid that activates Nrf2 pathway and inhibits NK-κβ pathway.	Improved 6-min walk distance in patients with connective tissue disease-associated pulmonary arterial hypertension (Sommer et al. 2021; Sitbon et al. 2019)	Phase II
16.	Ferinject or CosmoFer	Iron infusion	Well tolerated but had no clinical benefit in patients with idiopathic and heritable pulmonary arterial hypertension (Sommer et al. 2021)	Phase II

This table enlists the different drugs and biologics that are under clinical investigation for the treatment of pulmonary hypertension enzyme 2 (ACE2), and spironolactone have recently entered into clinical trials for evaluation of their efficacy against pulmonary hypertension and are presently recruiting for clinical trials. Besides these, inhibition of ROCK and kinases may be explored as a potential treatment for pulmonary hypertension (Sommer et al. 2021; Weiss et al. 2021).

Although several drug candidates are being tested for pulmonary hypertension, the process of drug evaluation is tedious. Besides, all the currently used medications have moderate to high levels of side-effects (Grinnan et al. 2019). Therefore, the knowledge on disease pathogenesis and novel biomarkers as identified by modern state-of-art omics techniques should be exploited for identification of new drug targets, repurposing of drugs and designing novel therapeutics.

10.15 Future Perspectives

Several important research advancements have taken place with regard to pulmonary hypertension, which have laid the foundation stone for further developments in this field. However, the medical fraternity is still far from curing pulmonary hypertension or reducing its high mortality rate. Newer screening technologies and biomarkers should be validated to enable early disease detection (Deshwal et al. 2021). Integration of multi-omics data and network analyses should be performed more for identifying treatable targets and disease modifying therapeutic strategies. Determining the role of metabolic and immune dysregulation to a deeper level may be of vital importance in understanding some of the global changes occurring in multiple organs in pulmonary hypertension patients (Gurtu and Michelakis 2015). Systems biology approaches should be extensively used to determine the possible role of miRNAs in disease pathogenesis, identify their targets, and validate miRNAs as potential therapeutic targets for pulmonary hypertension (Yuan et al. 2013). Techniques like lineage tracing and improved 3D imaging of human and experimental lung and heart tissues may aid in improving our understanding of the cellular involvement in disease pathobiology (Spiekerkoetter et al. 2019). These might aid in understanding the disease both at cellular and genetic level and thereby aid in effective disease phenotyping. Bayesian network modelling and supervised machine learning techniques based on interaction of several variables (phenotypic, clinical, and imaging features) should be explored for efficient pulmonary hypertension risk stratification and survival prediction (Sweatt et al. 2021). Exploring the different avenues mentioned above may improve diagnosis, treatment, and eventually lead to development of precision medicine for pulmonary hypertension.

References

Abdul-Salam VB, Russomanno G, Chien-Nien C, Mahomed AS, Yates LA, Wilkins MR, Zhao L, Gierula M, Dubois O, Schaeper U et al (2019) CLIC4/Arf6 Pathway. Circ Res 124(1):52–65

Al-Naamani N, Sagliani KD, Dolnikowski GG, Warburton RR, Toksoz D, Kayyali U, Hill NS, Fanburg BL, Roberts KE, Preston IR (2016) Plasma 12- and 15-hydroxyeicosanoids are predictors of survival in pulmonary arterial hypertension. Pulm Circ 6(2):224–233

Altschul E, Remy-Jardin M, Machnicki S, Sulica R, Moore JA, Singh A, Raoof S (2019) Imaging of pulmonary hypertension: pictorial essay. Chest 156(2):211–227

Aluja Jaramillo F, Gutierrez FR, Diaz Telli FG, Yevenes Aravena S, Javidan-Nejad C, Bhalla S (2018) Approach to pulmonary hypertension: from CT to clinical diagnosis. Radiographics 38(2):357–373

Anwar A, Ruffenach G, Mahajan A, Eghbali M, Umar S (2016) Novel biomarkers for pulmonary arterial hypertension. Respir Res 17(1):88

Ascha M, Renapurkar RD, Tonelli AR (2017) A review of imaging modalities in pulmonary hypertension. Ann Thorac Med 12(2):61–73

Atkinson C, Stewart S, Upton PD, Machado R, Thomson JR, Trembath RC, Morrell NW (2002) Primary pulmonary hypertension is associated with reduced pulmonary vascular expression of type II bone morphogenetic protein receptor. Circulation 105(14):1672–1678

Bendayan D, Shitrit D, Ygla M, Huerta M, Fink G, Kramer MR (2003) Hyperuricemia as a prognostic factor in pulmonary arterial hypertension. Respir Med 97(2):130–133

Benincasa G, Maron BA, Affinito O, D'Alto M, Franzese M, Argiento P, Schiano C, Romeo E, Bontempo P, Golino P et al (2022) Association between circulating CD4(+) T cell methylation signatures of network-oriented SOCS3 gene and hemodynamics in patients suffering pulmonary arterial hypertension. J Cardiovasc Transl Res

Benza RL, Gomberg-Maitland M, Demarco T, Frost AE, Torbicki A, Langleben D, Pulido T, Correa-Jaque P, Passineau MJ, Wiener HW et al (2015) Endothelin-1 pathway polymorphisms and outcomes in pulmonary arterial hypertension. Am J Respir Crit Care Med 192(11): 1345–1354

Bisserier M, Janostiak R, Lezoualc'h F, Hadri L (2020) Targeting epigenetic mechanisms as an emerging therapeutic strategy in pulmonary hypertension disease. Vasc Biol 2(1):R17–R34

Bogdan M, Humbert M, Francoual J, Claise C, Duroux P, Simonneau G, Lindenbaum A (1998) Urinary cGMP concentrations in severe primary pulmonary hypertension. Thorax 53(12): 1059–1062

Bossini-Castillo L, Campillo-Davo D, Lopez-Isac E, Carmona FD, Simeon CP, Carreira P, Callejas-Rubio JL, Castellvi I, Fernandez-Nebro A, Rodriguez-Rodriguez L et al (2017) An MIF promoter polymorphism is associated with susceptibility to pulmonary arterial hypertension in diffuse cutaneous systemic sclerosis. J Rheumatol 44(10):1453–1457

Bowers R, Cool C, Murphy RC, Tuder RM, Hopken MW, Flores SC, Voelkel NF (2004) Oxidative stress in severe pulmonary hypertension. Am J Respir Crit Care Med 169(6):764–769

Brittain EL, Talati M, Fessel JP, Zhu H, Penner N, Calcutt MW, West JD, Funke M, Lewis GD, Gerszten RE et al (2016) Fatty acid metabolic defects and right ventricular lipotoxicity in human pulmonary arterial hypertension. Circulation 133(20):1936–1944

Burkhoff D, Borlaug BA, Shah SJ, Zolty R, Tedford RJ, Thenappan T, Zamanian RT, Mazurek JA, Rich JD, Simon MA et al (2021) Levosimendan Improves Hemodynamics and Exercise Tolerance in PH-HFpEF: Results of the Randomized Placebo-Controlled HELP Trial. JACC Heart Fail 9(5):360–370

Chen Y, Ye C, Chen J, Lin D, Wang H, Wang S (2019) Association of the gene polymorphisms of BMPR2, ACVRL1, SMAD9 and their interactions with the risk of essential hypertension in the Chinese Han population. Biosci Rep 39(1)

Cracowski JL, Degano B, Chabot F, Labarere J, Schwedhelm E, Monneret D, Iuliano L, Schwebel C, Chaouat A, Reynaud-Gaubert M et al (2012) Independent association of urinary F2-isoprostanes with survival in pulmonary arterial hypertension. Chest 142(4):869–876

Damico R, Kolb TM, Valera L, Wang L, Housten T, Tedford RJ, Kass DA, Rafaels N, Gao L, Barnes KC et al (2015) Serum endostatin is a genetically determined predictor of survival in pulmonary arterial hypertension. Am J Respir Crit Care Med 191(2):208–218

Deshwal H, Weinstein T, Sulica R (2021) Advances in the management of pulmonary arterial hypertension. J Investig Med 69(7):1270–1280

Dromparis P, Michelakis ED (2012) F2-isoprostanes: an emerging pulmonary arterial hypertension biomarker and potential link to the metabolic theory of pulmonary arterial hypertension? Chest 142(4):816–820

Duarte JD, Kansal M, Desai AA, Riden K, Arwood MJ, Yacob AA, Stamos TD, Cavallari LH, Zamanian RT, Shah SJ et al (2018) Endothelial nitric oxide synthase genotype is associated with pulmonary hypertension severity in left heart failure patients. Pulm Circ 8(2): 2045894018773049

Durmus S, Atahan E, Avci Kilickiran B, Onal B, Cakatay U, Gelisgen R, Uzun H (2022) Significance of Cyclooxgenase-2 gene polymorphism and related miRNAs in pulmonary arterial hypertension. Clin Biochem 107:33–39

Elinoff JM, Mazer AJ, Cai R, Lu M, Graninger G, Harper B, Ferreyra GA, Sun J, Solomon MA, Danner RL (2020) Meta-analysis of blood genome-wide expression profiling studies in pulmonary arterial hypertension. Am J Physiol Lung Cell Mol Physiol 318(1):L98–L111

Ezenwa C, Soremekun OS, Nashiru O, Fatumo S (2020) Identification of differentially expressed genes present in the whole blood of Pulmonary Arterial Hypertension patients and control patients: An integrated bioinformatics approach. Informatics in Medicine Unlocked 20

Frost A, Badesch D, Gibbs JSR, Gopalan D, Khanna D, Manes A, Oudiz R, Satoh T, Torres F, Torbicki A (2019) Diagnosis of pulmonary hypertension. Eur Respir J 53(1)

Galie N, Humbert M, Vachiery JL, Gibbs S, Lang I, Torbicki A, Simonneau G, Peacock A, Vonk Noordegraaf A, Beghetti M et al (2016) 2015 ESC/ERS Guidelines for the diagnosis and treatment of pulmonary hypertension: The Joint Task Force for the Diagnosis and Treatment of Pulmonary Hypertension of the European Society of Cardiology (ESC) and the European Respiratory Society (ERS): Endorsed by: Association for European Paediatric and Congenital Cardiology (AEPC), International Society for Heart and Lung Transplantation (ISHLT). Eur Heart J 37(1):67–119

Germain M, Eyries M, Montani D, Poirier O, Girerd B, Dorfmuller P, Coulet F, Nadaud S, Maugenre S, Guignabert C et al (2013) Genome-wide association analysis identifies a susceptibility locus for pulmonary arterial hypertension. Nat Genet 45(5):518–521

Ghofrani HA, Wiedemann R, Rose F, Weissmann N, Schermuly RT, Quanz K, Grimminger F, Seeger W, Olschewski H (2002) Lung cGMP release subsequent to NO inhalation in pulmonary hypertension: responders versus nonresponders. Eur Respir J 19(4):664–671

Giannakoulas G, Mouratoglou SA, Gatzoulis MA, Karvounis H (2014) Blood biomarkers and their potential role in pulmonary arterial hypertension associated with congenital heart disease. A systematic review. Int J Cardiol 174(3):618–623

Gillmeyer KR, Lee MM, Link AP, Klings ES, Rinne ST, Wiener RS (2019) Accuracy of algorithms to identify pulmonary arterial hypertension in administrative data: a systematic review. Chest 155(4):680–688

Gorenflo M, Zheng C, Werle E, Fiehn W, Ulmer HE (2001) Plasma levels of asymmetrical dimethyl-L-arginine in patients with congenital heart disease and pulmonary hypertension. J Cardiovasc Pharmacol 37(4):489–492

Grigoryev DN, Mathai SC, Fisher MR, Girgis RE, Zaiman AL, Housten-Harris T, Cheadle C, Gao L, Hummers LK, Champion HC et al (2008) Identification of candidate genes in scleroderma-related pulmonary arterial hypertension. Transl Res 151(4):197–207

Grinnan D, Trankle C, Andruska A, Bloom B, Spiekerkoetter E (2019) Drug repositioning in pulmonary arterial hypertension: challenges and opportunities. Pulm Circ 9(1): 2045894019832226

Guignabert C, Tu L, Le Hiress M, Ricard N, Sattler C, Seferian A, Huertas A, Humbert M, Montani D (2013) Pathogenesis of pulmonary arterial hypertension: lessons from cancer. Eur Respir Rev 22(130):543–551

Guignabert C, Tu L, Girerd B, Ricard N, Huertas A, Montani D, Humbert M (2015) New molecular targets of pulmonary vascular remodeling in pulmonary arterial hypertension: importance of endothelial communication. Chest 147(2):529–537

Guo K, Xu L, Jin L, Wang H, Ren Y, Hu Y, Yu J, Cang J (2020) Bone morphogenetic protein 9, and its genetic variants contribute to susceptibility of idiopathic pulmonary arterial hypertension. Aging (Albany NY) 12(3):2123–2131

Gurtu V, Michelakis ED (2015) Emerging therapies and future directions in pulmonary arterial hypertension. Can J Cardiol 31(4):489–501

Harbaum L, Ghataorhe P, Wharton J, Jimenez B, Howard LSG, Gibbs JSR, Nicholson JK, Rhodes CJ, Wilkins MR (2019) Reduced plasma levels of small HDL particles transporting fibrinolytic proteins in pulmonary arterial hypertension. Thorax 74(4):380–389

Harbaum L, Rhodes CJ, Otero-Nunez P, Wharton J, Wilkins MR (2021) The application of 'omics' to pulmonary arterial hypertension. Br J Pharmacol 178(1):108–120

Hautefort A, Chesne J, Preussner J, Pullamsetti SS, Tost J, Looso M, Antigny F, Girerd B, Riou M, Eddahibi S et al (2017) Pulmonary endothelial cell DNA methylation signature in pulmonary arterial hypertension. Oncotarget 8(32):52995–53016

He W, Liu C, Liao J, Liu F, Lei H, Wei D, Ruan H, Kunwar B, Lu W, Wang J et al (2021) TIMP-1: a circulating biomarker for pulmonary hypertension diagnosis among chronic obstructive pulmonary disease patients. Front Med (Lausanne) 8:774623

Hemnes AR, Luther JM, Rhodes CJ, Burgess JP, Carlson J, Fan R, Fessel JP, Fortune N, Gerszten RE, Halliday SJ et al (2019) Human PAH is characterized by a pattern of lipid-related insulin resistance. JCI Insight 4(1)

Hemnes A, Rothman AMK, Swift AJ, Zisman LS (2020) Role of biomarkers in evaluation, treatment and clinical studies of pulmonary arterial hypertension. Pulm Circ 10(4): 2045894020957234

Hobbs AJ, Moyes AJ, Baliga RS, Ghedia D, Ochiel R, Sylvestre Y, Dore CJ, Chowdhury K, Maclagan K, Quartly HL et al (2019) Neprilysin inhibition for pulmonary arterial hypertension: a randomized, double-blind, placebo-controlled, proof-of-concept trial. Br J Pharmacol 176(9): 1251–1267

Hoeper MM, Hohlfeld JM, Fabel H (1999) Hyperuricaemia in patients with right or left heart failure. Eur Respir J 13(3):682–685

Hoeper MM, Ghofrani HA, Grunig E, Klose H, Olschewski H, Rosenkranz S (2017) Pulmonary hypertension. Dtsch Arztebl Int 114(5):73–84

Hsieh MW, Wang WT, Yeh JL, Lin CY, Kuo YR, Lee SS, Hou MF, Wu YC (2022) The potential application and promising role of targeted therapy in pulmonary arterial hypertension. Biomedicine 10(6)

Huang L, Zhang H, Liu Y, Long Y (2022) The role of gut and airway microbiota in pulmonary arterial hypertension. Front Microbiol 13:929752

Hudson J, Farkas L (2021) Epigenetic regulation of endothelial dysfunction and inflammation in pulmonary arterial hypertension. Int J Mol Sci 22(22)

Humbert M, Kovacs G, Hoeper MM, Badagliacca R, Berger RMF, Brida M, Carlsen J, Coats AJS, Escribano-Subias P, Ferrari P et al (2022a) 2022 ESC/ERS Guidelines for the diagnosis and treatment of pulmonary hypertension. Eur Heart J 43(38):3618–3731

Humbert M, Kovacs G, Hoeper MM, Badagliacca R, Berger RMF, Brida M, Carlsen J, Coats AJS, Escribano-Subias P, Ferrari P et al (2022b) 2022 ESC/ERS Guidelines for the diagnosis and treatment of pulmonary hypertension. Eur Respir J

Jasiewicz M, Moniuszko M, Pawlak D, Knapp M, Rusak M, Kazimierczyk R, Musial WJ, Dabrowska M, Kaminski KA (2016) Activity of the kynurenine pathway and its interplay with immunity in patients with pulmonary arterial hypertension. Heart 102(3):230–237

Jiao YR, Wang W, Lei PC, Jia HP, Dong J, Gou YQ, Chen CL, Cao J, Wang YF, Zhu YK (2019) 5-HTT, BMPR2, EDN1, ENG, KCNA5 gene polymorphisms and susceptibility to pulmonary arterial hypertension: a meta-analysis. Gene 680:34–42

Jose A, Apewokin S, Hussein WE, Ollberding NJ, Elwing JM, Haslam DB (2022) A unique gut microbiota signature in pulmonary arterial hypertension: a pilot study. Pulm Circ 12(1):e12051

Kaneko FT, Arroliga AC, Dweik RA, Comhair SA, Laskowski D, Oppedisano R, Thomassen MJ, Erzurum SC (1998) Biochemical reaction products of nitric oxide as quantitative markers of primary pulmonary hypertension. Am J Respir Crit Care Med 158(3):917–923

Kerstjens-Frederikse WS, Bongers EM, Roofthooft MT, Leter EM, Douwes JM, Van Dijk A, Vonk-Noordegraaf A, Dijk-Bos KK, Hoefsloot LH, Hoendermis ES et al (2013) TBX4 mutations (small patella syndrome) are associated with childhood-onset pulmonary arterial hypertension. J Med Genet 50(8):500–506

Kielstein JT, Bode-Boger SM, Hesse G, Martens-Lobenhoffer J, Takacs A, Fliser D, Hoeper MM (2005) Asymmetrical dimethylarginine in idiopathic pulmonary arterial hypertension. Arterioscler Thromb Vasc Biol 25(7):1414–1418

Kim S, Rigatto K, Gazzana MB, Knorst MM, Richards EM, Pepine CJ, Raizada MK (2020) Altered gut microbiome profile in patients with pulmonary arterial hypertension. Hypertension 75(4): 1063–1071

Komrokji R, Garcia-Manero G, Ades L, Prebet T, Steensma DP, Jurcic JG, Sekeres MA, Berdeja J, Savona MR, Beyne-Rauzy O et al (2018) Sotatercept with long-term extension for the treatment of anaemia in patients with lower-risk myelodysplastic syndromes: a phase 2, dose-ranging trial. Lancet Haematol 5(2):e63–e72

Lai J, Feng S, Xu S, Liu X (2022) Effects of oral anticoagulant therapy in patients with pulmonary diseases. Front Cardiovasc Med 9:987652

Lavoie JR, Ormiston ML, Perez-Iratxeta C, Courtman DW, Jiang B, Ferrer E, Caruso P, Southwood M, Foster WS, Morrell NW et al (2014) Proteomic analysis implicates translationally controlled tumor protein as a novel mediator of occlusive vascular remodeling in pulmonary arterial hypertension. Circulation 129(21):2125–2135

Lawrie A, Waterman E, Southwood M, Evans D, Suntharalingam J, Francis S, Crossman D, Croucher P, Morrell N, Newman C (2008) Evidence of a role for osteoprotegerin in the pathogenesis of pulmonary arterial hypertension. Am J Pathol 172(1):256–264

Leuchte HH, Baezner C, Baumgartner RA, Bevec D, Bacher G, Neurohr C, Behr J (2008) Inhalation of vasoactive intestinal peptide in pulmonary hypertension. Eur Respir J 32(5): 1289–1294

Lewis GD (2014) The emerging role of metabolomics in the development of biomarkers for pulmonary hypertension and other cardiovascular diseases (2013 Grover Conference series). Pulm Circ 4(3):417–423

Lewis GD, Ngo D, Hemnes AR, Farrell L, Domos C, Pappagianopoulos PP, Dhakal BP, Souza A, Shi X, Pugh ME et al (2016) Metabolic profiling of right ventricular-pulmonary vascular function reveals circulating biomarkers of pulmonary hypertension. J Am Coll Cardiol 67(2): 174–189

Li D, Sun Y, Kong X, Luan C, Yu Y, Chen F, Chen P (2018) Association between a single nucleotide polymorphism in the 3'-UTR of ARHGEF18 and the risk of nonidiopathic pulmonary arterial hypertension in Chinese population. Dis Markers 2018:2461845

Li M, Cai W, Chen Y, Dong L (2019) The CAV1 Gene 3' untranslated region single nucleotide polymorphisms are associated with the risk of pulmonary hypertension in Chinese Han chronic obstructive pulmonary patients. Genet Test Mol Biomarkers 23(9):634–643

Li N, Zhu L, Zhu C, Zhou H, Zheng D, Xu G, Shi H, Gao J, Li AJ, Wang Z et al (2021a) BMPR2 promoter methylation and its expression in valvular heart disease complicated with pulmonary artery hypertension. Aging (Albany NY) 13(22):24580–24604

Li A, He J, Zhang Z, Jiang S, Gao Y, Pan Y, Wang H, Zhuang L (2021b) Integrated bioinformatics analysis reveals marker genes and potential therapeutic targets for pulmonary arterial hypertension. Genes (Basel) 12(9)

Liu D, Wu WH, Mao YM, Yuan P, Zhang R, Ju FL, Jing ZC (2012) BMPR2 mutations influence phenotype more obviously in male patients with pulmonary arterial hypertension. Circ Cardiovasc Genet 5(5):511–518

Liu D, Yan Y, Chen JW, Yuan P, Wang XJ, Jiang R, Wang L, Zhao QH, Wu WH, Simonneau G et al (2017) Hypermethylation of BMPR2 promoter occurs in patients with heritable pulmonary arterial hypertension and inhibits BMPR2 expression. Am J Respir Crit Care Med 196(7): 925–928

Liu X, Mei M, Chen X, Lu Y, Dong X, Hu L, Hu X, Cheng G, Cao Y, Yang L et al (2019) Identification of genetic factors underlying persistent pulmonary hypertension of newborns in a cohort of Chinese neonates. Respir Res 20(1):174

Ma C, Li Y, Ma J, Liu Y, Li Q, Niu S, Shen Z, Zhang L, Pan Z, Zhu D (2011) Key role of 15-lipoxygenase/15-hydroxyeicosatetraenoic acid in pulmonary vascular remodeling and vascular angiogenesis associated with hypoxic pulmonary hypertension. Hypertension 58(4): 679–688

Machado RD, Pauciulo MW, Thomson JR, Lane KB, Morgan NV, Wheeler L, Phillips JA 3rd, Newman J, Williams D, Galie N et al (2001) BMPR2 haploinsufficiency as the inherited molecular mechanism for primary pulmonary hypertension. Am J Hum Genet 68(1):92–102

Malenfant S, Potus F, Fournier F, Breuils-Bonnet S, Pflieger A, Bourassa S, Tremblay E, Nehme B, Droit A, Bonnet S et al (2015) Skeletal muscle proteomic signature and metabolic impairment in pulmonary hypertension. J Mol Med (Berl) 93(5):573–584

Malhotra R, Paskin-Flerlage S, Zamanian RT, Zimmerman P, Schmidt JW, Deng DY, Southwood M, Spencer R, Lai CS, Parker W et al (2013) Circulating angiogenic modulatory factors predict survival and functional class in pulmonary arterial hypertension. Pulm Circ 3(2): 369–380

Mansoor JK, Schelegle ES, Davis CE, Walby WF, Zhao W, Aksenov AA, Pasamontes A, Figueroa J, Allen R (2014) Analysis of volatile compounds in exhaled breath condensate in patients with severe pulmonary arterial hypertension. PLoS One 9(4):e95331

Mathai SC, Hemnes AR, Manaker S, Anguiano RH, Dean BB, Saundankar V, Classi P, Nelsen AC, Gordon K, Ventetuolo CE (2019) Identifying patients with pulmonary arterial hypertension using administrative claims algorithms. Ann Am Thorac Soc 16(7):797–806

McGlinchey N, Johnson MK (2014) Novel serum biomarkers in pulmonary arterial hypertension. Biomark Med 8(8):1001–1011

McGoon MD, Miller DP (2012) REVEAL: a contemporary US pulmonary arterial hypertension registry. Eur Respir Rev 21(123):8–18

McGoon MD, Krichman A, Farber HW, Barst RJ, Raskob GE, Liou TG, Miller DP, Feldkircher K, Giles S (2008) Design of the REVEAL registry for US patients with pulmonary arterial hypertension. Mayo Clin Proc 83(8):923–931

Mei M, Cheng G, Sun B, Yang L, Wang H, Sun J, Zhou W (2016) EDN1 gene variant is associated with neonatal persistent pulmonary hypertension. Sci Rep 6:29877

Mumby S, Gambaryan N, Meng C, Perros F, Humbert M, Wort SJ, Adcock IM (2017) Bromodomain and extra-terminal protein mimic JQ1 decreases inflammation in human vascular endothelial cells: implications for pulmonary arterial hypertension. Respirology 22(1):157–164

Nagaya N, Uematsu M, Satoh T, Kyotani S, Sakamaki F, Nakanishi N, Yamagishi M, Kunieda T, Miyatake K (1999) Serum uric acid levels correlate with the severity and the mortality of primary pulmonary hypertension. Am J Respir Crit Care Med 160(2):487–492

Nagy BM, Nagaraj C, Meinitzer A, Sharma N, Papp R, Foris V, Ghanim B, Kwapiszewska G, Kovacs G, Klepetko W et al (2017) Importance of kynurenine in pulmonary hypertension. Am J Physiol Lung Cell Mol Physiol 313(5):L741–L751

Nakwan N, Mahasirimongkol S, Satproedprai N, Chaiyasung T, Kunhapan P, Charoenlap C, Singkhamanan K, Charalsawadi C (2022) Possible association between a polymorphism of EPAS1 gene and persistent pulmonary hypertension of the newborn: a case-control study. J Pediatr 98(4):383–389

Nozik-Grayck E, Woods C, Stearman RS, Venkataraman S, Ferguson BS, Swain K, Bowler RP, Geraci MW, Ihida-Stansbury K, Stenmark KR et al (2016) Histone deacetylation contributes to low extracellular superoxide dismutase expression in human idiopathic pulmonary arterial hypertension. Am J Physiol Lung Cell Mol Physiol 311(1):L124–L134

Papani R, Sharma G, Agarwal A, Callahan SJ, Chan WJ, Kuo YF, Shim YM, Mihalek AD, Duarte AG (2018) Validation of claims-based algorithms for pulmonary arterial hypertension. Pulm Circ 8(2):2045894018759246

Paulin R, Dromparis P, Sutendra G, Gurtu V, Zervopoulos S, Bowers L, Haromy A, Webster L, Provencher S, Bonnet S et al (2014) Sirtuin 3 deficiency is associated with inhibited mitochondrial function and pulmonary arterial hypertension in rodents and humans. Cell Metab 20(5): 827–839

Pullamsetti SS, Perros F, Chelladurai P, Yuan J, Stenmark K (2016) Transcription factors, transcriptional coregulators, and epigenetic modulation in the control of pulmonary vascular cell phenotype: therapeutic implications for pulmonary hypertension (2015 Grover Conference series). Pulm Circ 6(4):448–464

Qin X, Li T, Sun W, Guo X, Fang Q (2021) Proteomic analysis of pulmonary arterial hypertension. Ther Adv Chronic Dis 12:20406223211047304

Raja SG, Raja SM (2011) Treating pulmonary arterial hypertension: current treatments and future prospects. Ther Adv Chronic Dis 2(6):359–370

Rajagopal K, Bryant AJ, Sahay S, Wareing N, Zhou Y, Pandit LM, Karmouty-Quintana H (2021) Idiopathic pulmonary fibrosis and pulmonary hypertension: heracles meets the hydra. Br J Pharmacol 178(1):172–186

Ranchoux B, Bigorgne A, Hautefort A, Girerd B, Sitbon O, Montani D, Humbert M, Tcherakian C, Perros F (2017) Gut-lung connection in pulmonary arterial hypertension. Am J Respir Cell Mol Biol 56(3):402–405

Remy-Jardin M, Ryerson CJ, Schiebler ML, Leung ANC, Wild JM, Hoeper MM, Alderson PO, Goodman LR, Mayo J, Haramati LB et al (2021) Imaging of pulmonary hypertension in adults: a position paper from the Fleischner Society. Radiology 298(3):531–549

Rhodes CJ, Wharton J, Boon RA, Roexe T, Tsang H, Wojciak-Stothard B, Chakrabarti A, Howard LS, Gibbs JS, Lawrie A et al (2013) Reduced microRNA-150 is associated with poor survival in pulmonary arterial hypertension. Am J Respir Crit Care Med 187(3):294–302

Rhodes CJ, Ghataorhe P, Wharton J, Rue-Albrecht KC, Hadinnapola C, Watson G, Bleda M, Haimel M, Coghlan G, Corris PA et al (2017a) Plasma metabolomics implicates modified transfer RNAs and altered bioenergetics in the outcomes of pulmonary arterial hypertension. Circulation 135(5):460–475

Rhodes CJ, Wharton J, Ghataorhe P, Watson G, Girerd B, Howard LS, Gibbs JSR, Condliffe R, Elliot CA, Kiely DG et al (2017b) Plasma proteome analysis in patients with pulmonary arterial hypertension: an observational cohort study. Lancet Respir Med 5(9):717–726

Rhodes CJ, Batai K, Bleda M, Haimel M, Southgate L, Germain M, Pauciulo MW, Hadinnapola C, Aman J, Girerd B et al (2019) Genetic determinants of risk in pulmonary arterial hypertension: international genome-wide association studies and meta-analysis. Lancet Respir Med 7(3): 227–238

Rhodes CJ, Otero-Nunez P, Wharton J, Swietlik EM, Kariotis S, Harbaum L, Dunning MJ, Elinoff JM, Errington N, Thompson AAR et al (2020) Whole-blood RNA profiles associated with pulmonary arterial hypertension and clinical outcome. Am J Respir Crit Care Med 202(4): 586–594

Robbins IM, Morrow JD, Christman BW (2005) Oxidant stress but not thromboxane decreases with epoprostenol therapy. Free Radic Biol Med 38(5):568–574

Ross DJ, Hough G, Hama S, Aboulhosn J, Belperio JA, Saggar R, Van Lenten BJ, Ardehali A, Eghbali M, Reddy S et al (2015) Proinflammatory high-density lipoprotein results from oxidized lipid mediators in the pathogenesis of both idiopathic and associated types of pulmonary arterial hypertension. Pulm Circ 5(4):640–648

Rothman AM, Arnold ND, Pickworth JA, Iremonger J, Ciuclan L, Allen RM, Guth-Gundel S, Southwood M, Morrell NW, Thomas M et al (2016) MicroRNA-140-5p and SMURF1 regulate pulmonary arterial hypertension. J Clin Invest 126(7):2495–2508

Sanli C, Oguz D, Olgunturk R, Tunaoglu FS, Kula S, Pasaoglu H, Gulbahar O, Cevik A (2012) Elevated homocysteine and asymmetric dimethyl arginine levels in pulmonary hypertension associated with congenital heart disease. Pediatr Cardiol 33(8):1323–1331

Santos-Gomes J, Gandra I, Adao R, Perros F, Bras-Silva C (2022) An overview of circulating pulmonary arterial hypertension biomarkers. Front Cardiovasc Med 9:924873

Sarrion I, Milian L, Juan G, Ramon M, Furest I, Carda C, Cortijo Gimeno J, Mata Roig M (2015) Role of circulating miRNAs as biomarkers in idiopathic pulmonary arterial hypertension: possible relevance of miR-23a. Oxidative Med Cell Longev 2015:792846

Sasagawa S, Nishimura Y, Sawada H, Zhang E, Okabe S, Murakami S, Ashikawa Y, Yuge M, Kawaguchi K, Kawase R et al (2016) Comparative transcriptome analysis identifies CCDC80 as a novel gene associated with pulmonary arterial hypertension. Front Pharmacol 7:142

Savale L, Akagi S, Tu L, Cumont A, Thuillet R, Phan C, Le Vely B, Berrebeh N, Huertas A, Jais X et al (2021) Serum and pulmonary uric acid in pulmonary arterial hypertension. Eur Respir J 58(2)

Saygin D, Tabib T, Bittar HET, Valenzi E, Sembrat J, Chan SY, Rojas M, Lafyatis R (2020) Transcriptional profiling of lung cell populations in idiopathic pulmonary arterial hypertension. Pulm Circ 10(1)

Shitrit D, Bendayan D, Bar-Gil-Shitrit A, Huerta M, Rudensky B, Fink G, Kramer MR (2002a) Significance of a plasma D-dimer test in patients with primary pulmonary hypertension. Chest 122(5):1674–1678

Shitrit D, Bendayan D, Rudensky B, Izbicki G, Huerta M, Fink G, Kramer MR (2002b) Elevation of ELISA d-dimer levels in patients with primary pulmonary hypertension. Respiration 69(4): 327–329

Sitbon O, Gomberg-Maitland M, Granton J, Lewis MI, Mathai SC, Rainisio M, Stockbridge NL, Wilkins MR, Zamanian RT, Rubin LJ (2019) Clinical trial design and new therapies for pulmonary arterial hypertension. Eur Respir J 53(1)

Skoro-Sajer N, Mittermayer F, Panzenboeck A, Bonderman D, Sadushi R, Hitsch R, Jakowitsch J, Klepetko W, Kneussl MP, Wolzt M et al (2007) Asymmetric dimethylarginine is increased in chronic thromboembolic pulmonary hypertension. Am J Respir Crit Care Med 176(11): 1154–1160

Sommer N, Ghofrani HA, Pak O, Bonnet S, Provencher S, Sitbon O, Rosenkranz S, Hoeper MM, Kiely DG (2021) Current and future treatments of pulmonary arterial hypertension. Br J Pharmacol 178(1):6–30

Spiekerkoetter E, Goncharova EA, Guignabert C, Stenmark K, Kwapiszewska G, Rabinovitch M, Voelkel N, Bogaard HJ, Graham B, Pullamsetti SS et al (2019) Hot topics in the mechanisms of pulmonary arterial hypertension disease: cancer-like pathobiology, the role of the adventitia, systemic involvement, and right ventricular failure. Pulm Circ 9(4):2045894019889775

Stearman RS, Bui QM, Speyer G, Handen A, Cornelius AR, Graham BB, Kim S, Mickler EA, Tuder RM, Chan SY et al (2019) Systems analysis of the human pulmonary arterial hypertension lung transcriptome. Am J Respir Cell Mol Biol 60(6):637–649

Sweatt AJ, Hedlin HK, Balasubramanian V, Hsi A, Blum LK, Robinson WH, Haddad F, Hickey PM, Condliffe R, Lawrie A et al (2019) Discovery of distinct immune phenotypes using machine learning in pulmonary arterial hypertension. Circ Res 124(6):904–919

Sweatt AJ, Reddy R, Rahaghi FN, Al-Naamani N, American Thoracic Society Pulmonary Circulation Assembly Early Career Working G (2021) What's new in pulmonary hypertension clinical research: lessons from the best abstracts at the 2020 American Thoracic Society International Conference. Pulm Circ 11(3):20458940211040713

Takagi K, Kawamoto M, Higuchi T, Tochimoto A, Harigai M, Kawaguchi Y (2020) Single nucleotide polymorphisms of the HIF1A gene are associated with susceptibility to pulmonary arterial hypertension in systemic sclerosis and contribute to SSc-PAH disease severity. Int J Rheum Dis 23(5):674–680

Tang S, Liu Y, Liu B (2022) Integrated bioinformatics analysis reveals marker genes and immune infiltration for pulmonary arterial hypertension. Sci Rep 12(1):10154

Thenappan T, Ormiston ML, Ryan JJ, Archer SL (2018) Pulmonary arterial hypertension: pathogenesis and clinical management. BMJ 360:j5492

Thenappan T, Khoruts A, Chen Y, Weir EK (2019) Can intestinal microbiota and circulating microbial products contribute to pulmonary arterial hypertension? Am J Physiol Heart Circ Physiol 317(5):H1093–H1101

Torbicki A, Kurzyna M, Kuca P, Fijalkowska A, Sikora J, Florczyk M, Pruszczyk P, Burakowski J, Wawrzynska L (2003) Detectable serum cardiac troponin T as a marker of poor prognosis among patients with chronic precapillary pulmonary hypertension. Circulation 108(7):844–848

Varela-Chinchilla CD, Sanchez-Mejia DE, Trinidad-Calderon PA (2022) Congenital heart disease: the state-of-the-art on its pharmacological therapeutics. J Cardiovasc Dev Dis 9(7)

Wang Y, Huang X, Leng D, Li J, Wang L, Liang Y, Wang J, Miao R, Jiang T (2018) DNA methylation signatures of pulmonary arterial smooth muscle cells in chronic thromboembolic pulmonary hypertension. Physiol Genomics 50(5):313–322

Wei RQ, Zhang WM, Liang Z, Piao C, Zhu G (2022) Identification of signal pathways and hub genes of pulmonary arterial hypertension by bioinformatic analysis. Can Respir J 2022:1394088

Weiss A, Boehm M, Egemnazarov B, Grimminger F, Savai Pullamsetti S, Kwapiszewska G, Schermuly RT (2021) Kinases as potential targets for treatment of pulmonary hypertension and right ventricular dysfunction. Br J Pharmacol 178(1):31–53

Wu P, Zhu T, Tan Z, Chen S, Fang Z (2022) Role of gut microbiota in pulmonary arterial hypertension. Front Cell Infect Microbiol 12:812303

Xu M, Xu M, Han L, Yuan C, Mei Y, Zhang H, Chen S, Sun K, Zhu B (2017) Role for functional SOD2 polymorphism in pulmonary arterial hypertension in a Chinese population. Int J Environ Res Public Health 14(3)

Xu W, Comhair SAA, Chen R, Hu B, Hou Y, Zhou Y, Mavrakis LA, Janocha AJ, Li L, Zhang D et al (2019) Integrative proteomics and phosphoproteomics in pulmonary arterial hypertension. Sci Rep 9(1):18623

Yanagisawa R, Kataoka M, Inami T, Momose Y, Kawakami T, Takei M, Kimura M, Isobe S, Yamakado M, Fukuda K et al (2015) Usefulness of circulating amino acid profile and Fischer ratio to predict severity of pulmonary hypertension. Am J Cardiol 115(6):831–836

Yang D, Liu Z, Yang Z (2013) Ghrelin and its relation with N-terminal brain natriuretic peptide, endothelin-1 and nitric oxide in patients with idiopathic pulmonary hypertension. Cardiology 124(4):241–245

Yang X, Wang C, Lin Y, Zhang P (2022) Identification of crucial hub genes and differential T cell infiltration in idiopathic pulmonary arterial hypertension using bioinformatics strategies. Front Mol Biosci 9:800888

Yin C, Li K, Yu Y, Huang H, Yu Y, Wang Z, Yan J, Pu Y, Li Z, Li D et al (2018) Genome-wide association study identifies loci and candidate genes for non-idiopathic pulmonary hypertension in Eastern Chinese Han population. BMC Pulm Med 18(1):158

Yuan K, Orcholski M, Tian X, Liao X, de Jesus Perez VA (2013) MicroRNAs: promising therapeutic targets for the treatment of pulmonary arterial hypertension. Expert Opin Ther Targets 17(5):557–564

Zhang R, Sun ML, Fan YF, Jiang X, Zhao QH, He J, Wang L, Shailendra PK, Safdar Z, Jing ZC (2014) Plasma 15-F2t-isoprostane in idiopathic pulmonary arterial hypertension. Int J Cardiol 175(2):268–273

Zhang S, Yang T, Xu X, Wang M, Zhong L, Yang Y, Zhai Z, Xiao F, Wang C (2015) Oxidative stress and nitric oxide signaling related biomarkers in patients with pulmonary hypertension: a case control study. BMC Pulm Med 15:50

Zhang C, Zhang T, Lu W, Duan X, Luo X, Liu S, Chen Y, Li Y, Chen J, Liao J et al (2020) Altered airway microbiota composition in patients with pulmonary hypertension. Hypertension 76(5): 1589–1599

Zhang Y, Zervopoulos SD, Boukouris AE, Lorenzana-Carrillo MA, Saleme B, Webster L, Liu Y, Haromy A, Tabatabaei Dakhili SA, Ussher JR et al (2021) SNPs for genes encoding the mitochondrial proteins sirtuin3 and uncoupling protein 2 are associated with disease severity, type 2 diabetes, and outcomes in patients with pulmonary arterial hypertension and this is recapitulated in a new mouse model lacking both genes. J Am Heart Assoc 10(23):e020451

Zhang ZQ, Zhu SK, Wang M, Wang XA, Tong XH, Wan JQ, Ding JW (2022) New progress in diagnosis and treatment of pulmonary arterial hypertension. J Cardiothorac Surg 17(1):216

Zhao Y, Peng J, Lu C, Hsin M, Mura M, Wu L, Chu L, Zamel R, Machuca T, Waddell T et al (2014) Metabolomic heterogeneity of pulmonary arterial hypertension. PLoS One 9(2):e88727

Zhao YD, Chu L, Lin K, Granton E, Yin L, Peng J, Hsin M, Wu L, Yu A, Waddell T et al (2015) A biochemical approach to understand the pathogenesis of advanced pulmonary arterial hypertension: metabolomic profiles of arginine, sphingosine-1-phosphate, and heme of human lung. PLoS One 10(8):e0134958

Zheng D, Chen X, Li N, Sun L, Zhou Q, Shi H, Xu G, Liu J, Xu L, Duan S et al (2017) Differentially methylated regions in patients with rheumatic heart disease and secondary pulmonary arterial hypertension. Exp Ther Med 14(2):1367–1372

Zhuo Y, Zeng Q, Zhang P, Li G, Xie Q, Cheng Y (2017) VEGF promoter polymorphism confers an increased risk of pulmonary arterial hypertension in a Chinese population. Yonsei Med J 58(2): 305–311

Zolty R (2021) Novel experimental therapies for treatment of pulmonary arterial hypertension. J Exp Pharmacol 13:817–857

Check for updates

Infectious Pulmonary Diseases

11

Abstract

Different infectious pathogens like viruses, bacteria, and in rare occasion fungi infect the human respiratory tract leading to the onset of upper and lower respiratory tract infections. Infectious pulmonary diseases often display an array of no-specific symptoms like cough, fever, chest congestion, sputum production, dyspnoea, breathing difficulty, malaise, and fatigue. Microscopic examination, sputum culture, serological tests, antigen detection assays, nucleic acid amplification tests, and X-ray radiography are extensively used for disease diagnosis. Antibacterial drugs, anti-viral agents, and anti-fungal agents are used for treatment along with decongestants, expectorants, mucolytics, and antipyretics. In the recent past, several challenges have been encountered in tackling infectious pulmonary diseases including genetic evolution of microbial pathogens and development of antimicrobial drug resistance. Use of modern omics tools can prove to be beneficial in determining the genetic variations leading to rapid disease transmission and increased resistance to therapy. Omics tools shall also aid in identifying biomarkers for improved disease diagnosis and monitoring therapeutic efficacy.

Keywords

Respiratory tract infection · Pathogens · Bacteria · Virus · Fungus · Antibiotics · Anti-fungals · Antimicrobial resistance

11.1 Introduction

Infectious pulmonary diseases are mediated by pathogenic microorganisms like bacteria, viruses, and fungi. These pathogenic microorganisms may infect the upper or the lower respiratory tract or both. Immunosuppression, compromised

S. Saha et al., *Pulmonomics: Omics Approaches for Understanding Pulmonary Diseases*, https://doi.org/10.1007/978-981-99-3505-5_11

immunity, presence of comorbidities, life style habits (like alcoholism and smoking), and extremes of ages serve as risk factors for respiratory tract infection (File 2000). Respiratory tract infections are one of the most common forms of illness affecting individuals all over the world (Monto 2002). Acute lower respiratory tract infections are associated with significant mortality rate (Collaborators 2018). Disease characteristics, host health condition, socio-economic determinants, environmental factors, and malnutrition have been associated with mortality from severe lower respiratory tract infections (Sonego et al. 2015). The rapid genetic evolution of infectious pathogens, the emergence of novel strains with increased transmissibility and infectivity along with the development of antimicrobial drug resistance have also emerged as growing causes of concern in tackling and treating respiratory tract infections.

11.2 Different Types of Infectious Pulmonary Diseases

Classically, respiratory tract infections can be classified as upper respiratory tract infection and lower respiratory tract infection. In case of upper respiratory tract infection, parts of the upper respiratory tract including nose, sinuses, pharynx, larynx, and trachea are affected (https://en.wikipedia.org/wiki/Upper_respiratory_tract_infection). Common upper respiratory tract infections include common cold, sinusitis, tonsillitis, otitis media, pharyngitis, and laryngitis (https://en.wikipedia.org/wiki/Respiratory_disease). Upper respiratory tract infections are primarily caused by viruses and bacteria (https://www.ncbi.nlm.nih.gov/books/NBK532961/). Lower respiratory tract includes the trachea, bronchi, and the lungs. Lower respiratory tract infections include bronchitis and pneumonic illnesses such as COVID-19 and tuberculosis (https://en.wikipedia.org/wiki/Respiratory_disease; Mahashur 2018; Yuki et al. 2020).

Viruses and bacteria account for most of the respiratory tract infections (https://www.ncbi.nlm.nih.gov/books/NBK532961/; https://www.ncbi.nlm.nih.gov/books/NBK8142/). Influenza viruses A and B, parainfluenza viruses, coronaviruses, respiratory syncytial virus (RSV), rhinovirus, and adenovirus are among the common viral pathogens that infect the human respiratory tract (Liu et al. 2015; Woodhead et al. 2005). Common bacterial pathogens that cause respiratory tract infection include *Streptococcus pneumoniae, Haemophilus influenzae, Streptococcus pyogenes,* and *Moraxella catarrhalis* (Cappelletty 1998). Organisms like *Klebsiella pneumoniae, Mycobacterium tuberculosis, Corynebacterium diphtheriae, Corynebacterium ulcerans, Staphylococcus aureus,* and *Mycoplasma pneumoniae* are also frequently encountered causes of respiratory tract infections (Correia et al. 2021; Sun et al. 2022; Otshudiema et al. 2021; Moule and Cirillo 2020).

Fungal infection of the respiratory tract is rare but extremely severe. Such fungal infections are noted in immunocompromised individuals. Fungi like *Aspergillus* spp., *Cryptococcus* spp., *Pneumocystis* spp., *Histoplasma* spp., and *Candida* spp. often result in invasive, life-threatening conditions (Li et al. 2019). Table 11.1

Table 11.1 Different respiratory tract infections

Respiratory tract infection	Causative agents
Influenzae	*Haemophilus influenzae,* Influenzae viruses A and B
Sinusitis	*Streptococcus pneumoniae, Haemophilus influenzae, Moraxella catarrhalis, Staphylococcus aureus, Streptococcus pyogenes,* gram-negative organisms, and anaerobes (https://www.ncbi.nlm.nih.gov/books/NBK8142/)
Common cold	Rhinovirus, parainfluenza virus, influenzae viruses A and B, coronavirus OC43 or 229E, respiratory syncytial virus (RSV), adenovirus, enterovirus (Makela et al. 1998) Bacterial and fungal infections are rare (https://www.ncbi.nlm.nih.gov/books/NBK8142/; Makela et al. 1998)
Tonsillitis	Viral pathogens: Rhinovirus, influenzae virus, respiratory syncytial virus (RSV), adenovirus, coronavirus, Epstein–Barr virus, and cytomegalovirus Bacterial pathogens: *Streptococcus pyogenes,* Group A beta-hemolytic Streptococcus (GABHS), *Staphylococcus aureus, Streptococcus pneumoniae,* and *Haemophilus influenza* (https://www.ncbi.nlm.nih.gov/books/NBK544342/)
Otitis media	Viral pathogens: Respiratory syncytial virus (RSV), influenza virus, parainfluenza virus, rhinovirus, and adenovirus. Bacterial pathogens: *Streptococcus pneumoniae, Haemophilus influenza,* and *Moraxella catarrhalis* (https://www.ncbi.nlm.nih.gov/books/NBK470332/)
Pharyngitis	Viral pathogens: Rhinovirus, influenza virus, adenovirus, coronavirus, and parainfluenza virus Bacterial pathogens: Group A beta-hemolytic streptococci, Group B & C streptococci, *Chlamydia pneumoniae, Mycoplasma pneumoniae, Haemophilus influenzae, Neisseria meningitidis, Neisseria gonorrhoeae, Arcanobacterium haemolyticum, Fusobacterium necrophorum,* and *Corynebacterium diphtheriae* Fungal pathogens: *Candida* spp. (https://www.ncbi.nlm.nih.gov/books/NBK519550/)
Acute laryngitis	Viral pathogens: rhinovirus, parainfluenza virus, respiratory syncytial virus (RSV), coronavirus, adenovirus, and influenza virus Bacterial pathogens: *Streptococcus pneumoniae, Haemophilus influenzae,* and *Moraxella catarrhalis* (https://www.ncbi.nlm.nih.gov/books/NBK534871/)
Mumps	Paramyxovirus (Mumps Virus) (https://www.ncbi.nlm.nih.gov/books/NBK534785/; Rubin et al. 2015)
Measles or Rubeola	Paramyxovirus (*Morbillivirus* spp.) (https://www.ncbi.nlm.nih.gov/books/NBK448068/)
Diphtheria	*Corynebacterium diphtheriae* (Sharma et al. 2019)
Pertussis	*Bordetella pertussis* (Kilgore et al. 2016)
Bronchitis	Viral pathogens: Influenza virus A and B, parainfluenza virus, rhinovirus and respiratory syncytial virus (RSV) (https://www.ncbi.nlm.nih.gov/books/NBK448067/) Dominant bacterial species involved: *Streptococcus pneumonia,*

(continued)

Table 11.1 (continued)

Respiratory tract infection	Causative agents
	Staphylococcus aureus, and *Mycoplasma pneumonia* (https://www.ncbi.nlm.nih.gov/books/NBK482437/)
Tuberculosis	*Mycobacterium tuberculosis* (Bloom et al. 2017)
Pneumonia	Viral pathogens: Influenza virus, respiratory syncytial virus (rsv), parainfluenza virus, and adenovirus (Jain et al. 2015) Bacterial pathogens: *Streptococcus pneumoniae, Staphylococcus aureus,* Group A *Streptococcus, Klebsiella pneumoniae, Haemophilus influenzae, Moraxella catarrhalis,* anaerobes, and gram-negative organisms (https://www.ncbi.nlm.nih.gov/books/NBK534295/) Fungal pathogens: *Histoplasma* spp., *Blastomyces* spp., *and Coccidioides* spp. (Hage et al. 2012)
COVID-19	Severe acute respiratory syndrome coronavirus (SARS-CoV-2) (Rothan and Byrareddy 2020)
Aspergillosis	*Aspergillus fumigatus* (Li et al. 2019)
Cryptococcosis	*Cryptococcus neoformans* (Li et al. 2019)
Pneumocystis pneumonia (PCP)	*Pneumocystis jirovecii* (Li et al. 2019)

This table enlists the different types of respiratory tract infections along with their causative organisms

provides a list of the different respiratory tract infections along with the causative agents.

11.3 Clinical Features and Symptoms

The general symptoms of upper respiratory tract infection are non-specific and commonly include running nose, nasal congestion, sneezing, cold, cough, sore throat, headache, myalgia, malaise, fatigue, and low-grade fever (https://www.ncbi.nlm.nih.gov/books/NBK532961/; Goni et al. 2020). Generally symptoms appear after the incubation period for an infectious organism (around three days) following the exposure to the infectious agent. These symptoms usually persist for one to 2 weeks, but may also last up to 3 weeks (https://www.ncbi.nlm.nih.gov/books/NBK532961/).

The clinical symptoms associated with lower respiratory tract infection are mostly guided by the infectious agent and the severity of the infection. Fever, cough, mucopurulent sputum, shortness of breath, physical tiredness, chest retractions, and abnormal pulmonary auscultations are among the common symptoms (Hansen et al. 2020). Symptoms like haemoptysis, night sweats, and weight loss are observed in patients with pulmonary tuberculosis (https://www.ncbi.nlm.nih.gov/books/NBK441916/). High fever, rapid pulse rate, cough with phlegm, breathing difficulty have been reported in patients suffering from pneumonia (https://www.ncbi.nlm.nih.gov/books/NBK525774/). Difficult breathing, wheezing, malaise, yellowish sputum, and productive cough have been noted in acute bronchitis patients (https://www.ncbi.nlm.nih.gov/books/NBK448067/). Chest pain, breathing difficulty, fever, expectoration, haemoptysis, shortness of breath, and reduced oxygen saturation level are the symptoms noted in severe COVID-19 patients (Patgiri et al. 2022).

11.4 Diagnosis of Infectious Pulmonary Diseases

Different methods have been employed for diagnosis of infectious pulmonary diseases. Culture of nasal or sputum samples serves as the gold standard for the detection of bacterial respiratory pathogens. Culturing of the bacterial pathogens also allows for subsequent determination of antimicrobial susceptibility of the identified pathogens (Tenover 2011). Staining of nasal aspirates and sputum samples followed by microscopic examination is also performed for bacterial respiratory pathogens. Gram staining is done to identify the gram nature of the bacterial respiratory pathogens (https://www.ncbi.nlm.nih.gov/books/NBK562156/; Carroll 2002). Acid-fast staining is essentially performed for the identification of acid-fast bacteria like *Mycobacterium tuberculosis* (https://www.ncbi.nlm.nih.gov/books/NBK537121/). Viral pathogens were also previously detected using culture method. However, the culture method is largely limited by the fact that many viable pathogenic microorganisms are difficult to culture or are non-culturable (Das et al. 2018; Li et al. 2014).

Antigen detection assays, also referred to as rapid immunoassays (RIA) rely on detection of antigens directly from respiratory secretions and nasal swabs. Latex agglutination tests, horizontal flow immunoassays, lateral flow immunoassays, and optical immunoassays are the four common formats of RIA used for detection of current respiratory infection (Das et al. 2018; Weinberg and Walker 2005).

Serological tests are also widely used for detecting respiratory pathogen specific antibodies, especially for virus (Das et al. 2018). Serological tests are challenging for bacterial pathogens like *Mycoplasma pneumoniae* (Beersma et al. 2005).

Molecular biology techniques have revolutionized the diagnosis of infectious pulmonary diseases and have evolved as the new gold standard for disease diagnosis (Das et al. 2018). Different varieties of Nucleic Acid Amplification Tests (NAATs) like real-time PCR, nested multiplex RT-PCR, loop-mediated isothermal DNA amplification, and isothermal nucleic acid amplification are available for detection of respiratory pathogens. High complexity multiplex panel assays and moderate complexity multiplex integrated systems are also used to detect multiple respiratory pathogens simultaneously (Das et al. 2018).

Imaging techniques like chest X-ray radiography are widely used for diagnosing lower respiratory tract infections like pneumonia and tuberculosis. It also aids in monitoring disease progression and therapeutic efficacy (Parveen and Sathik 2011; Katz and Leung 1999). Chest radiography and CT scanning are also used for diagnosing COVID-19 (Jiang et al. 2020).

11.5 Treatment of Infectious Pulmonary Diseases

The nature of the infectious causative agent guides the treatment and medications prescribed for respiratory tract infection. In case of bacterial infection of the respiratory tract, primarily antibiotics, antitussives, mucolytics, expectorants, decongestants, and antipyretics are prescribed (https://www.ncbi.nlm.nih.gov/books/NBK8142/; Stanton et al. 2010). Antibiotics like macrolides, fluoroquinolones, beta lactam antibiotics, and aminoglycosides are commonly used (https://www.ncbi.nlm.nih.gov/books/NBK513321/). Combination therapies are also used especially in case of severe lower respiratory tract infection. Corticosteroids are also used for critically ill patients (Feldman and Richards 2018). However, judicial administration of antibiotic is vital owing to the tremendous rise in antimicrobial drug resistance. The treatment of viral respiratory infection is mostly supportive. Anti-viral drugs like ribavirin, zanamivir, oseltamivir, peramivir, and laninamivir are used (Liu et al. 2016; Abed and Boivin 2006). Antibiotics are administered in case of secondary bacterial infections (Manohar et al. 2020). Amphotericin B (AMB), fluconazole, triazoles, and echinocandins are the common anti-fungal agents that are widely used for the treatment of fungal respiratory tract infections (Li et al. 2019). However, these anti-fungals often fail to clear the infection due to emergence of anti-fungal drug resistance. Besides, the adverse effects and toxicity of these anti-fungal drugs are often a cause of concern in the patients (Pfaller 2012). Preventive measures in the form of vaccination are often used to immunize individuals and prevent incidences of respiratory tract infection (Woodhead et al. 2005; Woodhead 2011). Vaccination using different vaccines like pneumococcal vaccines, influenzae vaccines, meningococcal vaccines, COVID-19 vaccines, mumps vaccine, measles vaccine, pertussis vaccine, diphtheria vaccine, and Bacillus Calmette-Guérin (BCG) vaccine is often recommended as an important strategy for disease prevention (Whitney and Harper 2004; Ottolini 2000; Francis et al. 2022). Preventive measures other than vaccination like use of masks, improved nutritional uptake, and proper hygiene are also adopted in several cases (Goni et al. 2020; Brooks and Butler 2021; Gupta et al. 2009).

11.6 Challenges and Future Research Avenues for Infectious Pulmonary Diseases

Genetic evolution of microorganisms, especially viruses through mutation of the genome gives rise to new genetic variants of the infectious pathogens. Genetic evolution and variation in infectious agents often influence in pathogen transmissibility, antigenicity, virulence, and mechanisms of escaping host immune surveillance (Castonguay et al. 2021; Lauring and Hodcroft 2021). Modern state of art omics techniques should continue to be extensively explored for rapid determination of genetic mutation in microorganisms. Multi-omics data can also be correlated with alteration in transmissibility rate, infectivity, and immune escape mechanisms of the respiratory pathogens. Omics approaches can also help identifying host specific biomarkers for rapid disease diagnosis. Besides, the development of antimicrobial drug resistance is another major challenge encountered in the clinical management of infectious pulmonary diseases. Overuse and misuse of antimicrobial agents primarily contribute to antimicrobial drug resistance (Ventola 2015; Aslam et al. 2018). Emergence of multidrug resistant pathogenic strains and superbugs often contribute to failure of treatment in patients with acute lower respiratory tract infections and lead to increased mortality rate (Moellering Jr. 2004). Judicial use of antibiotics, anti-viral agents, and anti-fungals should be practised by clinicians to slow down the process of emergence and spread of drug resistance. Host-pathogen interactomics should be elaborately studied to identify novel therapeutic targets or pave the way for drug repurposing. The standard process of designing vaccines for effective immunization and prevention of respiratory tract infections is a tedious and complex. New vaccine platform technologies that can speed up the process of vaccine production should be extensively explored especially for tackling pandemic situations (Excler et al. 2021). Some of the major pulmonary infectious diseases like tuberculosis, COVID-19, and pneumonia have been dealt in detail in the subsequent chapters.

References

Abed Y, Boivin G (2006) Treatment of respiratory virus infections. Antivir Res 70(2):1–16

Aslam B, Wang W, Arshad MI, Khurshid M, Muzammil S, Rasool MH, Nisar MA, Alvi RF, Aslam MA, Qamar MU et al (2018) Antibiotic resistance: a rundown of a global crisis. Infect Drug Resist 11:1645–1658

Beersma MF, Dirven K, van Dam AP, Templeton KE, Claas EC, Goossens H (2005) Evaluation of 12 commercial tests and the complement fixation test for Mycoplasma pneumoniae-specific immunoglobulin G (IgG) and IgM antibodies, with PCR used as the "gold standard". J Clin Microbiol 43(5):2277–2285

Bloom BR, Atun R, Cohen T, Dye C, Fraser H, Gomez GB, Knight G, Murray M, Nardell E, Rubin E et al (2017) Tuberculosis. In: Holmes KK, Bertozzi S, Bloom BR, Jha P (eds) Major infectious diseases, Washington, DC

Brooks JT, Butler JC (2021) Effectiveness of mask wearing to control community spread of SARS-CoV-2. JAMA 325(10):998–999

Cappelletty D (1998) Microbiology of bacterial respiratory infections. Pediatr Infect Dis J 17(8 Suppl):S55–S61

Carroll KC (2002) Laboratory diagnosis of lower respiratory tract infections: controversy and conundrums. J Clin Microbiol 40(9):3115–3120

Castonguay N, Zhang W, Langlois MA (2021) Meta-analysis and structural dynamics of the emergence of genetic variants of SARS-CoV-2. Front Microbiol 12:676314

Collaborators GBDLRI (2018) Estimates of the global, regional, and national morbidity, mortality, and aetiologies of lower respiratory infections in 195 countries, 1990–2016: a systematic analysis for the Global Burden of Disease Study 2016. Lancet Infect Dis 18(11):1191–1210

Correia W, Dorta-Guerra R, Sanches M, Almeida Semedo CJB, Valladares B, de Pina-Araujo IIM, Carmelo E (2021) Study of the etiology of acute respiratory infections in children under 5 years at the Dr. Agostinho Neto Hospital, Praia, Santiago Island, Cabo Verde. Front Pediatr 9:716351

Das S, Dunbar S, Tang YW (2018) Laboratory diagnosis of respiratory tract infections in children - the State of the Art. Front Microbiol 9:2478

Excler JL, Saville M, Berkley S, Kim JH (2021) Vaccine development for emerging infectious diseases. Nat Med 27(4):591–600

Feldman C, Richards G (2018) Appropriate antibiotic management of bacterial lower respiratory tract infections. F1000Res:7

File TM (2000) The epidemiology of respiratory tract infections. Semin Respir Infect 15(3): 184–194

Francis AI, Ghany S, Gilkes T, Umakanthan S (2022) Review of COVID-19 vaccine subtypes, efficacy and geographical distributions. Postgrad Med J 98(1159):389–394

Goni MD, Hasan H, Wan-Arfah N, Naing NN, Deris ZZ, Arifin WN, Baaba AA, Aliyu A, Adam BM (2020) Health education intervention as an effective means for prevention of respiratory infections among hajj pilgrims: a review. Front Public Health 8:449

Gupta KB, Gupta R, Atreja A, Verma M, Vishvkarma S (2009) Tuberculosis and nutrition. Lung India 26(1):9–16

Hage CA, Knox KS, Wheat LJ (2012) Endemic mycoses: overlooked causes of community acquired pneumonia. Respir Med 106(6):769–776

Hansen LS, Lykkegaard J, Thomsen JL, Hansen MP (2020) Acute lower respiratory tract infections: symptoms, findings and management in Danish general practice. Eur J Gen Pract 26(1):14–20

Jain S, Self WH, Wunderink RG, Fakhran S, Balk R, Bramley AM, Reed C, Grijalva CG, Anderson EJ, Courtney DM et al (2015) Community-acquired pneumonia requiring hospitalization among U.S. Adults. N Engl J Med 373(5):415–427

Jiang ZZ, He C, Wang DQ, Shen HL, Sun JL, Gan WN, Lu JY, Liu XT (2020) The role of imaging techniques in management of COVID-19 in China: from diagnosis to monitoring and follow-up. Med Sci Monit 26:e924582

Katz DS, Leung AN (1999) Radiology of pneumonia. Clin Chest Med 20(3):549–562

Kilgore PE, Salim AM, Zervos MJ, Schmitt HJ (2016) Pertussis: microbiology, disease, treatment, and prevention. Clin Microbiol Rev 29(3):449–486

Lauring AS, Hodcroft EB (2021) Genetic variants of SARS-CoV-2-what do they mean? JAMA 325(6):529–531

Li L, Mendis N, Trigui H, Oliver JD, Faucher SP (2014) The importance of the viable but non-culturable state in human bacterial pathogens. Front Microbiol 5:258

Li Z, Lu G, Meng G (2019) Pathogenic fungal infection in the lung. Front Immunol 10:1524

Liu T, Li Z, Zhang S, Song S, Julong W, Lin Y, Guo N, Xing C, Xu A, Bi Z et al (2015) Viral etiology of acute respiratory tract infections in hospitalized children and adults in Shandong Province, China. Virol J 12:168

Liu Q, Zhou YH, Ye F, Yang ZQ (2016) Antivirals for respiratory viral infections: problems and prospects. Semin Respir Crit Care Med 37(4):640–646

Mahashur A (2018) Management of lower respiratory tract infection in outpatient settings: focus on clarithromycin. Lung India 35(2):143–149

Makela MJ, Puhakka T, Ruuskanen O, Leinonen M, Saikku P, Kimpimaki M, Blomqvist S, Hyypia T, Arstila P (1998) Viruses and bacteria in the etiology of the common cold. J Clin Microbiol 36(2):539–542

Manohar P, Loh B, Athira S, Nachimuthu R, Hua X, Welburn SC, Leptihn S (2020) Secondary bacterial infections during pulmonary viral disease: phage therapeutics as alternatives to antibiotics? Front Microbiol 11:1434

Moellering RC Jr (2004) The continuing challenge of lower respiratory tract infections. Clin Infect Dis 38(Suppl 4):S319–S321

Monto AS (2002) Epidemiology of viral respiratory infections. Am J Med 112(Suppl 6A):4S–12S

Moule MG, Cirillo JD (2020) Mycobacterium tuberculosis dissemination plays a critical role in pathogenesis. Front Cell Infect Microbiol 10:65

Otshudiema JO, Acosta AM, Cassiday PK, Hadler SC, Hariri S, Tiwari TSP (2021) Respiratory illness caused by Corynebacterium diphtheriae and C. ulcerans, and use of diphtheria antitoxin in the United States, 1996–2018. Clin Infect Dis 73(9):e2799–e2806

Ottolini MG (2000) Current and future use of vaccines for viral and bacterial respiratory tract infections. Curr Infect Dis Rep 2(2):121–129

Parveen NR, Sathik MM (2011) Detection of Pneumonia in chest X-ray images. J Xray Sci Technol 19(4):423–428

Patgiri PR, Rajendran V, Ahmed AB (2022) Clinico-epidemiological profiles of COVID-19 elderly patients in Guwahati City, Assam, India: a cross-sectional study. Cureus 14(4):e24043

Pfaller MA (2012) Antifungal drug resistance: mechanisms, epidemiology, and consequences for treatment. Am J Med 125(1 Suppl):S3–S13

Rothan HA, Byrareddy SN (2020) The epidemiology and pathogenesis of coronavirus disease (COVID-19) outbreak. J Autoimmun 109:102433

Rubin S, Eckhaus M, Rennick LJ, Bamford CG, Duprex WP (2015) Molecular biology, pathogenesis and pathology of mumps virus. J Pathol 235(2):242–252

Sharma NC, Efstratiou A, Mokrousov I, Mutreja A, Das B, Ramamurthy T (2019) Diphtheria. Nat Rev Dis Primers 5(1):81

Sonego M, Pellegrin MC, Becker G, Lazzerini M (2015) Risk factors for mortality from acute lower respiratory infections (ALRI) in children under five years of age in low and middle-income countries: a systematic review and meta-analysis of observational studies. PLoS One 10(1): e0116380

Stanton N, Francis NA, Butler CC (2010) Reducing uncertainty in managing respiratory tract infections in primary care. Br J Gen Pract 60(581):e466–e475

Sun H, Li S, Wang T, Chen Z (2022) Mycoplasma pneumoniae infection and persistent wheezing in young children: a retrospective case-control study. Front Pediatr 10:811086

Tenover FC (2011) Developing molecular amplification methods for rapid diagnosis of respiratory tract infections caused by bacterial pathogens. Clin Infect Dis 52(Suppl 4):S338–S345

Ventola CL (2015) The antibiotic resistance crisis: part 1: causes and threats. P T 40(4):277–283

Weinberg A, Walker ML (2005) Evaluation of three immunoassay kits for rapid detection of influenza virus A and B. Clin Diagn Lab Immunol 12(3):367–370

Whitney CG, Harper SA (2004) Lower respiratory tract infections: prevention using vaccines. Infect Dis Clin N Am 18(4):899–917

Woodhead M (2011) New guidelines for the management of adult lower respiratory tract infections. Eur Respir J 38(6):1250–1251

Woodhead M, Blasi F, Ewig S, Huchon G, Ieven M, Ortqvist A, Schaberg T, Torres A, van der Heijden G, Verheij TJ et al (2005) Guidelines for the management of adult lower respiratory tract infections. Eur Respir J 26(6):1138–1180

Yuki K, Fujiogi M, Koutsogiannaki S (2020) COVID-19 pathophysiology: a review. Clin Immunol 215:108427

12 Tuberculosis

Abstract

Tuberculosis is one of the major global health concerns associated with a high morbidity and mortality rate. Pulmonary tuberculosis may remain latent or manifest in the form of severe respiratory tract infection. Microbiological tests, molecular biology detection methods, and imaging modalities are primarily used for disease diagnosis. Delayed disease diagnosis and emergence of drug resistance to available therapeutic regime are major hinderances encountered in effectively controlling tuberculosis. This chapter discusses the scope of using advanced multi-omics approaches to tackle these problems. Probable host biomarkers in tuberculosis patients as identified by omics tools have also been discussed. Different databases and webservers designed for tuberculosis have been documented in this chapter. Present drug regimens and newly designed vaccines for controlling tuberculosis have also been highlighted. Lastly, the different futuristic research directions that can be investigated for tuberculosis have been mentioned.

Keywords

Mycobacterium tuberculosis · Granuloma · Ghon focus · Bilateral planar chest radiograph · Latent tuberculosis · Active tuberculosis · Multidrug-resistant TB (MDR-TB) · Extensively drug-resistant TB (XDR-TB)

12.1 Introduction

Tuberculosis is an infectious disease caused by *Mycobacterium tuberculosis. Mycobacterium tuberculosis* primarily acts as a pulmonary pathogen affecting the lungs but it can affect other systems like gastrointestinal (GI) system, liver, lymphoreticular system, musculoskeletal system, central nervous system, skin, and

S. Saha et al., *Pulmonomics: Omics Approaches for Understanding Pulmonary Diseases*, https://doi.org/10.1007/978-981-99-3505-5_12

reproductive system. Tuberculosis can affect individuals of all age groups across the globe and serves as a major cause of death especially in immunocompromised and immunosuppressed individuals. Global collaborative efforts have been initiated with the goal of eradicating tuberculosis (https://www.ncbi.nlm.nih.gov/books/NBK441916/). As per, the World Health Organization (WHO), there has been a 11% reduction in the incidence of tuberculosis and 9.2% decline in tuberculosis-related deaths (Jeremiah et al. 2022). However, tuberculosis still continues to remain associated with high morbidity and mortality rate, primarily due to delayed disease diagnosis, development of drug-resistant strains, and lack of host diagnostic and response biomarkers.

12.2 Etiopathology

Mycobacterium tuberculosis infection can lead to both primary and post-primary or secondary tuberculosis. Primary tuberculosis occurs typically in very young or very aged individuals or immunocompromised individuals. Such infections often lead to disseminated tuberculosis with miliary tuberculosis and extrapulmonary granulomas. Post primary or secondary tuberculosis classically initiates after systemic immunity is established in primary tuberculosis. In secondary tuberculosis, the host systemic immune response is evaded (Kanabalan et al. 2021; Hunter 2018).

The pathogenesis of tuberculosis is highly complicated and guided primarily by the complex interplay of host immunity and mycobacterial factors. The inhaled airborne infectious particles travel down the respiratory tract. The infectious droplets penetrate the terminal alveoli and are phagocytosed by alveolar macrophages and other phagocytes. *Mycobacterium tuberculosis* replicates within the alveolar macrophages. Thus, the macrophages not only protect the human host from the infectious agent but also provide a platform for the establishment of *Mycobacterium tuberculosis* infection at an early stage. Alveolar immune cells harbouring these microbes cross the alveolar barrier and undergo systemic dissemination. Intracellular replication and spread of the mycobacteria occur via body fluids (blood and lymph) to adjacent lymph nodes and extra pulmonary tissues. Development of both innate and adaptive immune responses is triggered by macrophages, whereas dendritic cells and CD4+ T cells primarily promote adaptive immune response in human host. Adaptive immune response typically develops 2–8 weeks post initial infection (Kanabalan et al. 2021; Ahmad 2011). Release of *Mycobacterium tuberculosis* in the extracellular milieu results in active bacterial replication and simultaneous release of pathogenic factors that interact with host factors. Activation of macrophages and adaptive immunity results in phago-lysosomal fusion and triggers cytokine mediated host bactericidal response, which eventually leads to inflammation and tissue destruction in lungs of active pulmonary tuberculosis patients. The dynamic equilibrium between *Mycobacterium tuberculosis* and host immunity in case of latent tuberculosis infection still remains largely unknown.

Granuloma formation is a hallmark of pulmonary tuberculosis in human host. Granuloma is compact, immunological structure composed of transformed macrophages, foamy macrophages, multinucleated giant cells, epithelioid cells, monocytes, neutrophils, and dendritic cells. Granulomas also have peripheral lymphatic cuffs comprising of plasma cells, B and T lymphocytes, which restricts mycobacterial replication. Granuloma formation is also surrounded by a plethora of immune cells including macrophages, neutrophils, dendritic cells, fibroblasts, T and B lymphocytes. Granuloma formation initiates shortly after infection to limit mycobacterial replication and minimize tissue damage at the site of infection. Granulomas are observed in both active and latent tuberculosis. Necrotic breakdown of granulomas and simultaneous enlargement of lesions occur in active tuberculosis infection. Release of *Mycobacterium tuberculosis* into the airway occurs upon granuloma cavitation and promotes disease transmission in humans. Mostly restrictive granulomas or fibrotic lesions are observed in latent tuberculosis. *Mycobacterium tuberculosis* may remain dormant within granulomas (Kanabalan et al. 2021; Ramakrishnan 2012; Martinot 2018). *Mycobacterium tuberculosis* manipulates host to promote early infiltration and enable mycobacterial survival within alveolar macrophages as obstructive lobular pneumonia. Mycobacterial antigens, host cells, and lipids within the foamy macrophages remain entrapped by the obstructed bronchioles. These early infiltrations spread to other sites within the lungs via the bronchi. Mostly, the infiltrates regress spontaneously, however, they might undergo necrosis or form fibro-caseous disease or post-primary granuloma or cavities. Granuloma formation is secondary to subclinical obstructive bronchopneumonia in case of post primary tuberculosis (Kumar 2016; Hunter 2016).

Mycobacterial virulence factors play a crucial role in pathogenesis of tuberculosis. These virulence factors mainly include genes encoding for cell surface proteins, lipid pathways proteins, and signal transduction regulators. Modulation of host immune response (including macrophages, granulomas) and host cellular response lies central to the pathogenicity of *Mycobacterium tuberculosis*. The pathogenicity of *Mycobacterium tuberculosis* in human system has been schematically represented in Fig. 12.1. *Mycobacterium tuberculosis* genome also suffers evolutionary changes to ensure transmissibility and efficient escape from human immune surveillance.

12.3 Clinical Features

Pulmonary tuberculosis is typically characterized by the presence of local granulomatous inflammation in the periphery of the lung (Ghon focus). It may also be associated with ipsilateral lymph node involvement, often referred as the Ghon complex. Tuberculosis may remain asymptomatic, which is commonly referred to as latent tuberculosis. Tuberculosis may otherwise clinically present itself as an acute form of lower respiratory tract infection. The classical symptoms of pulmonary tuberculosis primarily include chronic cough, sputum production, blood in sputum or hemoptysis, fever, night sweats, anaemia, weight loss, and appetite loss (Loddenkemper et al. 2015).

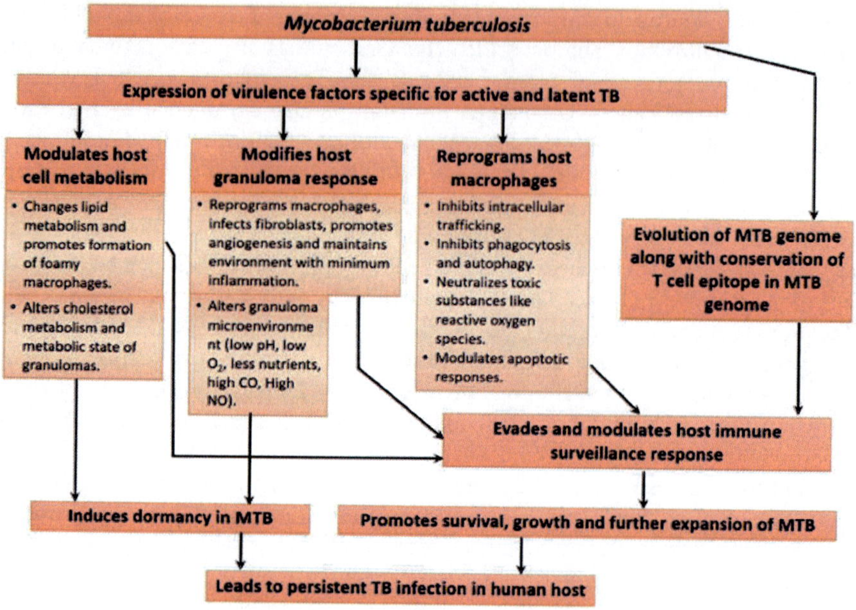

Fig. 12.1 Pathogenicity of *Mycobacterium tuberculosis* in human host. Host–mycobacterial interaction, *Mycobacterium tuberculosis* mediated reprogramming of host macrophages, modulation of host granuloma, alteration of host cell metabolism, and modulation of host immune responses lies central to the pathogenicity of *Mycobacterium tuberculosis*

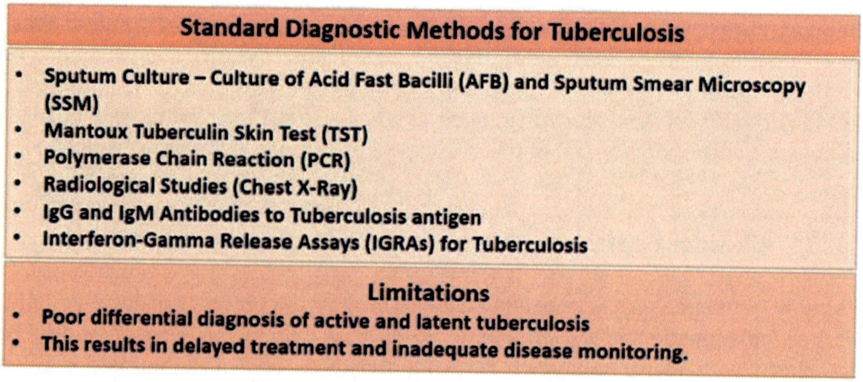

Fig. 12.2 Different techniques that are widely used for diagnosis of tuberculosis. The major techniques used for tuberculosis diagnosis have been mentioned along with the basic limitations of the available diagnostic techniques

The diagnosis of tuberculosis relies mainly on several microbiological and molecular biological techniques. Different methods widely used for the diagnosis of tuberculosis have been summarized in Fig. 12.2.

Sputum culture, sputum smear microscopy, and PCR-based nucleic acid amplification test (NAATs) are mostly used for the diagnosis of active tuberculosis. However, diagnosis of *Mycobacterium tuberculosis* by culture method takes several weeks to appear positive. PCR-based techniques often have very high specificity and relatively good sensitivity. Immunochromatographic IgG/IgM rapid test is also used for rapid detection of active tuberculosis. Other tests like tuberculin skin test and blood interferon-γ release assay (IGRA) can also be used to diagnose suspected cases of tuberculosis but takes 3–8 weeks to show positive results. IGRA, however, is used as a modern standard diagnostic tool for latent tuberculosis. Examination of drug resistance profile of the infectious *Mycobacterium tuberculosis* strain in active tuberculosis patients primarily depends on whole genome sequencing of the concerned strain. Imaging techniques also find wide application in diagnosis of tuberculosis. Bilateral planar chest X-ray radiography is the standard imaging tool for diagnosis of pulmonary tuberculosis. Ghon focus and Ghon complex can be visualized in chest radiograph. Apart from inflammatory infiltrates and lymphadenopathy, pleural effusion, multiple pulmonary nodules, cavity formation with necrosis, and tissue destruction in the upper lobes of the lungs can also be visualized in chest radiograph. Computed tomography (CT) provides better assessment of lymphadenopathy, endobronchial extension, and disease activity in tuberculosis patients, as compared to chest radiography (Loddenkemper et al. 2015; Ben-Selma et al. 2011; Suarez et al. 2019). The present diagnostic tools are associated with delayed tuberculosis diagnosis. Besides, they often fail to differentiate between latent and active tuberculosis. The combination of microscopic techniques, molecular biology detection methods, and imaging modalities can be used to improve tuberculosis diagnosis. However, identification of human biomarkers in tuberculosis patients can be of vital importance in expediting the process of disease diagnosis.

12.4 Biomarkers of Tuberculosis

Discovery of host specific factors, novel biomarkers, and therapeutic targets for tuberculosis is highly challenging. Presently, there is no validated host biomarkers for diagnosis of paediatric and adult pulmonary tuberculosis. Multi-omics approaches are of prime importance in unveiling vital protein–protein interactions involved in host immune modulation in tuberculosis (Ahamad et al. 2022). Such omics-based studies shall not only aid in improving our understanding of tuberculosis pathobiology and drug resistance mechanism but also provide valuable insights for identification of host diagnostic biomarkers, host response biomarkers, and novel therapeutic targets (Jakhar et al. 2020). The probable role of different state of art omics approaches like genomics, transcriptomics, proteomics, lipidomics, and metabolomics in tuberculosis biology, disease diagnosis, and treatment have been summarized in Fig. 12.3.

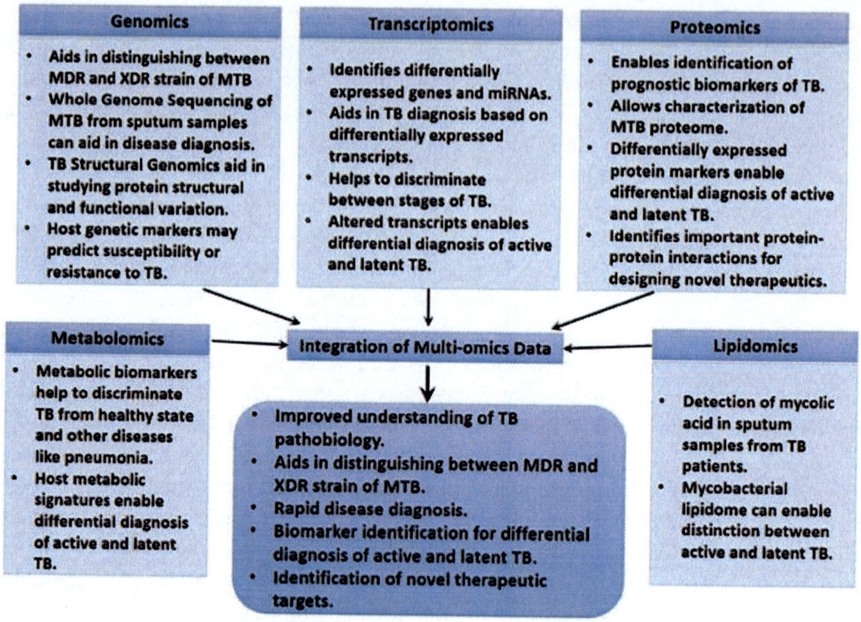

Fig. 12.3 Different omics approaches and integration of multi-omics data may enable improved understanding of TB pathobiology and lead to biomarker identification for differential diagnosis of active and latent TB, alongside identification of novel therapeutic targets

12.5 Genomics

Genomics finds enormous application in tuberculosis research. Whole genome sequencing (WGS) serves as a valuable tool to identify strains of *Mycobacterium tuberculosis* and identify novel drug resistance mutations. Apart from the mycobacterial aspect of tuberculosis, modern genomics techniques have aided in understanding the impact of human genetic variations and SNPs in risk assessment for latent or active pulmonary tuberculosis (Kanabalan et al. 2021; Ahamad et al. 2022). SNP in TGFβ (rs2317130) was found to be associated with disease severity in tuberculosis, and the CC genotype of rs2317130 was related to increased TGF-β1 and IL-17A production (Zhang et al. 2022). The G allele of rs4986790 and the T allele of rs4986791 of TLR4 were associated with reduced risk of active tuberculosis (Ortega et al. 2020). SNPs in NOD2 gene (namely rs1861759 and rs7194886) were significantly associated with augmented risk of tuberculosis. SNPs in CD14 (namely rs2569190 and rs2569191) were also found to influence the risk of tuberculosis depending on the allele (Cubillos-Angulo et al. 2021). Another functional SNP (rs13120371) in the 3′ UTR region of xCT gene was found to increase susceptibility to tuberculosis via interaction with miR-142-3p (Wang et al. 2020). A few other SNP biomarkers related to tuberculosis have been tabulated in Table 12.1.

Table 12.1 Different biomarkers of tuberculosis

Omics approach	Biomarkers	Biological sample	Comments
Genomics	SP110	Peripheral blood	SNPs in SP110 gene (rs9061, rs7580900, rs11556887, rs7580912, and rs2241525) were associated with latent tuberculosis infection and/or active tuberculosis disease. Also, rs9061 was found to be associated with plasma levels of TNFα in latent tuberculosis infection (Chang et al. 2018)
	IL10, IFN-γ		At IFN-γ T874A polymorphic site (rs2430561), A genotype was higher in latent tuberculosis infection as compared to healthy controls. The AA genotype of IL-101082 polymorphic site (rs1800896) was significantly associated with pulmonary tuberculosis (Hu et al. 2015)
	TNF		TNF rs1799964 and rs1800630 may serve as risk factors for tuberculosis in Chinese Tibetan population (Wu et al. 2019)
	IL6		The G allele at rs2069837 of *IL6 may confer protection from tuberculosis in Chinese Han population* (Wu et al. 2019)
	IL1B		The GG genotype of IL1B rs16944 polymorphism was more prevalent in controls than in tuberculosis patients (Wu et al. 2019)
Epigenomics	IL12B, IL12RB2, TYK2, IFNGR1, JAK1, JAK2	Peripheral blood mononuclear cells (PBMCs), CD4+ T cells	IL12B, IL12RB2, TYK2, IFNGR1, JAK1, and JAK2 genes were hypermethylated in

(continued)

Table 12.1 (continued)

Omics approach	Biomarkers	Biological sample	Comments
			tuberculosis patients (DiNardo et al. 2020)
	VDR	Genomic DNA	Hypermethylation of VDR was noted in tuberculosis patients (Jiang et al. 2017)
	IGSF8, TNK2-AS1, COL11A2, HYAL3, NAT6, KCNJ4, IER3, KLC2, FBRSL1, GNAS, H19, KMT5C, EIF5A, SOCS1, DXO, SIN3A, LTBP3, GABBR1, ZFP41, HLA-E, HLA-F, ZBTB22, C6orf47, B3GALT4, SLC29A1	Peripheral blood, peripheral blood mononuclear cells (PBMCs)	IGSF8, TNK2-AS1, COL11A2, HYAL3, NAT6, KCNJ4, IER3, KLC2, FBRSL1, GNAS, H19, KMT5C, EIF5A, SOCS1, DXO, SIN3A, LTBP3, GABBR1, ZFP41, HLA-E, HLA-F, ZBTB22, C6orf47, B3GALT4, and SLC29A1 were differentially methylated between tuberculosis patients and healthy subjects (Lyu et al. 2022)
	PARP9, miR-505, RASGRP4, GNG12, MRPS18B, RPTOR	Peripheral blood mononuclear cells (PBMCs)	PARP9 and miR-505 were hypomethylated, whereas RASGRP4 and GNG12 were hypermethylated in tuberculosis patients as compared to healthy subjects. Hypomethylation of MRPS18B and RPTOR genes was noted in tuberculosis patients having pleural involvement (Chen et al. 2020)
Transcriptomics	FCGR1A	Blood	FCGR1A was significantly increased in active tuberculosis patients as compared to latent tuberculosis infections before treatment, irrespective of genetic background or HIV infection (Sutherland et al. 2014).
			The combination of let-7a-5p, miR-589-5p,

(continued)

Table 12.1 (continued)

Omics approach	Biomarkers	Biological sample	Comments
	let-7a-5p, miR-589-5p, miR-196b-5p, SNORD104		miR-196b-5p, and SNORD104 can function as a highly sensitive (100%) classifier to differentiate tuberculosis from all other non-tuberculosis groups (de Araujo et al. 2019)
	Lactoferrin, CD64, GTPase 33A	Peripheral blood mononuclear cells (PBMCs)	Gene expression level of lactoferrin, CD64 and Ras-associated GTPase 33A could distinguish between tuberculosis patients, *Mycobacterium tuberculosis*-infected healthy donors and non-infected healthy volunteers (Jacobsen et al. 2007)
	ASUN, DHX29, PTPRC		PTPRC can be used to distinguish active tuberculosis patients from healthy subjects. ASUN may serve to distinguish between latent tuberculosis infections from healthy state. DHX29 can be utilized for identification of latent infections among active tuberculosis patients or healthy controls (Lee et al. 2016)
	circRNA_103017, circRNA_059914, circRNA_101128, circRNA_062400		Circular RNAs like circRNA_103017, circRNA_059914, and circRNA_101128 were increased, whereas circRNA_062400 was reduced in active tuberculosis patients (Fu et al. 2019)
	hsa-let-7e-5p, hsa-let-7d-5p, hsa-miR-450a-5p, hsa-miR-140-5p	Serum exosomes	MicroRNAs, namely hsa-let-7e-5p, hsa-let-7d-5p, hsa-miR-450a-5p, and hsa-miR-140-5p were specifically expressed in latent tuberculosis infection (Lyu et al. 2019)

(continued)

Table 12.1 (continued)

Omics approach	Biomarkers	Biological sample	Comments
	hsa-miR-1246, hsa-miR-2110, hsa-miR-370-3p, hsa-miR-28-3p, hsa-miR-193b-5p		MicroRNAs, namely hsa-miR-1246, hsa-miR-2110, hsa-miR-370-3p, hsa-miR-28-3p, and hsa-miR-193b-5p were specifically expressed in active tuberculosis patients (Lyu et al. 2019)
Proteomics	ADA	Serum	Serum ADA may function as a probable diagnostic biomarker candidate for diagnosis of tuberculosis and for evaluating treatment response in tuberculosis patients (Pandey et al. 2016; Soedarsono et al. 2020)
	SYWC, kallistatin, complement C9, gelsolin, testican-2, aldolase C		A Bayes classifier model comprising of 6 human proteins namely, SYWC, kallistatin, complement C9, gelsolin, testican-2, and aldolase C was developed that can be useful in diagnosis of active tuberculosis (De Groote et al. 2017)
	C reactive protein, transthyretin, IFN-γ, complement factor H, apolipoprotein-A1, inducible protein 10, serum amyloid A		A seven-protein serum biosignature comprising of C reactive protein, transthyretin, IFN-γ, complement factor H, apolipoprotein-A1, inducible protein 10, and serum amyloid A can be used for the diagnosis of tuberculosis irrespective of HIV infection (Chegou et al. 2016)
	S100A9, ORM2, SOD, IL-36a	Plasma	S100A9, ORM2, SOD, and IL-36a were significantly elevated in severe pulmonary tuberculosis patients. These proteins can be used to identify the different stages of tuberculosis (Liu et al. 2018)

(continued)

Table 12.1 (continued)

Omics approach	Biomarkers	Biological sample	Comments
	AGP1, ACT, CDH1		These proteins can efficiently discriminate pulmonary tuberculosis from latent tuberculosis with a specificity of 95.2% and sensitivity of 81.2% (Sun et al. 2018)
	Complement factor C9, IGFBP-2, CD79A, MXRA-7, NrCAM, CK-MB, C1qTNF3/ CTNFF3		The TB risk model 5 or TRM5 model (comprising of five proteins, namely complement factor C9, IGFBP-2, CD79A, MXRA-7, and NrCAM) and the 3-protein pair-ratio or 3PR (comprising of three proteins, namely C9, CK-MB, and C1qTNF3/CTNFF3) can be used for predicting tuberculosis progression (Penn-Nicholson et al. 2019)
	β-integrin, vitamin D binding protein (DBP), uteroglobin, profilin, cathelicidin antimicrobial peptide	Sputum	Five proteins namely, β-integrin, DBP, uteroglobin, profilin, and cathelicidin antimicrobial peptide could differentiate between active tuberculosis patients and non-tuberculosis patients (Bishwal et al. 2019)
Metabolomics	5-Oxoproline	Serum	5-Oxoproline levels were found to be reduced consistently in active tuberculosis patients (Che et al. 2013)
	Aspartate, glutamate, methionine, sulfoxide asparagine, glutamine, methionine		Serum levels of aspartate, glutamate, methionine, and sulphoxide were elevated, while serum levels of asparagine, glutamine, and methionine were reduced in active tuberculosis patients as compared to healthy subjects and latent tuberculosis infection (Cho et al. 2020)

(continued)

Table 12.1 (continued)

Omics approach	Biomarkers	Biological sample	Comments
	Ceramide, 12R-hydroxy-5Z, Z,10E,14Z-eicosatetraenoic acid [12 (R)-HETE], cholesterol sulphate, 4-formyl-4-methyl-5-cholesta-8-en-3-ol	Plasma	Ceramide, 12R-hydroxy-5Z, Z,10E,14Z-eicosatetraenoic acid [12 (R)-HETE], cholesterol sulphate, and 4-formyl-4-methyl-5- cholesta-8-en-3-ol were significantly increased in tuberculosis patients. These metabolites may be used for rapid and non-invasive diagnosis of tuberculosis (Lau et al. 2015)
	Diacetyl spermine, neopterin, sialic acid, N-acetyl hexosamine	Urine	Diacetyl spermine, neopterin, sialic acid, and N-acetyl hexosamine levels in urine may serve as potential non-invasive biomarkers for distinguishing active tuberculosis from healthy controls and other non-tuberculous pulmonary disease patients (Isa et al. 2018)
Metagenomics	*Haemophilus parainfluenzae, Roseburia inulinivorans, Roseburia hominis*	Stool	Classification model based on the abundance of *Haemophilus parainfluenzae, Roseburia inulinivorans,* and *Roseburia hominis* could discriminate pulmonary tuberculosis patients from healthy subjects (Hu et al. 2019).
	Prevotella spp., *Enterococcus* spp., Ruminococcaceae, Bifidobacteriaceae, *Faecalibacterium prausnitzii*		A decline in gut microbial diversity was noted in paediatric tuberculosis patients. Increased abundance of *Prevotella* spp. and *Enterococcus* spp., along with decreased abundance of Ruminococcaceae, Bifidobacteriaceae, and *Faecalibacterium prausnitzii* was noted in these patients (Li et al. 2019a)

(continued)

Table 12.1 (continued)

Omics approach	Biomarkers	Biological sample	Comments
	Erysipelotrichaceae spp., *Blautia* spp., *Anaerostipes* spp.		A distinct gut microbial flora characterized by the presence of *Erysipelotrichaceae* spp., *Blautia* spp. and *Anaerostipes* spp. was observed in tuberculosis patients as compared to healthy controls (Somboro et al. 2021)
	Actinobacteria, Proteobacteria, Bacteroidetes, *Prevotella* spp., *Lachnospira* spp.		Increased abundance of Actinobacteria and Proteobacteria along with reduced abundance of Bacteroidetes was noted in faecal samples of recurrent tuberculosis patients. *Prevotella* spp. and *Lachnospira* spp. were dramatically reduced in both new and recurrent tuberculosis patients as compared to healthy subjects (Luo et al. 2017)
	Streptococcus spp., *Fusobacterium* spp., *Mycobacterium* spp.	Broncho alveolar lavage (BAL)	Microbial diversity was reduced in tuberculosis patients. *Mycobacterium* abundance was increased, whereas abundance of *Streptococcus* spp. and *Fusobacterium* spp. was reduced in tuberculosis patients (Vazquez-Perez et al. 2020)
	Mogibacterium spp., *Moryella* spp., *Oribacterium* spp.	Sputum	Increased abundance of *Mogibacterium* spp., *Moryella* spp., and *Oribacterium* spp. was noted in tuberculosis patients (Cheung et al. 2013)

This table enlists various probable biomarker candidates of tuberculosis identified from human samples using different omics approaches

12.6 Epigenomics

Changes in DNA methylation pattern have been associated with dampened human immune response in tuberculosis patients. DNA hypermethylation of IL-2/STAT5 pathway, TNF/NF-κB pathway, and IFN-γ signalling pathway were noted in tuberculosis patients. DNA hypermethylation of these pathways resulted in reduced human immune responsiveness due to decreased production of IFN-γ, CXCL9, CXCL10, IL-6, TNF, and IL-1β (DiNardo et al. 2020). Tuberculosis has also been found to trigger premature cellular aging process by perturbing and enhancing DNA methylation (Bobak et al. 2022). Another study showed that tuberculosis patients harbouring WIPI2/GNG12 hypermethylation or MRPS18B/FOXO3 hypomethylation had lesser than one-year survival (Chen et al. 2020). A recent pilot study also highlighted the potential use of DNA methylation pattern as a diagnostic tool for differentiating between active tuberculosis cases and latent tuberculosis infections (Du et al. 2022). Assessment of global DNA methylation pattern in paediatric tuberculosis patients revealed significant difference in methylation profile between patients and control subjects. Exploring such methylation signatures for diagnostic purposes can aid in early detection of paediatric tuberculosis (Maruthai et al. 2018). Some of the differentially methylated genes noted in tuberculosis patients have been listed in Table 12.1.

12.7 Transcriptomics

Microarray and other transcriptomics tools have been extensively explored with the goal of identifying host specific biomarkers of tuberculosis. A high-throughput blood transcriptomics study identified 86 host transcript signatures that could efficiently distinguish between tuberculosis and other inflammatory or infectious diseases. Neutrophil-driven interferon inducible genes involved in type-1 IFN-αβ signalling and IFN-γ were largely expressed in active tuberculosis (Berry et al. 2010). A study on paediatric tuberculosis identified as 51 transcript signature that could successfully differentiate tuberculosis from other diseases in children (Anderson et al. 2014). In another study on adult tuberculosis in Africa, a 27-transcript signature was identified that could mediate differential diagnosis of active tuberculosis and latent tuberculosis infections. In the same study, a 44-transcript signature was also identified that could distinguish tuberculosis from other diseases (Kaforou et al. 2013). Apart from gene expression, recent RNA sequencing studies have also demonstrated the importance of non-coding RNAs (miRNAs, piRNAs, snoRNAs, and circular RNAs) in different stages of *Mycobacterium tuberculosis* infection (Kanabalan et al. 2021). A list of some of the host genes and non-coding RNAs that are differentially expressed in tuberculosis have been tabulated in Table 12.1.

12.8 Proteomics

Several proteomics studies have been conducted for identification of host biomarkers in tuberculosis. Large scale proteomics study showed that active tuberculosis patients have altered expression of proteins that are involved in lipid metabolism, tissue repair, and immune response. A panel comprising of eight to ten proteins were identified that could distinguish active tuberculosis cases from latent tuberculosis infections (Achkar et al. 2015). Another study highlighted that pulmonary tuberculosis results in enrichment of host proteins involved in immune mechanisms, inflammatory pathways, acute phase response, antimicrobial defense mechanisms, coagulation cascade, tissue healing, tissue remodelling, and apoptosis (De Groote et al. 2013). Plasma chemokine levels were found to be associated with disease severity and increased bacterial burden in pulmonary tuberculosis patients. CCL1, CCL3, CXCL1, CXCL2, CXCL9, and CXCL10 levels were significantly increased in pulmonary tuberculosis patients as compared to latent tuberculosis infections and healthy controls. CCL1, CCL3, CXCL1, CXCL10, and CXCL11 levels in pulmonary tuberculosis patients with bilateral or cavitary disease were significantly higher than that in pulmonary tuberculosis patients with unilateral or non-cavitary disease. Expression level of these chemokines was also positively correlated to the mycobacterial burden (Kumar et al. 2019). Plasma CCL3, CXCL8, and CXCL10 were identified as novel biomarkers for prediction of adverse treatment outcome in pulmonary tuberculosis patients (Kumar et al. 2021). Various other host protein signatures of tuberculosis have been documented in Table 12.1.

12.9 Metabolomics

Several metabolomics studies have been conducted using different biological samples like blood serum, plasma, sputum, breath, and urine from tuberculosis patients for identification of tuberculosis-specific metabolic signatures (Kanabalan et al. 2021). Profound alterations of metabolic pathways were observed from plasma of pulmonary tuberculosis patients. Vitamin A (retinol) metabolism, vitamin D3 (cholecalciferol) metabolism, bile acid biosynthesis, and purine metabolism were found to be altered in tuberculosis patients (Phuoc Long et al. 2022). Novel immunometabolic signatures were found to be significantly associated with tuberculosis progression in household contacts of TB index cases (HHCs). Cortisol, glutathione, tryptophan, and tRNA acylation networks have been implicated in progression of tuberculosis in HHCs (Duffy et al. 2019). Glycerophospholipid and arachidonic acid metabolic pathways were found to play regulatory role in tuberculosis (Kanabalan et al. 2021). An association of metabolic profiles with cytokine signalling was also established in tuberculosis patients (Weiner 3rd et al. 2012). *Mycobacterium tuberculosis* associated metabolites like acyl phosphatidylinositol mannoside (Ac1PIM1), lysophosphatidylinositol (Lyso-PI), and phosphatidylglycerol (PG) were significantly elevated in active tuberculosis patients

(Collins et al. 2018). Some of the human metabolic biomarkers obtained from tuberculosis patients have been enlisted in Table 12.1.

12.10 Metagenomics

Gut and pulmonary microbial dysbiosis have been studied for both adult and paediatric pulmonary tuberculosis (Liu et al. 2021). Alteration in gut microbiome in tuberculosis patients has also been associated with inflammation and changes in metabolic pathways (Somboro et al. 2021). Pulmonary microbiota in tuberculosis patients was found to be characterized by Proteobacteria and Bacteroidetes (Cheung et al. 2013). Increased abundance of *Streptococcus* spp., *Granulicatella* spp., and *Pseudomonas* spp., along with a decreased abundance of *Prevotella* spp., *Leptotrichia* spp., *Treponema* spp., *Catonella* spp., and *Coprococcus* spp. was observed in tuberculosis patients as compared to healthy volunteers. This study also highlighted that recurrent tuberculosis patients had lower *Treponema/Mycobacterium* ratio and higher *Pseudomonas/Mycobacterium* ratio as compared to new tuberculosis patients. Treatment failure tuberculosis patients had increased abundance of *Pseudomonas* spp. and higher *Pseudomonas/Mycobacterium* ratio as compared to new tuberculosis patients (Wu et al. 2013). Another study highlighted a reduction in the ratio of Firmicutes to Bacteroidetes in gut of tuberculosis patients as compared to healthy subjects (Huang et al. 2019). Various other alterations in gut and lung microbiome in tuberculosis patients have been documented in Table 12.1. The study of the microbial dysbiosis along the gut–lung axis and their probable role in immune modulation in tuberculosis may shed light on novel strategies based on modulation of microbiota for prevention and therapy of tuberculosis (Liu et al. 2021).

12.11 Bioinformatics

Tuberculosis is the most well-studied infectious disease both with respect to the causative agent, i.e. *M. tuberculosis*, and the host immune responses both experimentally and using in silico approaches. The emergence of drug-resistant MTB has increased the complexity of this disease, and thus several online databases (listed in Table 12.2) and prediction servers (tabulated in Table 12.3) have been developed across the globe to make it a serious effort to understand and tackle the disease.

12.12 Medical Imaging or Radiomics

Imaging in tuberculosis (TB) is primarily used to diagnose active TB patients and also to differentiate from lung cancer and other lung diseases. For this, conventional chest X-ray, computed tomography (CT), and even magnetic resonance imaging (MRI) are used for accurately detecting primary, post-primary tuberculosis

Table 12.2 Databases for *Mycobacterium tuberculosis*

Sl. no.	Name	Description	URL
1.	TBDB (TB Database or Tuberculosis Database)	Repository of TB genomic data, relevant to biomarker discovery, development of TB drugs and vaccines Also contains information on individual genes and operon, list of protein epitopes and sequencing data for 25 strains of *Mycobacterium tuberculosis* (Reddy et al. 2009)	http://tbdb.bu.edu/
2.	GenoMycDB	This database enables comparison across mycobacterial genes and genomes (Catanho et al. 2006)	http://www.dbbm.fiocruz.br/labwim/bioinfoteam/templates/archives/GenoMycDB/GenoMycDB.html
3.	Tbvar	Comprehensive database for genome variation for *Mycobacterium tuberculosis* (Joshi et al. 2014)	http://genome.igib.res.in/tbvar/
4.	Mycobrowser (Mycobacterial browser)	This is a repository of genomic and proteomic data for pathogenic strains of mycobacteria (Kapopoulou et al. 2011)	https://mycobrowser.epfl.ch/
5.	COMBAT-TB NeoDB	An integrated database made by compilation of omics data for *Mycobacterium tuberculosis* (Lose et al. 2020)	https://neodb.sanbi.ac.za/browser/
6.	CPLP-TB	Database of molecular epidemiological data of *Mycobacterium tuberculosis* (Perdigao et al. 2019)	http://cplp-tb.ff.ulisboa.pt/
7.	TB Portals	Collection of geographic, clinical, laboratory, radiological, and genomic data from patient cases of drug-resistant tuberculosis (https://tbportals.niaid.nih.gov/)	https://tbportals.niaid.nih.gov/
8.	TB Drug Target Database	Repository of anti-tubercular drugs. It also enlists different target proteins for the treatment of TB (https://www.bioinformatics.org/tbdtdb/)	https://www.bioinformatics.org/tbdtdb/
9.	TubercuList	Relational database of genome derived information for H37Rv strain of *Mycobacterium tuberculosis* (Lew et al. 2011)	http://genolist.pasteur.fr/TubercuList/index.html

(continued)

Table 12.2 (continued)

Sl. no.	Name	Description	URL
10.	SInCRe (Structural Interactome Computational Resource)	Database for *Mycobacterium tuberculosis* integrates sequence and structural information, thereby aiding in drug discovery (Metri et al. 2015)	http://proline.biochem.iisc.ernet.in/sincre/
11.	MUBII-TB-DB	Repository of mutations associated with drug resistance in *Mycobacterium tuberculosis* (Flandrois et al. 2014)	https://umr5558-proka.univ-lyon1.fr/mubii/mubii-select.cgi
12.	DRAGdb	Database containing mutational data across drug resistance-associated genes (DRAGs) in *Mycobacterium tuberculosis,* along with other microbes like ESKAPE pathogens (Ghosh and Saha 2020)	http://bicresources.jcbose.ac.in/ssaha4/drag/
13.	TB Drug Resistance Mutation Database (TBDReaMDB)	Repository of drug resistance-associated mutations in *Mycobacterium tuberculosis* (Sandgren et al. 2009)	http://www.tbdreamdb.com/
14.	SITVIT	Database of three major types of molecular markers (namely VNTRs, MIRUs and Spoligotypes) for *Mycobacterium tuberculosis* (Demay et al. 2012)	http://www.pasteur-guadeloupe.fr:8081/SITVIT_ONLINE/index.jsp
15.	SITVIT2	Repository of genotyping molecular markers (like 5ETRs, spoligotypes, 12-loci MIRU-VNTR, 15-loci MIRU-VNTR, and 24-loci MIRU-VNTR markers) of *Mycobacterium tuberculosis* (Couvin et al. 2019)	http://www.pasteur-guadeloupe.fr:8081/SITVIT2/
16.	MycoperonDB	Compilation of computationally predicted transcription units and operons of Mycobacteria (Ranjan et al. 2006)	http://cdfd.org.in/mycoperondb/about.html
17.	MycobacRV	Repository of vaccine candidates for mycobacteria (Chaudhuri et al. 2014)	https://mycobacteriarv.igib.res.in/
18.	*Mycobacterium tuberculosis* Pathway/Genome Databases	Knowledgebase of genomic data and metabolic pathways of *Mycobacterium tuberculosis* (https://mycobacterium.biocyc.org/)	https://mycobacterium.biocyc.org/
19.	MtbTnDB	Repository of *Mycobacterium tuberculosis* transposon	https://www.mtbtndb.app/

(continued)

Table 12.2 (continued)

Sl. no.	Name	Description	URL
		sequencing data (https://www.mtbtndb.app/)	
20.	TDR Targets	Database on drug targets, drugs, and bioactive compounds. It can aid in drug discovery for disease pathogens like *Mycobacterium tuberculosis* (Uran Landaburu et al. 2020)	https://tdrtargets.org/
21.	AntiTbPdb	Database of experimentally validated anti-mycobacterial or anti-tubercular peptides (Usmani et al. 2018)	https://webs.iiitd.edu.in/raghava/antitbpdb/index.html
22.	mycoDB.es	Knowledgebase of animal tuberculosis (Rodriguez-Campos et al. 2012)	https://www.visavet.es/mycodb/
23.	Collaborative Drug Discovery Tuberculosis database (CDD TB)	Database of small molecule compounds tested against *Mycobacterium tuberculosis* (Ekins et al. 2010)	www.collaborativedrug.com

This table enlists the name of the different data repositories available for *Mycobacterium tuberculosis* along with their description and URL

(reactivation), and miliary tuberculosis (Nachiappan et al. 2017; Jeong and Lee 2008). Some of the signs of tuberculosis in image analysis are lymphadenopathy (common in children), cavitation with a typical “tree-in-bud” appearance of distribution of nodules (a marker for active TB), and pulmonary tuberculomas with inflammatory tissue (Leung et al. 1992; Lee and Im 1995; Lee et al. 1993). High-resolution CT has been shown to be highly accurate in diagnosing active TB than chest X-ray, whereas MRI is superior in detecting and assessing central nervous system TB (Raniga et al. 2006; Trivedi et al. 2009; Skoura et al. 2015). ^{11}C-choline PET and ^{18}F-FDG PET were used to differentiate lung cancer and tuberculoma (granulomatous tissue) (Hara et al. 2003; Ordonez et al. 2020). In summary, chest radiograph followed by image analysis plays an important role in detecting and monitoring TB patients after drug treatment.

12.13 Multi-omics and Data Integration

Integrative omics approaches have been applied to study the mechanisms of hypervirulence in MTB, to identify novel serum/sputum biomarkers and drug targets in MTB and also to study host–pathogen interactions (Ahamad et al. 2022; Gomez-Gonzalez et al. 2019; Kontsevaya et al. 2021; Goff et al. 2020). Rajwani et al. have

Table 12.3 Web-servers and Prediction tools used for *Mycobacterium tuberculosis*

Sl. no.	Bioinformatics tools	Description	URL
1.	TB Profiler	Webserver allows analyses of *Mycobacterium tuberculosis* whole genome sequencing data for prediction of lineage and drug resistance (Phelan et al. 2019)	https://tbdr.lshtm.ac.uk/
2.	PhyResSE	Web tool predicting both lineage and drug resistance of *Mycobacterium tuberculosis* from whole genome sequencing (WGS) data (Feuerriegel et al. 2015)	https://bioinf.fz-borstel.de/mchips/phyresse/
3.	genTB	Machine learning-based tool for prediction of drug resistance from next generation sequencing data for *Mycobacterium tuberculosis* (Groschel et al. 2021)	https://gentb.hms.harvard.edu/
4.	Resistance Sniffer	Stand-alone software that predicts drug resistance in *Mycobacterium tuberculosis* isolates using complete or partial genome datasets (Muzondiwa et al. 2020)	http://resistance-sniffer.bi.up.ac.za/
5.	Tuberculosis Regulatory Network Analysis Tool (TBRNAT)	Collects and annotates information on regulation of *Mycobacterium tuberculosis* gene expression by transcription factors and provide users with an intuitive way to visualize and navigate the network (https://bioinformatics.niaid.nih.gov/tbrnat/)	https://bioinformatics.niaid.nih.gov/tbrnat/
6.	Mykrobe predictor	Software package that predicts antibiotic resistance from genome sequences of *Mycobacterium tuberculosis* and *Staphylococcus aureus* (https://www.mykrobe.com/)	https://www.mykrobe.com/
7.	MycoPrInt (Mycobacterial Protein–Protein Interaction Predictor)	Web interface predicts mycobacterial protein–protein interaction (PPI) using domain Interaction mapping (DIM) method	https://webs.iiitd.edu.in/raghava/mycoprint/

(continued)

Table 12.3 (continued)

Sl. no.	Bioinformatics tools	Description	URL
		(https://webs.iiitd.edu.in/raghava/mycoprint/)	
8.	CASTB (comprehensive analysis server for the *Mycobacterium tuberculosis* complex)	Webserver analyses whole genome sequencing (WGS) data and predicts phylogeny and drug resistance of *Mycobacterium tuberculosis* (Iwai et al. 2015)	http://castb.ri.ncgm.go.jp/CASTB/
9.	KvarQ	This program allows targeted, direct variant calling in FastQ reads of genomes from *Mycobacterium tuberculosis* and other bacteria. It also aids in spoligotyping, phylogenetic classification, and identification of drug resistance for *Mycobacterium tuberculosis* (Steiner et al. 2014)	https://www.swisstph.ch/en/about/mpi/tuberculosis-research/kvarq-targeted-and-direct-variant-calling-in-fastq-reads-of-bacterial-genomes/, https://github.com/kvarq/kvarq
10.	MIRU-VNTRplus	Web-based tool for analysing and comparing genotypic data of *Mycobacterium tuberculosis* (Weniger et al. 2010)	https://www.miru-vntrplus.org/MIRU/index.faces
11.	StackTB	This tool enables classification of lineage of *Mycobacterium tuberculosis* from MIRU-VNTR data (Thain et al. 2019)	https://shiny.rcg.sfu.ca/static/stacktb/
12.	MDRIpred	Webserver used for predicting inhibitors against drug-resistant *Mycobacterium tuberculosis* H37Rv strain in different phases like replicative phase and non-replicative phase (Singla et al. 2013)	https://webs.iiitd.edu.in/oscadd/mdri/
13.	eBooster	Webserver predicts EC_{50} value of compounds targeting EthR transcription repressor of *Mycobacterium tuberculosis* (https://webs.iiitd.edu.in/oscadd/ebooster/)	https://webs.iiitd.edu.in/oscadd/ebooster/
14.	GDoQ	Webserver that predicts inhibitors for *Mycobacterium tuberculosis*	https://webs.iiitd.edu.in/oscadd/gdoq/

(continued)

Table 12.3 (continued)

Sl. no.	Bioinformatics tools	Description	URL
		using QSAR and docking (Singla et al. 2011)	
15.	MtbVeb	Web portal that aids in designing vaccines for multidrug-resistant (MDR) and extremely drug-resistant (XDR) strains of *Mycobacterium tuberculosis* (Dhanda et al. 2016)	http://crdd.osdd.net/raghava/mtbveb/
16.	PhyTB	Web-based tool that aids in visualizing phylogenetic tree and identifying *Mycobacterium tuberculosis* clade-informative polymorphisms (Benavente et al. 2015)	http://pathogenseq.lshtm.ac.uk/phytblive/index.php
17.	PolyTB	Web-based tool that enables identification of genetic variations and polymorphisms. It also aids in analysing genomic diversity and distribution of strains of *Mycobacterium tuberculosis* (Coll et al. 2014)	http://pathogenseq.lshtm.ac.uk/polytblive/browser.php
18.	DaGO-Fun	Prediction tool that predicts function of uncharacterized proteins of *Mycobacterium tuberculosis* (Mazandu and Mulder 2013)	http://web.cbio.uct.ac.za/ITGOM/tools/functionprediction.php
19.	AuTuMN TB-modelling	Modular Software platform that simulates TB control interventions (http://www.tb-modelling.com/mdr_tb_at_retreatment/)	http://www.tb-modelling.com/mdr_tb_at_retreatment/
20.	TB-Insight	Collection of tools that enables prediction of lineage, classification, and analysis of strains of *Mycobacterium tuberculosis* (https://tbinsight.cs.rpi.edu/)	https://tbinsight.cs.rpi.edu/
21.	MycPermCheck	Prediction tool that aids in drug discovery for *Mycobacterium tuberculosis* by predicting the permeability of small molecules (Merget et al. 2013)	https://www.mycpermcheck.aksotriffer.pharmazie.uni-wuerzburg.de/

(continued)

Table 12.3 (continued)

Sl. no.	Bioinformatics tools	Description	URL
22.	TB DEPOT	Analytical tool that allows multidomain data analysis for tuberculosis (https://depot.tbportals.niaid.nih.gov/#/home)	https://depot.tbportals.niaid.nih.gov/#/home
23.	SNP-IT	This tool utilizes single nucleotide polymorphisms to identify subspecies and lineages of *Mycobacterium tuberculosis* (Lipworth et al. 2019)	https://github.com/samlipworth/snpit
24.	TB Portals-GAP	Tool for genomic analysis of *Mycobacterium tuberculosis* (https://tbportals.niaid.nih.gov/gap)	https://tbportals.niaid.nih.gov/gap
25.	TB Portals - RAP	Analytical tool for analysing TB imaging data (https://tbportals.niaid.nih.gov/rap)	https://tbportals.niaid.nih.gov/rap
26.	OSDD Chem	Web interface for large scale synthesis and screening and chemical compounds against TB, leishmania, and malaria (http://www.osdd.net/)	http://www.osdd.net/

This table enlists the different bioinformatics tools available for *Mycobacterium tuberculosis* along with their description and URL

identified the hypervirulent-specific mutations in FadE5, Rv0178, and pip MTB genes based on integrated genomic-transcriptomic and proteomic comparisons (Rajwani et al. 2022). Krishnan et al. have identified serum biomarkers of tuberculosis patients by combining miRNA expression, metabolite data, and serum cytokine or chemokine values and developed a machine-learning-based decision-tree algorithm (Krishnan et al. 2021). Studies were carried out in MTB to understand the mechanism of MTB inhibitors like 8-(2-cyclobuten-1-yl)octanoic acid (DA-CB) and identify their targets by combining the results of proteomics, metabolomics, and lipidomics (Sakallioglu et al. 2022). Host–pathogen interaction with respect to MTB is extensively studied to identify the host-directed therapies in tuberculosis (Abreu et al. 2020). There are a few challenges of OMICs-based host biomarkers/targets discovery in tuberculosis as the identified biomarkers may not be specific to tuberculosis but may just represent general host responses to any infectious diseases (Kanabalan et al. 2021).

12.14 Present Therapeutics

The treatment of tuberculosis relies on the use of combination of anti-tuberculosis drugs. Directly observed therapy (DOT) is the standard for tuberculosis treatment. Different therapeutic regime is followed for latent and active tuberculosis patients. Combination of rifampicin and isoniazid or prolonged monotherapy of isoniazid are the most preferred therapeutic regimes for latent tuberculosis (https://www.ncbi.nlm.nih.gov/books/NBK441916/). Monotherapy is never used for the treatment of active tuberculosis cases. Generally, combination of the different anti-tuberculosis medications (enlisted in Table 12.4) is used for the therapy of active tuberculosis patients. Isoniazid (INH), rifampicin (RIF), pyrazinamide (PZA), ethambutol (EMB), and streptomycin (SM) are the first-line drugs for tuberculosis (https://www.ncbi.nlm.nih.gov/books/NBK138747/). Fluoroquinolones (FQs), *p*-amino salicylic acid (PAS), ethionamide/prothionamide, and injectable aminoglycosides are second line of anti-tuberculosis drugs. The details of both first-line and second-line anti-tuberculosis drugs have been enlisted in Table 12.4. Certain drugs, namely clofazimine, linezolid, amoxicillin/clavulanic acid, imipenem/cilastatin, and clarithromycin have been grouped as third-line anti-tuberculosis drugs. These drugs have variable degrees of unproven efficacy against tuberculosis and are essentially used against extensively drug-resistant strains of tuberculosis (https://www.ncbi.nlm.nih.gov/books/NBK441916/).

Emergence and spread of drug-resistant strains of *Mycobacterium tuberculosis* are serious causes of concern which makes tuberculosis control extremely difficult. In case of multidrug-resistant tuberculosis (MDR-TB), the *Mycobacterium tuberculosis* strain is resistant to isoniazid and rifampicin. Practically, these patients cannot be treated using first line anti-tuberculosis drugs (Seung et al. 2015). The emergence of extensively drug-resistant tuberculosis (XDR-TB) have further worsened the scenario. These XDR-TB strains are not only resistant to isoniazid and rifampicin but also have additional resistance to fluoroquinolones and any of the second-line injectable anti-tuberculosis drugs (Roelens et al. 2021; Maitre et al. 2017). XDR-TB poses a formidable challenge towards WHO's End TB program which aims to put an end to tuberculosis epidemic by 2035.

Immunization using effective vaccines may serve as an alternative primary driving force towards reducing the global incidences of tuberculosis. Currently, only Bacillus Calmette-Guerin (BCG) vaccine is administered in many countries as part of the WHO's Expanded Programme on Immunization for controlling tuberculosis (Scriba et al. 2020). The details of BCG vaccine have been tabulated in Table 12.5. Several pre-infection or preventive vaccines as well as post-infection or therapeutic vaccine candidates have been developed and are being clinically investigated currently to maximize beneficial impact on public health and bring about rapid decline in tuberculosis morbidity and mortality. These vaccine candidates have been enlisted in Table 12.6.

Although, there has been a striving effort for designing preventive and therapeutic vaccines for tuberculosis, the vaccines are still in clinical trials. Besides, the rapid expansion of drug resistance continues to hinder the reduction in tuberculosis

Table 12.4 Drugs for treatment of tuberculosis

Sl. no.	Drugs	Line of drug	Mode of action	Adverse effects
1.	Isoniazid (INH)	First-line drug	Prodrug that gets activated by mycobacterial catalase-peroxidase enzyme. Inhibits mycolic acid synthesis and interferes with mycobacterial cell wall formation	Hepatotoxicity, neurotoxicity, aplastic anaemia (Sotgiu et al. 2015; https://www.ncbi.nlm.nih.gov/books/NBK557617/; Jnawali and Ryoo 2013)
2.	Rifampin or Rifampicin (RIF)	First-line drug	Inhibits mycobacterial transcriptional activity by binding to the β-subunit of bacterial RNA polymerase	Stomach upset, hepatotoxicity (Sotgiu et al. 2015; Jnawali and Ryoo 2013)
3.	Pyrazinamide (PZA)	First-line drug	Prodrug that gets converted by mycobacterial enzyme pyrazinamidase to the active form pyrazinoic acid. PZA is active at an acidic pH of 5.5. Mostly acts on semi dormant bacteria	Stomach upset, nausea, vomiting, fatigue, gastrointestinal intolerance, hypersensitive reactions (Sotgiu et al. 2015; Jnawali and Ryoo 2013; Zhang et al. 2003; Kwon et al. 2020)
4.	Ethambutol (EMB)	First-line drug	Bacteriostatic agent Inhibits polymerization of arabinan and interferes with synthesis of arabinogalactan component of cell wall	Stomach pain, nausea, vomiting, blurred vision, colour blindness, headache, swelling of face, breathlessness (Sotgiu et al. 2015; Jnawali and Ryoo 2013)
5.	Streptomycin (SM)	First-line drug	Aminoglycoside antibiotic. Inhibits bacterial protein synthesis by binding with 16srRNA and interfering with translation proof-reading	Ototoxicity, vestibular dysfunction, nephrotoxicity (Sotgiu et al. 2015; Jnawali and Ryoo 2013)
6.	Fluoroquinolones (FQs)	Second-line drug	Specifically inhibits DNA gyrase and topoisomerase IV, thus preventing DNA unwinding and interfering with bacterial DNA synthesis	Gastrointestinal intolerance, colitis, headache, dizziness, rashes (Jnawali and Ryoo 2013; Blondeau 2004)
7.	Aminoglycosides (kanamycin, amikacin, and capreomycin)	Second-line drug	Used as injectables for MDR-TB. Interferes with bacterial protein translation	Renal toxicity, ototoxicity, hearing problems (Sotgiu et al. 2015; Jnawali and Ryoo 2013)
8.	Ethionamide/ prothionamide	Second-line drug	Used for MDR-TB Act as prodrug and inhibit the mycolic acid synthesis pathway.	Anorexia, abdominal pain, nausea, vomiting, liver inflammation, depression, peripheral

(continued)

Table 12.4 (continued)

Sl. no.	Drugs	Line of drug	Mode of action	Adverse effects
				neuropathy, hypothyroidism (Jnawali and Ryoo 2013)
9.	p-Amino salicylic acid (PAS)	Second-line drug	Used for XDR-TB Also serves as a Prodrug Inhibits dihydrofolate reductase (DHFR) enzyme activity, thereby, interferes with folic acid synthesis	Gastrointestinal (GI) disturbances, liver inflammation, allergic reactions (Sotgiu et al. 2015; Jnawali and Ryoo 2013; Zheng et al. 2013)
10.	Cycloserine (CS)	Second-line drug	Used for MDR-TB and XDR-TB. Serves as cyclic analogue of D-alanine. Inhibits cell wall peptidoglycan synthesis.	Adverse psychiatric effects, allergic reactions, sleepiness, numbness, seizures, unsteadiness (Sotgiu et al. 2015; Jnawali and Ryoo 2013; Li et al. 2019b)

This table enlists the different medications widely used as treatment for tuberculosis

Table 12.5 Vaccine used for tuberculosis

Sl. No.	Vaccine	Description	Status	Effects
1.	Bacillus Calmette-Guerin (BCG) vaccine	Live attenuated strain of bovine tuberculosis bacillus, *Mycobacterium bovis*	In use since 1921	• Moderate efficacy in controlling severe forms of extra-pulmonary TB (meningeal and miliary TB) and pulmonary TB in young children. • Varied degree of effectiveness in preventing TB in adults and adolescents. • Ineffective in preventing global TB epidemic (Scriba et al. 2020; Lobo et al. 2021)

This table enlists the vaccine that is being used for conferring protection from tuberculosis

morbidity and mortality. Therefore, the modern omics technologies should be extensively utilized for identifying diagnostic biomarkers, host response factors, and novel drug targets for enabling timely disease diagnosis and effective tuberculosis management.

Table 12.6 Vaccines under investigation for tuberculosis

Sl. no.	Vaccine	Description	Vaccine type	Clinical trial stage
1.	MVA85A	Modified vaccinia Ankara vaccine expressing antigen 85A of *Mycobacterium tuberculosis*	Preventive vaccine	Phase IIb (Scriba et al. 2020; Kaufmann 2020; Hawn et al. 2014)
2.	Ad5Ag85A	Replication deficient adenoviral vector expressing antigen 85A of *Mycobacterium tuberculosis*	Preventive vaccine	Phase I (Scriba et al. 2020; Kaufmann 2020; Hawn et al. 2014)
3.	ChAdOx1.85A + MVA85A	Comprises of a prime boost with a chimpanzee Adenoviral (ChAd) vaccine expressing Antigen 85A of *Mycobacterium tuberculosis,* followed by a boost with modified vaccinia Ankara virus vaccine expressing antigen 85A of *Mycobacterium tuberculosis*	Preventive vaccine	Phase I (Kaufmann 2020)
4.	TB-FLU-04 L	Replication deficient H1N1 influenza virus strain expressing antigen 85A and ESAT-6 protein of *Mycobacterium tuberculosis*	Preventive vaccine, therapeutic vaccine	Phase IIa (Kaufmann 2020)
5.	RUTI	Killed vaccine comprising of *Mycobacterium tuberculosis* fragments	Preventive vaccine, Therapeutic vaccine	Phase IIa (Scriba et al. 2020; Kaufmann 2020; Hawn et al. 2014)
6.	*M. vaccae* (Vaccae, V7)	Killed *Mycobacterium vaccae*	Preventive vaccine, therapeutic vaccine	Phase III (Scriba et al. 2020; Kaufmann 2020; Hawn et al. 2014)
7.	GamTBvac	Protein subunit (Ag85A and ESAT6-	Preventive vaccine	Phase II (Scriba et al. 2020)

(continued)

Table 12.6 (continued)

Sl. no.	Vaccine	Description	Vaccine type	Clinical trial stage
		CFP10 fusion protein) vaccine		
8.	AEC/BC02	Protein subunit (Ag85A and ESAT6-CFP10 fusion protein) vaccine	Preventive vaccine	Phase I (Scriba et al. 2020)
9.	MIP (Immuvac)	Killed *Mycobacterium indicus pranii*	Preventive vaccine, Therapeutic vaccine	Phase III (Scriba et al. 2020; Kaufmann 2020)
10.	DAR-901	Killed *Mycobacterium obuense*	Preventive vaccine	Phase III (Scriba et al. 2020; Kaufmann 2020; Hawn et al. 2014)
11.	VPM1002	Recombinant BCG vaccine	Preventive vaccine, Therapeutic vaccine	Phase III (Scriba et al. 2020; Kaufmann 2020)
12.	MTBVAC	Genetically attenuated *Mycobacterium tuberculosis* vaccine	Preventive vaccine	Phase IIa completed (Scriba et al. 2020; Kaufmann 2020; Hawn et al. 2014)
13.	$M72/AS01_E$	Protein subunit (fusion protein comprised of MTB32A and MTB39A antigens) vaccine	Preventive vaccine	Phase IIb (Scriba et al. 2020)
14.	H56:IC31	Protein subunit (fusion protein comprised of ESAT6, Ag85B, and Rv2660 antigens) vaccine	Therapeutic vaccine	Phase I (Scriba et al. 2020; Kaufmann 2020)
15.	ID93:GLA-SE	Protein subunit (fusion protein comprised of Rv1813, Rv2608, Rv3619, and Rv3620 antigens) vaccine	Therapeutic vaccine	Phase I (Kaufmann 2020)

This table enlists the vaccines that are in clinical development for tuberculosis

12.15 Future Perspectives

Scientific advancements have unveiled several details about the pathobiology of tuberculosis. However, the intricate details of host–mycobacterial interaction that drives and distinguishes latent tuberculosis from active tuberculosis still remains unknown. Use of modern omics approaches and systems biology may be of immense significance in improving the mechanistic understanding of tuberculosis pathogenesis and vital host–pathogen interactions (Comas and Gagneux 2009). Omics-based identification of precise biomarkers and radiological signatures is very crucial for improved disease diagnosis, for predicting response and for determining the chances of tuberculosis relapse (Gill et al. 2022). Various diagnostic pathways also need to be evaluated for designing new diagnostic tests for rapid, timely, and accurate detection of active tuberculosis (Rylance et al. 2010). The role of miRNAs as immunomodulators in tuberculosis has been highlighted in recent times. Further studies should be conducted in this field to enable the use of miRNAs for differential diagnosis of active and latent tuberculosis as well as to explore miRNAs as therapeutic targets for tuberculosis (Sabir et al. 2018). Several advancements in imaging modalities shall also be explored for tuberculosis. Pathogen-specific positron emission tomography (PET) imaging may be used to evaluate changes in tuberculosis tissues. Non-invasive temporal monitoring and computational enhancements can be utilized to further reduce scan times and improve quality of data acquisition (Merchant et al. 2022). Mathematical models have been used for epidemiological studies of tuberculosis. An expansion of such mathematical modelling with inclusion of biological data including host–pathogen interaction can enhance the robustness and significance of such models (Comas and Gagneux 2009). Artificial intelligence algorithms and machine learning technologies based on clinical features, medical and molecular parameters can also be used to support timely diagnosis of tuberculosis (Merchant et al. 2022; Orjuela-Canon et al. 2022). Emergence of drug resistance is a huge problem for controlling tuberculosis. To overcome this challenge associated with cure of tuberculosis, the study of host–pathogen protein–protein interactions and alterations in metabolic pathways along with the study of functional biology of bacterial virulence factors needs to be further emphasized. The quest for new drug targets is challenging and relies largely on analysing their bactericidal activity. However, such drug selections should not be done on the sole basis of bactericidal activity. Newer methods need to be explored for drug sensitivity testing. Target specification, suitable optimization, and drug interaction studies should also be done to enhance specificity and activity of the novel drug candidates (Rylance et al. 2010; Bendre et al. 2021). Exploring these research areas may take us one step ahead towards fulfilling WHO's End TB program and mediating better cure to tuberculosis patients.

References

Abreu R, Giri P, Quinn F (2020) Host-pathogen interaction as a novel target for host-directed therapies in tuberculosis. Front Immunol 11:1553

Achkar JM, Cortes L, Croteau P, Yanofsky C, Mentinova M, Rajotte I, Schirm M, Zhou Y, Junqueira-Kipnis AP, Kasprowicz VO et al (2015) Host protein biomarkers identify active tuberculosis in HIV uninfected and co-infected individuals. EBioMedicine 2(9):1160–1168

Ahamad N, Gupta S, Parashar D (2022) Using omics to study leprosy, tuberculosis, and other mycobacterial diseases. Front Cell Infect Microbiol 12:792617

Ahmad S (2011) Pathogenesis, immunology, and diagnosis of latent Mycobacterium tuberculosis infection. Clin Dev Immunol 2011:814943

Anderson ST, Kaforou M, Brent AJ, Wright VJ, Banwell CM, Chagaluka G, Crampin AC, Dockrell HM, French N, Hamilton MS et al (2014) Diagnosis of childhood tuberculosis and host RNA expression in Africa. N Engl J Med 370(18):1712–1723

Benavente ED, Coll F, Furnham N, McNerney R, Glynn JR, Campino S, Pain A, Mohareb FR, Clark TG (2015) PhyTB: Phylogenetic tree visualisation and sample positioning for M. tuberculosis. BMC Bioinformatics 16:155

Bendre AD, Peters PJ, Kumar J (2021) Tuberculosis: past, present and future of the treatment and drug discovery research. Curr Res Pharmacol Drug Discov 2:100037

Ben-Selma W, Harizi H, Boukadida J (2011) Immunochromatographic IgG/IgM test for rapid diagnosis of active tuberculosis. Clin Vaccine Immunol 18(12):2090–2094

Berry MP, Graham CM, McNab FW, Xu Z, Bloch SA, Oni T, Wilkinson KA, Banchereau R, Skinner J, Wilkinson RJ et al (2010) An interferon-inducible neutrophil-driven blood transcriptional signature in human tuberculosis. Nature 466(7309):973–977

Bishwal SC, Das MK, Badireddy VK, Dabral D, Das A, Mahapatra AR, Sahu S, Malakar D, Singh II, Mazumdar H et al (2019) Sputum proteomics reveals a shift in vitamin D-binding protein and antimicrobial protein axis in tuberculosis patients. Sci Rep 9(1):1036

Blondeau JM (2004) Fluoroquinolones: mechanism of action, classification, and development of resistance. Surv Ophthalmol 49(Suppl 2):S73–S78

Bobak CA, Abhimanyu NH, Gandhi T, Grimm SL, Nishiguchi T, Koster K, Longlax SC, Dlamini Q, Kahari J et al (2022) Increased DNA methylation, cellular senescence and premature epigenetic aging in guinea pigs and humans with tuberculosis. Aging (Albany NY) 14(5): 2174–2193

Catanho M, Mascarenhas D, Degrave W, Miranda AB (2006) GenoMycDB: a database for comparative analysis of mycobacterial genes and genomes. Genet Mol Res 5(1):115–126

Chang SY, Chen ML, Lee MR, Liang YC, Lu TP, Wang JY, Yan BS (2018) SP110 polymorphisms are genetic markers for vulnerability to latent and active tuberculosis infection in Taiwan. Dis Markers 2018:4687380

Chaudhuri R, Kulshreshtha D, Raghunandanan MV, Ramachandran S (2014) Integrative immunoinformatics for Mycobacterial diseases in R platform. Syst Synth Biol 8(1):27–39

Che N, Cheng J, Li H, Zhang Z, Zhang X, Ding Z, Dong F, Li C (2013) Decreased serum 5-oxoproline in TB patients is associated with pathological damage of the lung. Clin Chim Acta 423:5–9

Chegou NN, Sutherland JS, Malherbe S, Crampin AC, Corstjens PL, Geluk A, Mayanja-Kizza H, Loxton AG, van der Spuy G, Stanley K et al (2016) Diagnostic performance of a seven-marker serum protein biosignature for the diagnosis of active TB disease in African primary healthcare clinic attendees with signs and symptoms suggestive of TB. Thorax 71(9):785–794

Chen YC, Hsiao CC, Chen TW, Wu CC, Chao TY, Leung SY, Eng HL, Lee CP, Wang TY, Lin MC (2020) Whole genome DNA methylation analysis of active pulmonary tuberculosis disease identifies novel epigenotypes: PARP9/miR-505/RASGRP4/GNG12 gene methylation and clinical phenotypes. Int J Mol Sci 21(9)

Cheung MK, Lam WY, Fung WY, Law PT, Au CH, Nong W, Kam KM, Kwan HS, Tsui SK (2013) Sputum microbiota in tuberculosis as revealed by 16S rRNA pyrosequencing. PLoS One 8(1): e54574

Cho Y, Park Y, Sim B, Kim J, Lee H, Cho SN, Kang YA, Lee SG (2020) Identification of serum biomarkers for active pulmonary tuberculosis using a targeted metabolomics approach. Sci Rep 10(1):3825

Coll F, Preston M, Guerra-Assuncao JA, Hill-Cawthorn G, Harris D, Perdigao J, Viveiros M, Portugal I, Drobniewski F, Gagneux S et al (2014) PolyTB: a genomic variation map for Mycobacterium tuberculosis. Tuberculosis (Edinb) 94(3):346–354

Collins JM, Walker DI, Jones DP, Tukvadze N, Liu KH, Tran VT, Uppal K, Frediani JK, Easley KA, Shenvi N et al (2018) High-resolution plasma metabolomics analysis to detect Mycobacterium tuberculosis-associated metabolites that distinguish active pulmonary tuberculosis in humans. PLoS One 13(10):e0205398

Comas I, Gagneux S (2009) The past and future of tuberculosis research. PLoS Pathog 5(10): e1000600

Couvin D, David A, Zozio T, Rastogi N (2019) Macro-geographical specificities of the prevailing tuberculosis epidemic as seen through SITVIT2, an updated version of the Mycobacterium tuberculosis genotyping database. Infect Genet Evol 72:31–43

Cubillos-Angulo JM, Fernandes CD, Araujo DN, Carmo CA, Arriaga MB, Andrade BB (2021) The influence of single nucleotide polymorphisms of NOD2 or CD14 on the risk of Mycobacterium tuberculosis diseases: a systematic review. Syst Rev 10(1):174

de Araujo LS, Ribeiro-Alves M, Leal-Calvo T, Leung J, Duran V, Samir M, Talbot S, Tallam A, Mello FCQ, Geffers R et al (2019) Reprogramming of small noncoding RNA populations in peripheral blood reveals host biomarkers for latent and active Mycobacterium tuberculosis infection. mBio 10(6)

De Groote MA, Nahid P, Jarlsberg L, Johnson JL, Weiner M, Muzanyi G, Janjic N, Sterling DG, Ochsner UA (2013) Elucidating novel serum biomarkers associated with pulmonary tuberculosis treatment. PLoS One 8(4):e61002

De Groote MA, Sterling DG, Hraha T, Russell TM, Green LS, Wall K, Kraemer S, Ostroff R, Janjic N, Ochsner UA (2017) Discovery and validation of a six-marker serum protein signature for the diagnosis of active pulmonary tuberculosis. J Clin Microbiol 55(10):3057–3071

Demay C, Liens B, Burguiere T, Hill V, Couvin D, Millet J, Mokrousov I, Sola C, Zozio T, Rastogi N (2012) SITVITWEB--a publicly available international multimarker database for studying Mycobacterium tuberculosis genetic diversity and molecular epidemiology. Infect Genet Evol 12(4):755–766

Dhanda SK, Vir P, Singla D, Gupta S, Kumar S, Raghava GP (2016) A Web-based platform for designing vaccines against existing and emerging strains of Mycobacterium tuberculosis. PLoS One 11(4):e0153771

DiNardo AR, Rajapakshe K, Nishiguchi T, Grimm SL, Mtetwa G, Dlamini Q, Kahari J, Mahapatra S, Kay A, Maphalala G et al (2020) DNA hypermethylation during tuberculosis dampens host immune responsiveness. J Clin Invest 130(6):3113–3123

Du Y, Gao X, Yan J, Zhang H, Cao X, Feng B, He Y, He Y, Guo T, Xin H et al (2022) Relationship between DNA methylation profiles and active tuberculosis development from latent infection: a pilot study in nested case-control design. Microbiol Spectr 10(3):e0058622

Duffy FJ, Weiner J 3rd, Hansen S, Tabb DL, Suliman S, Thompson E, Maertzdorf J, Shankar S, Tromp G, Parida S et al (2019) Immunometabolic signatures predict risk of progression to active tuberculosis and disease outcome. Front Immunol 10:527

Ekins S, Bradford J, Dole K, Spektor A, Gregory K, Blondeau D, Hohman M, Bunin BA (2010) A collaborative database and computational models for tuberculosis drug discovery. Mol BioSyst 6(5):840–851

Feuerriegel S, Schleusener V, Beckert P, Kohl TA, Miotto P, Cirillo DM, Cabibbe AM, Niemann S, Fellenberg K (2015) PhyResSE: a Web tool delineating mycobacterium tuberculosis antibiotic resistance and lineage from whole-genome sequencing data. J Clin Microbiol 53(6):1908–1914

Flandrois JP, Lina G, Dumitrescu O (2014) MUBII-TB-DB: a database of mutations associated with antibiotic resistance in Mycobacterium tuberculosis. BMC Bioinform 15:107
Fu Y, Wang J, Qiao J, Yi Z (2019) Signature of circular RNAs in peripheral blood mononuclear cells from patients with active tuberculosis. J Cell Mol Med 23(3):1917–1925
Ghosh ANS, Saha S (2020) Survey of drug resistance associated gene mutations in Mycobacterium tuberculosis, ESKAPE and other bacterial species. Sci Rep 10(1):8957
Gill CM, Dolan L, Piggott LM, McLaughlin AM (2022) New developments in tuberculosis diagnosis and treatment. Breathe (Sheff) 18(1):210149
Goff A, Cantillon D, Muraro Wildner L, Waddell SJ (2020) Multi-Omics technologies applied to tuberculosis drug discovery. Appl Sci:10
Gomez-Gonzalez PJ, Andreu N, Phelan JE, de Sessions PF, Glynn JR, Crampin AC, Campino S, Butcher PD, Hibberd ML, Clark TG (2019) An integrated whole genome analysis of Mycobacterium tuberculosis reveals insights into relationship between its genome, transcriptome and methylome. Sci Rep 9(1):5204
Groschel MI, Owens M, Freschi L, Vargas R Jr, Marin MG, Phelan J, Iqbal Z, Dixit A, Farhat MR (2021) GenTB: A user-friendly genome-based predictor for tuberculosis resistance powered by machine learning. Genome Med 13(1):138
Hara T, Kosaka N, Suzuki T, Kudo K, Niino H (2003) Uptake rates of 18F-fluorodeoxyglucose and 11C-choline in lung cancer and pulmonary tuberculosis: a positron emission tomography study. Chest 124(3):893–901
Hawn TR, Day TA, Scriba TJ, Hatherill M, Hanekom WA, Evans TG, Churchyard GJ, Kublin JG, Bekker LG, Self SG (2014) Tuberculosis vaccines and prevention of infection. Microbiol Mol Biol Rev 78(4):650–671
Hu Y, Wu L, Li D, Zhao Q, Jiang W, Xu B (2015) Association between cytokine gene polymorphisms and tuberculosis in a Chinese population in Shanghai: a case-control study. BMC Immunol 16:8
Hu Y, Feng Y, Wu J, Liu F, Zhang Z, Hao Y, Liang S, Li B, Li J, Lv N et al (2019) The gut microbiome signatures discriminate healthy from pulmonary tuberculosis patients. Front Cell Infect Microbiol 9:90
Huang SF, Yang YY, Chou KT, Fung CP, Wang FD, Su WJ (2019) Systemic proinflammation after Mycobacterium tuberculosis infection was correlated to the gut microbiome in HIV-uninfected humans. Eur J Clin Investig 49(5):e13068
Hunter RL (2016) Tuberculosis as a three-act play: a new paradigm for the pathogenesis of pulmonary tuberculosis. Tuberculosis (Edinb) 97:8–17
Hunter RL (2018) The pathogenesis of tuberculosis: the early infiltrate of post-primary (adult pulmonary) tuberculosis: a distinct disease entity. Front Immunol 9:2108
Isa F, Collins S, Lee MH, Decome D, Dorvil N, Joseph P, Smith L, Salerno S, Wells MT, Fischer S et al (2018) Mass spectrometric identification of urinary biomarkers of pulmonary tuberculosis. EBioMedicine 31:157–165
Iwai H, Kato-Miyazawa M, Kirikae T, Miyoshi-Akiyama T (2015) CASTB (the comprehensive analysis server for the Mycobacterium tuberculosis complex): a publicly accessible web server for epidemiological analyses, drug-resistance prediction and phylogenetic comparison of clinical isolates. Tuberculosis (Edinb) 95(6):843–844
Jacobsen M, Repsilber D, Gutschmidt A, Neher A, Feldmann K, Mollenkopf HJ, Ziegler A, Kaufmann SH (2007) Candidate biomarkers for discrimination between infection and disease caused by Mycobacterium tuberculosis. J Mol Med (Berl) 85(6):613–621
Jakhar S, Bitzer AA, Stromberg LR, Mukundan H (2020) Pediatric tuberculosis: the impact of “omics” on diagnostics development. Int J Mol Sci 21(19)
Jeong YJ, Lee KS (2008) Pulmonary tuberculosis: up-to-date imaging and management. AJR Am J Roentgenol 191(3):834–844
Jeremiah C, Petersen E, Nantanda R, Mungai BN, Migliori GB, Amanullah F, Lungu P, Ntoumi F, Kumarasamy N, Maeurer M et al (2022) The WHO Global Tuberculosis 2021 Report - not so good news and turning the tide back to End TB. Int J Infect Dis

Jiang C, Zhu J, Liu Y, Luan X, Jiang Y, Jiang G, Fan J (2017) The methylation state of VDR gene in pulmonary tuberculosis patients. J Thorac Dis 9(11):4353–4357

Jnawali HN, Ryoo S (2013) First– and second–line drugs and drug resistance. In: Mahboub BH, Vats MG (eds) Tuberculosis - current issues in diagnosis and management. IntechOpen, London

Joshi KR, Dhiman H, Scaria V (2014) tbvar: A comprehensive genome variation resource for Mycobacterium tuberculosis. Database (Oxford) 2014:bat083

Kaforou M, Wright VJ, Oni T, French N, Anderson ST, Bangani N, Banwell CM, Brent AJ, Crampin AC, Dockrell HM et al (2013) Detection of tuberculosis in HIV-infected and -uninfected African adults using whole blood RNA expression signatures: a case-control study. PLoS Med 10(10):e1001538

Kanabalan RD, Lee LJ, Lee TY, Chong PP, Hassan L, Ismail R, Chin VK (2021) Human tuberculosis and Mycobacterium tuberculosis complex: a review on genetic diversity, pathogenesis and omics approaches in host biomarkers discovery. Microbiol Res 246:126674

Kapopoulou A, Lew JM, Cole ST (2011) The MycoBrowser portal: a comprehensive and manually annotated resource for mycobacterial genomes. Tuberculosis (Edinb) 91(1):8–13

Kaufmann SHE (2020) Vaccination against tuberculosis: revamping BCG by molecular genetics guided by immunology. Front Immunol 11:316

Kontsevaya I, Lange C, Comella-Del-Barrio P, Coarfa C, DiNardo AR, Gillespie SH, Hauptmann M, Leschczyk C, Mandalakas AM, Martinecz A et al (2021) Perspectives for systems biology in the management of tuberculosis. Eur Respir Rev 30(160)

Krishnan S, Queiroz ATL, Gupta A, Gupte N, Bisson GP, Kumwenda J, Naidoo K, Mohapi L, Mave V, Mngqibisa R et al (2021) Integrative Multi-omics reveals serum markers of tuberculosis in advanced HIV. Front Immunol 12:676980

Kumar P (2016) Adult pulmonary tuberculosis as a pathological manifestation of hyperactive antimycobacterial immune response. Clin Transl Med 5(1):38

Kumar NP, Moideen K, Nancy A, Viswanathan V, Shruthi BS, Sivakumar S, Natarajan M, Kornfeld H, Babu S (2019) Plasma chemokines are biomarkers of disease severity, higher bacterial burden and delayed sputum culture conversion in pulmonary tuberculosis. Sci Rep 9(1):18217

Kumar NP, Moideen K, Nancy A, Viswanathan V, Thiruvengadam K, Nair D, Banurekha VV, Sivakumar S, Hissar S, Kornfeld H et al (2021) Plasma chemokines are baseline predictors of unfavorable treatment outcomes in pulmonary tuberculosis. Clin Infect Dis 73(9):e3419–e3427

Kwon BS, Kim Y, Lee SH, Lim SY, Lee YJ, Park JS, Cho YJ, Yoon HI, Lee CT, Lee JH (2020) The high incidence of severe adverse events due to pyrazinamide in elderly patients with tuberculosis. PLoS One 15(7):e0236109

Lau SK, Lee KC, Curreem SO, Chow WN, To KK, Hung IF, Ho DT, Sridhar S, Li IW, Ding VS et al (2015) Metabolomic profiling of plasma from patients with tuberculosis by use of untargeted mass spectrometry reveals novel biomarkers for diagnosis. J Clin Microbiol 53(12):3750–3759

Lee KS, Im JG (1995) CT in adults with tuberculosis of the chest: characteristic findings and role in management. AJR Am J Roentgenol 164(6):1361–1367

Lee KS, Song KS, Lim TH, Kim PN, Kim IY, Lee BH (1993) Adult-onset pulmonary tuberculosis: findings on chest radiographs and CT scans. AJR Am J Roentgenol 160(4):753–758

Lee SW, Wu LS, Huang GM, Huang KY, Lee TY, Weng JT (2016) Gene expression profiling identifies candidate biomarkers for active and latent tuberculosis. BMC Bioinformatics 17 (Suppl 1):3

Leung AN, Muller NL, Pineda PR, FitzGerald JM (1992) Primary tuberculosis in childhood: radiographic manifestations. Radiology 182(1):87–91

Lew JM, Kapopoulou A, Jones LM, Cole ST: TubercuList--10 years after. Tuberculosis (Edinb) 2011, 91(1):1-7

Li W, Zhu Y, Liao Q, Wang Z, Wan C (2019a) Characterization of gut microbiota in children with pulmonary tuberculosis. BMC Pediatr 19(1):445

Li Y, Wang F, Wu L, Zhu M, He G, Chen X, Sun F, Liu Q, Wang X, Zhang W (2019b) Cycloserine for treatment of multidrug-resistant tuberculosis: a retrospective cohort study in China. Infect Drug Resist 12:721–731

Lipworth S, Jajou R, de Neeling A, Bradley P, van der Hoek W, Maphalala G, Bonnet M, Sanchez-Padilla E, Diel R, Niemann S et al (2019) SNP-IT tool for identifying subspecies and associated lineages of Mycobacterium tuberculosis complex. Emerg Infect Dis 25(3):482–488

Liu Q, Pan L, Han F, Luo B, Jia H, Xing A, Li Q, Zhang Z (2018) Proteomic profiling for plasma biomarkers of tuberculosis progression. Mol Med Rep 18(2):1551–1559

Liu Y, Wang J, Wu C (2021) Microbiota and tuberculosis: a potential role of probiotics, and postbiotics. Front Nutr 8:626254

Lobo N, Brooks NA, Zlotta AR, Cirillo JD, Boorjian S, Black PC, Meeks JJ, Bivalacqua TJ, Gontero P, Steinberg GD et al (2021) 100 years of Bacillus Calmette-Guerin immunotherapy: from cattle to COVID-19. Nat Rev Urol 18(10):611–622

Loddenkemper R, Lipman M, Zumla A (2015) Clinical aspects of adult tuberculosis. Cold Spring Harb Perspect Med 6(1):a017848

Lose T, van Heusden P, Christoffels A (2020) COMBAT-TB-NeoDB: fostering tuberculosis research through integrative analysis using graph database technologies. Bioinformatics 36(3): 982–983

Luo M, Liu Y, Wu P, Luo DX, Sun Q, Zheng H, Hu R, Pandol SJ, Li QF, Han YP et al (2017) Alternation of gut microbiota in patients with pulmonary tuberculosis. Front Physiol 8:822

Lyu L, Zhang X, Li C, Yang T, Wang J, Pan L, Jia H, Li Z, Sun Q, Yue L et al (2019) Small RNA profiles of serum exosomes derived from individuals with latent and active tuberculosis. Front Microbiol 10:1174

Lyu M, Zhou J, Jiao L, Wang Y, Zhou Y, Lai H, Xu W, Ying B (2022) Deciphering a TB-related DNA methylation biomarker and constructing a TB diagnostic classifier. Mol Ther Nucleic Acids 27:37–49

Maitre T, Aubry A, Jarlier V, Robert J, Veziris N, MyRMA CNR (2017) Multidrug and extensively drug-resistant tuberculosis. Med Mal Infect 47(1):3–10

Martinot AJ (2018) Microbial offense vs host defense: who controls the TB granuloma? Vet Pathol 55(1):14–26

Maruthai K, Kalaiarasan E, Joseph NM, Parija SC, Mahadevan S (2018) Assessment of global DNA methylation in children with tuberculosis disease. Int J Mycobacteriol 7(4):338–342

Mazandu GK, Mulder NJ (2013) DaGO-Fun: tool for Gene Ontology-based functional analysis using term information content measures. BMC Bioinform 14:284

Merchant SA, Shaikh MJS, Nadkarni P (2022) Tuberculosis conundrum – current and future scenarios: a proposed comprehensive approach combining laboratory, imaging, and computing advances. World J Radiol 14(6):114–136

Merget B, Zilian D, Muller T, Sotriffer CA (2013) MycPermCheck: the Mycobacterium tuberculosis permeability prediction tool for small molecules. Bioinformatics 29(1):62–68

Metri R, Hariharaputran S, Ramakrishnan G, Anand P, Raghavender US, Ochoa-Montano B, Higueruelo AP, Sowdhamini R, Chandra NR, Blundell TL et al (2015) SInCRe-structural interactome computational resource for Mycobacterium tuberculosis. Database (Oxford) 2015:bav060

Muzondiwa D, Mutshembele A, Pierneef RE, Reva ON (2020) Resistance Sniffer: an online tool for prediction of drug resistance patterns of Mycobacterium tuberculosis isolates using next generation sequencing data. Int J Med Microbiol 310(2):151399

Nachiappan AC, Rahbar K, Shi X, Guy ES, Mortani Barbosa EJ Jr, Shroff GS, Ocazionez D, Schlesinger AE, Katz SI, Hammer MM (2017) Pulmonary tuberculosis: role of radiology in diagnosis and management. Radiographics 37(1):52–72

Ordonez AA, Wang H, Magombedze G, Ruiz-Bedoya CA, Srivastava S, Chen A, Tucker EW, Urbanowski ME, Pieterse L, Fabian Cardozo E et al (2020) Dynamic imaging in patients with tuberculosis reveals heterogeneous drug exposures in pulmonary lesions. Nat Med 26(4): 529–534

Orjuela-Canon AD, Jutinico AL, Awad C, Vergara E, Palencia A (2022) Machine learning in the loop for tuberculosis diagnosis support. Front Public Health 10:876949

Ortega E, Hernandez-Bazan S, Sanchez-Hernandez B, Licona-Limon I, Fuentes-Dominguez J (2020) Single nucleotide polymorphisms in TLR4 affect susceptibility to tuberculosis in Mexican population from the State of Veracruz. J Immunol Res 2020:2965697

Pandey R, Tamrakar D, Jaiswal S, Sharma A, Koju S, Duwal SR, Sharma I, Jayaswal RP, Pankaj PP (2016) Serum adenosine deaminase: a novel biomarker tool for the diagnosis of tuberculosis. Biosci Biotechnol Res Asia:13

Penn-Nicholson A, Hraha T, Thompson EG, Sterling D, Mbandi SK, Wall KM, Fisher M, Suliman S, Shankar S, Hanekom WA et al (2019) Discovery and validation of a prognostic proteomic signature for tuberculosis progression: A prospective cohort study. PLoS Med 16(4): e1002781

Perdigao J, Silva C, Diniz J, Pereira C, Machado D, Ramos J, Silva H, Abilleira F, Brum C, Reis AJ et al (2019) Clonal expansion across the seas as seen through CPLP-TB database: a joint effort in cataloguing Mycobacterium tuberculosis genetic diversity in Portuguese-speaking countries. Infect Genet Evol 72:44–58

Phelan JE, O'Sullivan DM, Machado D, Ramos J, Oppong YEA, Campino S, O'Grady J, McNerney R, Hibberd ML, Viveiros M et al (2019) Integrating informatics tools and portable sequencing technology for rapid detection of resistance to anti-tuberculous drugs. Genome Med 11(1):41

Phuoc Long N, Heo DY, Park S, Thi Hai Yen N, Cho YS, Shin JG, Oh JY, Kim DH (2022) Molecular perturbations in pulmonary tuberculosis patients identified by pathway-level analysis of plasma metabolic features. PLoS One 17(1):e0262545

Rajwani R, Galata C, Lee AWT, So PK, Leung KSS, Tam KKG, Shehzad S, Ng TTL, Zhu L, Lao HY et al (2022) A multi-omics investigation into the mechanisms of hyper-virulence in Mycobacterium tuberculosis. Virulence 13(1):1088–1100

Ramakrishnan L (2012) Revisiting the role of the granuloma in tuberculosis. Nat Rev Immunol 12(5):352–366

Raniga S, Parikh N, Arora A, Vaghani M, Vora PA, Vaidya V (2006) Is HRCT reliable in determining disease activity in pulmonary tuberculosis? Indian J Radiol Imaging 16:221–228

Ranjan S, Gundu RK, Ranjan A (2006) MycoperonDB: a database of computationally identified operons and transcriptional units in Mycobacteria. BMC Bioinformatics 7(Suppl 5):S9

Reddy TB, Riley R, Wymore F, Montgomery P, DeCaprio D, Engels R, Gellesch M, Hubble J, Jen D, Jin H et al (2009) TB database: an integrated platform for tuberculosis research. Nucleic Acids Res 37(Database issue):D499–D508

Rodriguez-Campos S, Gonzalez S, de Juan L, Romero B, Bezos J, Casal C, Alvarez J, Fernandez-de-Mera IG, Castellanos E, Mateos A et al (2012) A database for animal tuberculosis (mycoDB.es) within the context of the Spanish national programme for eradication of bovine tuberculosis. Infect Genet Evol 12(4):877–882

Roelens M, Battista Migliori G, Rozanova L, Estill J, Campbell JR, Cegielski JP, Tiberi S, Palmero D, Fox GJ, Guglielmetti L et al (2021) Evidence-based definition for extensively drug-resistant tuberculosis. Am J Respir Crit Care Med 204(6):713–722

Rylance J, Pai M, Lienhardt C, Garner P (2010) Priorities for tuberculosis research: a systematic review. Lancet Infect Dis 10(12):886–892

Sabir N, Hussain T, Shah SZA, Peramo A, Zhao D, Zhou X (2018) miRNAs in tuberculosis: new avenues for diagnosis and host-directed therapy. Front Microbiol 9:602

Sakallioglu IT, Maroli AS, De Lima LA, Marshall DD, Evans BW, Zinniel DK, Dussault PH, Barletta RG, Powers R (2022) Multi-omics investigation into the mechanism of action of an anti-tubercular fatty acid analogue. J Am Chem Soc 144(46):21157–21173

Sandgren A, Strong M, Muthukrishnan P, Weiner BK, Church GM, Murray MB (2009) Tuberculosis drug resistance mutation database. PLoS Med 6(2):e2

Scriba TJ, Netea MG, Ginsberg AM (2020) Key recent advances in TB vaccine development and understanding of protective immune responses against Mycobacterium tuberculosis. Semin Immunol 50:101431

Seung KJ, Keshavjee S, Rich ML (2015) Multidrug-resistant tuberculosis and extensively drug-resistant tuberculosis. Cold Spring Harb Perspect Med 5(9):a017863

Singla D, Anurag M, Dash D, Raghava GP (2011) A web server for predicting inhibitors against bacterial target GlmU protein. BMC Pharmacol 11:5

Singla D, Tewari R, Kumar A, Raghava GP (2013) Open source drug discovery C: designing of inhibitors against drug tolerant Mycobacterium tuberculosis (H37Rv). Chem Cent J 7(1):49

Skoura E, Zumla A, Bomanji J (2015) Imaging in tuberculosis. Int J Infect Dis 32:87–93

Soedarsono S, Prinasetyo K, Tanzilia M, Nugraha J (2020) Changes of serum adenosine deaminase level in new cases of pulmonary tuberculosis before and after intensive phase treatment. Lung India 37(2):126–129

Somboro AM, Diallo D, Holl JL, Maiga M (2021) The role of the microbiome in inflammation during tuberculosis. EBioMedicine 68:103435

Sotgiu G, Centis R, D'Ambrosio L, Migliori GB (2015) Tuberculosis treatment and drug regimens. Cold Spring Harb Perspect Med 5(5):a017822

Steiner A, Stucki D, Coscolla M, Borrell S, Gagneux S (2014) KvarQ: targeted and direct variant calling from fastq reads of bacterial genomes. BMC Genomics 15:881

Suarez I, Funger SM, Kroger S, Rademacher J, Fatkenheuer G, Rybniker J (2019) The diagnosis and treatment of tuberculosis. Dtsch Arztebl Int 116(43):729–735

Sun H, Pan L, Jia H, Zhang Z, Gao M, Huang M, Wang J, Sun Q, Wei R, Du B et al (2018) Label-free quantitative proteomics identifies novel plasma biomarkers for distinguishing pulmonary tuberculosis and latent infection. Front Microbiol 9:1267

Sutherland JS, Loxton AG, Haks MC, Kassa D, Ambrose L, Lee JS, Ran L, van Baarle D, Maertzdorf J, Howe R et al (2014) Differential gene expression of activating Fcgamma receptor classifies active tuberculosis regardless of human immunodeficiency virus status or ethnicity. Clin Microbiol Infect 20(4):O230–O238

Thain N, Le C, Crossa A, Ahuja SD, Meissner JS, Mathema B, Kreiswirth B, Kurepina N, Cohen T, Chindelevitch L (2019) Towards better prediction of Mycobacterium tuberculosis lineages from MIRU-VNTR data. Infect Genet Evol 72:59–66

Trivedi R, Saksena S, Gupta RK (2009) Magnetic resonance imaging in central nervous system tuberculosis. Indian J Radiol Imaging 19(4):256–265

Uran Landaburu L, Berenstein AJ, Videla S, Maru P, Shanmugam D, Chernomoretz A, Aguero F (2020) TDR Targets 6: driving drug discovery for human pathogens through intensive chemogenomic data integration. Nucleic Acids Res 48(D1):D992–D1005

Usmani SS, Kumar R, Kumar V, Singh S, Raghava GPS: AntiTbPdb: a knowledgebase of anti-tubercular peptides. Database (Oxford) 2018, 2018

Vazquez-Perez JA, Carrillo CO, Iniguez-Garcia MA, Romero-Espinoza I, Marquez-Garcia JE, Falcon LI, Torres M, Herrera MT (2020) Alveolar microbiota profile in patients with human pulmonary tuberculosis and interstitial pneumonia. Microb Pathog 139:103851

Wang W, Cai Y, Deng G, Yang Q, Tang P, Wu M, Yu Z, Yang F, Chen J, Werz O et al (2020) Allelic-specific regulation of xCT expression increases susceptibility to tuberculosis by modulating microRNA-mRNA interactions. mSphere 5(2)

Weiner J 3rd, Parida SK, Maertzdorf J, Black GF, Repsilber D, Telaar A, Mohney RP, Arndt-Sullivan C, Ganoza CA, Fae KC et al (2012) Biomarkers of inflammation, immunosuppression and stress with active disease are revealed by metabolomic profiling of tuberculosis patients. PLoS One 7(7):e40221

Weniger T, Krawczyk J, Supply P, Niemann S, Harmsen D (2010) MIRU-VNTRplus: a web tool for polyphasic genotyping of Mycobacterium tuberculosis complex bacteria. Nucleic Acids Res 38(Web Server issue):W326-331

Wu J, Liu W, He L, Huang F, Chen J, Cui P, Shen Y, Zhao J, Wang W, Zhang Y et al (2013) Sputum microbiota associated with new, recurrent and treatment failure tuberculosis. PLoS One 8(12):e83445

Wu S, Wang MG, Wang Y, He JQ (2019) Polymorphisms of cytokine genes and tuberculosis in two independent studies. Sci Rep 9(1):2507

Zhang Y, Wade MM, Scorpio A, Zhang H, Sun Z (2003) Mode of action of pyrazinamide: disruption of Mycobacterium tuberculosis membrane transport and energetics by pyrazinoic acid. J Antimicrob Chemother 52(5):790–795

Zhang S, Li G, Bi J, Guo Q, Fu X, Wang W, Liu S, Xiao G, Ou M, Zhang J et al (2022) Association between functional nucleotide polymorphisms up-regulating transforming growth factor beta1 expression and increased tuberculosis susceptibility. J Infect Dis 225(5):825–835

Zheng J, Rubin EJ, Bifani P, Mathys V, Lim V, Au M, Jang J, Nam J, Dick T, Walker JR et al (2013) para-Aminosalicylic acid is a prodrug targeting dihydrofolate reductase in Mycobacterium tuberculosis. J Biol Chem 288(32):23447–23456

COVID-19

13

Abstract

The global viral pandemic of COVID-19, caused by SARS-CoV-2 created an unprecedented demand for rapid diagnosis, containment, and treatment of the disease. However, the diverse clinical presentation of COVID-19 ranging from asymptomatic cases to ARDS and death in critically severe patients have complicated the overall scenario. Multi-omics approaches have played a vital role in understanding disease pathogenesis and tackling this global health emergency. Omics tools have unravelled mutant variants of SARS-CoV-2, highlighted host-viral interaction and identified probable host biomarkers that can be explored for disease diagnosis and designing novel anti-COVID-19 drugs. Several host biomarkers for COVID-19 identified by different omics techniques have been documented in this chapter. Besides, the different therapeutic strategies for the treatment of COVID-19 and various vaccines developed for mass immunization against SARS-CoV-2 have also been mentioned. Finally, the future research directions that can be explored for better diagnosis, control, and management of COVID-19 have been discussed.

Keywords

SARS-CoV-2 · COVID-19 · Asymptomatic · Multi-omics · Biomarkers · ACE2 · Computed tomography (CT) scans · Vaccines

13.1 Introduction

Coronavirus disease 2019 (COVID-19) pandemic has emerged as a serious global health concern in 2019. This catastrophic viral disease is caused by severe acute respiratory syndrome coronavirus 2 (SARS-CoV-2). SARS-CoV-2 has rapidly disseminated all over the world since the first case, due to the highly contagious

S. Saha et al., *Pulmonomics: Omics Approaches for Understanding Pulmonary Diseases*, https://doi.org/10.1007/978-981-99-3505-5_13

nature of the viral pathogen (Sharma et al. 2021). COVID-19 has resulted in approximately six million deaths all over the world (Cascella et al. 2022). The pandemic also exerted an adverse impact on the global economy (Kolahchi et al. 2021). Tremendous progress in scientific research has enabled better understanding of SARS-CoV-2 infection and has paved the way for limiting the spread of the disease via mass immunization. Besides, screening of immunomodulatory agents and repurposing of drugs have also enabled improved management of COVID-19. However, the emergence of novel mutant variants of SARS-CoV-2, that may lead to COVID-19 outbreaks in future, continues to serve as a serious cause of concern.

13.2 Pathogenesis

SARS-CoV-2 affects the respiratory tract and the lungs in majority of the cases. Lung damage accounts for death in a significant proportion of COVID-19 patients (Bosmuller et al. 2021). However, COVID-19 is not restricted to the pulmonary system, it is a multi-system disorder (Samprathi and Jayashree 2020). SARS-CoV-2 is transmitted via microdroplets or aerosols from infected individuals to others. Upon entering the body, the viral particle binds to host receptors and enters host cells either by receptor-mediated endocytosis or by membrane fusion. The viral spike (S) protein binds to angiotensin-converting enzyme 2 (ACE-2) coronavirus receptor. ACE2 is highly expressed by pulmonary epithelial cells, type 2 pneumocytes, alveolar macrophages, and dendritic cells. Upon binding ACE2, the viral S protein gets activated by a two-step protease cleavage that finally mediates viral and host membrane fusion and viral entry into the host cell. Thereafter viral contents are released within the host cell. The virus undergoes replication within the host, followed by subsequent RNA dependent RNA polymerase driven transcription. Subsequently, new viral proteins and viral particles are synthesized within the host cell. Newly formed viral particles leave the infected cell by exocytosis into the extracellular space. These particles are now capable of infecting adjacent host cells. During the initial phase of replication and propagation, a limited host immune response is triggered. During the pre-symptomatic period, although the viral load remains low, the individual remains capable of spreading the disease and is tested positive (Belouzard et al. 2012; Parasher 2021).

The virus gradually spreads to the upper respiratory tract. Symptoms like fever, dry cough, and malaise are noted in this stage. Host immune response gets triggered with the release of CXCL-10, IFNβ, and IFNγ from SARS-CoV-2 infected host cells. The heightened host immune response at this stage is adequate to contain the spread of SARS-CoV-2 infection. In most of the COVID-19 patients, the disease does not progress beyond this stage and patients recover from this stage (Parasher 2021; Tang et al. 2005).

Involvement of the lower respiratory tract results in the development of severe symptoms. The virus laden pneumocytes and other host cells release various inflammatory mediators and cytokines like IL-1, IL-6, IL-8, IL-12, TNF-α, IFNγ, IFNβ, CXCL-10, MCP-1, and MIP-1α. Such cytokine storm results in the recruitment of a plethora of immune cells, namely neutrophils, CD4 helper T cells, and CD8 cytotoxic T cells in the lung tissues. These immune cells not only function to combat the viral particles but also result in considerable inflammation and injury in the lung tissues. These infected host cells undergo apoptotic cell death and result in the release of new viral particles that infect adjacent alveolar epithelial cells. Persistent lung injury results in the loss of both type 1 and type 2 pneumocytes. Such persistent and diffuse alveolar damage ultimately leads to the development of acute respiratory distress syndrome (ARDS) in severe and critically severe COVID-19 patients (Cascella et al. 2022; Parasher 2021; Xu et al. 2020).

Vascular affection marked by intravascular thrombosis has been a prominent pathology in COVID-19, and the incidence is quite high especially in ICU admitted patients (Klok et al. 2020). The pulmonary vascular coagulopathy is differentially lung limited and medicated through complex immune mechanisms (McGonagle et al. 2020). Both thrombotic and haemorrhagic pathology can supervene and may add to hypoxemia from different mechanisms (Asakura and Ogawa 2021; Nitsure et al. 2020).

13.3 Clinical Features and Diagnosis

The time from exposure to SARS-CoV-2 and the onset of symptoms, also referred to as the incubation time is typically 5–6 days, but may also extend for 14 days. The clinical presentation and symptoms of COVID-19 may vary from patient to patient. Patients may be asymptomatic (having no symptoms at all) or may be mild, moderate, severe, and critically severe. The most common symptoms include fever, dry cough, and shortness of breath. Although initially COVID-19 was characterized by these triad of symptoms, later, the US Center for Disease Control and Prevention (CDC) added muscle pain, headache, chills, sore throat and other neurological manifestations like loss of taste or loss of smell to the existing triad list. Other symptoms of COVID-19 include diarrhoea, nausea, vomiting, malaise, fatigue, anorexia, sore throat, sneezing, rhinitis, pharyngalgia, sputum secretion, chest pain, and dermatological manifestations. Hemoptysis is the least common symptom noted in COVID-19 patients (Parasher 2021; da Rosa et al. 2021; Alimohamadi et al. 2020; He et al. 2020). Severe and critically severe COVID-19 patients have exacerbated respiratory symptoms, diarrhoea, abdominal pain, anorexia, and fatigue. These patients may further deteriorate to acute respiratory distress syndrome (ARDS), septic shock, metabolic acidosis, thrombosis, and organ failure. Presence of comorbidities like diabetes and hypertension further worsens the disease prognosis. Severe and critically severe COVID-19 is also associated with higher mortality rate (He et al. 2020; Stephenson et al. 2021).

RT-PCR, antibody detection method, serological tests, rapid antigen detection test, and imaging modalities like Chest X-ray and CT scanning are widely used for the diagnosis of COVID-19 (Allam et al. 2020). Nanopore sequencing or variant specific PCR is also conducted to identify the variant of SARS-CoV-2 responsible for causing COVID-19 (Dachert et al. 2022). Primarily clinical samples from the upper respiratory tract are collected in the form of nasopharyngeal swabs and oropharyngeal swabs. Samples from lower respiratory tract are obtained either as expectorated sputum or as bronchoalveolar lavage from mechanically ventilated patients (Parasher 2021). However, nasopharyngeal or oropharyngeal swabs are mostly used as clinical samples for detection of SARS-CoV-2. Amplification of viral genetic material from these clinical samples is done using RT-PCR. RT-PCR serves as the gold standard for the detection of COVID-19. The detection procedure relies on the amplification of viral ORF1ab, E gene, and N genes. RNase P internal gene is used as the internal control in RT-PCR. Positive control for such RT-PCR mostly includes synthetic dsDNA g-blocks corresponding to SARS-CoV-2 gene fragments (Yuce et al. 2021).

Antibody detection method based on lateral flow type assays (LFA) and enzyme-linked immunosorbent type assays (ELISA) detects the presence and concentration of anti-SARS-CoV-2 IgG and IgM in blood plasma or serum. However, presently there is no effective antibody test available for SARS-CoV-2 (Parasher 2021; Yuce et al. 2021). Certain serological tests are conducted using blood samples to predict the prognosis of COVID-19. Normal to reduced white blood cell (WBC) count is observed in COVID-19 patients with worse prognosis (Parasher 2021). Elevated levels of lactate dehydrogenase, C-reactive protein, creatine kinase, aspartate amino-transferase, alanine amino-transferase, D-dimer levels, and neutrophil-to-lymphocyte ratio are also observed in some patients (Cascella et al. 2022; Yang et al. 2020). Abnormalities in coagulation factors are also observed in severe COVID-19 patients (Parasher 2021). Rapid antigen detection tests are also widely used for the detection of SARS-CoV-2 infection. SARS-CoV-2 viral components, namely S glycoprotein, M protein, or released N protein are detected using rapid antigen test. Antigen test is conducted using LFA strips for rapid detection. Antigen tests may also be conducted by ELISA for increased sensitivity (Yuce et al. 2021).

Imaging modalities include chest radiography and computed tomography (CT) scanning. Chest X-ray is inconclusive during the initial stages of SARS-CoV-2 infection. However, with the progression of the disease, bilateral multifocal alveolar opacities can be visualized in chest radiographs. Such opacities may be associated with pleural effusion (Cascella et al. 2022). High-resolution CT (HRCT) is a highly sensitive tool for the diagnosis of COVID-19. Multifocal bilateral "ground-glass" appearance along with patchy peripheral distribution is observed especially in the lower lobes of COVID-19 patients. Reverse halo sign consisting of a focal opaque patch surrounded by a peripheral ring with consolidation is also noted in some patients. Other characteristics observed in COVID-19 patients include pleural effusion, cavity formation, calcification, and lymphadenopathy (Parasher 2021). COVID-19 pandemic has resulted in a huge demand for rapid diagnosis of the disease. Diagnosis of COVID-19 solely relied on RT-PCR. Rapid antigen detection test has been successful to some extent. CT scanning has also been helpful in evaluating the progression of COVID-19. The radiological changes are dynamic as they appear and evolve from GGO (ground glass opacities) to consolidation to fibrotic-like appearance and finally to resolve in most of the cases (Kwee and Kwee 2022). Despite the worthwhile developments in the available modalities, the diagnostic tools need to be improvised to accomplish the diagnosis of early and asymptomatic diseases in order to prevent spread of disease via asymptomatic carriers.

13.4 Significance of Multi-omics and Biomarkers for COVID-19

Although RT-PCR, rapid antigen test and CT scans are of immense significance in detection of COVID-19, each of these diagnostic tools has their own limitations (Ma et al. 2022a). The variation in clinical presentation of COVID-19 often complicates the diagnosis. Also, the asymptomatic nature of some COVID-19 patients plays a vital role in mediating the spread of this contagious disease. Identification of biomarkers that can diagnose and differentiate asymptomatic from symptomatic COVID-19 cases shall be of tremendous importance in preventing disease spread and in mediating early COVID-19 diagnosis. Cutting-edge multi-omics approaches including, genomics, transcriptomics, proteomics, metabolomics, and metagenomics can aid in improved understanding of pathobiology of SARS-CoV-2 infection and lead to identification of novel biomarkers. Genomics has been widely employed for viral genome sequence analysis and for identification of novel mutant variants of SARS-CoV-2. Structural proteomics aid in studying viral proteins (like spike proteins) that are involved in interaction with host proteins and modulating host immune response. Apart from viral characteristics and host–virus interactions, these multi-omics tools provide clue about host factors (genes, proteins, metabolites, and microbiome) that undergo changes upon SARS-CoV-2 infection. Such differentially expressed host bio-molecules may serve as biomarkers for rapid disease diagnosis, for stratification of disease based on severity and may also aid in monitoring disease prognosis (Samprathi and Jayashree 2020). Besides, novel therapeutic targets can be identified for designing drugs for COVID-19. The different roles of multi-omics approaches in enabling a better understanding and mediating an improved classification of COVID-19 have been schematically represented in Fig. 13.1. Various omics-based studies conducted using COVID-19 patient samples for the identification of host biomarkers have been discussed in the subsequent sections and have also been documented in Table 13.1.

13.5 Genomics

Genome wide association studies (GWAS) have associated several genetic variants with COVID-19 susceptibility. TMPRSS2 rs2070788 polymorphism was found to be associated with fatality in COVID-19 patients in the Indian population (Pandey et al. 2022). SNP in BIN1 (rs744373) was also associated with mortality in COVID-19 patients (Lehrer 2021). The T allele of rs35705950 SNP in MUC5B gene was found to confer protection from development of severe disease and hospitalization in COVID-19 patients (van Moorsel et al. 2021; Verma et al. 2022). SNPs in ACE2 (rs4646142 and rs6632677) have been identified as genetic biomarkers for

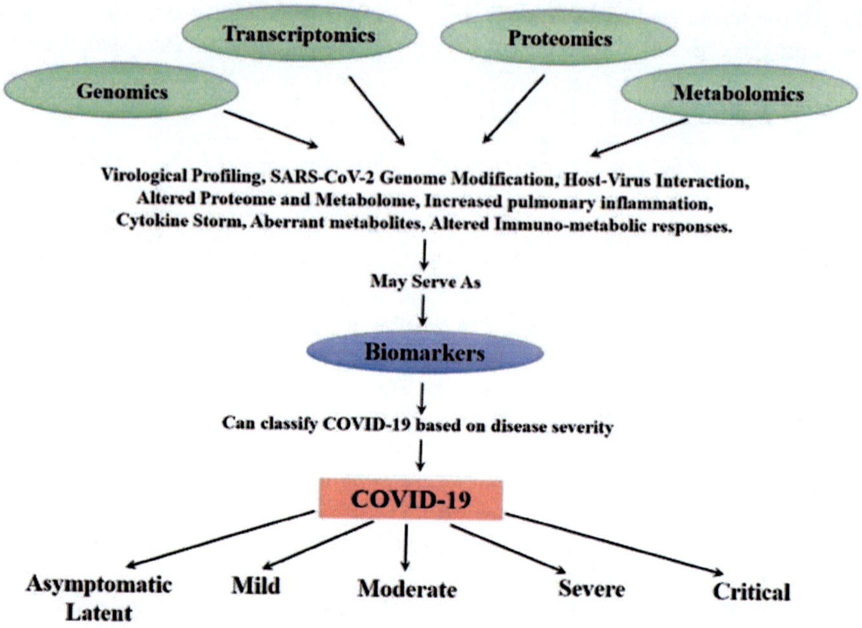

Fig. 13.1 Different omics approaches may enable biomarker identification for classification of COVID-19 based on disease severity. Genomics, transcriptomics, proteomics, and metabolomics can identify viral genome modifications, host–viral interaction, aberrant proteins, metabolites, inflammatory, and other immune mediators which may serve as biomarkers for classifying COVID-19 based on disease severity

COVID-19 with systemic inflammatory injury and cardiovascular risk (Ma et al. 2022b). The C allele of rs12252 in IFITM3 gene was found in greater abundance in severe COVID-19 cases than in moderate cases of COVID-19 (Zhang et al. 2020a). Missense variation in CD26 (rs13015258) was found to be associated with vulnerability to COVID-19 (Adli et al. 2022). DPP4 rs3788979 polymorphism was found to be associated with COVID-19 (Posadas-Sanchez et al. 2021). The rs10157379 CT and rs10754558 GG genotypes of NLRP3 were found to be positively associated with SARS-CoV-2 infection (Maes et al. 2022). Studies have highlighted the association of genetic variants in the 3p21.31 locus (LZTFL1, SLC6A20, CCR9, FYCO1, CXCR6, and XCR1) and 9q34.2 (ABO) with the severity of COVID-19 (Schmiedel et al. 2021; Fricke-Galindo and Falfan-Valencia 2021). The intergenic variant rs60200309 near DOCK2 gene in the 5q35 locus has also been associated with severe COVID-19, especially in young individuals (Ferreira et al. 2022). Some other SNPs associated with COVID-19 have been listed in Table 13.1.

Table 13.1 Different biomarkers of COVID-19

Omics approach	Biomarkers	Biological sample	Comments
Genomics	ACE2	Blood	SNPs of ACE2, namely rs2285666, rs2048683, rs879922, and rs4240157 showed significant association with COVID-19 disease severity in obese patients as compared to that in lean patients (Jalaleddine et al. 2022)
	VWA8, PDE8B, CTSC, THSD7B, STK39, FBXO34, RPL6P27, METTL21C		SNPs in eight genes, namely VWA8 (rs10507497), PDE8B (rs7715119), CTSC (rs72953026), THSD7B (rs7605851), STK39 (rs7595310), FBXO34 (rs10140801), RPL6P27 (rs11659676), and METTL21C (rs599976) showed a strong association with hospitalization in COVID-19 patients (Mousa et al. 2021)
	IL6, IFNG, MIP, CXCL16		SNPs in IL6 (rs1554606), IFNG (rs2069718), MIP (rs799187) and CXCL16 (rs8071286) were significantly associated with critical illness in COVID-19 patients (Alefishat et al. 2022)
	ACE2, TMPRSS2		SNPs in ACE2 (rs2285666) and TMPRSS2 (rs12329760) may serve as predictors of disease severity in COVID-19 patients (Abdelsattar et al. 2022)
	ACE2		SNPs in ACE2 (rs4646142, rs6632677, and rs2074192) were found to be associated with COVID-19 (Ma et al. 2022b)
Epigenomics	IFITM1, ISG20, NLRP3, MX1	Peripheral blood mononuclear cells (PBMCs)	Hypermethylation of IFITM1 and ISG20 along with hypo-methylation of NLRP3 and MX1 was noted in severe COVID-19 (Corley et al. 2021)

(continued)

Table 13.1 (continued)

Omics approach	Biomarkers	Biological sample	Comments
	IRF7, BCL6, MX1, TNF, IFI27, OAS2	Blood	IRF7, BCL6, MX1, TNF, IFI27 and OAS2 were differentially methylated in COVID-19 patients (Balnis et al. 2021)
	OAS1, OAS2, PARP9-DTX3L, IFIT3, IRF7, TRIM22, MX1, CD38, EPSTI1, LAT		OAS1, OAS2 were hyper-methylated while PARP9-DTX3L, IFIT3, IRF7, TRIM22, MX1, CD38, EPSTI1, LAT were hypo-methylated in COVID-19 patients and were associated with COVID progression (Barturen et al. 2022)
	TLR7		TLR7 was differentially methylated and consistently downregulated in male severe COVID-19 patients as compared to females (Gomez-Carballa et al. 2022)
	ACIN1, MEIS2, CARD14, CRHR1, HIVEP3, NCOA2, PIBF1, SAMD4A		Reduced methylation of ACIN1 (cg02037503 and cg23712970) and MEIS2 (cg06471042) was noted following SARS-CoV-2 infection. Hypermethylation of CARD14 (cg10846936), CRHR1 (cg09422970), HIVEP3 (cg11857452), NCOA2 (cg20282780), PIBF1 (cg00531853 and cg15128396), and SAMD4A (cg18499294) was observed following SARS-CoV-2 infection (Pang et al. 2022)
	AIM2, CCDC6, CDC42BPB, CEP85L, CELF4, CXCR2, DDO, HLA-C, KIFAIP3, LHX6, LCE1C, IFI44L, MOBKL2A, PM20D1, SGMS1, SORCS1, VIM, ZNF385D, UBAP2L, UBE2W		Differential methylation of AIM2, CCDC6, CDC42BPB, CEP85L, CELF4, CXCR2, DDO, HLA-C, KIFAIP3, LHX6, LCE1C, IFI44L, MOBKL2A, PM20D1, SGMS1, SORCS1, VIM, ZNF385D, UBAP2L and

(continued)

Table 13.1 (continued)

Omics approach	Biomarkers	Biological sample	Comments
			UBE2W was associated with clinical severity and respiratory failure in COVID-19 patients (Castro de Moura et al. 2021)
Transcriptomics	hsa-miR-4665-5p, hsa-miR-3190-3p, hsa-miR-331-3p, hsa-miR-4525, hsa-miR-431-5p, hsa-miR-6721-5p, hsa-miR-4661-5p, hsa-miR-548a-3p, hsa-miR-4745-5p, hsa-miR-3150b-3p, hsa-miR-7848-3p, hsa-miR-7110-3p, hsa-miR-1197, hsa-miR-4686, hsa-miR-374b-3p, hsa-miR-4473, hsa-miR-758-3p, hsa-miR-5004-3p, hsa-miR-146a-3p, hsa-miR-329-5p, hsa-miR-744-3p, hsa-miR-1285-5p, hsa-miR-6516-3p	Plasma	hsa-miR-4665-5p, hsa-miR-3190-3p, hsa-miR-331-3p, hsa-miR-4525, hsa-miR-431-5p, hsa-miR-6721-5p, hsa-miR-4661-5p, hsa-miR-548a-3p, hsa-miR-4745-5,p and hsa-miR-3150b-3p was upregulated in COVID-19 patients. hsa-miR-7848-3p, hsa-miR-7110-3p, hsa-miR-1197, hsa-miR-4686, hsa-miR-374b-3p, hsa-miR-4473, hsa-miR-758-3p, hsa-miR-5004-3p, hsa-miR-146a-3p, hsa-miR-329-5p, hsa-miR-744-3p, hsa-miR-1285-5p and hsa-miR-6516-3p were down regulated in severe COVID-19 patients (Fernandez-Pato et al. 2022)
	miR-776-3p, miR-1275, miR-4742-3p, miR-31-5p, miR-3215-3p		miR-776-3p, and miR-1275 were downregulated in COVID-19 patients. miR-4742-3p, miR-31-5p and miR-3215-3p were upregulated in COVID-19 patients (Farr et al. 2021)
	miR-335-5p, miR-221-3p, miR-146a-5p, miR-27b-3p, miR-130a-3p, miR-584-5p, miR-376a-3p, miR-361-5p, miR-140-3p, miR-25-3p, miR-16-5p, miR-424-5p, miR-425-5p, miR-30a-5p, miR-532-3p		miR-335-5p, miR-221-3p, miR-146a-5p, miR-27b-3p, miR-130a-3p, miR-584-5p, miR-376a-3p and miR-361-5p were downregulated, while, miR-140-3p, miR-25-3p, miR-16-5p, miR-424-5p, miR-425-5p, miR-30a-5p,

(continued)

Table 13.1 (continued)

Omics approach	Biomarkers	Biological sample	Comments
			and miR-532-3p were upregulated in COVID-19 patients as compared to community acquired pneumonia (Martinez-Fleta et al. 2021)
	CXCL5, CXCL12, CCL2, CCL4, CXCL10, IFIH1, IFI44, IFIT1, IL6, IL10, RPL41, RPL17, SLC25A6, CALM1, TUBA1A	Naso-pharyngeal swabs	CXCL5, CXCL12, CCL2, CCL4, CXCL10, IFIH1, IFI44, IFIT1, IL6 and IL10 were increased in COVID-19 patients, whereas RPL41, RPL17, SLC25A6, CALM1, and TUBA1A were reduced in COVID-19 patients (Jain et al. 2021)
	CCL2/MCP-1, CXCL10/IP-10, CCL3/MIP-1A, CCL4/MIP1B, CXCL1, CXCL2, CXCL6, CXCL8	Broncho alveolar lavage fluid (BALF)	CCL2/MCP-1, CXCL10/IP-10, CCL3/MIP-1A, CCL4/MIP1B, CXCL1, CXCL2, CXCL6, and CXCL8 were overexpressed in BALF of COVID-19 patients (Xiong et al. 2020)
	IL10, GDF11, NOG, CXCL10, NRG1, TIMP1, IL18, C5, AREG, CCL4, TNFSF10, IFNG, CXCL8, CMTM2	Peripheral blood mononuclear cells (PBMCs)	GDF11, NOG, CMTM2, IFNG, CXCL8, and CCL4 were reduced in COVID-19 patients, whereas IL10, CXCL10, TNFSF10, NRG1, IL18, TIMP1, AREG, and C5 were increased in COVID-19 patients (Xiong et al. 2020)
Proteomics	APOM	Serum	APOM is decreased in severe COVID-19 patients as compared to non-severe COVID-19 patients and healthy controls (Shen et al. 2020)
	SAA1, SAA2, SAA4, CRP, SERPINA3, APCS, PPBP, PF4		SAA1, SAA2, SAA4, CRP, SERPINA3, and APCS were upregulated in severe COVID-19 patients as compared to non-severe COVID-19 patients. PPBP and PF4 were downregulated in severe

(continued)

Table 13.1 (continued)

Omics approach	Biomarkers	Biological sample	Comments
			COVID-19 patients (Shen et al. 2020)
	IL-6, CRP, LRG1, S100A12, SAA1, SERPINA3, SFTPB, TIMP1, ASH1L, CETP, CRISP3, F13B, GSN, IGFALS, IGFPB3, PARP9, PSRC1, STOM		IL-6, CRP, LRG1, S100A12, SAA1, SERPINA3, SFTPB, and TIMP1 were increased, whereas ASH1L, CETP, CRISP3, F13B, GSN, IGFALS, IGFPB3, PARP9, PSRC1, and STOM were decreased in COVID-19 patients as compared to healthy subjects (D'Alessandro et al. 2020)
	APOA5, CES1, CFL1, HSP90AB1, ITGA7, NEXN		APOA5, CES1, CFL1, HSP90AB1, ITGA7, and NEXN were elevated in severe COVID-19 patients both in disease stage and recovery stage (Chen et al. 2021a)
	ATP6AP1, CD93, CHI3L1, CST3, CTSB, GIPC2, GRN, IGFBP6, LCN2, MB, MPO, SECTM1, SPP1, VIM, VSIG4, VWF, APOC1, APOC2, APOC3, APOC4, APOM, ARFIP1, IGFALS, IGFBP3, PRG4, SAA4, TTR		Increased expression of ATP6AP1, CD93, CHI3L1, CST3, CTSB, GIPC2, GRN, IGFBP6, LCN2, MB, MPO, SECTM1, SPP1, VIM, VSIG4, and VWF along with reduced expression of APOC1, APOC2, APOC3, APOC4, APOM, ARFIP1, IGFALS, IGFBP3, PRG4, SAA4, and TTR were associated with adverse prognosis in severe COVID-19 patients (Kimura et al. 2021)
	IL-6, CKAP4, Gal-9, IL-1ra, LILRB4, PD-L1	Plasma	Increased expression of IL-6, CKAP4, Gal-9, IL-1ra, LILRB4, and PD-L1 was associated with COVID-19 severity (Patel et al. 2021)
	Neurofibromin 2 (NF2)		NF2 was identified as a specific plasma biomarker for COVID-19 (Patel et al. 2021)

(continued)

Table 13.1 (continued)

Omics approach	Biomarkers	Biological sample	Comments
	D-dimer		Significant increase in plasma D-dimer in COVID-19 patients was associated with increased disease severity (Asakura and Ogawa 2021; Liao et al. 2020)
	IL-6, IL-10, CXCL10, CXCL11, CCL2, CCL7, CCL8, PD-L1, IL-18R1		IL-6, IL-10, CXCL10, CXCL11, CCL2, CCL7, CCL8, PD-L1, and IL-18R1 were found to be elevated in early stage of COVID-19 in ICU patients as compared to non-ICU COVID-19 patients (Haljasmagi et al. 2020)
	CASP8, TNFSF14, TGFB1, HGF		Apoptotic markers, namely CASP8, TNFSF14, TGFB1, and HGF were increased in COVID-19 patients as compared to healthy subjects (Haljasmagi et al. 2020)
Metabolomics	R-S lactoglutathione, glutamine, hypoxanthine, inosine, leukotriene D4 (LTD4)	Serum	R-S lactoglutathione and glutamine were downregulated, whereas inosine, hypoxanthine, and LTD4 were upregulated in COVID-19 patients (Dogan et al. 2021)
	Tryptophan, kynurenine, 3-hydroxy-DL-kynurenine, citrulline, ornithine		Tryptophan and citrulline were decreased, while kynurenine, 3-hydroxy-DL-kynurenine, and ornithine were increased in acute phase COVID-19 patients as compared to recovery phase COVID-19 patients (Ansone et al. 2021)
	Cysteine, isoleucine, glutamine, threonine, glyceric acid, citric acid, α-hydroxy isovaleric acid, α-hydroxy butyric acid, 2,3 dihydroxy butanoic acid, malic acid, glutamic acid, phenylalanine		Cysteine, isoleucine, glutamine, threonine, glyceric acid, and citric acid were reduced, whereas α-hydroxy isovaleric acid, α-hydroxy butyric acid, 2,3 dihydroxy butanoic acid,

(continued)

Table 13.1 (continued)

Omics approach	Biomarkers	Biological sample	Comments
			malic acid, glutamic acid, and phenylalanine were increased in COVID-19 patients (Paez-Franco et al. 2021)
	Methionine sulfoxide, Cystine, Creatine, Creatinine, Polyamines, Glucose, Free fatty acids		Methionine sulfoxide, cystine, creatine, creatinine, polyamines, glucose, and free fatty acids were increased in COVID-19 patients (Thomas et al. 2020)
	Choline, sphingolipids phosphocholine, 21-hydroxypregnenolone, kynurenate, kynurenine, 8-methoxykynurenate		Choline and its derivatives and sphingolipids were decreased in COVID-19 patients. Phosphocholine, 21-hydroxypregnenolone, kynurenate, kynurenine, and 8-methoxykynurenate were found to be increased in sera of COVID-19 patients (Shen et al. 2020)
	Glutamate, arginine, n-(l-arginino)-succinate, citrulline, ornithine, glutamine, 2-oxoglutarate, N-acetyl-L-glutamate, urea, fumarate		Glutamate, arginine, N-(l-arginino)-succinate, citrulline, ornithine, glutamine, 2-oxoglutarate, N-acetyl-L-glutamate, urea, and fumarate were decreased in COVID-19 patients as compared to healthy volunteers (Shen et al. 2020)
	Argininate, asymmetric dimethylarginine, symmetric dimethylarginine, homoarginine, N-acetyl arginine		Argininate, asymmetric dimethylarginine, symmetric dimethylarginine, homoarginine, and N-acetyl arginine were significantly reduced in non-severe COVID-19 patients (Shen et al. 2020)
	Kynurenine, tryptophan, glutamine, glutamate	Plasma	The ratio of kynurenine to tryptophan and the ratio of glutamine to glutamate were significantly elevated in COVID-19 patients (Kimhofer et al. 2020).

(continued)

Table 13.1 (continued)

Omics approach	Biomarkers	Biological sample	Comments
	3-hydroxibutirate, Linoleic acid, LPC (14:0 and 18:2), LPE (22:6), Kynurenic acid, Tryptophan		3-hydroxibutirate, linoleic acid, LPC (14:0 and 18:2), LPE (22:6), kynurenic acid, and tryptophan were identified as potential biomarkers of COVID-19 severity (Valdes et al. 2022)
	Methylpent-2-enal, 2,4-octadiene 1-chloroheptane, Nonanal	Exhaled Breath	Methylpent-2-enal, 2,4-octadiene 1-chloroheptane, and nonanal may serve as indicators of ARDS patients with COVID-19 (Grassin-Delyle et al. 2021)
Metagenomics	*Candidatus Saccharibacteria, Fibrobacteres, Spirochaetes*	Stool	Increased abundance of *Candidatus Saccharibacteria* was noted in SARS-CoV-2 infected patients. Reduced abundance of *Fibrobacteres* was observed in SARS-CoV-2 infected patients. Differential abundance of *Spirochaetes* was noted in SARS-CoV-2 infected patients as compared to asymptomatic cases (Liu et al. 2021a)
	Anaerostipes spp., *Tyzzerella* spp., *Coprobacillus* spp., *Lachnoclostridium* spp., *Unclassified_member of_Erysipelotrichaceae*		Increased abundance of *Anaerostipes spp., Tyzzerella spp., Coprobacillus* spp., *Lachnoclostridium* spp., and *unclassified member of_Erysipelotrichaceae* was noted in SARS-CoV-2 infected patients as compared to healthy subjects (Liu et al. 2021a)
	Enterococcus faecalis, Citrobacter freundii, Citrobacter unclassified, Haemophilus parainfluenzae, Saccharomyces cerevisiae		Enrichment of *Enterococcus faecalis, Citrobacter freundii, Citrobacter unclassified, Haemophilus parainfluenzae,* and *Saccharomyces cerevisiae*

(continued)

Table 13.1 (continued)

Omics approach	Biomarkers	Biological sample	Comments
			were observed in COVID-19 patients with fever (Zhou et al. 2021)
	Bacteroides cellulosilyticus, Bacteroides fragilis, Bacteroides thetaiotaomicron, Bacteroides xylanisolvens, Eubacterium ramulus, Erysipelotrichaceae bacterium_6_1_45		*Significant enrichment of Bacteroides cellulosilyticus, Bacteroides fragilis, Bacteroides thetaiotaomicron, Bacteroides xylanisolvens, Eubacterium ramulus, and Erysipelotrichaceae bacterium_6_1_45 was noted in* COVID-19 patients with non-fever *(*Zhou et al. 2021*)*
	Anelloviridae (Torque teno midi virus 8, TTV-like mini virus 19 and 26), *Cycloviridae* (Human associated cyclovirus 10), Human beta herpesvirus 6	Oropharyngeal swabs, nasopharyngeal swabs, tracheal aspirates	Increased abundance of members of *Anelloviridae* family (Torque teno midi virus 8, TTV-like mini virus 19 and 26), *Cycloviridae* family (human associated cyclovirus 10) and human beta herpesvirus 6 were observed in deceased and hospitalized COVID-19 patients as compared to ambulatory individuals (Isa et al. 2022)
	Rotavirus A, Measles morbillivirus, Alpha papilloma virus 10		The relative abundance of *Rotavirus A*, *Measles morbillivirus,* and *Alpha papilloma virus 10* was significantly increased in deceased COVID-19 patients as compared to hospitalized COVID-19 patients and ambulatory individuals (Isa et al. 2022)

This table enlists various important human biomarkers of COVID-19 identified using different omics approaches

13.6 Epigenomics

Several studies have associated differentially methylated CpGs with SARS-CoV-2 infection. Epigenetic analyses highlighted increased DNA methylation to be associated with severe cases of COVID-19. Hypermethylation of interferon-related genes and hypomethylation of genes involved in inflammation were documented in severe COVID-19 (Corley et al. 2021). Genes, namely MX1, OAS1, IRF7, IFIT3, IFI27, EPSTI1, FAM38A, PHOSPHO1, GTPBP2, CYSTM1, CMPK2, ARID5B, PARP9, DTX3L, TRIM22, and CD38 were found to be differentially methylated in SARS-CoV-2 positive patients (Konigsberg et al. 2021). Hypermethylation of ROC1, ZNF789, and H1F0 was found to be associated with increased mortality in COVID-19 patients (Bradic et al. 2022). Studies have also highlighted the role of DNA methylation in COVID-19 progression. High-throughput methylome profiling study showed that DNA methylation changes in B-lymphocytes, neutrophils, and CD8+ T-lymphocytes in COVID-19 patients lead to altered regulation of functional pathways associated with viral defense and autoimmune disorders (Barturen et al. 2022). Machine learning approaches based on Monte Carlo selection method identified epigenetic alterations of EPSTI1, SHROOM3, NACAP1, C19ORF35, and MX1 as potential biomarkers for distinguishing between COVID-19 patients and non-COVID subjects. The study also suggested that such changes in methylation pattern can also provide guidance for efficient detection and therapy of COVID-19 (Li et al. 2022a). Other epigenomic signatures associated with COVID-19 have been documented in Table 13.1.

13.7 Transcriptomics

Six long non-coding RNAs (TALAM1, DLEU2, UICLM CASC18, SNHG20, and GNAS) involved in the differentially expressed protein-coding genes-lncRNA network may serve as potential marker candidates for COVID-19 patients (Chakraborty et al. 2021). Another transcriptomics study showed that COVID-19 patients have increased expression of genes involved in keratinization of tears (Mastropasqua et al. 2021). Changes in expression pattern of various genes, namely NFKBIA, COL1A2, KANK4, FN1, FAP, COMP, FAM101B, TAGLN, ANKRD1, SPARC, ADAM19, CXCL10/11, OASL, OLFM4, FOS, IFI27, IFIT1, APOBEC3A, IFI44L, RSAD2, NDUFS1, HECTD1, SRSF6, CBX3, and DDX17 were noted upon COVID-19 infection (Jabeen et al. 2022). Expression patterns of cardio-metabolic microRNAs, namely miR-133a and miR-122 were found to be associated with clinical severity and mortality in COVID-19 patients (Gutmann et al. 2022). Differential expression of several serum miRNAs has also been noted in COVID-19 patients. MiRNAs like hsa-let-7d, hsa-miR-17, hsa-miR-34b, hsa-miR-93, hsa-miR-200b, hsa-miR-200c, and hsa-miR-223 were significantly downregulated, whereas hsa-miR-190a and hsa-miR-203 were significantly increased in COVID-19 patients as compared to healthy controls (Demiray et al. 2021). Transcriptomic analyses have also enabled clustering of critically ill COVID-19 patients based on the expression pattern of interferon-related genes or immune checkpoint genes (Lopez-Martinez et al. 2022). Differentially expressed host genes and miRNAs associated with COVID-19 have been enlisted in Table 13.1.

13.8 Proteomics

Multiple proteomics studies have been conducted to identify host biomarkers related to COVID-19, disease severity, and survivability in COVID-19 patients. Plasma proteomic study highlighted ADM, IL-6, MCP-3, TRAIL-R2, and PD-L1 as predictive indicators of death in COVID-19 patients. The same study also showed that proteins, namely ADM, CTSL1, HGF, IL-6, IL-27, KIM1, MERTK, MMP-1, MMP-12, OPG, TNFRSF10A, and TRAIL-R2 were overexpressed in hospitalized COVID-patients as compared to non-hospitalized COVID-19 patients (Bauer et al. 2021). Studies have also corelated differential expression profiles of APOA1, APOA2, APOC3, APOC1, APOD, APOE, APOL1, A2M, A1BG, ALB, AHSG, ACTB/ACTG1, B2M, CD44, CD14, CFB, CFD, CFH, CFI, C1R, C1S, C8A, CRP, CFHR1, CFHR2, CFHR5, C9, FGA, FGB, FGG, F12, F2, PLG, ITIH3, ITIH4, GSN, KLKB1, LBP, LGALS3BP, HP, LRG1, PIGR, SAA1, SAA2, LYZ, SERPINA1, SERPINA3, SERPINA10, SERPINC1, and TF with disease severity in COVID-19 patients (Messner et al. 2020; Demichev et al. 2021). Another high-throughput proteomic study correlated increased expression of ORM1, ORM2, CRP, CFI, S100A9, AZGP1, SERPINA3, LCP1 and decreased expression of FETUB, CETP, PI16 with COVID-19 severity (Shu et al. 2020). Serum proteomics-based comparative analysis of severe and non-severe COVID-19 patients revealed that proteins involved in humoral immune response, IFN signalling, acute phase response, inflammatory response, lipid metabolism, coagulation cascade, and platelet degranulation were mostly dysregulated in severe COVID-19 (Lee et al. 2021). Some of the other host proteomic signatures for COVID-19 have been documented in Table 13.1.

13.9 Metabolomics

Several high-throughput metabolomic studies have highlighted alteration in global metabolomic profile in COVID-19 patients. Plasma metabolomic study emphasized on the significance of the tryptophan-nicotinamide pathway in inflammatory response in COVID-19 patients. The study also stated cytosine and tryptophan metabolism as novel biomarkers of COVID-19 disease severity (Blasco et al. 2020). Plasma levels of diglycerides (particularly, DG (16:0/20:2/0:0)), free fatty acids (FAAs), and triglycerides (particularly, TG (14:0/22:1/22:3)) were increased in COVID-19 patients. Higher abundance of these lipids was associated with deterioration of COVID-19 patients. On the other hand, phosphatidylcholines (PCs) were gradually decreased in COVID-19 fatal cases (Wu et al. 2020). Studies have also correlated tryptophan metabolites and lipids with COVID-19 severity (Overmyer et al. 2021). Panel comprising of kynurenine/ tryptophan ratio, lysoPC (C26:0), and pyruvic acid was found to successfully distinguish between PCR positive non-hospitalized COVID-19 patients and non-COVID controls. Another metabolite panel comprising of C10:2, butyric acid, and pyruvic acid could distinguish PCR positive non-hospitalized COVID-19 patients from PCR positive hospitalized and intubated COVID-19 patients (Lopez-Hernandez et al. 2021). A meta-analysis highlighted cholesterol, D-mannose, tyrosine, L-phenylalanine, and bilirubin as potential metabolic biomarkers of COVID-19 (Pang et al. 2021). Some of the metabolite signatures noted in COVID-19 patients have been listed in Table 13.1.

13.10 Metagenomics

Functional dysbiosis of lung and gut microbiota has been documented in COVID-19 patients. Increased abundance of *Candida albicans* and *Pseudomonas phages Pf1* while, reduced abundance of *Bifidobacterium adolescentis* were associated with severe cluster of COVID-19 patients. Abnormal bursts of urea cycle in severe COVID-19 patients have been associated with *Klebsiella* spp. (Liu et al. 2022). Metagenomic study based on next-generation sequencing revealed an increased abundance of *Tannerella forsythia* in diabetic SARS-CoV-2 patients as compared to controls (Al-Emran et al. 2022). Another study revealed that *Bariatricus comes, Blautia_A obeum, Blautia_A wexlerae, Dorea formicigenerans, Faecalibacterium prausnitzii_D, Faecalibacterium sp900539945,* and *Fusicatenibacter saccharivorans* had significant strain loss in COVID-19 patients as compared to non-COVID-19 controls. Opportunistic pathogens, namely *Klebsiella quasivariicola*, *Klebsiella pneumoniae,* and *Escherichia coli* were found to be associated with the progression of COVID-19 (Ke et al. 2022). Changes in lung fungal flora were also observed in COVID-19 patients. *Cryptococcus spp.* was found to be dominant in the lungs of COVID-19 patients. *Aspergillus spp., Alternaria spp., Dipodascus spp., Mortierella spp., Naganishia spp., Diutina spp., Candida spp., Cladosporium spp., Issatchenkia spp.,* and *Wallemia spp.* were also noted in COVID-19 patients (de Oliveira et al. 2021). Some of the other microbial signatures specific to COVID-19 have been tabulated in Table 13.1.

13.11 Bioinformatics

A huge amount of COVID-19 data has been generated recently on several fronts like SARS-CoV-2 sequencing of strains, medical imaging, drug-vaccine development, clinical trials, and published literature. These datasets are compiled and curated into several databases as shown in Table 13.2. SARS-CoV-2 resources were available from the National Center for Biotechnology Information (NCBI), the European Molecular Biology Laboratory-European Bioinformatics Institute (EMBL-EBI), and DNA Data Bank of Japan (DDBJ). SARS-CoV-2 next-generation sequencing runs are available in these databases in Sequence Read Archive (SRA). These

Table 13.2 Databases related to SARS-CoV-2 and COVID-19

Sl. no.	Name	Description	URL
1.	LitCovid	A curated database of COVID-19 literature (Chen et al. 2021b)	https://www.ncbi.nlm.nih.gov/research/coronavirus/
2.	NCBI SARS-CoV-2 Resources	It includes SARS-CoV-2 SRA runs, nucleotide records, clinical trials, and PubMed records. It also allows assembled and raw read data submission (https://www.ncbi.nlm.nih.gov/sars-cov-2/)	https://www.ncbi.nlm.nih.gov/sars-cov-2/
3.	EMBL-EBI COVID-19 Data Portal	It serves as an easily accessible archive of viral sequences, host sequences, proteins, and images from microscopy (Harrison et al. 2021)	https://www.covid19dataportal.org/
4.	DDBJ-Research data and resource sites of SARS-CoV-2	It is a collection of information by DDBJ and its collaborators (https://www.ddbj.nig.ac.jp/covid19-e.html)	https://www.ddbj.nig.ac.jp/covid19-e.html
5.	CoV3D	A database of high-resolution coronavirus protein structures (Gowthaman et al. 2021)	https://cov3d.ibbr.umd.edu
6.	Database of COVID-19 vaccinations	A global public dataset of COVID-19 vaccination (Mathieu et al. 2021)	https://github.com/owid/covid-19-data/tree/master/public/data/vaccinations
7.	COMBATdb	A database for the COVID-19 Multi-Omics Blood ATlas (Wang et al. 2023)	https://db.combat.ox.ac.uk
8.	COVID19db	A database platform to discover potential drugs and targets of COVID-19 at transcriptomic scale (Zhang et al. 2022)	http://www.biomedical-web.com/covid19db
9.	DockCoV2	A drug database against SARS-CoV-2 (Chen et al. 2021c)	https://covirus.cc/drugs/
10.	COVID-19 biomarkers	Collection of COVID-19 biomarkers (Gogate et al. 2021)	https://data.oncomx.org/covid19

This table enlists the different databases that have been developed and are accessible online to end users

databases also allow submission of SARS-CoV-2 data. A global public dataset of COVID-19 vaccination has also been developed (Mathieu et al. 2021). COVID-19 host biomarkers were compiled and are available to the public (Gogate et al. 2021).

13.12 Medical Imaging or Radiomics

Chest imaging using CT has played an essential role in COVID-19 diagnosis and monitoring the disease severity. There are typical (peripheral, bilateral grounded-glass opacities (GGOs)) and atypical (presence of isolated lobar without GGO and discrete small nodules) features observed in chest CT images of COVID-19 patients which are well documented (Kanne et al. 2021). These chest CT features occur around 9–13 days post-infection. Several scoring and assessment systems were developed for lung CT images. Computational-based quantitative methods including artificial intelligence (AI) used the total lung involvement as well as the percentage of GGO and thus have higher accuracy than human semi-quantitative estimates (Pu et al. 2021). The AI based on CT imaging can correctly diagnose and also predict the severity of COVID-19 pneumonia (Mei et al. 2020; Zhang et al. 2020b). A few studies have reported that the CT severity score in COVID-19 patients showed high correlations with blood biomarkers (IL-6, C-reactive protein, lymphocyte, and neutrophil count) of disease severity (Broman et al. 2021; Zhang et al. 2020c). A web server named LHSPred can predict CT severity scores using six blood parameters (Bhattacharjee et al. 2022). A few open databases of COVID-19-related medical imaging like RSNA International COVID-19 Open Radiology Database (RICORD) and the National COVID-19 Chest Imaging Database (NCCID) allow archiving the chest image data, thereby supporting a better understanding of the COVID-19 viral infection and enabling development of AI-based models. In summary, CT and chest radiography can distinguish COVID-19 pneumonia from other types of pneumonia and lung diseases and also help to monitor the disease severity based on automated CT scores.

13.13 Multi-omics and Data Integration

Several studies have focused on the generation of multi-omics datasets including genomics, transcriptomics, proteomics, metabolomics, and lipidomic profiles from blood and lung samples of COVID-19 patients (COvid-19 Multi-omics Blood ATlas (COMBAT) Consortium 2022; Wu et al. 2021). These studies allow bioinformaticians and computational biologists to apply integrative multi-omics approaches to understand this complex disease and enable subtyping patients into asymptomatic, mild, severe, and critical. Multi-omics approaches based on multivariate analyses, supervised-based machine learning models, sparse decomposition of arrays (SDA) algorithm, and network-based approaches are applied to COVID-19 patients' datasets to understand COVID-19 infection, the host immune response, to identify correlation-based therapeutics targets and enable drug selection (Li et al. 2022b). Liu et al. used multivariate analyses such as an ordinal partial least-squares (PLS) regression approach from longitudinally assessed circulating proteins and 188 surface protein markers, transcriptome, and T cell receptor sequence to predict the severity outcomes in COVID-19 patients (Liu et al. 2021b). Supervised machine learning like the ExtraTrees classifier was used on a combination of four omics datasets of metabolites, lipids, proteins, and transcripts for the prediction of COVID-19 severity (Overmyer et al. 2021). A tensor-based approach, like the sparse decomposition of arrays (SDA) algorithm, was used in COVID-19 Multi-omics Blood ATlas (COMBAT) datasets to predict the disease severity and specificity on large cohorts (COvid-19 Multi-omics Blood ATlas (COMBAT) Consortium 2022). Stukalov et al. applied a network diffusion approach to study the interactome of SARS-CoV-2 and SARS-CoV virus proteins with host cellular proteins on a lung-derived human cell line based on proteomics, transcriptomics, and phosphor-proteomics approaches (Stukalov et al. 2021). An integrated analysis based on the correlation of clinical measurements, immune cells (CD4+ T Cell and B cell heterogeneity), plasma metabolomics, and proteomics of COVID-19 patients suggested that these parameters are associated with disease severity (Su et al. 2020). In summary, the integration of COVID-19 multi-omics datasets have led to a tremendous scientific impact in understanding this complex disease, which probably would not have been possible by a single omics-based study.

13.14 Present Therapeutics

One of the major challenges encountered in the COVID-19 pandemic was the lack of established therapeutic regime (Mehta et al. 2020). Various anti-virals, anti-parasitic agents, anti-malarial drugs, antibiotics, steroids, immunomodulatory agents, and biologics were considered for treatment of patients suffering from COVID-19. The different drugs used for treating COVID-19 patients have been tabulated in Table 13.3.

Oxygen supplementation, hospitalization, and mechanical ventilation are also provided to severe and critically severe COVID-19 patients (Alfano et al. 2022). However, till date there is no specific medication for COVID-19. The development and approval of several anti-COVID-19 vaccines have reduced the risk and instances of COVID-19 all over the world. Different types of COVID-19 vaccines like live attenuated vaccine, inactivated viral vaccine, non-replicating viral vector, virus-like particle (VLP) vaccine, mRNA vaccine, protein subunit vaccine, adjuvanted protein subunit vaccine, and DNA plasmid-based COVID-19 vaccine are available globally (Majumdar et al. 2021). The different vaccines that have been approved in various countries of the world have been documented in Table 13.4.

Although available drugs and vaccination strategies have largely reduced the overall global COVID-19 infections and deaths, there is still a need to screen novel biomarkers and therapeutic targets to tackle the spread of the newly emerging variants of SARS-CoV-2.

Table 13.3 Drugs used for treating COVID-19 patients

Sl. no.	Drugs	Description	Mode of action
1.	Remdesivir	Broad spectrum anti-viral drug (Adenosine nucleotide analog)	Prevents viral replication by inhibiting viral RNA dependent RNA polymerase (RdRp) Gets incorporated into viral RNA strand and leads to premature chain termination Interferes with viral proof-reading activity (Mehta et al. 2020; Majumdar et al. 2021; Mohanty et al. 2021; Barati et al. 2020; Jogalekar et al. 2021; Panoutsopoulos 2020; Choudhury et al. 2021; Hussain et al. 2021; Tiwari et al. 2020)
2.	Casirivimab/imdevimab	Combination of two human monoclonal antibodies against spike (S) protein of SARS-CoV-2	Exerts anti-viral effect because of its strong binding affinity for the receptor binding domain of the viral spike (S) protein Reduces viral load (Heustess et al. 2021; Alfano et al. 2022)
3.	Oral corticosteroids (dexamethasone or other steroids like methylprednisolone, prednisone, and hydrocortisone)	Oral steroids	Serves as immunosuppressants Interacts with cytoplasmic receptors, thereby altering mRNA transcription and reducing the expression of genes encoding inflammatory mediators Exerts anti-inflammatory, vasoconstrictive, and anti-fibrotic effects Inhibits production and release of pro-inflammatory cytokines, chemokines, and enzymes, thereby suppressing hyperactive inflammatory responses and cytokine storm Dexamethasone decreased mortality rate in critically ill COVID-19 patients requiring supplemental oxygen therapy or mechanical ventilation

(continued)

Table 13.3 (continued)

Sl. no.	Drugs	Description	Mode of action
			(Mehta et al. 2020; Majumdar et al. 2021; Jogalekar et al. 2021; Choudhury et al. 2021; Tiwari et al. 2020; Alfano et al. 2022; Owji et al. 2020; Rommasi et al. 2022)
4.	Inhaled corticosteroids (inhaled budesonide)	Inhaled steroids	Reduces the need for urgent medical care in COVID-19 patients and decreases the time for recovery (Ramakrishnan et al. 2021)
5.	Chloroquine (CQ), hydroxychloroquine (HCQ)	Amino- quinolone anti-malarial drugs	Blocks glycosylation and binding of ACE2 receptor to viral spike (S) protein, thereby interfering with viral entry Increases endosomal pH, thereby preventing fusion of virus and endosomes Inhibits autophagy Exerts anti-viral effect by interfering with viral nucleic acid replication, glycosylation of viral particles, virus assembly, transport of viral particles, and viral release Has immunomodulatory properties Reduces secretion of proinflammatory cytokines Inhibits the release of IL-6, IL-1β, and TNF-α Interferes with platelet aggregation and blood clotting proteins, thus exerting anti-thrombotic effects (Mehta et al. 2020; Majumdar et al. 2021; Mohanty et al. 2021; Barati et al. 2020; Jogalekar et al. 2021; Choudhury et al. 2021; Hussain et al. 2021; Tiwari et al. 2020; Yadav 2021)
6.	Azithromycin	Macrolide antibiotic (broad spectrum)	Interferes with the interaction of the virus and host cell receptor CD147,

(continued)

Table 13.3 (continued)

Sl. no.	Drugs	Description	Mode of action
			thus blocking CD147 mediated viral entry into host cells May also interfere with interaction of viral spike (S) protein with host cell ACE2 receptor Acts as lipophilic acidotrophic weak base that increases endosomal pH and trans Golgi network pH, thereby affecting uncoating of enveloped viral, endocytosis, and shedding of viral genomes from lysosomes, ultimately resulting in reduced viral replication Possess anti-inflammatory properties Serves as an immunomodulatory agent by reducing production of proinflammatory cytokines like IL-6, IL-10, IL-12, and TNF-α Reduces inflammation by suppressing IP-10 via MAPK-JNK/ERK and NF-κβ/p65 signalling pathways May aid in reducing viral load within the host by promoting production of anti-viral proteins like type I and type III interferons (IFNβ and IFNλ1), toll-like receptor 3 (TLR3), retinoic inducible gene I (RIG-I), RIG-I-like helicase. and melanoma differentiation-associated protein 5 (MDA5) in bronchial epithelial cells May promote host cell sensitivity to viral infection by upregulating Pathogen Recognition Receptors (PRRs)

(continued)

Table 13.3 (continued)

Sl. no.	Drugs	Description	Mode of action
			May also exert anti-viral effect by activating IKK and TANK1 signalling pathways or by modulating IL-28 and IL-29 receptor complexes or by binding to IFNAR1 complex and inducing STAT1/2, IRF7, IRF9, and ISGF3 production (Mehta et al. 2020; Majumdar et al. 2021; Khoshnood et al. 2022; Al-Masaeed et al. 2021)
7.	Ivermectin	Broad spectrum anti-parasitic drug (macrocyclic lactone derivative)	Halts growth of SARS-CoV-2 in cell cultures Prevents viral RNA synthesis Exerts anti-viral activity by binding to importin α/β complex, thereby destabilizing the heterodimer complex and hampering the translocation of SARS-CoV-2 proteins into host cell nucleus (Majumdar et al. 2021; Panoutsopoulos 2020; Choudhury et al. 2021; Tiwari et al. 2020)
8.	Famotidine	Antacid (Histamine-2 receptor antagonist)	May bind to 3-chymotrypsin-like protease (3CLpro) and inhibit replication and proliferation of SARS-CoV-2 within the host cell May prevent inflammation and cytokine storm (Majumdar et al. 2021; Panoutsopoulos 2020; Choudhury et al. 2021; Rommasi et al. 2022)
9.	Tocilizumab and Sarilumab	Monoclonal antibody against interleukin-6 receptor (IL- 6R)	Attenuates fatal inflammatory response by inhibiting the production of cytokines and acute phase proteins. Blocks interaction of IL-6 with IL-6R, thereby

(continued)

Table 13.3 (continued)

Sl. no.	Drugs	Description	Mode of action
			blocking downstream signalling cascade associated with inflammation (Majumdar et al. 2021; Choudhury et al. 2021; Tiwari et al. 2020; Heustess et al. 2021; Rommasi et al. 2022)
10.	Nafamostat and camostat	Serine protease inhibitors	Blocks SARS-CoV-2 viral entry into lung alveolar cells, by inhibiting the transmembrane serine protease, TMPRSS2 Reduces SARS-CoV-2 viral replication (Majumdar et al. 2021; Mohanty et al. 2021; Jogalekar et al. 2021; Choudhury et al. 2021; Tiwari et al. 2020; Heustess et al. 2021)
11.	Baricitinib, fedratinib, and ruxolitinib	Inhibitors of Janus Kinase (JAK)	Blocks JAK-STAT signalling pathways Has anti-inflammatory actions Reduces IL-6 and TNF-α Controls and reduces cytokine storm Baricitinib, additionally, blocks clathrin-dependent endocytosis mediated entry of virus (Majumdar et al. 2021; Choudhury et al. 2021; Tiwari et al. 2020; Owji et al. 2020; Rommasi et al. 2022)
12.	Lopinavir/ ritonavir	Anti-retroviral aspartic protease inhibitors	Blocks production of active viral peptides due to inhibitory action on coronavirus endo-peptidase C30 (CEP-C30) (Mehta et al. 2020; Majumdar et al. 2021; Barati et al. 2020; Jogalekar et al. 2021; Choudhury et al. 2021; Hussain et al. 2021; Tiwari et al. 2020; Heustess et al. 2021; Vellingiri et al. 2020)
13.	Favipiravir (T-705)		Exerts anti-viral effect by competitive inhibition of

(continued)

Table 13.3 (continued)

Sl. no.	Drugs	Description	Mode of action
		Broad spectrum anti-viral drug (purine analog prodrug)	viral RNA-dependent RNA polymerase (RdRp) Interferes with and slows down viral replication and viral mutagenicity Decreases viral load Leads to termination of viral protein synthesis upon getting incorporated into the viral RNA strand (Mehta et al. 2020; Majumdar et al. 2021; Mohanty et al. 2021; Barati et al. 2020; Jogalekar et al. 2021; Choudhury et al. 2021; Hussain et al. 2021; Tiwari et al. 2020; Heustess et al. 2021; Al-Masaeed et al. 2021)
14.	Umifenovir (arbidol)	Broad spectrum anti-viral drug (indole carboxylic acid derivative)	Prevents fusion of virus with host cell membrane by interfering with the binding of the viral spike (S) protein with host cell ACE2 receptor Inhibits viral replication (Mohanty et al. 2021; Jogalekar et al. 2021; Choudhury et al. 2021; Hussain et al. 2021; Tiwari et al. 2020)
15.	Oseltamivir	Anti-retroviral drug	Functions by blocking the active site of 3C-like protease (Mohanty et al. 2021; Jogalekar et al. 2021; Al-Masaeed et al. 2021)
16.	Anakinra	Interleukin-1 receptor (IL-1R) Antagonist	Serves as an immunosuppressive agent by preventing the interaction of IL-1α and IL-1β with their cognate receptor IL-1R Aids in alleviating cytokine release. Blocks IL-1 mediated hyaluronic acid production (Tiwari et al. 2020; Heustess et al. 2021; Rommasi et al. 2022)

(continued)

Table 13.3 (continued)

Sl. no.	Drugs	Description	Mode of action
17.	Ribavirin	Anti-viral drug (Guanosine analogue)	Serves as competitive inhibitor of viral RNA-dependent RNA polymerase (RdRp) Blocks viral replication (Mohanty et al. 2021; Jogalekar et al. 2021; Hussain et al. 2021)
18.	Gimsilumab, lenzilumab, and namilumab	Monoclonal antibody targeting granulocyte macrophage-colony stimulating factor (GM-CSF)	Prevents inflammation at an early stage. Blocks production of IL-6 and other pro-inflammatory cytokines by targeting GM-CSF and blocking its interaction with its cognate cell surface receptor (Majumdar et al. 2021; Choudhury et al. 2021; Owji et al. 2020)
19.	Darunavir-cobicistat	Anti-retroviral drug (Protease inhibitor)	Interferes with viral replication by inhibiting viral protease (Barati et al. 2020; Jogalekar et al. 2021; Tiwari et al. 2020)
20.	Naproxen	Non-steroidal anti-inflammatory drugs (NSAIDs)	Exerts anti-inflammatory effect by inhibiting cyclooxygenase (both COX-1 and COX-2) enzyme complex (Rommasi et al. 2022)
21.	Thalidomide	Synthetic L-glutamic acid derivative	Serves as an anti-fibrotic, immunomodulatory agent Exerts anti-inflammatory effect by reducing NF-κβ activity, reducing infiltration of inflammatory cells, and inhibiting release of chemokines and cytokines like CCL5, IL-6, and TNF-α (Majumdar et al. 2021; Mohanty et al. 2021; Owji et al. 2020)
22.	Fluvoxamine	Anti-depressant drug (Selective serotonin reuptake inhibitor (SSRI) and a σ-1 receptor (S1R) agonist)	May serve as an immunomodulatory agent Exerts anti-inflammatory effect by leading to the activation of S1R, which in turn blocks the transcriptional activation of

(continued)

Table 13.3 (continued)

Sl. no.	Drugs	Description	Mode of action
			X-box protein 1 (an important regulator of IL-6, IL-8, IL-1β, and IL-12 production), thereby shutting down the inflammatory response from the endoplasmic reticulum (ER) May exert antiplatelet activity and confer protection from thrombosis (Majumdar et al. 2021; Marzolini et al. 2022; Reis et al. 2022)
23.	APN01	Soluble human recombinant ACE2	Inhibits the binding of viral spike protein to host cell surface ACE2 receptor and thus, prevents viral entry into host cells (Majumdar et al. 2021)
24.	*N*-acetylcysteine	Pleiotropic molecule having antioxidant properties	Has anti-inflammatory properties. Acts on the mechanisms involved in prothrombotic state noted in severe COVID-19 (Di Marco et al. 2022) Exhibits mucolytic activity (Jorge-Aaron and Rosa-Ester 2020) NAC reduces CRP levels and improves oxygenation parameters and length of hospital stay in hospitalized COVID-19 patients (Avdeev et al. 2022)
25.	Doxycycline	Broad spectrum antibiotic	Has anti-inflammatory properties. This antibiotic prevents disease progression in COVID-19 patients and prevents incidences of ICU admission (Dhar et al. 2023)

COVID-19 is treated mainly by anti-viral drugs, agents to tackle the cytokine storm (steroid) and anti-thrombosis medications apart from different supportive measures. This table enlists the different drugs that have been used for curing COVID-19 patients

Table 13.4 Vaccines approved for COVID-19

Sl. no.	Vaccine	Other names	Description
1.	Abdala	CIGB-66	Protein subunit vaccine (https://covid19.trackvaccines.org/vaccines/approved/#vaccine-list; Accessed 4 Oct 2022)
2.	Ad5-nCoV-IH	–	Non-replicating viral vector vaccine (https://covid19.trackvaccines.org/vaccines/approved/#vaccine-list; Accessed 4 Oct 2022)
3.	Aurora-CoV	EpiVacCorona-N	Protein subunit vaccine (https://covid19.trackvaccines.org/vaccines/approved/#vaccine-list; Accessed 4 Oct 2022)
4.	Comirnaty (Pfizer/BioNTech)	Tozinameran, BNT162b2	mRNA vaccine (https://covid19.trackvaccines.org/vaccines/approved/#vaccine-list; Accessed 4 Oct 2022)
5.	Comirnaty Bivalent Original/Omicron BA.1	BNT162b2 (B.1.1.529), BNT162b2 Bivalent (WT/OMI BA.1)	mRNA vaccine (https://covid19.trackvaccines.org/vaccines/approved/#vaccine-list; Accessed 4 Oct 2022)
6.	Comirnaty Bivalent Original/Omicron BA.4/BA.5	BNT162b2 Bivalent (WT/OMI BA.4/BA.5)	mRNA vaccine (https://covid19.trackvaccines.org/vaccines/approved/#vaccine-list; Accessed 4 Oct 2022)
7.	Convidecia	Ad5-nCoV	Non-replicating viral vector vaccine (https://covid19.trackvaccines.org/vaccines/approved/#vaccine-list; Accessed 4 Oct 2022)
8.	Corbevax	BECOV2A	Protein subunit vaccine (https://covid19.trackvaccines.org/vaccines/approved/#vaccine-list; Accessed 4 Oct 2022)
9.	CoronaVac	–	Inactivated COVID-19 vaccine (https://covid19.trackvaccines.org/vaccines/approved/#vaccine-list; Accessed 4 Oct 2022)
10.	Covaxin	BBV152	Inactivated COVID-19 vaccine (https://covid19.trackvaccines.org/vaccines/approved/#vaccine-list; Accessed 4 Oct 2022)
11.	Covifenz	CoVLP, MT-2766, Plant-based VLP	Virus like particle (VLP) vaccine (https://covid19.trackvaccines.org/vaccines/approved/#vaccine-list; Accessed 4 Oct 2022)
12.	Covilo	BBIBP-CorV (Vero Cells)	Inactivated virus-based COVID-19 vaccine (https://covid19.trackvaccines.org/vaccines/approved/#vaccine-list; Accessed 4 Oct 2022)

(continued)

Table 13.4 (continued)

Sl. no.	Vaccine	Other names	Description
13.	COVIran Barekat	COVID-19 Inactivated Vaccine	Inactivated virus-based COVID-19 vaccine (https://covid19.trackvaccines.org/vaccines/approved/#vaccine-list; Accessed 4 Oct 2022)
14.	Covishield (Oxford/ AstraZeneca formulation)	–	Non-replicating viral vector vaccine (https://covid19.trackvaccines.org/vaccines/approved/#vaccine-list; Accessed 4 Oct 2022)
15.	COVOVAX (Novavax formulation)	–	Protein subunit vaccine (https://covid19.trackvaccines.org/vaccines/approved/#vaccine-list; Accessed 4 Oct 2022)
16.	EpiVacCorona	–	Protein subunit vaccine (https://covid19.trackvaccines.org/vaccines/approved/#vaccine-list; Accessed 4 Oct 2022)
17.	FAKHRAVAC (MIVAC)	–	Inactivated virus-based COVID-19 vaccine (https://covid19.trackvaccines.org/vaccines/approved/#vaccine-list; Accessed 4 Oct 2022)
18.	GEMCOVAC-19	Gemcovac	mRNA vaccine (https://covid19.trackvaccines.org/vaccines/approved/#vaccine-list; Accessed 4 Oct 2022)
19.	Inactivated (Vero Cells)	–	Inactivated virus-based COVID-19 vaccine (https://covid19.trackvaccines.org/vaccines/approved/#vaccine-list; Accessed 4 Oct 2022)
20.	iNCOVACC	BBV154	Non-replicating viral vector vaccine (https://covid19.trackvaccines.org/vaccines/approved/#vaccine-list; Accessed 4 Oct 2022)
21.	Janssen (Johnson & Johnson)	Ad26COVS1, Ad26.COV2.S, JNJ-78436735, Jcovden	Non-replicating viral vector vaccine (https://covid19.trackvaccines.org/vaccines/approved/#vaccine-list; Accessed 4 Oct 2022)
22.	KoviVac	–	Inactivated virus-based COVID-19 vaccine (https://covid19.trackvaccines.org/vaccines/approved/#vaccine-list; Accessed 4 Oct 2022)
23.	KCONVAC	SARS-CoV-2 Vaccine (Vero Cells), KconecaVac	Inactivated virus-based COVID-19 vaccine (https://covid19.trackvaccines.org/vaccines/approved/#vaccine-list; Accessed 4 Oct 2022)
24.	MVC-COV1901	–	Protein subunit vaccine (https://covid19.trackvaccines.org/vaccines/approved/#vaccine-list; Accessed 4 Oct 2022)

(continued)

Table 13.4 (continued)

Sl. no.	Vaccine	Other names	Description
25.	Noora vaccine	COVID-19 Recombinant RBD Protein Vaccine	Protein subunit vaccine (https://covid19.trackvaccines.org/vaccines/approved/#vaccine-list; Accessed 4 Oct 2022)
26.	Nuvaxovid	NVX-CoV2373	Protein subunit vaccine (https://covid19.trackvaccines.org/vaccines/approved/#vaccine-list; Accessed 4 Oct 2022)
27.	QazVac	QazCovid-in	Inactivated virus-based COVID-19 vaccine (https://covid19.trackvaccines.org/vaccines/approved/#vaccine-list; Accessed 4 Oct 2022)
28.	Razi Cov Pars	–	Protein subunit vaccine (https://covid19.trackvaccines.org/vaccines/approved/#vaccine-list; Accessed 4 Oct 2022)
29.	Recombinant SARS-CoV-2 Vaccine (CHO Cell)	Recombinant COVID-19 Vaccine (CHO cell, NVSI-06-08)	Protein subunit vaccine (https://covid19.trackvaccines.org/vaccines/approved/#vaccine-list; Accessed 4 Oct 2022)
30.	SKYCovione	GBP510	Protein subunit vaccine (https://covid19.trackvaccines.org/vaccines/approved/#vaccine-list; Accessed 4 Oct 2022)
31.	Soberana 02	FINLAY-FR-2, Pastu Covac, Pastocovac	Protein subunit vaccine (https://covid19.trackvaccines.org/vaccines/approved/#vaccine-list; Accessed 4 Oct 2022)
32.	Soberana Plus	FINLAY-FR-1A	Protein subunit vaccine (https://covid19.trackvaccines.org/vaccines/approved/#vaccine-list; Accessed 4 Oct 2022)
33.	Spikevax	mRNA-1273, Elasomeran, Moderna COVID-19 vaccine	mRNA vaccine (https://covid19.trackvaccines.org/vaccines/approved/#vaccine-list; Accessed 4 Oct 2022)
34.	Spikevax Bivalent Original/Omicron BA.1	mRNA-1273.214	mRNA vaccine (https://covid19.trackvaccines.org/vaccines/approved/#vaccine-list; Accessed 4 Oct 2022)
35.	Spikevax Bivalent Original/Omicron BA.4/BA.5	mRNA-1273.222	mRNA vaccine (https://covid19.trackvaccines.org/vaccines/approved/#vaccine-list; Accessed 4 Oct 2022)
36.	SpikoGen	COVAX-19	Protein subunit vaccine (https://covid19.trackvaccines.org/vaccines/approved/#vaccine-list; Accessed 4 Oct 2022)
37.	Sputnik V	Gam-COVID-Vac, rAd5	Non-replicating viral vector vaccine (https://covid19.trackvaccines.org/vaccines/approved/#vaccine-list; Accessed 4 Oct 2022)

(continued)

Table 13.4 (continued)

Sl. no.	Vaccine	Other names	Description
38.	Sputnik Light	–	Non-replicating viral vector vaccine (https://covid19.trackvaccines.org/vaccines/approved/#vaccine-list; Accessed 4 Oct 2022)
39.	TAK-919 (Moderna formulation)	–	mRNA vaccine (https://covid19.trackvaccines.org/vaccines/approved/#vaccine-list; Accessed 4 Oct 2022)
40.	TAK-019 (Novavax formulation)	–	Protein subunit vaccine (https://covid19.trackvaccines.org/vaccines/approved/#vaccine-list; Accessed 4 Oct 2022)
41.	Turkovac	ERUCOV-VAC	Inactivated virus-based vaccine (https://covid19.trackvaccines.org/vaccines/approved/#vaccine-list; Accessed 4 Oct 2022)
42.	V-01	–	Protein subunit vaccine (https://covid19.trackvaccines.org/vaccines/approved/#vaccine-list; Accessed 4 Oct 2022)
43.	Vaxzevria	AZD1222, ChAdOx1 nCoV-19	Non-replicating viral vector vaccine (https://covid19.trackvaccines.org/vaccines/approved/#vaccine-list; Accessed 4 Oct 2022)
44.	VLA2001	–	Inactivated virus-based vaccine (https://covid19.trackvaccines.org/vaccines/approved/#vaccine-list; Accessed 4 Oct 2022)
45.	Zifivax	RBD-Dimer, ZF2001	Adjuvanted protein subunit vaccine (https://covid19.trackvaccines.org/vaccines/approved/#vaccine-list; Accessed 4 Oct 2022)
46.	ZyCoV-D	–	DNA plasmid-based COVID-19 vaccine (https://covid19.trackvaccines.org/vaccines/approved/#vaccine-list; Accessed 4 Oct 2022)

This table enlists the different COVID-19 vaccines that have been approved worldwide

13.15 Future Perspectives

Cutting-edge omics approaches have been extensively utilized by the scientific community to combat this transboundary COVID-19 crisis. However, certain gaps are yet to be addressed. Diagnosis of asymptomatic COVID-19 patients has been difficult and is a serious cause of concern for disease transmission. Identification of suitable diagnostic biomarkers or diagnostic tools for detection of asymptomatic COVID-19 cases is of great significance for monitoring subclinical infection. Future research should also focus on identifying appropriate serosurveys for both vaccinated and unvaccinated population (Pettifor et al. 2022). Artificial intelligence empowered machine learning should be explored for automated detection of COVID-19 using volumetric CT scans (Majumdar et al. 2021). Besides, virological and immunological events occurring in the initial stage of infection and prolonged period of recovery should be studied in detail for identification of suitable prognostic biomarkers. Most of the COVID-19 research have focussed on pulmonary complications, but COVID-19 affects other organs like intestines, heart, kidney, which have not been explored in detail like respiratory complications. State of art omics approaches should also be used to study and dissect multisystem complications in COVID-19 (Wang et al. 2021). Improved understanding of host and viral interaction using modern omics tools may aid in identifying novel therapeutic targets and designing drugs specific for COVID-19. The clinical significance of miRNA as therapeutic targets should also be studied. Computational pharmacology based on machine learning and deep learning algorithms should be explored for speeding up the process of designing novel therapeutics by overcoming the existing limitations of experimental screening of chemical libraries (Kanapeckaite et al. 2022). Understanding the alterations in viral structural and functional proteins, changes in host–viral interaction, disease transmission, and clinical manifestation upon emergence of novel mutant variants of SARS-CoV-2 is also of prime importance in preventing new waves of COVID-19 across the world. Mathematical simulation models should also be constructed for predicting transmission and infection dynamics for novel mutant strains of SARS-CoV-2 (Jayatilaka et al. 2022). In a nutshell, addressing and exploring these above-mentioned research avenues can aid in effectively combating viral transmission, SARS-CoV-2 infection and bring about significant reduction in COVID-19 mortality rate.

References

Abdelsattar S, Kasemy ZA, Ewida SF, Abo-Elsoud RAA, Zytoon AA, Abdelaal GA, Abdelgawad AS, Khalil FO, Kamel HFM (2022) ACE2 and TMPRSS2 SNPs as determinants of susceptibility to, and severity of, a COVID-19 infection. Br J Biomed Sci 79:10238

Adli A, Rahimi M, Khodaie R, Hashemzaei N, Hosseini SM (2022) Role of genetic variants and host polymorphisms on COVID-19: from viral entrance mechanisms to immunological reactions. J Med Virol 94(5):1846–1865

Alefishat E, Mousa M, Albreiki M, Jelinek HF, Al Halwachi Z, Khalili M, Waasia F, Uddin M, Al Kaabi N, Mahboub B et al (2022) Genetic variants and serum profiles of cytokines in COVID-19 severity. Shock

Al-Emran HM, Rahman S, Hasan MS, Ul Alam R, Islam OK, Anwar A, Jahid MIK, Hossain A (2022) Microbiome analysis revealing microbial interactions and secondary bacterial infections in COVID-19 patients comorbidly affected by Type 2 diabetes. J Med Virol

Alfano G, Morisi N, Frisina M, Ferrari A, Fontana F, Tonelli R, Franceschini E, Meschiari M, Donati G, Guaraldi G (2022) Awaiting a cure for COVID-19: therapeutic approach in patients with different severity levels of COVID-19. Infez Med 30(1):11–21

Alimohamadi Y, Sepandi M, Taghdir M, Hosamirudsari H (2020) Determine the most common clinical symptoms in COVID-19 patients: a systematic review and meta-analysis. J Prev Med Hyg 61(3):E304–E312

Allam M, Cai S, Ganesh S, Venkatesan M, Doodhwala S, Song Z, Hu T, Kumar A, Heit J, Study Group C et al (2020) COVID-19 diagnostics, tools, and prevention. Diagnostics (Basel) 10(6)

Al-Masaeed M, Alghawanmeh M, Al-Singlawi A, Alsababha R, Alqudah M (2021) An examination of COVID-19 medications' effectiveness in managing and treating COVID-19 patients: a comparative review. Healthcare (Basel) 9(5)

Ansone L, Briviba M, Silamikelis I, Terentjeva A, Perkons I, Birzniece L, Rovite V, Rozentale B, Viksna L, Kolesova O et al (2021) Amino acid metabolism is significantly altered at the time of admission in hospital for severe COVID-19 patients: findings from longitudinal targeted metabolomics analysis. Microbiol Spectr 9(3):e0033821

Asakura H, Ogawa H (2021) COVID-19-associated coagulopathy and disseminated intravascular coagulation. Int J Hematol 113(1):45–57

Avdeev SN, Gaynitdinova VV, Merzhoeva ZM, Berikkhanov ZG (2022) N-acetylcysteine for the treatment of COVID-19 among hospitalized patients. J Infect 84(1):94–118

Balnis J, Madrid A, Hogan KJ, Drake LA, Chieng HC, Tiwari A, Vincent CE, Chopra A, Vincent PA, Robek MD et al (2021) Blood DNA methylation and COVID-19 outcomes. Clin Epigenetics 13(1):118

Barati F, Pouresmaieli M, Ekrami E, Asghari S, Ziarani FR, Mamoudifard M (2020) Potential drugs and remedies for the treatment of COVID-19: a critical review. Biol Proced Online 22:15

Barturen G, Carnero-Montoro E, Martinez-Bueno M, Rojo-Rello S, Sobrino B, Porras-Perales O, Alcantara-Dominguez C, Bernardo D, Alarcon-Riquelme ME (2022) Whole blood DNA methylation analysis reveals respiratory environmental traits involved in COVID-19 severity following SARS-CoV-2 infection. Nat Commun 13(1):4597

Bauer W, Weber M, Diehl-Wiesenecker E, Galtung N, Prpic M, Somasundaram R, Tauber R, Schwenk JM, Micke P, Kappert K (2021) Plasma proteome fingerprints reveal distinctiveness and clinical outcome of SARS-CoV-2 infection. Viruses 13(12)

Belouzard S, Millet JK, Licitra BN, Whittaker GR (2012) Mechanisms of coronavirus cell entry mediated by the viral spike protein. Viruses 4(6):1011–1033

Bhattacharjee S, Saha B, Bhattacharyya P, Saha S (2022) LHSPred: a web based application for predicting lung health severity. Biomed Signal Process Control 77:103745

Blasco H, Bessy C, Plantier L, Lefevre A, Piver E, Bernard L, Marlet J, Stefic K, Benz-de Bretagne I, Cannet P et al (2020) The specific metabolome profiling of patients infected by SARS-COV-2 supports the key role of tryptophan-nicotinamide pathway and cytosine metabolism. Sci Rep 10(1):16824

Bosmuller H, Matter M, Fend F, Tzankov A (2021) The pulmonary pathology of COVID-19. Virchows Arch 478(1):137–150

Bradic M, Taleb S, Thomas B, Chidiac O, Robay A, Hassan N, Malek J, Ait Hssain A, Abi Khalil C (2022) DNA methylation predicts the outcome of COVID-19 patients with acute respiratory distress syndrome. J Transl Med 20(1):526

Broman N, Rantasarkka K, Feuth T, Valtonen M, Waris M, Hohenthal U, Rintala E, Karlsson A, Marttila H, Peltola V et al (2021) IL-6 and other biomarkers as predictors of severity in COVID-19. Ann Med 53(1):410–412

Cascella M, Rajnik M, Aleem A, Dulebohn SC, Di Napoli R (2022) Features, Evaluation, and Treatment of Coronavirus (COVID-19). In: StatPearls, Treasure Island, FL

Castro de Moura M, Davalos V, Planas-Serra L, Alvarez-Errico D, Arribas C, Ruiz M, Aguilera-Albesa S, Troya J, Valencia-Ramos J, Velez-Santamaria V et al (2021) Epigenome-wide association study of COVID-19 severity with respiratory failure. EBioMedicine 66:103339

Chakraborty C, Sharma AR, Bhattacharya M, Zayed H, Lee SS (2021) Understanding gene expression and transcriptome profiling of COVID-19: an initiative towards the mapping of protective immunity genes against SARS-CoV-2 infection. Front Immunol 12:724936

Chen Y, Yao H, Zhang N, Wu J, Gao S, Guo J, Lu X, Cheng L, Luo R, Liang X et al (2021a) Proteomic analysis identifies prolonged disturbances in pathways related to cholesterol metabolism and myocardium function in the COVID-19 recovery stage. J Proteome Res 20(7): 3463–3474

Chen Q, Allot A, Lu Z (2021b) LitCovid: an open database of COVID-19 literature. Nucleic Acids Res 49(D1):D1534–D1540

Chen TF, Chang YC, Hsiao Y, Lee KH, Hsiao YC, Lin YH, Tu YE, Huang HC, Chen CY, Juan HF (2021c) DockCoV2: a drug database against SARS-CoV-2. Nucleic Acids Res 49(D1):D1152–D1159

Choudhury A, Mukherjee G, Mukherjee S (2021) Chemotherapy vs. Immunotherapy in combating nCOVID19: an update. Hum Immunol 82(9):649–658

Corley MJ, Pang APS, Dody K, Mudd PA, Patterson BK, Seethamraju H, Bram Y, Peluso MJ, Torres L, Iyer NS et al (2021) Genome-wide DNA methylation profiling of peripheral blood reveals an epigenetic signature associated with severe COVID-19. J Leukoc Biol 110(1):21–26

COvid-19 Multi-omics Blood ATlas (COMBAT) Consortium (2022) A blood atlas of COVID-19 defines hallmarks of disease severity and specificity. Cell 185(5):916–938 e958

da Rosa Mesquita R, Francelino Silva Junior LC, Santos Santana FM, Farias de Oliveira T, Campos Alcantara R, Monteiro Arnozo G, Rodrigues da Silva Filho E, Galdino Dos Santos AG, Oliveira da Cunha EJ, Salgueiro de Aquino SH et al: Clinical manifestations of COVID-19 in the general population: systematic review. Wien Klin Wochenschr 2021, 133(7–8):377–382

Dachert C, Muenchhoff M, Graf A, Autenrieth H, Bender S, Mairhofer H, Wratil PR, Thieme S, Krebs S, Grzimek-Koschewa N et al (2022) Rapid and sensitive identification of omicron by variant-specific PCR and nanopore sequencing: paradigm for diagnostics of emerging SARS-CoV-2 variants. Med Microbiol Immunol 211(1):71–77

D'Alessandro A, Thomas T, Dzieciatkowska M, Hill RC, Francis RO, Hudson KE, Zimring JC, Hod EA, Spitalnik SL, Hansen KC (2020) Serum proteomics in COVID-19 patients: altered coagulation and complement status as a function of IL-6 level. J Proteome Res 19(11): 4417–4427

de Oliveira GLV, Oliveira CNS, Pinzan CF, de Salis LVV, Cardoso CRB (2021) Microbiota modulation of the gut-lung axis in COVID-19. Front Immunol 12:635471

Demichev V, Tober-Lau P, Lemke O, Nazarenko T, Thibeault C, Whitwell H, Rohl A, Freiwald A, Szyrwiel L, Ludwig D et al (2021) A time-resolved proteomic and prognostic map of COVID-19. Cell Syst 12(8):780–794 e787

Demiray A, Sarı T, Çalışkan A, Nar R, Aksoy L, Akbubak IH (2021) Serum microRNA signature is capable of predictive and prognostic factor for SARS-COV-2 virulence. Turkish J Biochem 46(3)

Dhar R, Kirkpatrick J, Gilbert L, Khanna A, Modi MM, Chawla RK, Dalal S, Maturu VN, Stern M, Keppler OT et al (2023) Doxycycline for the prevention of progression of COVID-19 to severe disease requiring intensive care unit (ICU) admission: a randomized, controlled, open-label, parallel group trial (DOXPREVENT.ICU). PLoS One 18(1):e0280745

Di Marco F, Foti G, Corsico AG (2022) Where are we with the use of N-acetylcysteine as a preventive and adjuvant treatment for COVID-19? Eur Rev Med Pharmacol Sci 26(2):715–721

Dogan HO, Senol O, Bolat S, Yildiz SN, Buyuktuna SA, Sariismailoglu R, Dogan K, Hasbek M, Hekim SN (2021) Understanding the pathophysiological changes via untargeted metabolomics in COVID-19 patients. J Med Virol 93(4):2340–2349

Farr RJ, Rootes CL, Rowntree LC, Nguyen THO, Hensen L, Kedzierski L, Cheng AC, Kedzierska K, Au GG, Marsh GA et al (2021) Altered microRNA expression in COVID-19 patients enables identification of SARS-CoV-2 infection. PLoS Pathog 17(7):e1009759

Fernandez-Pato A, Virseda-Berdices A, Resino S, Ryan P, Martinez-Gonzalez O, Perez-Garcia F, Martin-Vicente M, Valle-Millares D, Brochado-Kith O, Blancas R et al (2022) Plasma miRNA profile at COVID-19 onset predicts severity status and mortality. Emerg Microbes Infect 11(1): 676–688

Ferreira LC, Gomes CEM, Rodrigues-Neto JF, Jeronimo SMB (2022) Genome-wide association studies of COVID-19: Connecting the dots. Infect Genet Evol 106:105379

Fricke-Galindo I, Falfan-Valencia R (2021) Genetics insight for COVID-19 susceptibility and severity: a review. Front Immunol 12:622176

Gogate N, Lyman D, Bell A, Cauley E, Crandall KA, Joseph A, Kahsay R, Natale DA, Schriml LM, Sen S et al (2021) COVID-19 biomarkers and their overlap with comorbidities in a disease biomarker data model. Brief Bioinform 22(6)

Gomez-Carballa A, Pardo-Seco J, Pischedda S, Rivero-Calle I, Butler-Laporte G, Richards JB, Viz-Lasheras S, Martinon-Torres F, Salas A (2022) network G: Sex-biased expression of the TLR7 gene in severe COVID-19 patients: Insights from transcriptomics and epigenomics. Environ Res 215(Pt 2):114288

Gowthaman R, Guest JD, Yin R, Adolf-Bryfogle J, Schief WR, Pierce BG (2021) CoV3D: a database of high resolution coronavirus protein structures. Nucleic Acids Res 49(D1):D282–D287

Grassin-Delyle S, Roquencourt C, Moine P, Saffroy G, Carn S, Heming N, Fleuriet J, Salvator H, Naline E, Couderc LJ et al (2021) Metabolomics of exhaled breath in critically ill COVID-19 patients: a pilot study. EBioMedicine 63:103154

Gutmann C, Khamina K, Theofilatos K, Diendorfer AB, Burnap SA, Nabeebaccus A, Fish M, McPhail MJW, O'Gallagher K, Schmidt LE et al (2022) Association of cardiometabolic microRNAs with COVID-19 severity and mortality. Cardiovasc Res 118(2):461–474

Haljasmagi L, Salumets A, Rumm AP, Jurgenson M, Krassohhina E, Remm A, Sein H, Kareinen L, Vapalahti O, Sironen T et al (2020) Longitudinal proteomic profiling reveals increased early inflammation and sustained apoptosis proteins in severe COVID-19. Sci Rep 10(1):20533

Harrison PW, Lopez R, Rahman N, Allen SG, Aslam R, Buso N, Cummins C, Fathy Y, Felix E, Glont M et al (2021) The COVID-19 Data Portal: accelerating SARS-CoV-2 and COVID-19 research through rapid open access data sharing. Nucleic Acids Res 49(W1):W619–W623

He X, Cheng X, Feng X, Wan H, Chen S, Xiong M (2020) Clinical symptom differences between mild and severe COVID-19 patients in China: a meta-analysis. Front Public Health 8:561264

Heustess AM, Allard MA, Thompson DK, Fasinu PS (2021) Clinical management of COVID-19: a review of pharmacological treatment options. Pharmaceuticals (Basel) 14(6)

Hussain I, Hussain A, Alajmi MF, Rehman MT, Amir S (2021) Impact of repurposed drugs on the symptomatic COVID-19 patients. J Infect Public Health 14(1):24–38

Isa P, Taboada B, Garcia-Lopez R, Boukadida C, Ramirez-Gonzalez JE, Vazquez-Perez JA, Hernandez-Teran A, Romero-Espinoza JA, Munoz-Medina JE, Grajales-Muniz C et al (2022) Metagenomic analysis reveals differences in the co-occurrence and abundance of viral species in SARS-CoV-2 patients with different severity of disease. BMC Infect Dis 22(1):792

Jabeen A, Ahmad N, Raza K (2022) Global gene expression and docking profiling of COVID-19 infection. Front Genet 13:870836

Jain R, Ramaswamy S, Harilal D, Uddin M, Loney T, Nowotny N, Alsuwaidi H, Varghese R, Deesi Z, Alkhajeh A et al (2021) Host transcriptomic profiling of COVID-19 patients with mild, moderate, and severe clinical outcomes. Comput Struct Biotechnol J 19:153–160

Jalaleddine N, Bouzid A, Hachim M, Sharif-Askari NS, Mahboub B, Senok A, Halwani R, Hamoudi RA, Al Heialy S (2022) ACE2 polymorphisms impact COVID-19 severity in obese patients. Sci Rep 12(1):21491

Jayatilaka R, Patel R, Brar M, Tang Y, Jisrawi NM, Chishtie F, Drozd J, Valluri SR (2022) A mathematical model of COVID-19 transmission. Mater Today Proc 54:101–112

Jogalekar MP, Veerabathini A, Patel AB (2021) COVID-19: antiviral agents and enzyme inhibitors/receptor blockers in development. Exp Biol Med (Maywood) 246(13):1533–1540

Jorge-Aaron RM, Rosa-Ester MP (2020) N-acetylcysteine as a potential treatment for COVID-19. Future Microbiol 15:959–962

Kanapeckaite A, Mazeikiene A, Geris L, Burokiene N, Cottrell GS, Widera D (2022) Computational pharmacology: new avenues for COVID-19 therapeutics search and better preparedness for future pandemic crises. Biophys Chem 290:106891

Kanne JP, Bai H, Bernheim A, Chung M, Haramati LB, Kallmes DF, Little BP, Rubin GD, Sverzellati N (2021) COVID-19 imaging: what we know now and what remains unknown. Radiology 299(3):E262–E279

Ke S, Weiss ST, Liu YY (2022) Dissecting the role of the human microbiome in COVID-19 via metagenome-assembled genomes. Nat Commun 13(1):5235

Khoshnood S, Shirani M, Dalir A, Moradi M, Haddadi MH, Sadeghifard N, Birjandi FS, Yashmi I, Heidary M (2022) Antiviral effects of azithromycin: a narrative review. Biomed Pharmacother 147:112682

Kimhofer T, Lodge S, Whiley L, Gray N, Loo RL, Lawler NG, Nitschke P, Bong SH, Morrison DL, Begum S et al (2020) Integrative modeling of quantitative plasma lipoprotein, metabolic, and amino acid data reveals a multiorgan pathological signature of SARS-CoV-2 infection. J Proteome Res 19(11):4442–4454

Kimura Y, Nakai Y, Shin J, Hara M, Takeda Y, Kubo S, Jeremiah SS, Ino Y, Akiyama T, Moriyama K et al (2021) Identification of serum prognostic biomarkers of severe COVID-19 using a quantitative proteomic approach. Sci Rep 11(1):20638

Klok FA, Kruip M, van der Meer NJM, Arbous MS, Gommers D, Kant KM, Kaptein FHJ, van Paassen J, Stals MAM, Huisman MV et al (2020) Incidence of thrombotic complications in critically ill ICU patients with COVID-19. Thromb Res 191:145–147

Kolahchi Z, De Domenico M, Uddin LQ, Cauda V, Grossmann I, Lacasa L, Grancini G, Mahmoudi M, Rezaei N (2021) COVID-19 and its global economic impact. Adv Exp Med Biol 1318:825–837

Konigsberg IR, Barnes B, Campbell M, Davidson E, Zhen Y, Pallisard O, Boorgula MP, Cox C, Nandy D, Seal S et al (2021) Host methylation predicts SARS-CoV-2 infection and clinical outcome. Commun Med (Lond) 1(1):42

Kwee TC, Kwee RM (2022) Chest CT in COVID-19: what the radiologist needs to know. Radiographics 42(1):E32

Lee JS, Han D, Kim SY, Hong KH, Jang MJ, Kim MJ, Kim YG, Park JH, Cho SI, Park WB et al (2021) Longitudinal proteomic profiling provides insights into host response and proteome dynamics in COVID-19 progression. Proteomics 21(11–12):e2000278

Lehrer S (2021) Rheinstein PH: BIN1 rs744373 SNP and COVID-19 mortality. World Acad Sci J 3(2)

Li Z, Mei Z, Ding S, Chen L, Li H, Feng K, Huang T, Cai YD (2022a) Identifying methylation signatures and rules for COVID-19 with machine learning methods. Front Mol Biosci 9:908080

Li CX, Gao J, Zhang Z, Chen L, Li X, Zhou M, Wheelock AM (2022b) Multiomics integration-based molecular characterizations of COVID-19. Brief Bioinform 23(1)

Liao D, Zhou F, Luo L, Xu M, Wang H, Xia J, Gao Y, Cai L, Wang Z, Yin P et al (2020) Haematological characteristics and risk factors in the classification and prognosis evaluation of COVID-19: a retrospective cohort study. Lancet Haematol 7(9):e671–e678

Liu Y, Zhang H, Tang X, Jiang X, Yan X, Liu X, Gong J, Mew K, Sun H, Chen X et al (2021a) Distinct metagenomic signatures in the SARS-CoV-2 infection. Front Cell Infect Microbiol 11:706970

Liu C, Martins AJ, Lau WW, Rachmaninoff N, Chen J, Imberti L, Mostaghimi D, Fink DL, Burbelo PD, Dobbs K et al (2021b) Time-resolved systems immunology reveals a late juncture linked to fatal COVID-19. Cell 184(7):1836–1857 e1822

Liu Q, Su Q, Zhang F, Tun HM, Mak JWY, Lui GC, Ng SSS, Ching JYL, Li A, Lu W et al (2022) Multi-kingdom gut microbiota analyses define COVID-19 severity and post-acute COVID-19 syndrome. Nat Commun 13(1):6806

Lopez-Hernandez Y, Monarrez-Espino J, Oostdam AH, Delgado JEC, Zhang L, Zheng J, Valdez JJO, Mandal R, Gonzalez FLO, Moreno JCB et al (2021) Targeted metabolomics identifies high performing diagnostic and prognostic biomarkers for COVID-19. Sci Rep 11(1):14732

Lopez-Martinez C, Martin-Vicente P, Gomez de Ona J, Lopez-Alonso I, Gil-Pena H, Cuesta-Llavona E, Fernandez-Rodriguez M, Crespo I, Salgado Del Riego E, Rodriguez-Garcia R et al (2022) Transcriptomic clustering of critically ill COVID-19 patients. Eur Respir J

Ma J, Deng Y, Zhang M, Yu J (2022a) The role of multi-omics in the diagnosis of COVID-19 and the prediction of new therapeutic targets. Virulence 13(1):1101–1110

Ma Y, Li Q, Chen J, Liu S, Liu S, He X, Ling Y, Zheng J, Corpe C, Lu H et al (2022b) Angiotensin-converting enzyme 2 SNPs as common genetic loci and optimal early identification genetic markers for COVID-19. Pathogens 11(8)

Maes M, Tedesco Junior WLD, Lozovoy MAB, Mori MTE, Danelli T, Almeida ERD, Tejo AM, Tano ZN, Reiche EMV, Simao ANC (2022) In COVID-19, NLRP3 inflammasome genetic variants are associated with critical disease and these effects are partly mediated by the sickness symptom complex: a nomothetic network approach. Mol Psychiatry 27(4):1945–1955

Majumdar S, Verma R, Saha A, Bhattacharyya P, Maji P, Surjit M, Kundu M, Basu J, Saha S (2021) Perspectives about modulating host immune system in targeting SARS-CoV-2 in India. Front Genet 12:637362

Martinez-Fleta P, Vera-Tome P, Jimenez-Fernandez M, Requena S, Roy-Vallejo E, Sanz-Garcia A, Lozano-Prieto M, Lopez-Sanz C, Vara A, Lancho-Sanchez A et al (2021) A differential signature of circulating miRNAs and cytokines between COVID-19 and community-acquired pneumonia uncovers novel physiopathological mechanisms of COVID-19. Front Immunol 12: 815651

Marzolini C, Marra F, Boyle A, Khoo S, Back DJ (2022) Fluvoxamine for the treatment of COVID-19. Lancet Glob Health 10(3):e331

Mastropasqua L, Toto L, Chiricosta L, Diomede F, Gugliandolo A, Silvestro S, Marconi GD, Sinjari B, Vecchiet J, Cipollone F et al (2021) Transcriptomic analysis revealed increased expression of genes involved in keratinization in the tears of COVID-19 patients. Sci Rep 11(1):19817

Mathieu E, Ritchie H, Ortiz-Ospina E, Roser M, Hasell J, Appel C, Giattino C, Rodes-Guirao L (2021) A global database of COVID-19 vaccinations. Nat Hum Behav 5(7):947–953

McGonagle D, O'Donnell JS, Sharif K, Emery P, Bridgewood C (2020) Immune mechanisms of pulmonary intravascular coagulopathy in COVID-19 pneumonia. Lancet Rheumatol 2(7):e437–e445

Mehta N, Mazer-Amirshahi M, Alkindi N, Pourmand A (2020) Pharmacotherapy in COVID-19; A narrative review for emergency providers. Am J Emerg Med 38(7):1488–1493

Mei X, Lee HC, Diao KY, Huang M, Lin B, Liu C, Xie Z, Ma Y, Robson PM, Chung M et al (2020) Artificial intelligence-enabled rapid diagnosis of patients with COVID-19. Nat Med 26(8): 1224–1228

Messner CB, Demichev V, Wendisch D, Michalick L, White M, Freiwald A, Textoris-Taube K, Vernardis SI, Egger AS, Kreidl M et al (2020) Ultra-high-throughput clinical proteomics reveals classifiers of COVID-19 infection. Cell Syst 11(1):11–24 e14

Mohanty SS, Sahoo CR, Padhy RN (2021) Targeting some enzymes with repurposing approved pharmaceutical drugs for expeditious antiviral approaches against newer strains of COVID-19. AAPS PharmSciTech 22(6):214

Mousa M, Vurivi H, Kannout H, Uddin M, Alkaabi N, Mahboub B, Tay GK, Alsafar HS (2021) Partnership UC-C: genome-wide association study of hospitalized COVID-19 patients in the United Arab Emirates. EBioMedicine 74:103695

Nitsure M, Sarangi B, Shankar GH, Reddy VS, Walimbe A, Sharma V, Prayag S (2020) Mechanisms of hypoxia in COVID-19 patients: a pathophysiologic reflection. Indian J Crit Care Med 24(10):967–970

Overmyer KA, Shishkova E, Miller IJ, Balnis J, Bernstein MN, Peters-Clarke TM, Meyer JG, Quan Q, Muehlbauer LK, Trujillo EA et al (2021) Large-scale multi-omic analysis of COVID-19 severity. Cell Syst 12(1):23–40 e27

Owji H, Negahdaripour M, Hajighahramani N (2020) Immunotherapeutic approaches to curtail COVID-19. Int Immunopharmacol 88:106924

Paez-Franco JC, Torres-Ruiz J, Sosa-Hernandez VA, Cervantes-Diaz R, Romero-Ramirez S, Perez-Fragoso A, Meza-Sanchez DE, German-Acacio JM, Maravillas-Montero JL, Mejia-Dominguez

NR et al (2021) Metabolomics analysis reveals a modified amino acid metabolism that correlates with altered oxygen homeostasis in COVID-19 patients. Sci Rep 11(1):6350

Pandey RK, Srivastava A, Singh PP, Chaubey G (2022) Genetic association of TMPRSS2 rs2070788 polymorphism with COVID-19 case fatality rate among Indian populations. Infect Genet Evol 98:105206

Pang Z, Zhou G, Chong J, Xia J (2021) Comprehensive meta-analysis of COVID-19 global metabolomics datasets. Meta 11(1)

Pang APS, Higgins-Chen AT, Comite F, Raica I, Arboleda C, Went H, Mendez T, Schotsaert M, Dwaraka V, Smith R et al (2022) Longitudinal study of DNA methylation and epigenetic clocks prior to and following test-confirmed COVID-19 and mRNA vaccination. Front Genet 13: 819749

Panoutsopoulos AA (2020) Known drugs and small molecules in the battle for COVID-19 treatment. Genes Dis 7(4):528–534

Parasher A (2021) COVID-19: current understanding of its pathophysiology, clinical presentation and treatment. Postgrad Med J 97(1147):312–320

Patel H, Ashton NJ, Dobson RJB, Andersson LM, Yilmaz A, Blennow K, Gisslen M, Zetterberg H (2021) Proteomic blood profiling in mild, severe and critical COVID-19 patients. Sci Rep 11(1): 6357

Pettifor A, DiPrete BL, Shook-Sa BE, Premkumar L, Kuczynski K, Dittmer D, Aiello A, Wallet S, Maile R, Tan J et al (2022) A prospective study of asymptomatic SARS-CoV-2 infection among individuals involved in academic research under limited operations during the COVID-19 pandemic. PLoS One 17(4):e0267353

Posadas-Sanchez R, Sanchez-Munoz F, Guzman-Martin CA, Hernandez-Diaz Couder A, Rojas-Velasco G, Fragoso JM, Vargas-Alarcon G (2021) Dipeptidylpeptidase-4 levels and DPP4 gene polymorphisms in patients with COVID-19. Association with disease and with severity. Life Sci 276:119410

Pu J, Leader JK, Bandos A, Ke S, Wang J, Shi J, Du P, Guo Y, Wenzel SE, Fuhrman CR et al (2021) Automated quantification of COVID-19 severity and progression using chest CT images. Eur Radiol 31(1):436–446

Ramakrishnan S, Nicolau DV Jr, Langford B, Mahdi M, Jeffers H, Mwasuku C, Krassowska K, Fox R, Binnian I, Glover V et al (2021) Inhaled budesonide in the treatment of early COVID-19 (STOIC): a phase 2, open-label, randomised controlled trial. Lancet Respir Med 9(7):763–772

Reis G, Dos Santos Moreira-Silva EA, Silva DCM, Thabane L, Milagres AC, Ferreira TS, Dos Santos CVQ, de Souza Campos VH, Nogueira AMR, de Almeida A et al (2022) Effect of early treatment with fluvoxamine on risk of emergency care and hospitalisation among patients with COVID-19: the TOGETHER randomised, platform clinical trial. Lancet Glob Health 10(1): e42–e51

Rommasi F, Nasiri MJ, Mirsaeidi M (2022) Immunomodulatory agents for COVID-19 treatment: possible mechanism of action and immunopathology features. Mol Cell Biochem 477(3): 711–726

Samprathi M, Jayashree M (2020) Biomarkers in COVID-19: an up-to-date review. Front Pediatr 8: 607647

Schmiedel BJ, Rocha J, Gonzalez-Colin C, Bhattacharyya S, Madrigal A, Ottensmeier CH, Ay F, Chandra V, Vijayanand P (2021) COVID-19 genetic risk variants are associated with expression of multiple genes in diverse immune cell types. Nat Commun 12(1):6760

Sharma A, Ahmad Farouk I, Lal SK (2021) COVID-19: a review on the novel coronavirus disease evolution, transmission, detection, control and prevention. Viruses 13(2)

Shen B, Yi X, Sun Y, Bi X, Du J, Zhang C, Quan S, Zhang F, Sun R, Qian L et al (2020) Proteomic and metabolomic characterization of COVID-19 patient sera. Cell 182(1):59–72 e15

Shu T, Ning W, Wu D, Xu J, Han Q, Huang M, Zou X, Yang Q, Yuan Y, Bie Y et al (2020) Plasma proteomics identify biomarkers and pathogenesis of COVID-19. Immunity 53(5):1108–1122 e1105

Stephenson E, Reynolds G, Botting RA, Calero-Nieto FJ, Morgan MD, Tuong ZK, Bach K, Sungnak W, Worlock KB, Yoshida M et al (2021) Single-cell multi-omics analysis of the immune response in COVID-19. Nat Med 27(5):904–916

Stukalov A, Girault V, Grass V, Karayel O, Bergant V, Urban C, Haas DA, Huang Y, Oubraham L, Wang A et al (2021) Multilevel proteomics reveals host perturbations by SARS-CoV-2 and SARS-CoV. Nature 594(7862):246–252

Su Y, Chen D, Yuan D, Lausted C, Choi J, Dai CL, Voillet V, Duvvuri VR, Scherler K, Troisch P et al (2020) Multi-omics resolves a sharp disease-state shift between mild and moderate COVID-19. Cell 183(6):1479–1495 e1420

Tang NL, Chan PK, Wong CK, To KF, Wu AK, Sung YM, Hui DS, Sung JJ, Lam CW (2005) Early enhanced expression of interferon-inducible protein-10 (CXCL-10) and other chemokines predicts adverse outcome in severe acute respiratory syndrome. Clin Chem 51(12):2333–2340

Thomas T, Stefanoni D, Reisz JA, Nemkov T, Bertolone L, Francis RO, Hudson KE, Zimring JC, Hansen KC, Hod EA et al (2020) COVID-19 infection alters kynurenine and fatty acid metabolism, correlating with IL-6 levels and renal status. JCI Insight 5(14)

Tiwari N, Upadhyay J, Ansari MN, Joshi R (2020) Novel beta-Coronavirus (SARS-CoV-2): current and future aspects of pharmacological treatments. Saudi Pharm J 28(10):1243–1252

Valdes A, Moreno LO, Rello SR, Orduna A, Bernardo D, Cifuentes A (2022) Metabolomics study of COVID-19 patients in four different clinical stages. Sci Rep 12(1):1650

van Moorsel CHM, van der Vis JJ, Duckworth A, Scotton CJ, Benschop C, Ellinghaus D, Ruven HJT, Quanjel MJR, Grutters JC (2021) The MUC5B promoter polymorphism associates with severe COVID-19 in the European population. Front Med (Lausanne) 8:668024

Vellingiri B, Jayaramayya K, Iyer M, Narayanasamy A, Govindasamy V, Giridharan B, Ganesan S, Venugopal A, Venkatesan D, Ganesan H et al (2020) COVID-19: a promising cure for the global panic. Sci Total Environ 725:138277

Verma A, Minnier J, Wan ES, Huffman JE, Gao L, Joseph J, Ho YL, Wu WC, Cho K, Gorman BR et al (2022) A MUC5B gene polymorphism, rs35705950-T, confers protective effects against COVID-19 hospitalization but not severe disease or mortality. Am J Respir Crit Care Med 206(10):1220–1229

Wang X, Xu G, Liu X, Liu Y, Zhang S, Zhang Z (2021) Multiomics: unraveling the panoramic landscapes of SARS-CoV-2 infection. Cell Mol Immunol 18(10):2313–2324

Wang D, Kumar V, Burnham KL, Mentzer AJ, Marsden BD, Knight JC (2023) COMBATdb: a database for the COVID-19 Multi-omics Blood ATlas. Nucleic Acids Res 51(D1):D896–D905

Wu D, Shu T, Yang X, Song JX, Zhang M, Yao C, Liu W, Huang M, Yu Y, Yang Q et al (2020) Plasma metabolomic and lipidomic alterations associated with COVID-19. Natl Sci Rev 7(7): 1157–1168

Wu P, Chen D, Ding W, Wu P, Hou H, Bai Y, Zhou Y, Li K, Xiang S, Liu P et al (2021) The trans-omics landscape of COVID-19. Nat Commun 12(1):4543

Xiong Y, Liu Y, Cao L, Wang D, Guo M, Jiang A, Guo D, Hu W, Yang J, Tang Z et al (2020) Transcriptomic characteristics of bronchoalveolar lavage fluid and peripheral blood mononuclear cells in COVID-19 patients. Emerg Microbes Infect 9(1):761–770

Xu Z, Shi L, Wang Y, Zhang J, Huang L, Zhang C, Liu S, Zhao P, Liu H, Zhu L et al (2020) Pathological findings of COVID-19 associated with acute respiratory distress syndrome. Lancet Respir Med 8(4):420–422

Yadav UCS (2021) Vaccines and drugs under clinical trials for prevention and treatment of COVID-19. Virus 32(1):13–19

Yang AP, Liu JP, Tao WQ, Li HM (2020) The diagnostic and predictive role of NLR, d-NLR and PLR in COVID-19 patients. Int Immunopharmacol 84:106504

Yuce M, Filiztekin E, Ozkaya KG (2021) COVID-19 diagnosis – a review of current methods. Biosens Bioelectron 172:112752

Zhang Y, Qin L, Zhao Y, Zhang P, Xu B, Li K, Liang L, Zhang C, Dai Y, Feng Y et al (2020a) Interferon-induced transmembrane protein 3 genetic variant rs12252-C associated with disease severity in coronavirus disease 2019. J Infect Dis 222(1):34–37

Zhang K, Liu X, Shen J, Li Z, Sang Y, Wu X, Zha Y, Liang W, Wang C, Wang K et al (2020b) Clinically applicable AI system for accurate diagnosis, quantitative measurements, and prognosis of COVID-19 pneumonia using computed tomography. Cell 181(6):1423–1433 e1411

Zhang G, Zhang J, Wang B, Zhu X, Wang Q, Qiu S (2020c) Analysis of clinical characteristics and laboratory findings of 95 cases of 2019 novel coronavirus pneumonia in Wuhan, China: a retrospective analysis. Respir Res 21(1):74

Zhang W, Zhang Y, Min Z, Mo J, Ju Z, Guan W, Zeng B, Liu Y, Chen J, Zhang Q et al (2022) COVID19db: a comprehensive database platform to discover potential drugs and targets of COVID-19 at whole transcriptomic scale. Nucleic Acids Res 50(D1):D747–D757

Zhou Y, Shi X, Fu W, Xiang F, He X, Yang B, Wang X, Ma WL (2021) Gut microbiota dysbiosis correlates with abnormal immune response in moderate COVID-19 patients with fever. J Inflamm Res 14:2619–2631

Pneumonia

14

Abstract

Pneumonia is an acute lower respiratory tract infection affecting the bronchioles, alveoli, and distal airways. Such infections are primarily caused by bacteria, viruses, and mycoplasma, but may also be caused by fungal pathogens. Medical imaging and routine culture of blood and sputum are widely used for disease diagnosis. Primarily antibiotics and antimicrobial agents are used for the treatment of pneumonia. However, the disease continues to be associated with high morbidity and high mortality rate. New age omics techniques have been widely explored to address the complications associated with pneumonia. This chapter documents the various probable molecular and metabolic signatures of the different types of pneumonia as identified by multi-omics approaches. Microbial dysbiosis has also been observed in pneumonia patients. The different research avenues that can be explored in future for improving the diagnosis and management of pneumonia have been discussed at the end of this chapter.

Keywords

Community-acquired pneumonia (CAP) · Hospital-acquired pneumonia (HAP) · Ventilator-acquired pneumonia (VAP) · Bronchopneumonia · Lobar pneumonia · CURB-65 pneumonia severity score · Chest radiograph · Biomarkers · Antibiotics

14.1 Introduction

Pneumonia may be defined as a severe lower respiratory tract infection where the lung parenchyma gets affected. This deadly disease is characterized by the inflammation of alveoli and distal airways. The disease is mostly caused by bacteria but may also be caused by viruses, mycoplasma, and fungal pathogens (https://www.ncbi.nlm.nih.gov/books/NBK534295/; Torres et al. 2021). The global burden of

S. Saha et al., *Pulmonomics: Omics Approaches for Understanding Pulmonary Diseases*, https://doi.org/10.1007/978-981-99-3505-5_14

pneumonia among infants, children, and elderly individuals has made it a major health concern. Children less than 5 years and elderly adults greater than 65 years of age or having comorbidities are at a higher risk of pneumonia (McAllister et al. 2019; Weir et al. 2015; Torres et al. 2013). Suboptimal breastfeeding, malnutrition, ambient particulate matter, and household air pollution are the major risk factors for community-acquired pneumonia in children (Bradley et al. 2011). Comorbidities like COPD, cardiovascular disease, diabetes mellitus, and chronic liver disease often increase the risk of community-acquired pneumonia in adults (Torres et al. 2013). Men, in general, are at a higher risk of having community-acquired pneumonia as compared to women (Barbagelata et al. 2020). Immunocompromised and immunosuppressed individuals are at a higher risk of hospital-acquired pneumonia (Schnell et al. 2010; https://www.ncbi.nlm.nih.gov/books/NBK557843/). Pneumonia continues to be one of the leading causes of hospitalization and mortality among infants and elderly adults (Walker et al. 2013; Hespanhol and Barbara 2020; Shi et al. 2020; https://www.ncbi.nlm.nih.gov/books/NBK430749/).

14.2 Classification of Pneumonia

Pneumonia can be classified either on the basis of source of acquiring the disease or on the basis of the causative organism or on the basis of the area of the lung affected. Based on the source, pneumonia can be community-acquired pneumonia (CAP), hospital-acquired pneumonia (HAP), and health care-associated pneumonia (HCAP). Community-acquired pneumonia (CAP) refers to those pneumonia cases where patients have no history of hospitalization during the month of the onset of symptoms. Aspiration pneumonia accounts for 5–15% of all CAP cases (Mandell and Niederman 2019). The causative agents of CAP include bacteria like *Streptococcus pneumoniae*, *Haemophilus influenzae*, *Mycoplasma pneumoniae*, *Legionella pneumophila*, and other respiratory viruses (Cilloniz et al. 2011). Hospital-acquired pneumonia (HAP), on the other hand, is acquired within at least 2 days of hospitalization. HAP patients generally do not present any suspicion of disease incubation prior to hospitalization. HAP is mostly caused by methicillin-resistant *Staphylococcus aureus* (MRSA), methicillin-susceptible *Staphylococcus aureus* (MSSA), *Enterobacterales* spp., *Pseudomonas aeruginosa*, or *Acinetobacter* spp. (Torres et al. 2017). Ventilator-associated pneumonia (VAP) is a form of HAP that occurs typically more than 48 hours after endotracheal intubation. Health care-associated pneumonia (HCAP) is generally acquired in non-hospital health care units (Torres et al. 2021).

Based on causative microbial agent, pneumonia can be bacterial pneumonia, viral pneumonia, mycoplasma pneumonia, and fungal pneumonia. Bacterial pneumonia can be categorized as typical and atypical pneumonia. Typical pneumonia is primarily caused by *Streptococcus pneumoniae*, *Haemophilus influenzae*, *Klebsiella pneumoniae*, *Pseudomonas aeruginosa*, *Staphylococcus aureus*, Group A streptococci, *Moraxella catarrhalis*, anaerobes and aerobic gram-negative bacteria. Microbes like *Legionella* spp., *Chlamydia pneumoniae*, and *Chlamydia psittaci* are responsible for causing atypical pneumonia (https://www.ncbi.nlm.nih.gov/books/NBK513321/). Viral pneumonia can be caused by viruses like influenza virus, respiratory syncytial virus (RSV), parainfluenza virus (PIV), adenovirus, human metapneumovirus (HMPV), rhinovirus, human coronaviruses (HCoV), middle east respiratory syndrome coronavirus (MERS-CoV), and severe acute respiratory syndrome coronavirus 2 (SARS-CoV-2). Pneumonia can also be caused by other viruses like varicella-zoster virus (VZV), cytomegalovirus (CMV), and herpes simplex virus (HSV) (Dandachi and Rodriguez-Barradas 2018; https://www.ncbi.nlm.nih.gov/books/NBK513286/). Mycoplasma pneumonia, commonly caused by *Mycoplasma pneumoniae* gets transmitted via respiratory droplets. *Mycoplasma pneumoniae* serves as a common cause of community-acquired pneumonia (CAP) (https://www.ncbi.nlm.nih.gov/books/NBK430780/). Fungal pneumonia is rare (Davies 1994). Fungal pneumonia can be caused by different fungi like *Aspergillus* spp., *Cryptococcus neoformans*, *Pneumocystis jirovecii*, *Histoplasma. capsulatum*, *Blastomyces dermatitidis*, *Coccidioides immitis*, and *Candida* spp. (Pound et al. 2002; Li et al. 2019; Castillo et al. 2016; Pitcher and Zuckerman 2016). Opportunistic fungal pneumonia is caused by *Fusarium* spp., *Penicillium* spp., and dematiaceous fungi (Connolly Jr. et al. 1999).

Based on the area of the lung affected, pneumonia can be histologically or anatomically classified as bronchopneumonia or lobar pneumonia. In case of bronchopneumonia, the infection occurs around the bronchi and bronchioles. The infection gradually spreads locally within the lungs. Multiple discrete foci of consolidation are observed within the lungs especially, the lower lobe of the lungs. These patches of consolidation represent accumulation of neutrophils in the bronchi and the alveoli. In case of lobar pneumonia, there is an acute exudative inflammation of the whole lobe. Uniform consolidation of an entire lobe is observed. Most of the lobar pneumonia cases are caused by *Streptococcus pneumoniae*. There are four classical stages of inflammation in lobar pneumonia. In the first stage, congestion occurs in the first 24 hours where the lungs appear to be red, heavy, and boggy. Intra-alveolar oedema, vascular engorgement, several bacteria, and few neutrophils are observed microscopically. In the second stage of inflammation, red hepatization begins about 2–3 days after consolidation and persists about 2–4 days. The infected lung appears dry, red-pink, granular, airless and exhibits firm liver-like consistency. The oedema fluid gets replaced by fibrin strands. Presence of exudate makes the alveolar septa less prominent. Exudates containing neutrophils and ingested bacteria along with extravasation of erythrocytes, desquamation of epithelial cells, and presence of fibrin strands in the alveoli are visualized microscopically. The third stage called grey hepatization occurs 2–3 days after red hepatization and extends for 4–8 days. Presence of fibrinopurulent exudate, progressive disintegration of red blood cells, and hemosiderin accounts for the greyish appearance of the lung. Appearance of macrophages is also observed in this stage. The resolution and restoration of lung architecture occur in the last stage. This occurs by the eighth day. Solid fibrinous content gets liquified by enzymatic action, and the aeration gets restored eventually. Predominance of macrophages containing engulfed neutrophils and debris is also observed in this stage (https://www.ncbi.nlm.nih.gov/books/NBK534295/; Lim 2020).

14.3 Pathogenesis

The airways and pulmonary system are continuously exposed to environmental pathogens. Healthy airways are not sterile and harbours unique microbiome. Microbes and pathogens enter the respiratory tract and the lungs via micro-aspiration process or through hematogenous spread (https://www.ncbi.nlm.nih.gov/books/NBK430749/). Various defense mechanisms including anatomical barriers, cough reflex, and mucociliary clearance system function to expel the microbial pathogens. These immune responses help to restore the normal respiratory microflora and prevent infectious agents from colonizing the respiratory tract. Pneumonia occurs when the host is immunocompromised or when the infectious pathogen evades the host immune response or when the infectious microbial inoculum is large. Entry, invasion, multiplication, and subsequent propagation of the microbial pathogen in the lung parenchyma activate the host immune and inflammatory response. The alveolar macrophages primarily exert immune response against lower airway bacteria. The complex interplay of host–microbe interaction and host immune response leads to the onset of the symptoms of pneumonia (https://www.ncbi.nlm.nih.gov/books/NBK534295/; Torres et al. 2021; https://www.ncbi.nlm.nih.gov/books/NBK430780/). IL-1 and TNF result in fever in pneumonia patients. Chemokines like G-CSF and IL-8 mediate chemotaxis and maturation of neutrophils in bacterial pneumonia. Overproduction of cytokines leads to leakage of the alveolar-capillary membrane, which ultimately results in loss of compliance and shortness of breath in pneumonia patients. Excessive cytokine and chemokine production can lead to sepsis and multiorgan failure in pneumonia patients. At times, RBCs can cross the alveolar-capillary barrier and result in haemoptysis (https://www.ncbi.nlm.nih.gov/books/NBK534295/; https://www.ncbi.nlm.nih.gov/books/NBK513321/). Pneumocytes are primarily targeted in viral pneumonia. Involvement of CD4 and CD8 cells has also been noted in the release of immune products which leads to increased vascular permeability and oedema in patients suffering from viral pneumonia (https://www.ncbi.nlm.nih.gov/books/NBK557843/; https://www.ncbi.nlm.nih.gov/books/NBK513286/).

14.4 Clinical Features and Diagnosis

Pneumonia is characterized by symptoms like cough, sputum production, high fever, chills, breathlessness, dyspnoea, pleuritic chest pain, high pulse rate, fatigue, headache, muscle pain, and arthralgias. Less common symptoms include diarrhoea, drowsiness, stomach pain, and confusion. Bronchial breath sounds and crackles are observed in chest auscultation of pneumonia patients. However, the symptoms vary largely based on the degree of disease severity. Hypoxia, confusion, sepsis, and multiorgan failure are noted in severe pneumonia patients. Severe pneumonia is also associated with increased mortality rate (https://www.ncbi.nlm.nih.gov/books/NBK534295/; Torres et al. 2021; https://www.ncbi.nlm.nih.gov/books/NBK525774/). Extrapulmonary complications are also noted in pneumonia patients. Common extrapulmonary complications include sepsis, cardiovascular complications (like myocarditis, acute coronary syndrome, depression of left ventricular function, ischemia, arrhythmias, and infarction), dementia, decline in cognitive function and depression (Dremsizov et al. 2006; Giuliano and Baker 2020; Corrales-Medina et al. 2013; Iwashyna et al. 2010; Shah et al. 2013; Davydow et al. 2013).

Pneumonia patients can be stratified based on the disease severity. CURB-65 pneumonia severity score or expanded CURB-65 aids in classifying pneumonia patients. There are five factors, namely confusion, uremia (BUN more than 20 mg/dL), respiratory rate (more than 30 per min), hypotension (SBP lower than 90 and DBP lower than 60), and age more than 65 years. One point is allotted for each factor. A cumulative score of 2 or more indicates the need for hospitalization. On the other hand, a score of 4 or more is indicative of admission to intensive care unit (ICU) and the requirement for intense therapy (https://www.ncbi.nlm.nih.gov/books/NBK534295/).

Pneumonia can be primarily detected by the physicians based on the clinical presentation and symptoms. Imaging tools like chest radiography and CT scan find huge application in the diagnosis of pneumonia. Chest X-ray provides information about the features of pneumonia along with the site and extent of infection. Presence of new infiltrates, lobes involved, pleural effusion, cavitation, and pulmonary opacities are all detected from chest radiographs. Pulmonary infiltrates on simple posterior–anterior (PA) and lateral images are signatures for CAP. Lateral projection images are crucial to enhance the performance of chest radiography in the diagnosis of CAP (Torres et al. 2021). Presence of pleural fluid or multi-lobar pneumonia indicates severe pneumonia (Metlay et al. 2019; Self et al. 2013). Observation of infiltrates in chest radiograph alone serves as detection criteria for HAP. Detection of new infiltrates by anterior–posterior projection in the supine position in chest radiograph aids in the diagnosis of VAP. However, the sensitivity and specificity of chest X-ray are inadequate for the detection of VAP (Torres et al. 2021). Computed tomography (CT) scan can detect CAP with a specificity and sensitivity of 77% and 66%, respectively. Chest CT scans may serve as a complementary diagnostic tool to chest radiograph for the detection of CAP (Laursen et al. 2013; Claessens et al. 2015). Lung ultrasonography (USG) may serve as an accurate and complementary detection tool for pneumonia. Thoracic USG can detect lung consolidations in CAP patients (D'Amato et al. 2017; Long et al. 2017). Alterations in the outer regions of the lungs can also be documented from lung USG (https://www.ncbi.nlm.nih.gov/books/NBK525774/). Lung USG can diagnose pneumonia with a sensitivity and specificity of 88% and 89%, respectively. Ventilator-associated pneumonia lung ultrasound score (VPLUS) have been devised to diagnose VAP with a specificity of 69% and a sensitivity of 71% (Bouhemad et al. 2018).

Complete blood count (CBC) along with measurement of parameters like ESR and acute phase reactants can aid in monitoring the inflammation and degree of disease severity. Leukocytosis with a leftward shift is noted as an abnormality in pneumonia patients. Leukopenia, on the other hand, is indicative of poor prognosis (https://www.ncbi.nlm.nih.gov/books/NBK534295/). Oxygen saturation level in pneumonia patients should be monitored using pulse oximetry. Blood, urine, sputum, and coughed up mucous can also be used for laboratory microbiological investigations (https://www.ncbi.nlm.nih.gov/books/NBK525774/). Blood culture, sputum culture, and urine antigen testing are performed for hospitalized patients for determining the course of antibiotic therapy. Sputum culture and gram staining of sputum samples are of vital importance in specific diagnosis of lobar pneumonia (Musher et al. 2004; Fukuyama et al. 2014). Molecular detection methods based on multiplex PCR can help to identify the causative microbial pathogens from biological samples like sputum (Torres et al. 2016; Schulte et al. 2014). Invasive techniques like thoracocentesis, pleural biopsy, bronchoscopy, or direct transthoracic aspiration from pneumonia patients can help to get material for microscopic examination, culture, or genetic tests. Open lung biopsy is the most specific diagnostic test for pneumonia; however, it is rarely conducted (https://www.ncbi.nlm.nih.gov/books/NBK534295/).

14.5 Biomarkers of Pneumonia

Biomarkers not only aid in improving our understanding of disease pathogenesis but have also been increasingly helpful in identifying individuals susceptible to pneumonia. Biomarkers may also serve as adjunctive tools for disease diagnosis. Besides, biomarkers play a vital role in guiding the nature and tenure of antibiotic therapy alongside monitoring disease prognosis upon therapeutic administration. Various omics approaches, namely genomics, epigenomics, transcriptomics, proteomics, metabolomics, and metagenomics have been extensively utilized for the identification of biomarkers for pneumonia. Some of the host molecular, metabolic, and microbial signatures of pneumonia as observed by omics studies have been discussed below and also tabulated in Table 14.1.

14.6 Genomics

Large-scale genome-wide association study (GWAS) showed that SNP of FER gene (rs4957796) was strongly associated with survival in pneumonia patients having sepsis (Rautanen et al. 2015). The GT genotype of SNP of TNFRSF1B (rs1061622) was found to be associated with lower mortality rate in pneumonia patients as compared to the GG or TT genotype (Sole-Violan et al. 2010). The −174 GG genotype of IL6 was found to protect community-acquired pneumonia patients from acute respiratory distress syndrome (ARDS) (Martin-Loeches et al. 2012). The A allele of IL10 rs1800896 conferred protection from severe inflammation, sepsis, and septic shock in pneumonia patients (Smelaya et al. 2016). On the other hand, genetic polymorphisms in the promoter region of TIM1, namely rs9313422

Table 14.1 Different biomarkers of pneumonia

Omics approach	Biomarkers	Biological sample	Comments
Genomics	HSP70-2	Blood	The +1267 genotype of HSP70-2 serves as a strong predictor of septic shock in community-acquired pneumonia (CAP) patients (Waterer et al. 2003)
	PTX3		The AG genotype of rs1840680 of PTX3 was associated with lower risk of severe community-acquired pneumonia (Zeng et al. 2021)
	ACE		SNP of ACE (rs4340) was significantly associated with the risk of *Mycoplasma pneumoniae* pneumonia (MPP) in Chinese children (Zhao et al. 2017)
	IL6		SNP of IL6 (rs1800795) was significantly associated with the risk of *Mycoplasma pneumoniae* pneumonia (MPP) in Chinese children (Zhao et al. 2017)
	NOS3		SNP of NOS3 (rs1799983) was significantly associated with the risk of *Mycoplasma pneumoniae* pneumonia (MPP) in Chinese children (Zhao et al. 2017)
	CFTR	Genomic DNA	SNP of CFTR (rs113827944) was significantly associated with both pneumonia susceptibility and severity in individuals of European ancestry (Chen et al. 2021)
	HBB		

(continued)

Table 14.1 (continued)

Omics approach	Biomarkers	Biological sample	Comments
			SNP of HBB (rs334) was significantly associated with both pneumonia susceptibility and severity in individuals of African ancestry (Chen et al. 2021)
Transcriptomics	HLA-DOA, HLA-DMA, HLA-DMB, ICOS, ICOSLG, IL2RA, CD1, CD3, CD28, CD40LG	Blood	Reduced expression of HLA-DOA, HLA-DMA, HLA-DMB, ICOS, ICOSLG, IL2RA, CD1, CD3, CD28 and CD40LG was observed in ventilator-associated pneumonia (VAP) patients (Almansa et al. 2018)
	miR-223-3p	Plasma	miR-223-3p can serve as potential biomarker for predicting sepsis due to pneumonia (Zhang et al. 2019)
	miR-146a-5p, miR-16-5p		Increased expression of miR-146a-5p and miR-16-5p was associated with lower mortality in community-acquired pneumonia (CAP) patients. Thus, these miRNAs can serve as potential biomarkers of good prognosis in hospitalized CAP patients (Galvan-Roman et al. 2020)
	LBR, DHCR24, SC5D, SQLE, IL10, IL10RB, C1QA, C1QB, C2, CR1, CDK1, CCNA1, CCNB1, HLA-DOA, HLA-DPB1, HLA-DRA, HLA-DMA,	Circulating monocytes	LBR, DHCR24, SC5D, SQLE, IL10, IL10RB, C1QA, C1QB, C2, CR1, CDK1, CCNA1, and CCNB1 are upregulated, while HLA-DOA, HLA-DPB1, HLA-DRA, and HLA-DMA are downregulated in circulating monocytes during the acute stage of CAP (Brands et al. 2021)
	IGHV1-69, FLT1, CH17-472G23.1,	Bronchoalveolar lavage (BAL)	IGHV1-69, FLT1, CH17-472G23.1,

(continued)

Table 14.1 (continued)

Omics approach	Biomarkers	Biological sample	Comments
	ATP1B2, FCER2, MUC21, IL13, FCRLB, INHBA, CLEC5A, FAM124A		ATP1B2, FCER2, MUC21, IL13, FCRLB, INHBA, CLEC5A, and FAM124A were upregulated in children with severe *Mycoplasma pneumoniae* pneumonia (MPP) as compared to mild MPP (Wang et al. 2017)
	OSTN-AS1, IL22RA2, COL3A1, CD79A, AICDA, CD19, C1orf141, TNFRSF13C, IGKV2-29, RP11-731F5.2, IGHV4-4, KIRREL, DNASE1L3, COL6A2		OSTN-AS1, IL22RA2, COL3A1, CD79A, AICDA, CD19, C1orf141, TNFRSF13C, IGKV2-29, RP11-731F5.2, IGHV4-4, KIRREL, DNASE1L3, and COL6A2 were downregulated in children with severe *Mycoplasma pneumoniae* pneumonia (MPP) as compared to mild MPP (Wang et al. 2017)
	miR-17-5p, miR-193a-5p	Extracellular vesicles derived from Bronchoalveolar Lavage Fluid (BALF)	Increased expression of miR-17-5p and miR-193a-5p in extracellular vesicles derived from bronchoalveolar lavage fluid may serve as diagnostic biomarkers of pneumonia (Sun et al. 2022)
Proteomics	FCGBP	Serum	Serum FCGBP may serve as a biomarker of severe *Mycoplasma pneumoniae* pneumonia (MPP) (Liu et al. 2022)
	AGT, FGB, CRP, SERPINA1		AGT, FGB, CRP, and SERPINA1 were significantly elevated in refractory *Mycoplasma pneumoniae* pneumonia patients as compared to non-refractory *Mycoplasma*

(continued)

Table 14.1 (continued)

Omics approach	Biomarkers	Biological sample	Comments
			pneumoniae pneumonia patients (Yu et al. 2017)
	Haptoglobin, Immunoglobulin kappa chain, Apo-AI, transthyretin.	Blood Plasma	Haptoglobin and immunoglobulin kappa chain were increased while Apo-AI and transthyretin were decreased in complicated pneumococcal pneumonia patients (Tsai et al. 2009)
	IL-1, IL-4, IL-6, IL-8, IL-9, IL-15, eotaxin, b-FGF, G-CSF, GM-CSF, TNF-α		IL-1, IL-4, IL-6, IL-8, IL-9, IL-15, eotaxin, b-FGF, G-CSF, GM-CSF, and TNF-α were significantly increased in the plasma of severe community-acquired pneumonia (CAP) patients (Haugen et al. 2015)
	S100A8, LTF, ACTN1	Bronchoalveolar lavage fluid (BALF)	Increased expression of S100A8, LTF, and ACTN1 can serve to discriminate ventilator-associated pneumonia (VAP) positive patients as compared to VAP negative group (Nguyen et al. 2013)
	SERPINA1, ITIH4, IGLV6-57, HIST2H3A, HIST1H4, HIST1H2BL, COL1A1, ANXA2, COL2A1, ACAN, DSP, XP32, DSC1, HAPLN1, DSG1	Serum small extracellular vesicles	SERPINA1, ITIH4, IGLV6-57, HIST2H3A, HIST1H4, and HIST1H2BL were upregulated in children suffering from pneumonia as compared to healthy controls. COL1A1, ANXA2, COL2A1, ACAN, DSP, XP32, DSC1, HAPLN1, and DSG1 were downregulated in pneumonia patients as compared to the healthy group (Cheng et al. 2022)
Metabolomics	Sphinganine, *p*-cresol sulphate,	Serum	Sphinganine, *p*-cresol sulphate, and

(continued)

Table 14.1 (continued)

Omics approach	Biomarkers	Biological sample	Comments
	dehydroepiandrosterone sulphate (DHEA-S), lactate		dehydroepiandrosterone sulphate (DHEA-S) were significantly reduced in CAP patients as compared to healthy subjects. Serum lactate and sphinganine levels were positively correlated while DHEA-S was inversely correlated with CURB-65, pneumonia severity index (PSI) and Acute Physiology and Chronic Health Evaluation II (APACHE II) scores (Ning et al. 2018)
	N2, N2-dimethylguanosine, N6-carbamoyl threonyl adenosine, pseudouridine, N1-methyladenosine	Plasma	N2, N2-dimethylguanosine, N6-carbamoyl threonyl adenosine, pseudouridine, and N1-methyladenosine were elevated in non-survivors of pneumonia and sepsis as compared to survivors (Seymour et al. 2013)
	1-Linoleoylglycero-phosphoethanolamine, caffeine, paraxanthine, 2-Oleoylglycero-phosphoethanolamine		1-Linoleoylglycero-phosphoethanolamine, caffeine, paraxanthine, and 2-oleoylglycero-phosphoethanolamine were reduced in non-survivors of pneumonia and sepsis as compared to survivors (Seymour et al. 2013).
	N-acetyl serine, Erythritol, pyridoxate, C-glycosyl tryptophan, Alpha-CEHC glucuronide, N1-methyl-2-pyridone-5-carboxamide, kynurenate, cortisol, urea, erythronate, mannitol, glycerate, fumarate, xylonate,		N-acetyl serine, erythritol, pyridoxate, C-glycosyl tryptophan, alpha-CEHC glucuronide, N1-methyl-2-pyridone-5-carboxamide, kynurenate, cortisol, urea, erythronate, mannitol, glycerate, fumarate, xylonate,

(continued)

Table 14.1 (continued)

Omics approach	Biomarkers	Biological sample	Comments
	glycocholenate sulphate, taurocholate sulphate, 10-heptadecenoate (17:1n7)		glycocholenate sulphate, taurocholate sulphate, and 10-heptadecenoate (17:1n7) were elevated in non-survivors of pneumonia and sepsis as compared to survivors (Seymour et al. 2013)
	Isopropanol, citrate, taurine, 2-oxoglutarate, glutamine, alanine, serine, galactose, lactic acid, glucose, pyroglutamic acid, galactopyranoside, glyceric acid, fructose, beta-alanine, glycerol, glycine, 3-hydroxyl butanoic acid, phenylalanine, aspartate, phenylalanine, formate, 3-hydroxyisovalerate, fumarate, *O*-phosphocholine, adipate, choline, 2-hydroxyisovalerate, proline, ornithine, heptadecanoic acid, phosphoric acid, hexadecanoic acid, octadecenoic acid, octadecane, quinic acid		Isopropanol, citrate, taurine, 2-oxoglutarate, glutamine, alanine, serine, galactose, lactic acid, glucose, pyroglutamic acid, galactopyranoside, glyceric acid, fructose, beta-alanine, glycerol, glycine, 3-hydroxyl butanoic acid, and phenylalanine were reduced in patients suffering from H1N1 influenza pneumonia as compared to ventilated ICU controls. Aspartate, phenylalanine, formate, 3-hydroxyisovalerate, fumarate, *O*-phosphocholine, adipate, choline, 2-hydroxy isovalerate, proline, ornithine, heptadecanoic acid, phosphoric acid, hexadecanoic acid, octadecenoic acid, octadecane, and quinic acid were increased in H1N1 influenza pneumonia patients as compared to ventilated ICU controls (Banoei et al. 2017)
	Citrate, fumarate, 3-methyl,2-isovalerate, alanine, tyrosine, methionine, histidine, 4-hydroxybutyrate, uric acid, asparagine, myoinositol, lysine,		Citrate, fumarate, 3-methyl,2-isovalerate, alanine, tyrosine, methionine, histidine, 4-hydroxybutyrate, uric acid, asparagine, myoinositol, lysine,

(continued)

Table 14.1 (continued)

Omics approach	Biomarkers	Biological sample	Comments
	arabinonic acid, threonine, aspartic acid, threonic acid, acetoacetate, beta-alanine, formate, dimethylamine, carnitine, glycine, gulonic acid, pentadecane, 2-amino butanoic acid, alkane, quinic acid, benzoic acid		arabinonic acid, threonine, aspartic acid, and threonic acid were decreased in H1N1 pneumonia as compared to positive bacterial culture pneumonia. Acetoacetate, beta-alanine, formate, dimethylamine, carnitine, glycine, gulonic acid, pentadecane, 2-amino butanoic acid, alkane, quinic acid, and benzoic acid were increased in H1N1 pneumonia as compared to positive bacterial culture pneumonia (Banoei et al. 2017)
	2-Aminobutyrate, acetoacetate, 2-hydroxybutyrate, arginine, 3-hydroxybutyrate, methionine, pentadecane, 4-amino benzoic acid, hydroxylamine, 2-oxoglutarate, dimethylamine isopropanol, carnitine, 2-hydroxisovalerate, lactate, phenylalanine, acetate, tyrosine, threonic acid, dodecane, decanoic acid, 2-amino butanoic acid, valine, glycerol		2-Aminobutyrate, acetoacetate, 2-hydroxybutyrate, arginine, 3-hydroxybutyrate, methionine, pentadecane, 4-amino benzoic acid and hydroxylamine were increased in H1N1 pneumonia non-survivors as compared to survivors. 2-oxoglutarate, dimethylamine isopropanol, carnitine, 2-hydroxisovalerate, lactate, phenylalanine, acetate, tyrosine, threonic acid, dodecane, decanoic acid, 2-amino butanoic acid, valine, and glycerol were decreased in H1N1 pneumonia non-survivors as compared to survivors (Banoei et al. 2017)

(continued)

Table 14.1 (continued)

Omics approach	Biomarkers	Biological sample	Comments
	Uric acid, hypoxanthine, glutamic acid, L-tryptophan, adenosine-5′-diphosphate (ADP)		Increased plasma levels of uric acid, hypoxanthine, and glutamic acid were noted in pneumonia patients. L-tryptophan and ADP were reduced in plasma of pneumonia patients (Laiakis et al. 2010)
	Uric acid, L-histidine	Urine	Uric acid and L-histidine were significantly reduced in the urine of pneumonia patients (Laiakis et al. 2010)
	3-Succinoylpyridine		3-Succinoylpyridine level was elevated in all ventilator-associated pneumonia (VAP) patients. This increased urinary level of 3-succinoylpyridine was more pronounced in case of *Pseudomonas aeruginosa* VAP (Jongers et al. 2022).
Metagenomics	*Rothia* spp., *Lactobacillus* spp., *Streptococcus* spp.	Oropharyngeal samples	The nasal microbial diversity is significantly reduced in pneumonia patients. Domination of *Rothia* spp., *Lactobacillus* spp., and *Streptococcus* spp. was noted in pneumonia patients (de Steenhuijsen Piters et al. 2016)
	Proteobacteria	Sputum	Dominance of *Proteobacteria* was noted in HIV patients with pneumonia (Ueckermann and Ehlers 2021)
	Streptococcus pneumoniae, *Haemophilus influenzae*, *Moraxella catarrhalis*	Hypopharyngeal aspirates	Colonization of neonatal airways with *Streptococcus pneumoniae*, *Haemophilus influenzae*, and *Moraxella catarrhalis* was found to be associated with

(continued)

Table 14.1 (continued)

Omics approach	Biomarkers	Biological sample	Comments
			increased risk of pneumonia and bronchiolitis in early stage of life (Vissing et al. 2013)
	Pseudomonas aeruginosa, *Staphylococcus aureus*, *Klebsiella pneumoniae*, *Haemophilus influenzae*	Bronchoalveolar lavage fluid (BALF)	Reduced alpha diversity was noted in pneumonia patients. Increased abundance of *Pseudomonas*, *Klebsiella*, *Staphylococcus*, and *Haemophilus* was noted in ventilator-associated pneumonia (VAP) patients with positive culture (Fenn et al. 2022)
	Burkholderia spp., *Bacillales*, *Staphylococcus aureus*, *Pseudomonadales* spp.	Tracheal aspirates	*Burkholderia* spp., *Bacillales*, *Staphylococcus aureus*, and *Pseudomonadales* spp. were identified as vital players involved in the development of ventilator-associated pneumonia (VAP) (Zakharkina et al. 2017)
	Burkholderia spp., *Alcaligenes* spp., *Pseudomonas* spp., *Massilia* spp., *Flavobacterium* spp., *Enterobacter* spp.	Deep endotracheal secretions	*Burkholderia* spp., *Alcaligenes* spp., *Pseudomonas* spp., *Massilia* spp., *Flavobacterium* spp., and *Enterobacter* spp. in the lower respiratory tract of *Pseudomonas aeruginosa* ventilator-associated pneumonia (VAP) patients were found to be associated with severity of VAP (Qi et al. 2018)

This table enlists various host biomarkers of pneumonia identified using different omics approaches

(G > C) and rs41297579 (G > A) were found to increase the risk of community-acquired pneumonia (CAP) in children (Liu and Xu 2020). SNP of CYP1A1 (rs2606345) was also found to be strongly associated with susceptibility to CAP (Salnikova et al. 2013). Another study identified rs4340 polymorphism of CD143 as a risk factor for pneumonia (Wang et al. 2015). Two SNPs of HMGB1, namely rs1412125 and rs2249825 were found to be associated with susceptibility, severity, and inflammatory response in pneumonia (Song et al. 2019). SNPs of CRP (rs1205), IL-6 (rs1800797), and IL-10 (rs1800872) were also found to be associated with the susceptibility and severity of CAP (Chou et al. 2016). The C allele of rs1800795 of IL6 has been related to severe shock or sepsis or severe inflammatory response in CAP patients (Smelaya et al. 2016). Besides, functional expression quantitative trait locus (eQTL)-SNPs in ACE (rs4316 and rs4353) and CD276 (rs8032531) were identified to be significantly linked with susceptibility to *Mycoplasma pneumoniae* pneumonia in children (Dong et al. 2022). Other SNPs associated with pneumonia have been listed in Table 14.1.

14.7 Epigenomics

There is a scarcity of knowledge regarding the changes in host DNA methylation pattern in pneumonia patients, especially community-acquired pneumonia (CAP) patients. Upregulation of a histone demethylase KDM6B and downregulation of demethylases with opposing roles (like C14orf169, KDM1A, and KDM5D) were found to be associated with increased severity in CAP patients, suggesting a potential role of altered methylation in severe CAP patients (Hopp et al. 2018). Epigenetic studies performed in circulating monocytes from acute stage CAP patients highlighted differential methylation pattern in acute stage of CAP. Higher levels of methylation were observed in DNase hypersensitive sites on Chromosome 22 located within the genomic region dense in C to U RNA-editing cytidine deaminases and apolipoprotein B mRNA editing enzymes. On the contrary, reduced methylation was noted in DNase hypersensitive sites on Chromosome 8 located in the exon 4 of PARP10 (Brands et al. 2021).

14.8 Transcriptomics

New age transcriptomics tools have been used to study the differential gene expression in pneumonia patients as compared to healthy subjects and highlight the differences in gene expression among the various types of pneumonia. Comparative gene expression analysis highlighted 75% similarity in blood leukocyte response in community-acquired pneumonia (CAP) and hospital-acquired pneumonia (HAP) patients. Increased expression of genes involved in cell–cell junction remodelling, cell-cell adhesion, and diapedesis was noted in HAP patients (van Vught et al. 2016). A study conducted with extracellular vesicles (EVs) derived from sera highlighted differences of miRNA expression of CAP patients relative to healthy subjects and sepsis patients. miR-193a-5p, miR-542-3p, and miR-1246 were significantly reduced in both CAP and sepsis patients as compared to healthy volunteers. Although expression of miR-193a-5p and miR-542-3p did not show significant difference between CAP and sepsis patients, miR-1246 was significantly reduced in sepsis patients as compared to CAP patients (Hermann et al. 2020). Another gene expression study identified five genes, namely CCL4, TIMP metallopeptidase inhibitor 1, intercellular adhesion molecule 1, plasminogen activator, urokinase receptor and cathepsin B as key genes to be associated with pneumonia caused by Gram-positive bacteria (Jia et al. 2018). Other differentially expressed host transcriptomic signatures associated with pneumonia have been listed in Table 14.1.

14.9 Proteomics

Plasma proteomic study highlighted the importance of three proteins, namely haptoglobin, tumour necrosis factor receptor 2, or IL-10 and tissue inhibitor of metalloproteinases 1 in classifying paediatric pneumonia patients based on bacterial, malarial, and viral aetiologies (Valim et al. 2016). Another study showed elevation of YKL-40 in plasma of community-acquired pneumonia (CAP) patients. Plasma YKL-40 may serve as a diagnostic biomarker of CAP and may aid in the detection of disease severity in CAP patients (Wang et al. 2013). Serum YKL-40 and CCL18 level were identified to predict pneumonia caused by atypical pathogens and assess disease severity (Spoorenberg et al. 2018). Persistent increased level of CRP was associated with increased risk of ventilator-associated pneumonia (VAP). The same study also highlighted the significance of procalcitonin in the diagnosis of VAP (Ramirez et al. 2008). Studies have also shown the role of TREM-1 in CAP diagnosis and prognosis (Tejera et al. 2007). Another study showed increased serum levels of acute phase proteins, namely IL-1RA, IL-6, IL-8, IL-10, and MCP in pneumonia patients at the time of hospitalization (Endeman et al. 2011). HP, THBS1, and SAA1/2 were identified as prognostic biomarker candidates for determining poor outcome in infant pneumonia patients suffering from sepsis (Luo et al. 2022). Plasma proteomic studies have also highlighted an impairment of lipid metabolism in hospital-acquired pneumonia (HAP) patients having sepsis. APOA4, APOB, APOC1, APOL1, SAA4, and PON1 were found to be reduced in sepsis due to HAP (Sharma et al. 2019). Other proteomic biomarkers of pneumonia have been listed in Table 14.1.

14.10 Metabolomics

Several metabolomics studies were conducted to determine the changes in the global metabolome in pneumonia patients. Increased concentration of androstenedione, pregnanediol, 11-b-OH-androsterone, tetra hydrocortisone, cortolone-3-glucuronide, and 18-hydroxycortisol were noted in urine samples obtained from pneumococcal paediatric community-acquired pneumonia patients (Del Borrello et al. 2020). Another study showed significant increase of lysophosphatidic acid, lysophosphatidylcholine, phosphatidic acid, phosphatidylcholine, glycerol phosphoserines, phosphatidylethanolamines, phosphatidate, 1-Acyl-sn-glycero-3-phosphocholine, palmityl-CoA, and diacylglycerol in plasma samples derived from *Mycoplasma pneumoniae* pneumonia patients (Li et al. 2022). Some of the other host metabolic signatures of pneumonia have been documented in Table 14.1.

14.11 Metagenomics

Clinical metagenomics based on next generation sequencing (NGS) yields detection of greater number of pathogens and therefore ensures improved diagnosis and better clinical outcomes in pneumonia (Xie et al. 2021). Metagenomics NGS (mNGS) also enables antibiotic de-escalation in pneumonia patients and improves patient care (Zhou et al. 2021; Zhan et al. 2021). NGS by both Illumina and Nanopore were used to detect pathogens in BALF of suspected community-acquired pneumonia (CAP) patients. Superior detection of fungi was achieved using Nanopore, whereas bacteria and *Chlamydia psittaci* were better detected by Illumina. *Candida albicans* was the predominantly detected fungi in BALF of CAP patients. Viruses detected in CAP patients include human alpha herpesvirus 1, Epstein–Barr virus, and human cytomegalovirus. *Klebsiella aerogenes*, *Streptococcus pseudopneumoniae*, and *Streptococcus milleri* were among the bacterial species detected in CAP patients (Zhang et al. 2022). *Candida* spp. and *Staphylococcus* spp. were identified as probable prognostic biomarker candidates for community-acquired pneumonia (CAP) (Aishwarya and Gunasekaran 2022). Few other microbial biomarkers of pneumonia have been tabulated in Table 14.1. Understanding the significance of respiratory microbial dysbiosis in pneumonia may pave the way for the development of targeted host-directed therapy based on microbial modulation and thereby may lead to improved clinical outcomes in pneumonia patients (Roquilly et al. 2019).

14.12 Bioinformatics

There are a few dedicated patient databases of pneumonia for identifying patients that need medical emergency and for developing strategies of treatment. The excellence network for Community-Acquired Pneumonia (CAPNETZ) database started in 2001 is the most comprehensive CAP database worldwide (Pletz et al. 2022). CAPNETZ has specific inclusion criteria (age $\geq$ 18 years, pulmonary infiltrate on chest X-ray, clinical symptoms of cough) and exclusion criteria (nosocomial pneumonia, human immunodeficiency virus-infected or had active tuberculosis) (Braeken et al. 2021). CAP (community-acquired pneumonia) and HAP (hospital-acquired pneumonia) patients' data were compiled in a Pneumonia Research Partnership to Assess WHO Recommendations (PREPARE) database and includes data from 285,839 children with pneumonia (244,323 in the hospital and 41,516 in the community) (Martin et al. 2022). A database of adult pneumonia patients in Dubai, United Arab Emirates (UAE) was developed for identification of the significant healthcare burden of this disease in the UAE (Al Dallal et al. 2021). The gene expression profiles in pneumonia patients are compiled in BioGPS and are available at http://biogps.org/dataset/tag/pneumonia/ (Wu et al. 2016). These databases are useful in the management and treatment of the disease.

14.13 Medical Imaging or Radiomics

Along with the clinical features, chest X-ray plays a fundamental role in the diagnosis of typical and atypical pneumonia (Sharma et al. 2007). Computed tomography is only required to suggest specific pathogens and to differentiate them from other diseases (Nambu et al. 2014). Typical pneumonia X-ray images include peri-bronchial nodules, especially tree-in-bud appearance, and concavity of the opacities, whereas in atypical pneumonia there is a lack of typical findings of lobar consolidation (Franquet 2001; Dueck et al. 2021). The chest X-ray image data of pneumonia from the Radiological Society of North America (RSNA) were used to develop an ensemble-based deep learning model that can predict this disease with high (86.5%) accuracy (Kundu et al. 2021). These AI-based models can play an important role in the diagnosis and management of pneumonia.

14.14 Multi-omics and Data Integration

Multi-omics data analyses in pneumonia are mainly focused on paediatric patients. Imaging and serum markers like white blood cell (WBC) and C-reactive protein (CRP) were integrated for the diagnosis of Neonatal Infectious Pneumonia (Li 2020). In another study, an association of oropharyngeal microbiome and metabolome was observed in paediatric patients with influenza A virus pneumonia (Hu et al. 2022). There is a need for more research work to explore the power of the integration of multi-omics approaches to identify clinical biomarkers and drug targets of this age-old disease.

14.15 Present Therapeutics

Primarily, antibiotics are used as the mainstay therapy for pneumonia. The antibiotic used is largely guided by several host and pathogen factors along with disease severity (https://www.ncbi.nlm.nih.gov/books/NBK534295/; Torres et al. 2021; https://www.ncbi.nlm.nih.gov/books/NBK525774/). The different antibiotics used for the treatment of pneumonia; especially bacterial pneumonia have been tabulated in Table 14.2.

The antibiotic therapy mostly lasts for a period of 5–7 days. However, in case of delayed response to the therapy, the antibiotic therapy can be extended. Severe pneumonia patients are often treated with oral or injectable steroids. Oxygen supplementation and mechanical ventilation are given to critically severe pneumonia patients. Virostatic agents are at times used for viral pneumonia (https://www.ncbi.nlm.nih.gov/books/NBK534295/; https://www.ncbi.nlm.nih.gov/books/NBK52

Table 14.2 Antibiotics for treatment of pneumonia

Sl. no.	Class of antibiotics	Antibiotic names	Mode of action
1.	Macrolides	Azithromycin, clarithromycin	Blocks bacterial protein synthesis by binding to 50S ribosomal subunit (Mantero et al. 2017; https://www.ncbi.nlm.nih.gov/books/NBK551495/)
2.	Tetracycline	Doxycycline	Binds to bacterial 30S ribosomal subunit and prevents the binding of amino acyl tRNA to the ribosome acceptor site, thus blocking bacterial protein translation (Mantero et al. 2017; https://www.ncbi.nlm.nih.gov/books/NBK549905/)
3.	Fluoroquinolones	Levofloxacin, moxifloxacin, ciprofloxacin	Interacts with bacterial DNA gyrase and topoisomerase IV. Binds to enzyme DNA complex and inhibits bacterial DNA replication (Mantero et al. 2017; Hooper 2000; Hooper 2001)
4.	Beta-lactam antibiotics	Amoxicillin	Disrupts synthesis of peptidoglycan layer of bacterial cell wall (Mantero et al. 2017; Suarez and Gudiol 2009; https://www.ncbi.nlm.nih.gov/books/NBK545311/)
5.	Combination of beta-lactam antibiotics and beta-lactamase inhibitor	Amoxicillin/clavulanate	Amoxicillin blocks synthesis of peptidoglycan layer of bacterial cell wall Clavulanate inhibits beta-lactamases and prevents development of antimicrobial resistance to amoxicillin (Mantero et al. 2017; https://www.ncbi.nlm.nih.gov/books/NBK545311/)
6.	Cephalosporins (second generation)	Cefuroxime	Binds to peptidoglycan transpeptidase [also called penicillin-binding proteins (PBPs)] and prevents peptidoglycan crosslinking Blocks bacterial cell wall synthesis Has increased coverage against *H. influenza*, as compared to first-generation cephalosporins (Mantero et al. 2017; https://www.ncbi.nlm.nih.gov/books/NBK551517/)
7.	Cephalosporins (third generation)	Ceftriaxone, cefdinir, ceftazidime, cefpodoxime	Mode of action is similar to second-generation cephalosporins, but has increased coverage extended to gram-negative bacteria (https://www.ncbi.nlm.nih.gov/books/NBK551517/ ; Lupia et al. 2020)

(continued)

Table 14.2 (continued)

Sl. no.	Class of antibiotics	Antibiotic names	Mode of action
8.	Cephalosporins (fourth generation)	Cefepime	Mode of action is similar to third-generation cephalosporins, but, has better drug penetrability to outer membrane of gram-negative bacteria and increased coverage extended to beta-lactamase producing gram-negative bacilli (https://www.ncbi.nlm.nih.gov/books/NBK551517/; Lupia et al. 2020)
9.	Sulphonamides	Co-trimoxazole (sulfamethoxazole-trimethoprim)	Inhibits bacterial folate metabolism and folate production (Ovung and Bhattacharyya 2021) Used for pneumonia caused by yeast like fungus *Pneumocystis jirovecii* (Hughes and Killmar 1996)
10.	Lincosamide antibiotic (Lincomycin)	Clindamycin	Binds to 23S RNA of the bacterial 50S ribosomal subunit, thereby inhibiting bacterial protein synthesis (Duncan and Jeans 1965; https://www.ncbi.nlm.nih.gov/books/NBK519574/; Smieja 1998)
11.	Aminoglycosides	Amikacin	Binds to amino acyl site of 16S ribosomal RNA within the bacterial 30S ribosomal subunit, thereby inhibiting bacterial protein synthesis (Mantero et al. 2017; Krause et al. 2016)

This table enlists the different antimicrobial therapy used for pneumonia

5774/). Follow-up of pneumonia patients is very crucial for monitoring and preventing recurrence of pneumonia.

The emergence of antibiotic resistance has been a major issue in tackling pneumonia. The development of resistance to β-lactam was handled by combinatorial use of β-lactamase inhibitors or by use of new generations of β-lactam (Torres et al. 2021). However, genetic evolution of microbes and concomitant emergence of drug resistance are continuous processes which continue to create hurdles in combating pneumonia.

Susceptible or vulnerable individuals are often subjected to vaccination to confer protection against pneumonia. The different vaccines used to prevent the occurrence of viral and bacterial pneumonia have been documented in Table 14.3.

The advent of pneumococcal vaccines had played a significant role in the prevention of pneumonia. The pneumococcal vaccine plays a vital role in triggering active immune response against conjugate and capsular polysaccharides of *Streptococcus pneumoniae*. Immunity develops within 2–3 weeks post vaccination and lasts for about 5 years. Reimmunization is done after the immune response wanes off

Table 14.3 Vaccines used to confer protection from pneumonia caused by bacteria and virus

Sl. no.	Microorganism	Disease	Vaccine	Description	Vaccine types and names
1.	*Haemophilus influenzae* Type b	Hib infection	Hib Vaccine	Polysaccharide-protein conjugate vaccine polyribosyl ribitol phosphate (PRP) is conjugated to protein to elicit enhanced immune response than PRP alone It is a type of inactivated bacterial vaccine	*Monovalent Vaccines* [ActHIB (PRP-T), Hiberix (PRP-T), PedvaxHIB (PRP-OMB)] *Combination Vaccines* [Pentacel and Vaxelis] (Gilsdorf 2021; Congdon et al. 2021)
2.	Influenza viruses	Influenza (flu)	Flu vaccine	Inactivated viral vaccine, live attenuated viral vaccine	*Intramuscular injection of trivalent or quadrivalent inactivated viral vaccine (TIV/QIV)* [IIV3, IIV4, RIV4] *Nasal spray of live attenuated viral vaccine* [LAIV, Q/LAIV] (Schotsaert and Garcia-Sastre 2017; Heo et al. 2018; Mohn et al. 2018)
3.	Measles virus	Measles	Measles vaccine	Live attenuated viral vaccine	*Standalone Measles vaccine* *Combination Vaccines* [MR vaccine (Measles and rubella combined vaccine), MMR vaccine (mumps, measles and rubella combined vaccine), MMRV vaccine (Mumps, measles, rubella and varicella combined vaccine)] (Madhi et al. 2008; https://www.ncbi.nlm.nih.gov/books/NBK554450/)
4.	*Bordetella pertussis*	Whooping cough (Pertussis)	Pertussis vaccine	Whole-cell vaccines, acellular vaccines.	*Diphtheria, tetanus, and pertussis (DTaP) vaccines* Daptacel, Infanrix, Kinrix, Pediarix, Pentacel, Vaxelis, Quadracel *Tetanus, diphtheria, and pertussis (Tdap) vaccines*

					Adacel, Boostrix (Madhi et al. 2008; Robbins et al. 2009; https://www.ncbi.nlm.nih.gov/books/NBK545173/; https://www.ncbi.nlm.nih.gov/books/NBK572038/)
5.	Varicella-zoster virus (VZV)	Chickenpox	Chickenpox/ Varicella vaccine	Live weakened viral vaccine	*MMRV vaccine* (Mumps, measles, rubella, and varicella combined vaccine) (https://www.ncbi.nlm.nih.gov/books/NBK441946/ ; Mohsen and McKendrick 2003)
6.	*Streptococcus pneumoniae*	Pneumococcal disease	Pneumococcal vaccine	Pneumococcal conjugate vaccine—purified pneumococcal capsular polysaccharide conjugated to a carrier protein, pneumococcal polysaccharide vaccine (PPSV)—derived from capsular polysaccharide	*Pneumococcal conjugate vaccines* [PCV13, PCV15, PCV20] *Pneumococcal polysaccharide vaccine* [PPSV23] (Berical et al. 2016; https://www.ncbi.nlm.nih.gov/books/NBK507794/)

This table enlists the different vaccines developed for the different viruses and bacteria that can cause pneumonia (https://www.ncbi.nlm.nih.gov/books/NBK507794/). This vaccination shows efficacy in preventing pneumococcal community-acquired pneumonia (CAP) among the elderly individuals, infants, and young children (Berical et al. 2016; Heo et al. 2022).

Although vaccinations have helped to tackle pneumonia to a large extent, the disease still continues to be associated with a high mortality rate. Therefore, exploring the multi-omics tools for identifying novel biomarkers and drug targets of pneumonia can be of prime importance in better management of the disease.

14.16 Future Perspectives

Advancements in omics tools have elaborated our knowledge about the pathogenesis and various other facets of pneumonia. However, several knowledge gaps are yet to be bridged, and various scientific questions still remain to be answered about the different types of pneumonia. The knowledge about changes in epigenetic modification of host genes in pneumonia patients is extremely limited (Hopp et al. 2018). High-throughput epigenomic tools should be elaborately used to understand the association of alteration in gene methylation and acetylation with pneumonia and its severity. Multi-layered omics studies can help in elucidating the defense mechanism against pneumonia and assist in mechanistically connecting lung infection with extrapulmonary patho-physiologies. Comparative analyses of host immune response towards the different pathogens that cause pneumonia can highlight common as well as pathogen specific host responses, which can be exploited for designing novel therapeutics. Multi-omics studies can also aid in identifying biomarkers of disease severity, disease prognosis, disease relapse, and mortality in pneumonia patients. Efforts should also be made for designing mathematical models and prediction tools for predicting the susceptibility or resistance to recurrent pneumonia based on clinical information obtained from electronic health records and various reliable biological omics profiles (Dela Cruz et al. 2018). Deep learning approaches may be used for the diagnosis of pathogen specific pneumonia based on radiographic findings and radiological patterns. Integration of clinical findings and omics data can further enrich such approaches. Emergence of drug resistance is a huge problem encountered while tackling pneumonia. Identification of novel drug targets through omics studies and development of new antibiotics can aid in combating drug resistance. Misuse and overuse of antibiotics primarily lead to emergence of drug resistance. Better understanding of the different inflammatory phenotypes associated with pneumonia can aid in designing personalized therapy for curing pneumonia and minimize the misuse and overuse of antibiotics (Pletz et al. 2022). Thus, it may be summarized that the above-mentioned futuristic research goals shall pave the way for better diagnosis, risk stratification, and management of pneumonia and bring about reduction in pneumonia associated mortality rate.

References

Aishwarya S, Gunasekaran K (2022) Meta-analysis of the microbial biomarkers in the gut-lung crosstalk in COVID-19, community-acquired pneumonia and Clostridium difficile infections. Lett Appl Microbiol 75(5):1293–1306

Al Dallal SAM, Farghaly M, Ghorab A, Elaassar M, Haridy H, Awad N, Chickballapur Ramachandrachar B, Natarajan A (2021) Real-world evaluation of costs of illness for pneumonia in adult patients in Dubai-A claims database study. PLoS One 16(9):e0256856

Almansa R, Nogales L, Martin-Fernandez M, Batlle M, Villareal E, Rico L, Ortega A, Lopez-Campos G, Andaluz-Ojeda D, Ramirez P et al (2018) Transcriptomic depression of immunological synapse as a signature of ventilator-associated pneumonia. Ann Transl Med 6(21):415

Banoei MM, Vogel HJ, Weljie AM, Kumar A, Yende S, Angus DC, Winston BW (2017) Canadian critical care translational biology G: plasma metabolomics for the diagnosis and prognosis of H1N1 influenza pneumonia. Crit Care 21(1):97

Barbagelata E, Cilloniz C, Dominedo C, Torres A, Nicolini A, Solidoro P (2020) Gender differences in community-acquired pneumonia. Minerva Med 111(2):153–165

Berical AC, Harris D, Dela Cruz CS, Possick JD (2016) Pneumococcal vaccination strategies. an update and perspective. Ann Am Thorac Soc 13(6):933–944

Bouhemad B, Dransart-Raye O, Mojoli F, Mongodi S (2018) Lung ultrasound for diagnosis and monitoring of ventilator-associated pneumonia. Ann Transl Med 6(21):418

Bradley JS, Byington CL, Shah SS, Alverson B, Carter ER, Harrison C, Kaplan SL, Mace SE, McCracken GH Jr, Moore MR et al (2011) The management of community-acquired pneumonia in infants and children older than 3 months of age: clinical practice guidelines by the Pediatric Infectious Diseases Society and the Infectious Diseases Society of America. Clin Infect Dis 53(7):e25–e76

Braeken DCW, Essig A, Panning M, Hoerster R, Nawrocki M, Dalhoff K, Suttorp N, Welte T, Pletz MW, Witzenrath M et al (2021) Shift in bacterial etiology from the CAPNETZ cohort in patients with community-acquired pneumonia: data over more than a decade. Infection 49(3):533–537

Brands X, Haak BW, Klarenbeek AM, Butler J, Uhel F, Qin W, Otto NA, Jakobs ME, Faber DR, Lutter R et al (2021) An epigenetic and transcriptomic signature of immune tolerance in human monocytes through multi-omics integration. Genome Med 13(1):131

Castillo CG, Kauffman CA, Miceli MH (2016) Blastomycosis. Infect Dis Clin N Am 30(1): 247–264

Chen HH, Shaw DM, Petty LE, Graff M, Bohlender RJ, Polikowsky HG, Zhong X, Kim D, Buchanan VL, Preuss MH et al (2021) Host genetic effects in pneumonia. Am J Hum Genet 108(1):194–201

Cheng J, Ji D, Yin Y, Wang S, Song K, Pan Q, Zhang Q, Yang L (2022) Proteomic profiling of serum small extracellular vesicles reveals immune signatures of children with pneumonia. Transl Pediatr 11(6):891–908

Chou SC, Ko HW, Lin YC (2016) CRP/IL-6/IL-10 single-nucleotide polymorphisms correlate with the susceptibility and severity of community-acquired pneumonia. Genet Test Mol Biomarkers 20(12):732–740

Cilloniz C, Ewig S, Polverino E, Marcos MA, Esquinas C, Gabarrus A, Mensa J, Torres A (2011) Microbial aetiology of community-acquired pneumonia and its relation to severity. Thorax 66(4):340–346

Claessens YE, Debray MP, Tubach F, Brun AL, Rammaert B, Hausfater P, Naccache JM, Ray P, Choquet C, Carette MF et al (2015) Early chest computed tomography scan to assist diagnosis and guide treatment decision for suspected community-acquired pneumonia. Am J Respir Crit Care Med 192(8):974–982

Congdon M, Hong H, Young RR, Cunningham CK, Enane LA, Arscott-Mills T, Banda FM, Chise M, Motlhatlhedi K, Feemster K et al (2021) Effect of Haemophilus influenzae type b and 13-valent pneumococcal conjugate vaccines on childhood pneumonia hospitalizations and deaths in Botswana. Clin Infect Dis 73(2):e410–e416

Connolly JE Jr, McAdams HP, Erasmus JJ, Rosado-de-Christenson ML (1999) Opportunistic fungal pneumonia. J Thorac Imaging 14(1):51–62
Corrales-Medina VF, Musher DM, Shachkina S, Chirinos JA (2013) Acute pneumonia and the cardiovascular system. Lancet 381(9865):496–505
D'Amato M, Rea G, Carnevale V, Grimaldi MA, Saponara AR, Rosenthal E, Maggi MM, Dimitri L, Sperandeo M (2017) Assessment of thoracic ultrasound in complementary diagnosis and in follow up of community-acquired pneumonia (cap). BMC Med Imaging 17(1):52
Dandachi D, Rodriguez-Barradas MC (2018) Viral pneumonia: etiologies and treatment. J Investig Med 66(6):957–965
Davies SF (1994) Fungal pneumonia. Med Clin North Am 78(5):1049–1065
Davydow DS, Hough CL, Levine DA, Langa KM, Iwashyna TJ (2013) Functional disability, cognitive impairment, and depression after hospitalization for pneumonia. Am J Med 126(7): 615–624 e615
de Steenhuijsen Piters WA, Huijskens EG, Wyllie AL, Biesbroek G, van den Bergh MR, Veenhoven RH, Wang X, Trzcinski K, Bonten MJ, Rossen JW et al (2016) Dysbiosis of upper respiratory tract microbiota in elderly pneumonia patients. ISME J 10(1):97–108
Del Borrello G, Stocchero M, Giordano G, Pirillo P, Zanconato S, Da Dalt L, Carraro S, Esposito S, Baraldi E (2020) New insights into pediatric community-acquired pneumonia gained from untargeted metabolomics: a preliminary study. Pediatr Pulmonol 55(2):418–425
Dela Cruz CS, Wunderink RG, Christiani DC, Cormier SA, Crothers K, Doerschuk CM, Evans SE, Goldstein DR, Khatri P, Kobzik L et al (2018) Future research directions in pneumonia. NHLBI Working Group Report. Am J Respir Crit Care Med 198(2):256–263
Dong Y, Gao Y, Luo C, Wu N, Cheng Z, Qiu A, Zhou Y, Zhang W, Chu M, Chang Q (2022) Novel functional eQTL-SNPs associated with susceptibility to mycoplasma pneumoniae pneumonia in children. Front Public Health 10:899045
Dremsizov T, Clermont G, Kellum JA, Kalassian KG, Fine MJ, Angus DC (2006) Severe sepsis in community-acquired pneumonia: when does it happen, and do systemic inflammatory response syndrome criteria help predict course? Chest 129(4):968–978
Dueck NP, Epstein S, Franquet T, Moore CC, Bueno J (2021) Atypical pneumonia: definition, causes, and imaging features. Radiographics 41(3):720–741
Duncan IB, Jeans B (1965) Lincomycin in hospital practice. Can Med Assoc J 93(13):685–691
Endeman H, Meijvis SC, Rijkers GT, van Velzen-Blad H, van Moorsel CH, Grutters JC, Biesma DH (2011) Systemic cytokine response in patients with community-acquired pneumonia. Eur Respir J 37(6):1431–1438
Fenn D, Abdel-Aziz MI, van Oort PMP, Brinkman P, Ahmed WM, Felton T, Artigas A, Povoa P, Martin-Loeches I, Schultz MJ et al (2022) Composition and diversity analysis of the lung microbiome in patients with suspected ventilator-associated pneumonia. Crit Care 26(1):203
Franquet T (2001) Imaging of pneumonia: trends and algorithms. Eur Respir J 18(1):196–208
Fukuyama H, Yamashiro S, Kinjo K, Tamaki H, Kishaba T (2014) Validation of sputum Gram stain for treatment of community-acquired pneumonia and healthcare-associated pneumonia: a prospective observational study. BMC Infect Dis 14:534
Galvan-Roman JM, Lancho-Sanchez A, Luquero-Bueno S, Vega-Piris L, Curbelo J, Manzaneque-Pradales M, Gomez M, de la Fuente H, Ortega-Gomez M, Aspa J (2020) Usefulness of circulating microRNAs miR-146a and miR-16-5p as prognostic biomarkers in community-acquired pneumonia. PLoS One 15(10):e0240926
Gilsdorf JR (2021) Hib vaccines: their impact on Haemophilus influenzae type B disease. J Infect Dis 224(12 Suppl 2):S321–S330
Giuliano KK, Baker D (2020) Sepsis in the context of nonventilator hospital-acquired pneumonia. Am J Crit Care 29(1):9–14
Haugen J, Chandyo RK, Brokstad KA, Mathisen M, Ulak M, Basnet S, Valentiner-Branth P, Strand TA (2015) Cytokine concentrations in plasma from children with severe and non-severe community acquired pneumonia. PLoS One 10(9):e0138978

Heo JY, Song JY, Noh JY, Choi MJ, Yoon JG, Lee SN, Cheong HJ, Kim WJ (2018) Effects of influenza immunization on pneumonia in the elderly. Hum Vaccin Immunother 14(3):744–749

Heo JY, Seo YB, Choi WS, Kim EJ, Jeong HW, Lee J, Yoon JG, Noh JY, Cheong HJ, Kim WJ et al (2022) Effectiveness of pneumococcal vaccination against pneumococcal pneumonia hospitalization in older adults: a prospective, Test-negative study. J Infect Dis 225(5):836–845

Hermann S, Brandes F, Kirchner B, Buschmann D, Borrmann M, Klein M, Kotschote S, Bonin M, Reithmair M, Kaufmann I et al (2020) Diagnostic potential of circulating cell-free microRNAs for community-acquired pneumonia and pneumonia-related sepsis. J Cell Mol Med 24(20): 12054–12064

Hespanhol V, Barbara C (2020) Pneumonia mortality, comorbidities matter? Pulmonology 26(3): 123–129

Hooper DC (2000) Mechanisms of action and resistance of older and newer fluoroquinolones. Clin Infect Dis 31(Suppl 2):S24–S28

Hooper DC (2001) Mechanisms of action of antimicrobials: focus on fluoroquinolones. Clin Infect Dis 32(Suppl 1):S9–S15

Hopp L, Loeffler-Wirth H, Nersisyan L, Arakelyan A, Binder H (2018) Footprints of sepsis framed within community acquired pneumonia in the blood transcriptome. Front Immunol 9:1620

Hu Q, Liu B, Fan Y, Zheng Y, Wen F, Yu U, Wang W (2022) Multi-omics association analysis reveals interactions between the oropharyngeal microbiome and the metabolome in pediatric patients with influenza A virus pneumonia. Front Cell Infect Microbiol 12:1011254

Hughes WT, Killmar J (1996) Monodrug efficacies of sulfonamides in prophylaxis for Pneumocystis carinii pneumonia. Antimicrob Agents Chemother 40(4):962–965

Iwashyna TJ, Ely EW, Smith DM, Langa KM (2010) Long-term cognitive impairment and functional disability among survivors of severe sepsis. JAMA 304(16):1787–1794

Jia R, Yang J, Cui Y, Guo D, Li T (2018) Gene expression analysis for pneumonia caused by Gram-positive bacterial infection. Exp Ther Med 15(4):3989–3996

Jongers B, Hotterbeekx A, Bielen K, Vervliet P, Boddaert J, Lammens C, Fransen E, Baggerman G, Covaci A, Goossens H et al (2022) Identification of potential urinary metabolite biomarkers of Pseudomonas aeruginosa ventilator-associated pneumonia. Biomark Insights 17: 11772719221099131

Krause KM, Serio AW, Kane TR, Connolly LE: Aminoglycosides: an overview. Cold Spring Harb Perspect Med 2016, 6(6)

Kundu R, Das R, Geem ZW, Han GT, Sarkar R (2021) Pneumonia detection in chest X-ray images using an ensemble of deep learning models. PLoS One 16(9):e0256630

Laiakis EC, Morris GA, Fornace AJ, Howie SR (2010) Metabolomic analysis in severe childhood pneumonia in the Gambia, West Africa: findings from a pilot study. PLoS One 5(9)

Laursen CB, Sloth E, Lambrechtsen J, Lassen AT, Madsen PH, Henriksen DP, Davidsen JR, Rasmussen F (2013) Diagnostic performance of chest X-ray for the diagnosis of community acquired pneumonia in acute admitted patients with respiratory symptoms. Scand J Trauma Resusc Emerg Med:21

Li Y (2020) Combined Imaging and Serum Bioinformatics Analysis for Neonatal Infectious Pneumonia. J Med Imaging Health Inform 10(6):1385–1388(1384)

Li Z, Lu G, Meng G (2019) Pathogenic fungal infection in the lung. Front Immunol 10:1524

Li J, Luu LDW, Wang X, Cui X, Huang X, Fu J, Zhu X, Li Z, Wang Y, Tai J (2022) Metabolomic analysis reveals potential biomarkers and the underlying pathogenesis involved in Mycoplasma pneumoniae pneumonia. Emerg Microbes Infect 11(1):593–605

Lim WS (2020) pneumonia—overview. Reference Module Biomed Sci:B978-0-12-801238-3.11636-8

Liu Y, Xu HB (2020) Genetic polymorphisms of rs9313422 G>C and rs41297579 G>A at the promoter of TIM-1 gene contribute to the risk of community-acquired pneumonia in children. J Clin Lab Anal 34(3):e23095

Liu J, Shen R, Feng L, Cheng S, Chen J, Xiao T, Zhao S (2022) Proteomics study of Mycoplasma pneumoniae pneumonia reveals the Fc fragment of the IgG-binding protein as a serum biomarker and implicates potential therapeutic targets. Front Med 16(3):378–388

Long L, Zhao HT, Zhang ZY, Wang GY, Zhao HL (2017) Lung ultrasound for the diagnosis of pneumonia in adults: a meta-analysis. Medicine (Baltimore) 96(3):e5713

Luo T, Yan H, Li X, Deng Y, Huang J, Li L, Xiao Z, Lu X (2022) Proteomic analysis identified potential age-associated prognostic biomarkers in pneumonia-derived paediatric sepsis. Proteomics Clin Appl 16(3):e2100036

Lupia T, Corcione S, Mornese Pinna S, De Rosa FG (2020) New cephalosporins for the treatment of pneumonia in internal medicine wards. J Thorac Dis 12(7):3747–3763

Madhi SA, Levine OS, Hajjeh R, Mansoor OD, Cherian T (2008) Vaccines to prevent pneumonia and improve child survival. Bull World Health Organ 86(5):365–372

Mandell LA, Niederman MS (2019) Aspiration pneumonia. N Engl J Med 380(7):651–663

Mantero M, Tarsia P, Gramegna A, Henchi S, Vanoni N, Di Pasquale M (2017) Antibiotic therapy, supportive treatment and management of immunomodulation-inflammation response in community acquired pneumonia: review of recommendations. Multidiscip Respir Med 12:26

Martin H, Falconer J, Addo-Yobo E, Aneja S, Arroyo LM, Asghar R, Awasthi S, Banajeh S, Bari A, Basnet S et al (2022) Assembling a global database of child pneumonia studies to inform WHO pneumonia management algorithm: methodology and applications. J Glob Health 12:04075

Martin-Loeches I, Sole-Violan J, Rodriguez de Castro F, Garcia-Laorden MI, Borderias L, Blanquer J, Rajas O, Briones ML, Aspa J, Herrera-Ramos E et al (2012) Variants at the promoter of the interleukin-6 gene are associated with severity and outcome of pneumococcal community-acquired pneumonia. Intensive Care Med 38(2):256–262

McAllister DA, Liu L, Shi T, Chu Y, Reed C, Burrows J, Adeloye D, Rudan I, Black RE, Campbell H et al (2019) Global, regional, and national estimates of pneumonia morbidity and mortality in children younger than 5 years between 2000 and 2015: a systematic analysis. Lancet Glob Health 7(1):e47–e57

Metlay JP, Waterer GW, Long AC, Anzueto A, Brozek J, Crothers K, Cooley LA, Dean NC, Fine MJ, Flanders SA et al (2019) Diagnosis and treatment of adults with community-acquired pneumonia. An Official Clinical Practice Guideline of the American Thoracic Society and Infectious Diseases Society of America. Am J Respir Crit Care Med 200(7):e45–e67

Mohn KG, Smith I, Sjursen H, Cox RJ (2018) Immune responses after live attenuated influenza vaccination. Hum Vaccin Immunother 14(3):571–578

Mohsen AH, McKendrick M (2003) Varicella pneumonia in adults. Eur Respir J 21(5):886–891

Musher DM, Montoya R, Wanahita A (2004) Diagnostic value of microscopic examination of Gram-stained sputum and sputum cultures in patients with bacteremic pneumococcal pneumonia. Clin Infect Dis 39(2):165–169

Nambu A, Ozawa K, Kobayashi N, Tago M (2014) Imaging of community-acquired pneumonia: roles of imaging examinations, imaging diagnosis of specific pathogens and discrimination from noninfectious diseases. World J Radiol 6(10):779–793

Nguyen EV, Gharib SA, Palazzo SJ, Chow YH, Goodlett DR, Schnapp LM (2013) Proteomic profiling of bronchoalveolar lavage fluid in critically ill patients with ventilator-associated pneumonia. PLoS One 8(3):e58782

Ning P, Zheng Y, Luo Q, Liu X, Kang Y, Zhang Y, Zhang R, Xu Y, Yang D, Xi W et al (2018) Metabolic profiles in community-acquired pneumonia: developing assessment tools for disease severity. Crit Care 22(1):130

Ovung A, Bhattacharyya J (2021) Sulfonamide drugs: structure, antibacterial property, toxicity, and biophysical interactions. Biophys Rev 13(2):259–272

Pitcher JH, Zuckerman JB (2016) Coccidioides immitis infection with involvement of the airway. Conn Med 80(9):539–541

Pletz MW, Jensen AV, Bahrs C, Davenport C, Rupp J, Witzenrath M, Barten-Neiner G, Kolditz M, Dettmer S, Chalmers JD et al (2022) Unmet needs in pneumonia research: a comprehensive approach by the CAPNETZ study group. Respir Res 23(1):239

Pound MW, Drew RH, Perfect JR (2002) Recent advances in the epidemiology, prevention, diagnosis, and treatment of fungal pneumonia. Curr Opin Infect Dis 15(2):183–194

Qi X, Qu H, Yang D, Zhou L, He YW, Yu Y, Qu J, Liu J (2018) Lower respiratory tract microbial composition was diversified in Pseudomonas aeruginosa ventilator-associated pneumonia patients. Respir Res 19(1):139

Ramirez P, Garcia MA, Ferrer M, Aznar J, Valencia M, Sahuquillo JM, Menendez R, Asenjo MA, Torres A (2008) Sequential measurements of procalcitonin levels in diagnosing ventilator-associated pneumonia. Eur Respir J 31(2):356–362

Rautanen A, Mills TC, Gordon AC, Hutton P, Steffens M, Nuamah R, Chiche JD, Parks T, Chapman SJ, Davenport EE et al (2015) Genome-wide association study of survival from sepsis due to pneumonia: an observational cohort study. Lancet Respir Med 3(1):53–60

Robbins JB, Schneerson R, Keith JM, Miller MA, Kubler-Kielb J, Trollfors B (2009) Pertussis vaccine: a critique. Pediatr Infect Dis J 28(3):237–241

Roquilly A, Torres A, Villadangos JA, Netea MG, Dickson R, Becher B, Asehnoune K (2019) Pathophysiological role of respiratory dysbiosis in hospital-acquired pneumonia. Lancet Respir Med 7(8):710–720

Salnikova LE, Smelaya TV, Moroz VV, Golubev AM, Rubanovich AV (2013) Functional polymorphisms in the CYP1A1, ACE, and IL-6 genes contribute to susceptibility to community-acquired and nosocomial pneumonia. Int J Infect Dis 17(6):e433–e442

Schnell D, Mayaux J, de Bazelaire C, Legoff J, Feuillet S, Scieux C, Andreu-Gallien J, Darmon M, Baruchel A, Schlemmer B et al (2010) Risk factors for pneumonia in immunocompromised patients with influenza. Respir Med 104(7):1050–1056

Schotsaert M, Garcia-Sastre A (2017) Inactivated influenza virus vaccines: the future of TIV and QIV. Curr Opin Virol 23:102–106

Schulte B, Eickmeyer H, Heininger A, Juretzek S, Karrasch M, Denis O, Roisin S, Pletz MW, Klein M, Barth S et al (2014) Detection of pneumonia associated pathogens using a prototype multiplexed pneumonia test in hospitalized patients with severe pneumonia. PLoS One 9(11): e110566

Self WH, Courtney DM, McNaughton CD, Wunderink RG, Kline JA (2013) High discordance of chest X-ray and computed tomography for detection of pulmonary opacities in ED patients: implications for diagnosing pneumonia. Am J Emerg Med 31(2):401–405

Seymour CW, Yende S, Scott MJ, Pribis J, Mohney RP, Bell LN, Chen YF, Zuckerbraun BS, Bigbee WL, Yealy DM et al (2013) Metabolomics in pneumonia and sepsis: an analysis of the GenIMS cohort study. Intensive Care Med 39(8):1423–1434

Shah FA, Pike F, Alvarez K, Angus D, Newman AB, Lopez O, Tate J, Kapur V, Wilsdon A, Krishnan JA et al (2013) Bidirectional relationship between cognitive function and pneumonia. Am J Respir Crit Care Med 188(5):586–592

Sharma S, Maycher B, Eschun G (2007) Radiological imaging in pneumonia: recent innovations. Curr Opin Pulm Med 13(3):159–169

Sharma NK, Ferreira BL, Tashima AK, Brunialti MKC, Torquato RJS, Bafi A, Assuncao M, Azevedo LCP, Salomao R (2019) Lipid metabolism impairment in patients with sepsis secondary to hospital acquired pneumonia, a proteomic analysis. Clin Proteomics 16:29

Shi T, Denouel A, Tietjen AK, Lee JW, Falsey AR, Demont C, Nyawanda BO, Cai B, Fuentes R, Stoszek SK et al (2020) Global and regional burden of hospital admissions for pneumonia in older adults: a systematic review and meta-analysis. J Infect Dis 222(Suppl 7):S570–S576

Smelaya TV, Belopolskaya OB, Smirnova SV, Kuzovlev AN, Moroz VV, Golubev AM, Pabalan NA, Salnikova LE (2016) Genetic dissection of host immune response in pneumonia development and progression. Sci Rep 6:35021

Smieja M (1998) Current indications for the use of clindamycin: a critical review. Can J Infect Dis 9(1):22–28

Sole-Violan J, de Castro F, Garcia-Laorden MI, Blanquer J, Aspa J, Borderias L, Briones ML, Rajas O, Carrondo IM, Marcos-Ramos JA et al (2010) Genetic variability in the severity and outcome of community-acquired pneumonia. Respir Med 104(3):440–447

Song W, Tan H, Wang S, Zhang Y, Ding Y (2019) Association of high mobility group box protein B1 gene polymorphisms with pneumonia susceptibility and severity. Genet Test Mol Biomarkers 23(1):3–11
Spoorenberg SMC, Vestjens SMT, Voorn GP, van Moorsel CHM, Meek B, Zanen P, Rijkers GT, Bos WJW, Grutters JC (2018) Ovidius study g: Course of SP-D, YKL-40, CCL18 and CA 15-3 in adult patients hospitalised with community-acquired pneumonia and their association with disease severity and aetiology: a post-hoc analysis. PLoS One 13(1):e0190575
Suarez C, Gudiol F (2009) Beta-lactam antibiotics. Enferm Infecc Microbiol Clin 27(2):116–129
Sun Y, Xian Y, Duan Z, Wan Z, Li J, Liao Y, Bi X, Wu Z, Wang L, Zhang K (2022) Diagnostic potential of microRNAs in extracellular vesicles derived from bronchoalveolar lavage fluid for pneumonia-A preliminary report. Cell 11(19)
Tejera A, Santolaria F, Diez ML, Aleman-Valls MR, Gonzalez-Reimers E, Martinez-Riera A, Milena-Abril A (2007) Prognosis of community acquired pneumonia (CAP): value of triggering receptor expressed on myeloid cells-1 (TREM-1) and other mediators of the inflammatory response. Cytokine 38(3):117–123
Torres A, Peetermans WE, Viegi G, Blasi F (2013) Risk factors for community-acquired pneumonia in adults in Europe: a literature review. Thorax 68(11):1057–1065
Torres A, Lee N, Cilloniz C, Vila J, Van der Eerden M (2016) Laboratory diagnosis of pneumonia in the molecular age. Eur Respir J 48(6):1764–1778
Torres A, Niederman MS, Chastre J, Ewig S, Fernandez-Vandellos P, Hanberger H, Kollef M, Li Bassi G, Luna CM, Martin-Loeches I et al (2017) International ERS/ESICM/ESCMID/ALAT guidelines for the management of hospital-acquired pneumonia and ventilator-associated pneumonia: Guidelines for the management of hospital-acquired pneumonia (HAP)/ventilator-associated pneumonia (VAP) of the European Respiratory Society (ERS), European Society of Intensive Care Medicine (ESICM), European Society of Clinical Microbiology and Infectious Diseases (ESCMID) and Asociacion Latinoamericana del Torax (ALAT). Eur Respir J 50(3)
Torres A, Cilloniz C, Niederman MS, Menendez R, Chalmers JD, Wunderink RG, van der Poll T (2021) Pneumonia. Nat Rev Dis Primers 7(1):25
Tsai MH, Lin TY, Hsieh SY, Chiu CY, Chiu CH, Huang YC (2009) Comparative proteomic studies of plasma from children with pneumococcal pneumonia. Scand J Infect Dis 41(6-7):416–424
Ueckermann V, Ehlers M (2021) Late breaking abstract - dysbiosis in the lung microbiome of patients with hiv and pneumonia is associated with worse disease severity and mortality scores. 58(Suppl 65):OA1662
Valim C, Ahmad R, Lanaspa M, Tan Y, Acacio S, Gillette MA, Almendinger KD, Milner DA Jr, Madrid L, Pelle K et al (2016) Responses to bacteria, virus, and malaria distinguish the etiology of pediatric clinical pneumonia. Am J Respir Crit Care Med 193(4):448–459
van Vught LA, Scicluna BP, Wiewel MA, Hoogendijk AJ, Klein Klouwenberg PM, Franitza M, Toliat MR, Nurnberg P, Cremer OL, Horn J et al (2016) Comparative analysis of the host response to community-acquired and hospital-acquired pneumonia in critically Ill patients. Am J Respir Crit Care Med 194(11):1366–1374
Vissing NH, Chawes BL, Bisgaard H (2013) Increased risk of pneumonia and bronchiolitis after bacterial colonization of the airways as neonates. Am J Respir Crit Care Med 188(10): 1246–1252
Walker CLF, Rudan I, Liu L, Nair H, Theodoratou E, Bhutta ZA, O'Brien KL, Campbell H, Black RE (2013) Global burden of childhood pneumonia and diarrhoea. Lancet 381(9875):1405–1416
Wang HL, Hsiao PC, Tsai HT, Yeh CB, Yang SF (2013) Usefulness of plasma YKL-40 in management of community-acquired pneumonia severity in patients. Int J Mol Sci 14(11): 22817–22825
Wang H, Zhang K, Qin H, Yang L, Zhang L, Cao Y (2015) Genetic association between CD143 rs4340 polymorphism and pneumonia risk: a meta analysis. Medicine (Baltimore) 94(30):e883
Wang K, Gao M, Yang M, Meng F, Li D, Lu R, Wang Y, Zhuang H, Li M, Cheng G et al (2017) Transcriptome analysis of bronchoalveolar lavage fluid from children with severe Mycoplasma

pneumoniae pneumonia reveals novel gene expression and immunodeficiency. Hum Genomics 11(1):4

Waterer GW, ElBahlawan L, Quasney MW, Zhang Q, Kessler LA, Wunderink RG (2003) Heat shock protein 70-2+1267 AA homozygotes have an increased risk of septic shock in adults with community-acquired pneumonia. Crit Care Med 31(5):1367–1372

Weir DL, Majumdar SR, McAlister FA, Marrie TJ, Eurich DT (2015) The impact of multimorbidity on short-term events in patients with community-acquired pneumonia: prospective cohort study. Clin Microbiol Infect 21(3):264:e267-264 e213

Wu C, Jin X, Tsueng G, Afrasiabi C, Su AI (2016) BioGPS: building your own mash-up of gene annotations and expression profiles. Nucleic Acids Res 44(D1):D313–D316

Xie F, Duan Z, Zeng W, Xie S, Xie M, Fu H, Ye Q, Xu T, Xie L (2021) Clinical metagenomics assessments improve diagnosis and outcomes in community-acquired pneumonia. BMC Infect Dis 21(1):352

Yu JL, Song QF, Xie ZW, Jiang WH, Chen JH, Fan HF, Xie YP, Lu G (2017) iTRAQ-based quantitative proteomics study in patients with refractory Mycoplasma pneumoniae pneumonia. Jpn J Infect Dis 70(5):571–578

Zakharkina T, Martin-Loeches I, Matamoros S, Povoa P, Torres A, Kastelijn JB, Hofstra JJ, de Wever B, de Jong M, Schultz MJ et al (2017) The dynamics of the pulmonary microbiome during mechanical ventilation in the intensive care unit and the association with occurrence of pneumonia. Thorax 72(9):803–810

Zeng Q, Tang T, Huang B, Bu S, Xiao Y, Dai Y, Wei Z, Huang L, Jiang S (2021) rs1840680 single nucleotide polymorphism in Pentraxin 3: a potential protective biomarker of severe community-acquired pneumonia. J Int Med Res 49(4):3000605211010621

Zhan Y, Xu T, He F, Guan WJ, Li Z, Li S, Xie M, Li X, Chen R, Cheng L et al (2021) Clinical evaluation of a metagenomics-based assay for pneumonia management. Front Microbiol 12: 751073

Zhang W, Jia J, Liu Z, Si D, Ma L, Zhang G (2019) Circulating microRNAs as biomarkers for Sepsis secondary to pneumonia diagnosed via Sepsis 3.0. BMC Pulm Med 19(1):93

Zhang J, Gao L, Zhu C, Jin J, Song C, Dong H, Li Z, Wang Z, Chen Y, Yang Z et al (2022) Clinical value of metagenomic next-generation sequencing by Illumina and Nanopore for the detection of pathogens in bronchoalveolar lavage fluid in suspected community-acquired pneumonia patients. Front Cell Infect Microbiol 12:1021320

Zhao J, Zhang W, Shen L, Yang X, Liu Y, Gai Z (2017) Association of the ACE, GSTM1, IL-6, NOS3, and CYP1A1 polymorphisms with susceptibility of mycoplasma pneumoniae pneumonia in Chinese children. Medicine (Baltimore) 96(15):e6642

Zhou H, Larkin PMK, Zhao D, Ma Q, Yao Y, Wu X, Wang J, Zhou X, Li Y, Wang G et al (2021) Clinical impact of metagenomic next-generation sequencing of bronchoalveolar lavage in the diagnosis and management of pneumonia: a multicenter prospective observational study. J Mol Diagn 23(10):1259–1268

Lung Cancer

15

Abstract

Lung cancer has a complex pathophysiology and is one of the leading causes of death all over the world. Cutting edge imaging modalities and omics approaches have improved the diagnosis, stratification, and staging of the disease. Different pathological drivers and probable biomarker candidates of lung cancer identified using different advanced omics tools have been documented. Several databases and bioinformatics tools designed to improve our understanding of the disease and enable downstream analyses have been mentioned in this chapter. The potential use of computational modelling and multi-omics data analyses in lung cancer have been discussed. The present therapeutic regime for lung cancer including surgical resection, chemotherapy, immunotherapy, and targeted therapy have also been highlighted in this chapter. Finally, the different research avenues that can be explored in the field of lung cancer have been discussed.

Keywords

Non-small cell lung cancer (NSCLC) · Small cell lung cancer (SCLC) · TNM staging · Metastasis · Tumour biomarkers · Chemotherapy · Immunotherapy · Targeted therapy

15.1 Introduction

Lung cancer, also called bronchogenic carcinoma denotes tumour arising within the bronchi or in the lung parenchyma. Lung cancer is one of the most commonly diagnosed cancers across the globe. It has a poor prognosis and is associated with extremely high mortality rate. The disease arises due to the complex interplay of genetic and environmental factors. Smoking serves as the primary cause of lung cancer. Other risk factors include exposure to carcinogens and other occupational

S. Saha et al., *Pulmonomics: Omics Approaches for Understanding Pulmonary Diseases*, https://doi.org/10.1007/978-981-99-3505-5_15

hazards (like asbestos, radon, chromium, nickel, arsenic, and polycyclic aromatic hydrocarbons). Exposure to these carcinogenic agents results in genetic mutation and aberrant protein production. This eventually results in disrupted cell cycle, carcinogenesis, and dysplasia of lung epithelium. Mutations in genes like MYC, BCL2, p53, EGFR, KRAS, and p16 have been implicated in lung cancer. There has been a significant increase in the incidence of lung cancer in the past decade (https://www.ncbi.nlm.nih.gov/books/NBK482357/). Besides, lack of cure and high death rate have always kept the disease under limelight.

15.2 Major Subtypes and Stages of Lung Cancer

Lung cancer can be primarily classified into non-small cell lung carcinoma (NSCLC) and small cell lung carcinoma (SCLC). Other rare types of lung cancer include bronchial gland carcinomas, carcinoid tumours, and sarcomatoid carcinomas (Lu et al. 2017). NSCLC accounts for majority (~85%) of the lung cancer cases, while SCLC accounts for 10–15% of the lung cancer cases. Lung adenocarcinoma, squamous cell lung cancer, and large cell anaplastic carcinoma are the major subtypes of NSCLC (Zappa and Mousa 2016). Adenosquamous carcinoma of the lung and pulmonary enteric adenocarcinoma are rare subtypes of NSCLC (Li and Lu 2018; Gong et al. 2021). The generalized classification of lung cancer into the major types have been schematically shown in Fig. 15.1.

In 2021, the World Health Organization (WHO) classified thoracic tumours with ICD-O codes. The list of lung cancers provided in 2021 WHO classification have been summarized in Fig. 15.2. Morphological analyses serve as the basis for WHO

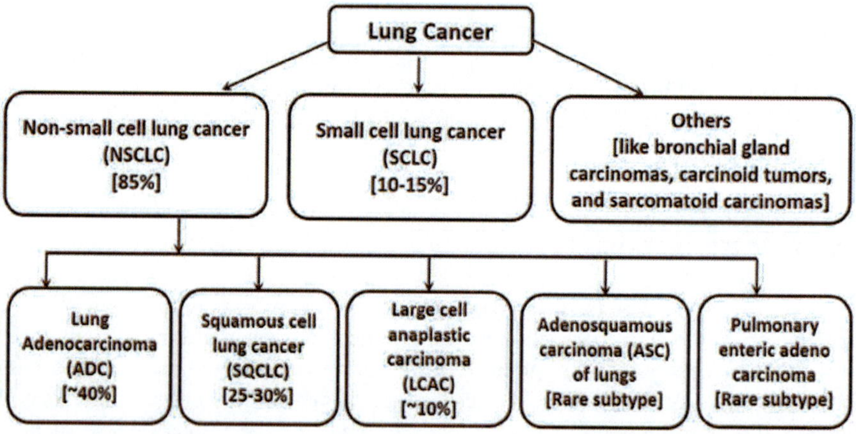

Fig. 15.1 Classification of lung cancer into major types and subtypes. The incidences of the different types of lung cancers have been expressed as % of all lung cancers (Zappa and Mousa 2016; Li and Lu 2018; Gong et al. 2021)

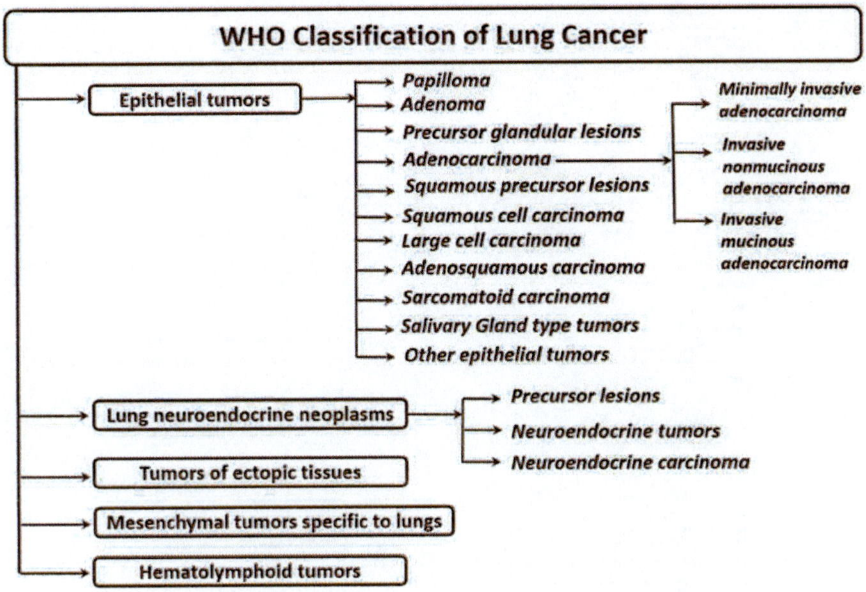

Fig. 15.2 2021 World Health Organization (WHO) Classification of lung tumours. Morphology, immunohistochemistry, and molecular abnormalities serve as the major criteria for classification of lung cancers by WHO (2021) (Nicholson et al. 2022)

classification and are often supported by immunohistochemistry and molecular biology data (Nicholson et al. 2022).

The anatomic extent of lung cancer aids in assigning prognosis and guiding therapy. The TNM staging system is widely used to describe the extent and stage of lung cancer. The TNM staging system comprises of three components, namely the features of the primary tumour (T), involvement of regional lymph nodes (N), and metastases (M) (Lababede and Meziane 2018). The different clinical stages of lung cancer and the details of the TNM staging system have been provided in Fig. 15.3. The TNM staging is primarily used for NSCLC. This staging system can also be used for SCLC. However, since SCLC is a systemic disease, a more direct and straightforward staging is successfully followed (https://www.ncbi.nlm.nih.gov/books/NBK482357/).

15.3 Clinical Features and Diagnosis

Lung cancer patients do not exhibit any specific symptoms. Clinical symptoms often appear at an advanced stage in most cases. Most of the symptoms arise due to local impact of the tumour. Coughing, hemoptysis, and chest pain are often noted in lung cancer patients. Pleural involvement is often noted in lung cancer. Superior vena cava syndrome with facial oedema, Pancoast syndrome, shoulder pain, Horner

A. Stages of Lung Cancer based on TNM Classification

Clinical Stages of Lung Cancer	TNM groups	Clinical Stages of Lung Cancer	TNM groups
Occult Cancer	TX, N0, M0	Stage IIB	T1a/T1b/T1c/T2a/T2b, N1, M0, OR, T3, N0, M0
Stage 0	Tis, N0, M0	Stage IIIA	T1a/T1b/T1c/T2a/T2b, N2, M0, OR, T3, N1, M0, OR, T4, N0/N1, M0
Stage IA_1	T1mi/T1a, N0, M0		
Stage IA_2	T1b, N0, M0	Stage IIIB	T1a/T1b/T1c/T2a/T2b, N3, M0, OR, T3/T4, N2, M0
Stage IA_3	T1c, N0, M0	Stage IIIC	T3/T4, N3, M0
Stage IB	T2a, N0, M0	Stage IVA	Any T, Any N, M1a/M1b
Stage IIA	T2b, N0, M0	Stage IVB	Any T, Any N, M1c

B. Detailed TNM (Tumor, Node, Metastasis) Classification

Primary Tumor (T)		Features
TX		Tumor cells in bronchial washings or sputum, but, no primary tumor noted by bronchoscopy or imaging
T0		No primary tumor noted
Tis		Carcinoma *in situ* (CIS)
T1		Primary Tumor Size < 3 cm
	T1mi	Minimally invasive tumor
	T1a	Primary Tumor Size ≤ 1 cm Superficial tumor may be limited to tracheal or bronchial walls (central airways)
	T1b	1 cm < Primary Tumor Size ≤ 2 cm
	T1c	2 cm < Primary Tumor Size ≤ 3 cm
T2		3 cm < Primary Tumor Size ≤ 5 cm, OR, Involvement of main bronchus (but not carina) Invasion of visceral pleura, OR, Atelectasis to the hilum
	T2a	3 cm < Primary Tumor Size ≤ 4 cm
	T2b	4 cm < Primary Tumor Size ≤ 5 cm
T3		5 cm < Primary Tumor Size ≤ 7 cm, OR, Involvement of the pericardium, phrenic nerve or chest wall, OR, Presence of separate tumor nodules in the same lobe
	T3Inv	Involvement of the pericardium, phrenic nerve or chest wall
	T3Satell	Presence of separate tumor nodules in the same lobe
T4		Primary Tumor Size > 7 cm
	T4Inv	Invading diaphragm, mediastinum, carina, trachea, heart, recurrent laryngeal nerve, esophagus, or vertebral body
	T4Ipsi nod	Separate tumor nodules in the ipsilateral lobes

Node (N)		Features
NX		Involvement of lymph nodes could not be assessed
N0		No involvement of regional lymph nodes
N1		Involvement of ipsilateral hilar and/or peribronchial lymph nodes observed
	N1a	Single station N1 nodes noted
	N1b	Multiple station N1 nodes noted
N2		Involvement of ipsilateral mediastinal and/or subcarinal lymph nodes noted
	N2a1	Single station N2 nodes with no N1 involvement
	N2a2	Single station N2 nodes with atleast one N1 nodal involvement
	N2b	Involvement of multiple station N2 nodes
N3		Involvement of contralateral mediastinal and/or hilar lymph nodes observed, OR, Involvement of scalene or supraclavicular lymph nodes

Metastasis (M)		Features
MX		Distant metastasis could not be assessed
M0		No distant metastasis noted
M1		Distant metastasis noted
	M1a	Tumor in contralateral lung, OR, Malignant pericardial or pleural effusion, OR, Presence of pleural or pericardial nodules along with tumor
	M1b	Single extra-thoracic metastasis
	M1c	Several extra-thoracic metastases

Fig. 15.3 Different stages of lung cancer based on TNM system of classification. (**a**) Lung Cancer Staging. This staging is primarily used for non-small cell lung cancer (NSCLC). Small cell lung cancer (SCLC) staging can be conducted using this TNM staging system, but a straightforward classification system is followed for SCLC since it is a systemic disease (https://www.ncbi.nlm.nih.gov/books/NBK482357/). (**b**) Details of the TNM (Tumour, Node and Metastasis) system of cancer classification (Lababede and Meziane 2018)

syndrome, para-neoplastic syndrome, stroke-like symptoms, and kidney stones are noted in patients due to metastases of lung neoplasms to different organs like bones, brain, liver, and kidneys. Adrenal metastasis is rarely symptomatic (https://www.ncbi.nlm.nih.gov/books/NBK482357/).

Lung cancer is primarily diagnosed using radiological and invasive techniques. Intravenous (IV) contrast-enhanced CT scan and positron emission tomography (PET) scan are widely used to screen every patient who are suspected to suffer from lung cancer. However, none of these techniques are sufficient to determine the stage of lung cancer. Invasive techniques are commonly used to confirm malignancy, determine tumour stage, and confirm histological differentiation of lung cancer. Bronchoscopic endobronchial ultrasound-trans bronchial needle aspiration (TBNA), endoscopic-TBNA, mediastinoscopy, and thoracoscopy or video-assisted thoracoscopy (VATS) are the common invasive techniques widely used in the evaluation and diagnosis of lung cancer (https://www.ncbi.nlm.nih.gov/books/NBK482357/). In spite of the considerable progress in diagnostic approaches, the diagnosis of lung cancer at an initial stage still remains beyond reach.

15.4 Biomarkers of Lung Cancer

Omics technologies have a wide application in improving the understanding of tumorigenesis, identifying biomarkers and drug targets (Lu and Zhan 2018). Despite the improvement in available diagnostic tools, there is still a need of diagnostic biomarkers for early diagnosis of lung cancer. Identification of specific onco-markers shall also aid in efficient molecular subtyping and stratification of lung cancer into different types and stages. Apart from diagnostic biomarkers, predictive biomarkers and prognostic biomarkers can be of immense importance in guiding treatment and monitoring clinical outcome. PD-L1 serves as the historical biomarker of lung cancer which paved the way for anti-PD-1 drugs as second-line treatment for lung cancer (Bourbonne et al. 2022). Other established potential biomarker candidates associated with lung cancer include CTLA-4, BRAF, ALK, KRAS, c-MET, CEA, EGFR, p53, Rb, CA-125, ERCC1/RRM1, CYFR A 21–21, chromogranin A, neuron-specific enolase (NSE), retinol-binding protein (RBP), and α1-antitrypsin (Zappa and Mousa 2016; Scott and Salgia 2008). Advancement of omics techniques has revolutionized biomarker research. Different potential biomarkers for lung cancer as identified by the various modern omics tools have been tabulated in Table 15.1.

15.5 Genomics

Genome wide association studies have enabled the identification of several lung cancer associated single-nucleotide polymorphisms (SNPs) and specific lung cancer susceptibility regions (Weissfeld et al. 2015). The G allele of rs10412613 in PPP2R1A was associated with reduced risk of lung cancer (Zhang et al. 2017). Variants in or near CHRNA2, BRCA2, CYP2A6 were associated with lung cancer (Byun et al. 2018). SNPs in long non-coding RNA genes were found to be associated with non-small cell lung cancer (NSCLC). TT genotype of rs498238, CC genotype of rs16901995, and variant genotypes in rs219741 have been associated with risk of NSCLC in Chinese population (Wang et al. 2019). SNPs of miR-196a2 (rs11614913) and miR-146a (rs2910164) were also found to be significantly associated with risk of lung cancer (Xiao et al. 2018). SNP of N-acetyltransferase 2 (NAT2) gene (rs1801280) was associated with non-small cell lung carcinoma (Lawi et al. 2022). Another independent genomics study highlighted a strong relation between SNP rs13383928 and asbestos-related lung cancer (Meng et al. 2021). SNPs in DNA repair genes like ERCC2 (rs1799793, rs13181), XRCC1 (rs25487), and XRCC3 (rs861539) have been associated with susceptibility to lung cancer in Saudi population (Alsagaby et al. 2022). SNP in DDO (rs9384742), PEX5L (rs9825224), DOCK2 (rs261083), VAV2 (rs2519996), and EPHB1 (rs36215) have also been significantly related to survival in NSCLC patients (Chen et al. 2022; Du et al. 2021). Some of the other SNPs related to lung cancer have been enlisted in Table 15.1.

Table 15.1 Different biomarkers of lung cancer

Omics approach	Biomarkers	Biological sample	Comments
Genomics	ERCC2	Peripheral Blood	Presence of SNPs of ERCC2 (rs13181 and rs1799793) was significantly associated with death risk in lung cancer patients (Zhang et al. 2020a)
	SLC31A1, FDX1, ATP7B		Minor alleles of SLC31A1 (rs10981694) and FDX1 (rs10488764) were significantly associated with an increased risk of lung cancer. Minor alleles of rs9535826 and rs9535828 in ATP7B gene were significantly associated with decreased risk of lung cancer (Yun et al. 2022)
	HMGB1		AG and GG genotypes of rs1360485 in HMGB1 were associated with decreased risk of female lung adenocarcinoma. Polymorphism rs1412125 was associated with the risk of lung adenocarcinoma and small cell lung cancer (Jiang et al. 2018)
	CYP2R1		SNP rs10741657 in the CYP2R1 gene was related to prognosis in elderly NSCLC patients, not receiving chemotherapy (Kong et al. 2020)
	PTEN		Small cell lung cancer (SCLC) patients with AA genotype of PTEN SNP rs2299939 was associated with better survival in response to early and late radiotherapy (Wang et al. 2020)
	RBFOX1		SNPs rs4787050 and rs8045980 in RBFOX1 gene were significantly associated with

(continued)

Table 15.1 (continued)

Omics approach	Biomarkers	Biological sample	Comments
			susceptibility to lung cancer (Li et al. 2019)
	MDM2		Haplotype-tagging SNPs of MDM2, namely rs1690924, rs1846402, rs2291857, rs3730581, and rs3730635 were associated with risk of lung cancer (Yin et al. 2020)
	EIF2AK3, HSPA5, DDIT3		SNPs of EIF2AK3 (rs867529), HSPA5 (rs391957), and DDIT3 (rs697221) were associated with risk of lung cancer (Liu et al. 2022)
	FANCB		A/G genotype of FANCB SNP rs754552650 was associated with reduced risk of lung cancer (Su et al. 2022)
	BIRC5		SNP of BIRC5 gene (rs8073069) may serve as a predictive biomarker of treatment outcome in NSCLC patients (Szczyrek et al. 2022)
Epigenomics	CLDN1, TP63, TBX5, TCF21, ADHFE1, HNF1B	Lung tissues	Differential DNA methylation noted in CLDN1, TP63, TBX5, TCF21, ADHFE1, and HNF1B may serve as novel predictive biomarkers for the diagnosis of squamous cell lung cancer (Shi et al. 2017)
	mir-34b		DNA methylation status of mir-34b may serve as an indicator of invasive phenotype of lung cancer (Watanabe et al. 2012)
	miR-9-3, miR-193a		miR-9-3 and miR-193a were methylated in NSCLC patients. Methylation of miR-9-3 was associated with

(continued)

Table 15.1 (continued)

Omics approach	Biomarkers	Biological sample	Comments
			shorter disease-free survival in lung squamous cell carcinoma patients (Heller et al. 2012)
	NID2	Lung tissues, Broncho alveolar lavage (BAL), blood plasma	Aberrant methylation of NID2 may serve as a useful predictive biomarker of NSCLC (Feng et al. 2018)
	BCAT1, CDO1, TRIM58, ZNF177	Bronchial aspirates, sputum	Hypermethylation of BCAT1, CDO1, TRIM58, and ZNF177 may serve as epigenetic biomarkers for early and rapid diagnosis of lung cancer (Diaz-Lagares et al. 2016)
	HOXA9, KRTAP8–1, CCND1, TULP2	Tumour tissues	Hypermethylation of HOXA9 and hypomethylation of KRTAP8–1, CCND1, and TULP2 were noted in lung adenocarcinoma patients (Shen et al. 2019).
	SLIT2, CDO1, TCF21, PCHD17, IRX1, HSPB6, TBX5		SLIT2, CDO1, TCF21, PCHD17, IRX1, HSPB6, and TBX5 are hypermethylated in NSCLC patients (Sun et al. 2021)
	ID4, DCL1, BNIP3, H2AFX, CACNA1G, TIMP3, RUNX3, SCGB3A1, SFRP4, HTLF, SFRP5		ID4, DCL1, BNIP3, H2AFX, CACNA1G, and TIMP3 methylation rates were different between squamous cell lung carcinoma and lung adenocarcinoma. Methylation of RUNX3, SCGB3A1, SFRP4, and DLC1 was related to the extent of the disease, whereas methylation of HTLF, SFRP5, and TIMP3 were related to overall survival in lung cancer patients (Castro et al. 2010)
	HOXA2, HOXA10		Methylation status of HOXA2 and HOXA10

(continued)

Table 15.1 (continued)

Omics approach	Biomarkers	Biological sample	Comments
			may serve as prognostic indicator of squamous cell carcinoma (Heller et al. 2013)
	AGTR1		AGTR1 hypermethylation may serve as a promising diagnostic biomarker of lung squamous cell carcinoma (Chen et al. 2017)
	SHOX2	Bronchial aspirates	Hypermethylation of SHOX2 may aid in identifying lung carcinoma patients (Schmidt et al. 2010)
Transcriptomics	CLDN1, TP63, TBX5, TCF21, ADHFE1, HNF1B	Lung tissues	CLDN1and TP63 are overexpressed while, TBX5, TCF21, ADHFE1, and HNF1B are reduced in squamous cell lung cancer patients (Shi et al. 2017)
	SPAG5, POLH, KIF23, RAD54L, SGCG, NLRC4, MMRN1, SFTPD		SPAG5, POLH, KIF23, and RAD54L were upregulated in NSCLC patients; whereas SGCG, NLRC4, MMRN1, and SFTPD were downregulated in NSCLC patients (Valk et al. 2010)
	RN7SL494P		RN7SL494P may serve as a promising biomarker for predicting nodal metastasis and evaluating prognosis in patients with primary lung adenocarcinoma (Zhu et al. 2019)
	hsa-miR-205		hsa-miR-205 serves as a highly specific and sensitive biomarker for squamous cell lung carcinoma (Lebanony et al. 2009)
	miR-31		miR-31 serves as a predictive marker of lymph node metastases and survival in lung

(continued)

Table 15.1 (continued)

Omics approach	Biomarkers	Biological sample	Comments
			adenocarcinoma patients (Meng et al. 2013)
	miR-487b		miR-487b was significantly down regulated in tissue specimens from lung cancer patients as compared to normal lung tissues from smokers (Xi et al. 2013)
	miR-126, miR-205, miR-222, miR-520 g	Serum	Increased expression of miR-126, miR-205, miR-222, and miR-520 g have been noted in asbestos-related NSCLC (Santarelli et al. 2019)
Proteomics	SERPINA4, PON1	Serum	Meta-marker model comprising of two proteins, SERPINA4 and PON1 has better sensitivity and specificity than single biomarker and this meta-marker model enables improved differential diagnosis of lung cancer (Kim et al. 2016)
	CRP-SAA complex		Increased expression of CRP-SAA complex serves as a prognostic biomarker for early-stage lung cancer (Zhang et al. 2015)
	Zyxin (ZYX)	Plasma	Plasma level of zyxin may serve as a potential biomarker for early diagnosis of NSCLC (Kim et al. 2015)
	CXCL8, CXCL10		Blood plasma levels of CXCL8 and CXCL10 could serve as a signature for predicting response of NSCLC patients to ICI therapy (Harel et al. 2022)
	FTL, MAPK1IP1L, FGB, RAB33B, RAB15	Urine	Urinary levels of FTL, MAPK1IP1L, FGB, RAB33B, and RAB15A were elevated in lung cancer patients. A

(continued)

Table 15.1 (continued)

Omics approach	Biomarkers	Biological sample	Comments
			combination of these 5 urinary biomarkers aids in distinguishing lung cancer from normal controls and other types of tumour (Zhang et al. 2018)
	LRG1	Urinary exosomes, lung tissues	Increased LRG1 expression was observed in urinary exosomes and lung tissues of NSCLC patients. Urine exosomal LRG1 may aid in non-invasive diagnosis of NSCLC (Li et al. 2011)
	PPIA, TAGLN, TAGLN2	Lung tissues	Overexpression of PPIA, TAGLN, and TAGLN2 is observed in lung tissues of lung adenocarcinoma patients. These proteins can serve as biomarkers for early diagnosis of lung adenocarcinoma (Rho et al. 2009)
	HPT, PRDX2, AAT		Haptoglobin (HPT) is upregulated, whereas Peroxiredoxin-2 (PRDX2) and Alpha-1 antitrypsin (AAT) are downregulated in lung tissue samples derived from NSCLC patients (Najafi et al. 2020)
	KRT6A, KRT6B, KRT6C	Exhaled breath condensate (EBC)	KRT6A, KRT6B, and KRT6C were significantly increased in EBC derived from lung cancer patients, and the expression of these KRTs was positively correlated with the tumour size (Lopez-Sanchez et al. 2017)
	S100A11, ANXA1, ENO1, FABP5		Increased expression of S100A11, ANXA1, ENO1, and FABP5 may serve as novel proteomic signatures of lung cancer (Ma et al. 2021a)

(continued)

Table 15.1 (continued)

Omics approach	Biomarkers	Biological sample	Comments
Metabolomics	Glucose, methanol	Sputum, exhaled breath condensate (EBC)	Absence of glucose was noted in sputum samples from lung cancer patients. Reduced level of methanol was observed in EBC of lung cancer patients. Sputum glucose level and EBC methanol level may serve as metabolic signatures of lung cancer for early disease detection and monitoring prognosis (Ahmed et al. 2016)
	Propionate, ethanol, acetate, acetone	Exhaled breath condensate (EBC)	Propionate, ethanol, acetate, and acetone were elevated in EBCs of lung cancer patients (Ahmed et al. 2016)
	N-acetyl sugars, acetate, propionate, glycoprotein, lysine, formate	Sputum	N-acetyl sugars, acetate, propionate, glycoprotein, lysine, and formate were reduced in sputum samples of lung cancer patients (Ahmed et al. 2016)
	Glucose, N1, N12-diacetylspermine, adenosine monophosphate		Glucose, N1, N12-diacetylspermine, and adenosine monophosphate levels in the sputum of early-stage non-small cell lung cancer patients were significantly elevated post-surgical resection (Ahmed et al. 2022)
	Hippurate, trigonelline, β-hydroxy isovalerate, α-hydroxyisobutyrate, N-acetyl glutamine, creatinine	Urine	Hippurate and trigonelline were reduced, whereas N-acetyl glutamine, β-hydroxy isovalerate, α-hydroxyisobutyrate, and creatinine were elevated in urine of lung cancer patients as compared to that in controls. These urine metabolites may thus serve as putative biomarkers of lung cancer (Carrola et al. 2011)
	HDL, VLDL, LDL, lactate, pyruvate, glucose,	Blood plasma	VLDL, LDL, lactate, and pyruvate were increased

(continued)

Table 15.1 (continued)

Omics approach	Biomarkers	Biological sample	Comments
	citrate, formate, acetate, methanol, alanine, glutamine, histidine, tyrosine, valine		while HDL, glucose, citrate, formate, acetate, methanol, alanine, glutamine, histidine, tyrosine, and valine were decreased in lung cancer patients (Rocha et al. 2011)
	Serine, sphingomyelin SM 42:4, sarcosine, glutamine, lyso phosphatidyl choline 18:0, sphingomyelins (SM 42:2, SM 35:1, and SM 38:3), ceramide Cer (d18:1/24:1), phosphatidylcholine (PC 30:0 and PC 30:1)		The study identified plasma levels of serine, sphingomyelin SM 42:4, sarcosine, glutamine, and lysophosphatidyl choline 18:0 as potential biomarker panel that can be useful in early-stage diagnosis of lung cancer SM 42:2, SM 35:1, Cer (d18:1/24:1), PC 30:0, PC 30:1, and SM 38:3 expression may aid in diagnostic and prognostic purposes in lung cancer patients (Yang et al. 2020)
	Glucose, lactate, phospholipids		Metabolic profiling of plasma may enable detection of lung cancer. Glucose level was increased while lactate and phospholipids were reduced in lung cancer patients (Louis et al. 2016)
	Carnitine, cadaverine		Carnitine level was elevated in all stages of NSCLC, whereas cadaverine was only elevated in stage IIIB and IV NSCLC patients (Madama et al. 2021)
	β-hydroxy butyric acid, LysoPC 20:3, PC C40:6, citric acid, fumaric acid		β-hydroxy butyric acid, LysoPC 20:3, PC C40:6, citric acid, and fumaric acid levels in plasma may serve to detect early-stage (I/II) NSCLC patients (Zhang et al. 2020b)
Metagenomics	Phylum like *Chloroflexi* and *Gemmatimonadetes;*	Sub solid nodules (SSN),	Organisms namely, *Cloacibacterium* spp.,

(continued)

Table 15.1 (continued)

Omics approach	Biomarkers	Biological sample	Comments
	Cloacibacterium spp., *Subdoligranulum* spp., *Mycobacterium* spp.	solid nodules (SN)	*Subdoligranulum* spp., and *Mycobacterium* spp. as well as organisms belonging to the phyla *Chloroflexi* and *Gemmatimonadetes* were increased significantly in sub solid nodules as compared to solid nodules in lung adenocarcinomas (Ma et al. 2021b)
	Streptococcus viridans, Granulicatella adiacens	Sputum	High abundance of *Streptococcus viridans* and *Granulicatella adiacens* was noted in lung cancer patients. Abundance of *Granulicatella adiacens* was correlated with six other bacterial species (*Enterococcus* sp. 130, *Streptococcus intermedius, Escherichia coli, Acinetobacter junii, Streptococcus viridans,* and *Streptococcus* sp. 6, in lung cancer patients and could be related to the stage of lung cancer (Cameron et al. 2017)
	Capnocytophaga spp., *Selenomonas* spp., *Veillonella* spp., *Neisseria* spp.	Saliva	Salivary *Capnocytophaga* spp., *Selenomonas* spp., *Veillonella* spp., and *Neisseria* spp. were significantly altered in squamous cell lung carcinoma and lung adenocarcinoma patients as compared to control subjects. Increased abundance of *Capnocytophaga* spp., *Selenomonas* spp., *Veillonella* spp. and reduced abundance of *Neisseria* spp. were noted in lung cancer patients (Yan et al. 2015)

(continued)

Table 15.1 (continued)

Omics approach	Biomarkers	Biological sample	Comments
	Haemophilus influenzae, *Enterobacter* spp., *Escherichia coli*, *Mycobacteria* spp., *Candida albicans*	Bronchial aspirates	Increased abundance of *Haemophilus influenzae*, *Enterobacter* spp., *Escherichia coli*, and *Mycobacteria* spp. was noted in lung cancer patients. *Candida albicans* was also observed in bronchial aspirates of lung cancer patients (Laroumagne et al. 2011)
	Veillonella spp., *Megasphaera* spp.	Broncho alveolar lavage fluid (BALF)	Increased relative abundance of *Veillonella* spp. and *Megasphaera* spp. was noted in lung cancer patients. These two genera may serve as predictive biomarkers of lung cancer (Lee et al. 2016)
	Streptococcus mutans, *Enterococcus casseliflavus*, *Acidobacteria* spp., *Granulicella* spp., *Leuconostoc lactis*, *Eubacterium siraeum*, *Streptococcus oligofermentans*, *Megasphaera micronuciformis*	Stool	*Streptococcus mutans*, *Enterococcus casseliflavus*, *Acidobacteria* spp., and *Granulicella* spp. were enriched in lung cancer patients who responded to treatment *Leuconostoc lactis*, *Eubacterium siraeum*, *Streptococcus oligofermentans*, and *Megasphaera micronuciformis* were enriched in lung cancer patients who are non-responders to treatment (Zhao et al. 2021a)

This table enlists various important biomarkers of lung cancer identified using different omics approaches

15.6 Epigenomics

Different high-throughput genome wide methylation studies have highlighted that hyper-methylation of CDKN2A/p16INK4a, PTEN, DAL1, RASSF1A, GSTP1, APC, SHOX2, RARβ, RUNX3, CDH1, CDH13, TSLC1, DAPK, FHIT, ASC/TMS1, KIF1A, NRCAM, CYGB, LOX, DOK1, CTSZ, OGDHL, NISCH, BRMS1, PAK3, MSX1, BNC1, and MGMT are commonly noted in lung cancer patients (Langevin et al. 2015; Ansari et al. 2016). Another genome wide methylation study identified 105 differentially methylated genes in lung tumours (Bakulski et al. 2019). ANKRD18A was identified as a novel epigenetic marker of lung cancer. Downregulation of ANKRD18A expression was associated with hypermethylation of ANKRD18A in lung cancer tissues (Liu et al. 2012). A study has shown that genes involved in apoptotic pathway, namely TNFRSF6/Fas, TRAIL-R1/DR4, and CASP8 are hypermethylated in small cell lung cancer (SCLC) patients (Hoang and Landi 2022; Hopkins-Donaldson et al. 2003). Another global methylation study revealed differential methylation of NEUROD1, HAND1, ZNF423, and REST as well as defective differentiation of neuroendocrine cells in SCLC patients (Kalari et al. 2013). Studies have also associated hypermethylation of TGFBR2, MLH1, MSH2, and hypomethylation of MAGE with non-small cell lung cancer (NSCLC) (Hoang and Landi 2022). Other epigenetic signatures of lung carcinoma have been documented in Table 15.1.

15.7 Transcriptomics

Gene expression studies using high-throughput omics techniques have identified several genes and miRNAs that are differentially expressed in lung cancer. Some of these have been listed in Table 15.1. Microarray-based study conducted in lung adenocarcinoma patients have highlighted differentially expressed genes to be associated with mitochondrial ATP synthesis, DNA replication, cell surface receptor linked signal transduction, enzyme-linked receptor protein signalling pathway, cytoskeletal regulation, and phosphatidylinositol signalling system. Differentially expressed genes like JAK1, NADH, STAT3, RAF1, PRIM1, MCM3, and MAPK1 have been implicated in the development of lung adenocarcinoma (Xu et al. 2016). A recent study highlighted that a panel of 14 differentially expressed long non-coding RNAs, including, HAGLR, TP53TG1, C14orf132, ADAMTS9-AS2, MCM3AP-AS1, LINC00261, LINC00312, LINC00968, TP73-AS1, LINC00673, SOX2-OT, LOC344887, LOC730101, and AFAP1-AS1 could aid in early diagnosis of NSCLC (Sulewska et al. 2022). Reduced expression of miR-138, miR-30a, miR-204, miR-107, miR-32, miR-224, miR-148b, miR-125b, miR-145, miR-200c, miR-375 and increased expression of miR-126, miR-150, miR-197, miR-21, miR-141 have been associated with poor clinical outcome in lung cancer patients (Ansari et al. 2016).

15.8 Proteomics

Recent advancements in onco-proteomics have aided in identification of several novel signatures, probable biomarker candidates, and therapeutic targets of lung cancer (Cheung and Juan 2017). Besides, models based on proteomic datasets have been proposed for classification of the different progression stage of lung adenocarcinoma and lung squamous cell carcinoma (Ren et al. 2018). Quantitative proteomic study identified GSTP1, HSPB1, and CKB as novel potential biomarker candidates for early detection of lung squamous cell cancer. Downregulation of GSTP1 was also implicated in human bronchial epithelial carcinogenesis (Baran and Brzezianska-Lasota 2021; Zeng et al. 2012). Serum levels of Apolipoprotein C-I (Apo C-I), haptoglobin alpha-1 chain, and S100A4 were also identified to diagnose NSCLC with a sensitivity of 96.56% and specificity of 94.79% (Yang et al. 2009). Proteomic analysis has also revealed SELENBP1, carbonic dehydratase, heat shock 20KD-like protein, and SM22α to be reduced and alpha enolase to be increased in NSCLC patients. This study also highlighted basaloid carcinoma as a unique subtype of NSCLC based on proteomic profile. (Li et al. 2004). ERO1L and 14-3-3 sigma were identified as a potential metastasis-related biomarker in lung adenocarcinoma patients and squamous cell lung carcinoma patients, respectively (Baran and Brzezianska-Lasota 2021). Some other proteomic signatures of lung cancer have been tabulated in Table 15.1.

15.9 Metabolomics

NMR-based plasma metabolomics study in lung cancer patients showed an increase in glycolysis, glutaminolysis, gluconeogenesis along with a simultaneous reduction in Krebs cycle and lipid catabolism (Rocha et al. 2011). Serum metabolomic profiling of newly diagnosed small cell lung cancer patients highlighted that metabolites involved in tri-carboxylic acid cycle, lipid metabolism, amino acid metabolism, and ketone body metabolism were significantly altered in patients as compared to healthy subjects (Pedersen et al. 2021). Glucose and glutamine metabolism were found to be significantly altered in squamous cell lung carcinoma patients as compared to lung adenocarcinoma patients or healthy controls (Sellers et al. 2019). High-throughput metabolomic studies conducted using nuclear magnetic resonance (NMR) and liquid chromatography mass spectrometry (LC-MS) revealed that metabolic profile of sputum and exhaled breath condensate (EBC) in early-stage non-small cell lung cancer patients suffered significant alterations after surgical resection (Ahmed et al. 2022). Modern omics tools have revolutionized our understanding of metabolic dysregulation in lung cancer. More than 150 metabolites have been related to lung cancers (Madama et al. 2021). A few of these metabolites have been listed in Table 15.1.

15.10 Metagenomics

Study of alteration in microbial profile in lung cancer provides a novel insight into pulmonary carcinogenesis. Characterization of the microbiota in NSCLC patients revealed that the microbial profile in tumorous and healthy tissue samples from the same patient were more similar than those derived from different patients (Dumont-Leblond et al. 2021). Proteobacteria were found to be enriched in lung cancer patients. Greater diversity of Proteobacteria was noted in squamous cell carcinoma than adenocarcinoma. Presence of several Enterobacteriaceae was associated with worse survival in squamous cell carcinoma (Gomes et al. 2019). Metagenomic analysis of lung cancerous site revealed an overall reduced microbial diversity in lung cancer site. Enrichment of *Streptococcus* spp. was noted in the lung cancer sites (Mur et al. 2018). Another small-scale study highlighted the abundance of *Granulicatella* spp., *Abiotrophia* spp., and *Streptococcus* spp. in sputum of lung cancer patients as compared to healthy controls (Hosgood 3rd et al. 2014). Other microbial signatures associated with different types of lung cancer have been listed in Table 15.1.

15.11 Bioinformatics

There are several web-based repositories of cancer containing genomics, transcriptomics, proteomics, mutations, and imaging datasets, from where lung cancer data can also be assessed. The list of databases related to cancer from where lung cancer data can be retrieved and analysed, have been tabulated in Table 15.2. Lung cancer high-throughput omics datasets can be browsed from BioGPS available at http://biogps.org/dataset/tag/lung%20cancer/ and can also be visualized using a plugin like "Gene expression/activity chart" (Wu et al. 2016).

15.12 Medical Imaging or Radiomics

Imaging plays a crucial role in TNM staging in lung cancer patients as shown in Fig. 15.2, and thus the image findings of T stage and their implications have an important role in treatment decisions (Hollings and Shaw 2002). Chest X-ray (CXR), computed tomography (CT), positron emission tomography (PET), and Magnetic Resonance Imaging (MRI) are used to determine T stage imaging of lung cancer patients, but CT is more commonly used to obtain a correct staging (Panunzio and Sartori 2020). However, when there is difficulty in locating the tumour, FDG PET/CT is recommended for accurately identifying the viable tumour (Purandare and Rangarajan 2015). The chest images of lung patients are archived in PadChest, Lung Image Database Consortium (LIDC), and The Cancer Imaging Archive (TCIA) (Clark et al. 2013; Bustos et al. 2020; McNitt-Gray et al. 2007). A few Artificial Intelligence (AI)-based software were developed on chest radiographs for detection of lung nodules and to improve radiologist performance (Sim et al. 2020;

Table 15.2 Different databases and bioinformatics-based analysis tools available for lung cancer

Sl. no.	Name	Description	URL
1.	The Cancer Genome Atlas (TCGA)	Repository of genomics data for different types of cancers including lung cancer (Cancer Genome Atlas Research N et al. 2013)	https://www.cancer.gov/about-nci/organization/ccg/research/structural-genomics/tcga
2.	Lung Cancer Explorer (LCE)	Knowledgebase of gene expression data and clinical data for lung cancer (Cai et al. 2019)	http://lce.biohpc.swmed.edu/
3.	Genomic Data Commons (GDC)	Knowledgebase of cancer genomics data from various types of cancer like lung cancer (Jensen et al. 2017)	https://gdc.cancer.gov/
4.	Proteomic Data Commons (PDC)	Repository of proteomic and proteogenomic datasets for various types of cancer including lung cancer (https://pdc.cancer.gov/pdc/)	https://pdc.cancer.gov/pdc/
5.	The Cancer Imaging Archive (TCIA)	Collection of medical images (like MRI, CT scans, and digital histopathology) for different types of cancer including lung cancer (Clark et al. 2013)	https://www.cancerimagingarchive.net/collections/
6.	miRCancer (microRNA Cancer Association) Database	Repository of expression profiles of different microRNAs in different types of human cancers including lung cancer (Xie et al. 2013)	http://mircancer.ecu.edu/
7.	Integrative OncoGenomics (IntOGen)	Collection of cancer driver genes and important somatic mutations associated with different types of cancer including lung cancer (Martinez-Jimenez et al. 2020)	https://www.intogen.org/search
8.	International Cancer Genome Consortium (ICGC)	Knowledgebase of cancer genomics data for different types of cancer including lung cancer (Zhang et al. 2011)	https://dcc.icgc.org/
9.	TGDBs	Collection of proto-oncogenes and tumour suppressor genes associated with major types of cancers including lung cancer (http://www.tumor-gene.org/tgdf.html)	http://www.tumor-gene.org/tgdf.html
10.	DriverDBv3	Knowledgebase of omics data (gene expression, miRNA expression, methylation	http://driverdb.tms.cmu.edu.tw/

(continued)

Table 15.2 (continued)

Sl. no.	Name	Description	URL
		pattern, mutational data, copy number variations, and clinical data) for different types of cancer including lung cancer (Liu et al. 2020)	
11.	COSMIC	Comprehensive repository of somatic mutations associated with different types of carcinomas including lung carcinoma (Tate et al. 2019)	https://cancer.sanger.ac.uk/cosmic/
12.	The Cancer Immunome Database	Comprehensive repository of next generation sequencing data (NGS) data for 20 solid cancers including lung cancer (Charoentong et al. 2017)	https://tcia.at/home
13.	National Cancer Database (NCDB)	Hospital-based repository of patient characteristics, tumour histology, disease stages, therapy, cancer recurrence, and survival for different types of cancer, including lung cancer (Bilimoria et al. 2008)	https://www.facs.org/quality-programs/cancer-programs/national-cancer-database/
14.	Surveillance, Epidemiology, and End Results (SEER) Program	Collection of data on cancer incidence, demographics, tumour stages, treatment strategy, and survival rate of cancer patients in US population (Jairam and Park 2019)	https://seer.cancer.gov/
15.	NCG7.0 (Network of Cancer Genes)	Manually curated repository of different cancer genes and healthy drivers (Repana et al. 2019)	http://ncg.kcl.ac.uk/
16.	European Genome-phenome Archive (EGA)	Archive of genetic, phenotypic, and clinical data for cancers (including lung cancer) and other diseases (Freeberg et al. 2022)	https://ega-archive.org/
17.	ONCOMINE	Repository of differential gene and miRNA expression datasets across major types of cancer (Rhodes et al. 2004)	https://www.oncomine.com/
18.	PubMeth	Database of various methylations associated with cancer (Ongenaert et al. 2008)	http://www.pubmeth.org/
19.	TARGET (Tumour Alterations Relevant	Repository of genes that are altered in different types of cancer and are also indicative of	https://software.broadinstitute.org/cancer/cga/target

(continued)

Table 15.2 (continued)

Sl. no.	Name	Description	URL
	for Genomics Driven Therapy)	response or resistance to cancer therapy (https://software.broadinstitute.org/cancer/cga/target)	
20.	dbPepNeo	Manually curated repository of neo-antigen peptides for human cancers including lung cancer (Tan et al. 2020)	http://www.biostatistics.online/dbPepNeo/index.php
21.	CSCdb	Repository of marker genes and miRNAs associated with cancer stem cells (CSCs) from different types of cancer including lung cancer (Shen et al. 2016)	http://bioinformatics.ustc.edu.cn/cscdb
22.	MoKCa database (Mutations, Oncogenes, Knowledge & Cancer)	Knowledgebase of oncogenes, tumour suppressor genes, kinases, and other proteins that are implicated in cancer (Richardson et al. 2009)	http://strubiol.icr.ac.uk/extra/mokca/
23.	BioXpress v3.0	Database of genes and miRNAs that are differentially expressed in different types of cancer (Dingerdissen et al. 2018)	https://hive.biochemistry.gwu.edu/bioxpress
24.	SomamiR	Database of cancer-associated mutations in miRNA and their target sites (Bhattacharya et al. 2013; Bhattacharya and Cui 2016)	https://compbio.uthsc.edu/SomamiR/
25.	canSAR	Knowledgebase that aids in translational research and drug discovery for cancers (including lung cancer) and other diseases (Coker et al. 2019)	https://cansarblack.icr.ac.uk/
26.	Genomics of Drug Sensitivity in Cancer (GDSC)	Knowledgebase that helps in discovery of therapeutic biomarkers in cancer cells (Yang et al. 2013)	https://www.cancerrxgene.org/
27.	CancerDR (Cancer Drug Resistance Database)	It is platform for cancer therapeutics. It contains mutational data and drug profiles for anti-cancer drugs in a single platform that aids in correlating mutational status of drug targets with efficacy of the anti-cancer drugs (Kumar et al. 2013)	https://webs.iiitd.edu.in/raghava/cancerdr/

(continued)

Table 15.2 (continued)

Sl. no.	Name	Description	URL
28.	Cancer Program Scientific Tools and Resources	Collection of tools and resources that enables analysis of different omics data obtained from different types of cancer (https://www.broadinstitute.org/cancer/cancer-program-scientific-tools-and-resources)	https://www.broadinstitute.org/cancer/cancer-program-scientific-tools-and-resources
29.	Genome Data Analysis Center (GDAC)	Enables analysis of cancer genomics data derived from TCGA (https://gdac.broadinstitute.org/)	https://gdac.broadinstitute.org/
30.	cBioPortal	Web portal that enables analysis and visualization of cancer genomics data (Cerami et al. 2012; Gao et al. 2013)	https://www.cbioportal.org/
31.	BCSCdb	A repository of cancer stem cell biomarkers (Firdous et al. 2022)	http://dibresources.jcbose.ac.in/ssaha4/bcscdb
32.	UCSC Cancer Browser: UCSC Xena	Web-based tool that enables analysis, integration and visualization of cancer genomics data (https://xena.ucsc.edu/welcome-to-ucsc-xena/)	http://xena.ucsc.edu/welcome-to-ucsc-xena/

This table enlists the different databases, web servers, and prediction tools designed for lung cancer

Tam et al. 2021). Currently, AI based on chest radiographs has been adopted in digital pathology for automated analysis in early detection and customized treatment planning with high accuracy (Cellina et al. 2022).

15.13 Multi-omics and Data Integration

Multi-omics approaches were applied in studying lung cancer, mainly for lung cancer stage prediction, prognosis, classification of lung adenocarcinoma into subtypes, and survival analysis. The datasets of transcriptome and microbiome were used to predict the lung cancer stage by Random Forest algorithm (Li et al. 2022). Machine learning techniques were also used with radiomic and genomic features to predict the prognosis of lung cancer (Emaminejad et al. 2016). Integrative analyses of multi-omics datasets based on mRNA expression, lncRNA expression, miRNA expression, DNA methylation, and somatic mutation were carried out for the classification of lung adenocarcinoma using the R package "MOVICS" and clustering analysis (Ruan et al. 2022; Zhao et al. 2021b). Integration of histopathological images, genomics, transcriptomics, and proteomics was used to predict the survival of lung adenocarcinoma patients (Chen et al. 2021). In another lung cancer survival study, multi-omics analysis was performed using machine learning models which could distinguish high-risk and low-risk patients (Takahashi et al. 2020). Overall, machine learning tools were mainly used for data integration of lung cancer multi-omics datasets.

15.14 Present Therapeutics

Surgical resection is the standard care advised for early-stage lung cancer patients, who are capable of tolerating thoracic surgery. Surgery involves removal of a lobe (lobectomy) or section of lungs harbouring the tumour. Modern surgical techniques like video-assisted thoracoscopic surgery (VATS) are also used as minimal invasive tools for lung resection (Zappa and Mousa 2016; Jones and Baldwin 2018). An adjuvant therapy is generally recommended after surgical resection to kill remaining cancer cells, prevent relapse, and prolong post-surgical survival in lung cancer patients. Radiation therapy, chemotherapy, immunotherapy, and targeted therapy are commonly used as post-operative adjuvant therapy (Zappa and Mousa 2016). Conventional external beam radiation therapy, stereotactic body radiation therapy, high dose rate (HDR) brachytherapy, and low dose rate (LDR) brachytherapy are commonly used for lung cancer patients. Modern radiation therapy techniques like CyberKnife, TomoTherapy, and robot-assisted brachytherapy are also used for early-stage lung cancer patients (Parashar et al. 2013).

Chemotherapy is not only used as a post-surgical adjuvant, but also widely used as a first-line therapy for stage IV lung cancer (Zappa and Mousa 2016). The common chemotherapeutic agents used for the treatment of non-small cell lung cancer (NSCLC) and small cell lung cancer (SCLC) have been tabulated in Table 15.3. Although chemotherapies provide major palliative benefits, they are associated with risk of adverse effects. The chemotherapeutic regime should be altered in case of serious adverse effects and stopped if it fails to reduce the tumour (Zappa and Mousa 2016).

Cancer immunotherapy primarily functions to boost the immune system and promote the recognition of cancer cells as foreign elements. Malignant cells generally utilize certain immune checkpoints to overcome immune surveillance and create tolerance in the immune system. Some of these crucial checkpoints are PD-1, PD-L1, and CTLA-4. These checkpoints are primarily targeted by cancer immunotherapy (https://www.ncbi.nlm.nih.gov/books/NBK482357/). A list of the different immunotherapeutic agents used for lung cancer have been summarized in Table 15.4.

Apart from immune checkpoints, certain driver mutations in genes like EGFR, KRAS, ALK, and ROS1 have also been implicated in lung cancer. It was hypothesized that blocking these mutation pathways may lead to improvement in health and survival in lung cancer patients (https://www.ncbi.nlm.nih.gov/books/NBK482357/). This paved the way for the use of targeted therapy in lung cancer. The different targeted therapeutic agents used for lung cancer have been listed in Table 15.5.

Table 15.3 Chemotherapeutic agents for treatment of lung cancer

Types of lung cancer	Sl. no.	Drug or drug combinations	Mode of action
Non-Small Cell Lung Cancer (NSCLC)	1.	Cisplatin	Platinum-containing compound Forms DNA adducts, prevents repair of damaged DNA, and subsequently induces death of cancer cells by apoptosis or necrosis (Gonzalez et al. 2001; Fuertes et al. 2003; Fennell et al. 2016)
	2.	Carboplatin	Platinum-containing compound Leads to intrastrand and interstrand crosslinking of DNA. This leads to inhibition of DNA synthesis during S phase of cell cycle and eventually induces cell death (Green et al. 1992; https://www.ncbi.nlm.nih.gov/books/NBK548565/; Wang et al. 2010)
	3.	Paclitaxel	It is an anti-microtubule agent Promotes polymerization of tubulin into microtubules and blocks microtubule dissociation, thereby blocking cell cycle progression and mitosis. Thus it primarily acts by inhibiting cancer cell growth and proliferation (Zhu and Chen 2019; https://www.ncbi.nlm.nih.gov/books/NBK536917/)
	4.	Albumin-bound paclitaxel (Nab-Paclitaxel)	It is a solvent-free, albumin-bound nanoparticle formulation of paclitaxel that exploits transcytosis mediated increased delivery of albumin to cancer cells Prevents depolymerization of microtubules and arrests cell cycle, thereby, preventing cancer cell growth Has better response rate and tolerability as compared to paclitaxel (Adrianzen Herrera et al. 2019; Yardley 2013; Miele et al. 2009)
	5.	Docetaxel	An anti-mitotic drug that binds to tubulin to promote stable microtubule assembly Prevents depolymerization of microtubules Eventually induces cell cycle arrest and cell death (Georgoulias 2002; Pircher et al. 2013)
	6.	Gemcitabine	It is a potent deoxycytidine analog. The active metabolite gemcitabine triphosphate (dFdCTP) gets incorporated into growing DNA strand and competitively inhibits DNA elongation. It then leads to chain termination and DNA fragmentation and induces apoptosis-mediated cancer cell death (de Sousa and Monteiro 2014; Ciccolini et al. 2016)
	7.	Vinorelbine	It is an anti-microtubule agent. Prevents metaphasic cell division by binding

(continued)

Table 15.3 (continued)

Types of lung cancer	Sl. no.	Drug or drug combinations	Mode of action
			to microtubular proteins in the mitotic spindle and interfering with chromosomal segregation during mitosis (Nobili et al. 2020; Capasso 2012)
	8.	Etoposide	Forms complex with topoisomerase II and DNA, thereby inhibiting DNA synthesis. This complex induces breaks in DNA strand and prevents repair by Topoisomerase II. Breakage of DNA prevents entry into mitotic phase of cell division and results in cell death. Functions mostly at G2 and S phase of cell cycle (Furuse 1992; Baldwin and Osheroff 2005; https://www.ncbi.nlm.nih.gov/books/NBK557864/)
	9.	Pemetrexed	It is a multi-targeted anti-folate drug Prevents survival and growth of cells by inhibiting several folate-dependent enzymes involved in the formation of purine and pyrimidine nucleotide precursors of DNA and RNA (Hanauske et al. 2001; Adjei 2004)
Small Cell Lung Cancer (SCLC)	1.	Cisplatin and etoposide	Cisplatin is a platinum-containing compound that crosslinks DNA, inhibits DNA synthesis, and eventually induces cancer cell death Etoposide prevents repair of DNA breakage by binding with topoisomerase II, thereby arresting cell cycle and ultimately, leading to cell death (Fuertes et al. 2003; https://www.ncbi.nlm.nih.gov/books/NBK557864/; Loehrer Sr. et al. 1988)
	2.	Carboplatin and etoposide	Carboplatin is a platinum-containing compound that leads to inter- and intra strand DNA crosslinking, thus blocking DNA synthesis and eventually inducing cancer cell death Etoposide prevents DNA repair by topoisomerase II, arrests cell cycle and ultimately, results in cell death (https://www.ncbi.nlm.nih.gov/books/NBK548565/; https://www.ncbi.nlm.nih.gov/books/NBK557864/; Bishop 1992)
	3.	Cisplatin and irinotecan	Cisplatin is a platinum-containing compound that interferes with DNA replication by promoting DNA crosslinking and eventually, leads to cancer cell death Irinotecan serves as an antineoplastic topoisomerase I inhibitor. Binds to topoisomerase I-DNA complex, prevents re-ligation of DNA strands, and interferes

(continued)

Table 15.3 (continued)

Types of lung cancer	Sl. no.	Drug or drug combinations	Mode of action
			with moving of the replication fork. This halts the replication and leads to apoptotic cell death (Fuertes et al. 2003; Kubota et al. 2014; Lara Jr. et al. 2009; https://www.ncbi.nlm.nih.gov/books/NBK554441/)
	4.	Carboplatin and irinotecan	Carboplatin is a platinum-containing compound that leads to DNA crosslinking, thereby inhibiting DNA synthesis and eventually inducing cancer cell death. Irinotecan inhibits topoisomerase I, interferes with replication, and leads to apoptotic cell death (https://www.ncbi.nlm.nih.gov/books/NBK548565/; https://www.ncbi.nlm.nih.gov/books/NBK554441/; Rutledge et al. 2014)
	5.	Topotecan and lurbinectedin	Topotecan exerts its effects during S phase. It acts as an uncompetitive inhibitor by binding to the topoisomerase I-DNA complex. It inhibits DNA re-ligation, arrests DNA replication, and eventually triggers apoptotic cell death Lurbinectedin normalizes tumour microenvironment (TME) and kills macrophages in tumours by preventing the production of tumour growth factors and by blocking transcription (Patel et al. 2021; Das et al. 2021; Kollmannsberger et al. 1999)

This table enlists the different chemotherapy drugs used for the two major types of lung cancers

Alongside the different adjuvant therapies for lung cancer, specialist supportive and palliative care are of prime importance in improving patient outcomes and survivability (Jones and Baldwin 2018).

In spite of the tremendous advancement in the therapy of lung cancer, till date there is no complete cure to this disease. Often resistance to standard chemotherapeutic agents is observed in lung cancer patients. Besides, lung cancer still continues to have a high mortality rate. Therefore, state-of-art omics tools should be exhaustively explored not only for biomarker identification but also for identifying novel drug targets and designing new therapeutic strategies for lung cancer.

15.15 Future Perspectives

Advanced omics approaches have brought about significant improvement in the field of pulmonary onco-research. However, several avenues still remain to be explored. Early diagnosis of lung cancer is extremely challenging. Omics approaches should

Table 15.4 Biologics used as immunotherapy for lung cancer

Sl. no.	Name	Description	Types of lung cancer	Comments
1.	Nivolumab	Monoclonal antibody targeting PD-1	Non-Small Cell Lung Cancer (NSCLC)	Approved by FDA for early-stage NSCLC and also metastatic NSCLC that has progressed after platinum-based chemotherapy (Kazandjian et al. 2016a; https://www.fda.gov/drugs/resources-information-approved-drugs/fda-approves-neoadjuvant-nivolumab-and-platinum-doublet-chemotherapy-early-stage-non-small-cell-lung)
2.	Pembrolizumab (MK-3475)	Monoclonal antibody targeting PD-1	Non-Small Cell Lung Cancer (NSCLC)	Approved by FDA for unresectable Stage III NSCLC patients as well as metastatic NSCLC patients (Pai-Scherf et al. 2017; Akinboro et al. 2022; https://www.fda.gov/drugs/fda-expands-pembrolizumab-indication-first-line-treatment-nsclc-tps-1; Sul et al. 2016)
3.	Cemiplimab	Antibody-directed against programmed death 1 (PD-1) Receptor	Non-Small Cell Lung Cancer (NSCLC)	FDA approved cemiplimab for treatment of advanced NSCLC patients having have high PD-L1 expression in tumours (Akinboro et al. 2022)
4.	Atezolizumab	Anti-PD-L1 monoclonal antibody	Non-Small Cell Lung Cancer (NSCLC) Small Cell Lung Cancer (SCLC)	FDA approved for use in NSCLC patients (with high PD-L1 expression) after surgery (Akinboro et al. 2022; de Marinis et al. 2022) Has also been approved by FDA for extensive stage SCLC (Mathieu et al. 2021)
5.	Durvalumab	Anti-PD-L1 monoclonal antibody	Non-Small Cell Lung Cancer (NSCLC) Small Cell Lung Cancer (SCLC)	Approved for use in unresectable Stage III NSCLC patients after chemo-radiation (Durvalumab n.d.; Vaddepally et al. 2020) Has also been approved by FDA for extensive stage SCLC (Mathieu et al. 2021)
6.	Ipilimumab	Anti-cytotoxic T-lymphocyte	Non-Small Cell Lung	Use of ipilimumab along with nivolumab was approved by FDA for use in

(continued)

Table 15.4 (continued)

Sl. no.	Name	Description	Types of lung cancer	Comments
		antigen 4 (CTLA-4) monoclonal antibody	Cancer (NSCLC)	metastatic or recurrent NSCLC (Vellanki et al. 2021)

This table enlists the different FDA approved biologics used for the immunotherapy for different forms of lung cancer

be used extensively for identification of non-invasive biomarkers that can diagnose and distinguish between the different types of lung cancer (Nooreldeen and Bach 2021). More advancement in the field of radiomics should be encouraged for decoding tumour biology from imaging data and for developing imaging surrogates for genetic testing (Lee et al. 2020). Screening of driver mutation involved in cancer growth and lung metastasis is crucial. Besides, identification of novel molecular targets and response markers is of prime importance in guiding anti-angiogenic therapy. Squamous cell lung cancer has often been more difficult to target than lung adenocarcinoma, therefore, efforts need to be made for development of effective treatment. Resistance to chemotherapeutic agents and targeted therapy is also an area of concern. Strategies of bypassing and overcoming such resistances should be studied for targeting the different resistance-related pathways (Forde and Ettinger 2013; Minguet et al. 2016). The prospect and effectiveness of particle therapy and proton therapy in lung cancer also need to be evaluated (Grutters et al. 2010). Machine learning-based lung cancer prediction models should be designed for improving diagnosis-related decision making, understanding disease progression, and guiding therapy (Kadir and Gleeson 2018; Shaikh and Rao 2022; Bekisz and Geris 2020). In recent times, association of allergy and lung cancer have been mentioned by several epidemiological studies. The exact immunoregulatory mechanisms driving such allergo-oncological association needs to be explored (Majumdar and Saha 2020). Exploring these different research avenues can help us to move one step towards early diagnosis and precision medicine designing for lung cancer.

Table 15.5 Targeted therapy for lung cancer

Sl. no.	Name	Description	Types of lung cancer	Comments
1.	Bevacizumab	Anti-vascular endothelial growth factor (VEGF) monoclonal antibody (drug that targets angiogenesis)	Non-Small Cell Lung Cancer (NSCLC)	Approved by FDA, in combination with carboplatin and paclitaxel for the treatment of unresectable metastatic/ advanced recurrent non-squamous NSCLC (Cohen et al. 2007)
2.	Ramucirumab	Human monoclonal antibody that targets VEGFR2 on blood vessel endothelial cells (drug that targets angiogenesis)	Non-Small Cell Lung Cancer (NSCLC)	Approved by FDA in combination with erlotinib for first-line therapy of metastatic NSCLC (Nakagawa et al. 2019; https://www.fda.gov/drugs/resources-information-approved-drugs/fda-approves-ramucirumab-plus-erlotinib-first-line-metastatic-nsclc)
3.	Erlotinib	Epidermal growth factor receptor (EGFR) inhibitor	Non-Small Cell Lung Cancer (NSCLC)	FDA approved for use in patients with advanced or metastatic NSCLC with EGFR mutations (Khozin et al. 2014)
4.	Afatinib	Epidermal growth factor receptor (EGFR) inhibitor	Non-Small Cell Lung Cancer (NSCLC)	Approved by FDA for metastatic NSCLC patients with tumours having EGFR mutations (Dungo and Keating 2013)
5.	Gefitinib	Epidermal growth factor receptor (EGFR) inhibitor	Non-Small Cell Lung Cancer (NSCLC)	FDA approved for patients with metastatic NSCLC having EGFR gene mutations (Kazandjian et al. 2016b)
6.	Osimertinib	Epidermal growth factor receptor (EGFR) inhibitor	Non-Small Cell Lung Cancer (NSCLC)	FDA approved it as adjuvant therapy post-surgery for NSCLC patients with EGFR exon 19 deletion or exon 21 mutation (Koch et al. 2021) Also approved by FDA as first-line therapy for metastatic NSCLC patients with most

(continued)

Table 15.5 (continued)

Sl. no.	Name	Description	Types of lung cancer	Comments
				common EGFR mutations (Scott 2018)
7.	Dacomitinib	Epidermal growth factor receptor (EGFR) inhibitor	Non-Small Cell Lung Cancer (NSCLC)	Approved by FDA as first-line therapy for metastatic NSCLC patients with tumours having EGFR mutations (Shirley 2018)
8.	Amivantamab	Bi-specific antibody targeting both EGFR and MET	Non-Small Cell Lung Cancer (NSCLC)	Approved by FDA for NSCLC patients with tumours having EGFR exon 20 insertion mutations (Olivier and Prasad 2022)
9.	Mobocertinib	Epidermal growth factor receptor (EGFR) inhibitor	Non-Small Cell Lung Cancer (NSCLC)	Approved by FDA for metastatic NSCLC with EGFR exon 20 insertion mutations (Duke et al. 2022)
10.	Necitumumab	Epidermal growth factor receptor (EGFR) inhibitor (Monoclonal antibody targeting EGFR)	Non-Small Cell Lung Cancer (NSCLC)	Approved by FDA for metastatic squamous NSCLC (Fala 2016)
11.	Sotorasib	KRAS inhibitor	Non-Small Cell Lung Cancer (NSCLC)	Approved by FDA for metastatic NSCLC having a particular G12C mutation in KRAS gene (Nakajima et al. 2022)
12.	Crizotinib	Anaplastic lymphoma kinase (ALK) inhibitor and ROS1 inhibitor	Non-Small Cell Lung Cancer (NSCLC)	Approved by FDA for metastatic NSCLC with ALK positive or ROS1 positive tumours (Kazandjian et al. 2016c)
13.	Lorlatinib	Anaplastic lymphoma kinase (ALK) inhibitor and ROS1 inhibitor	Non-Small Cell Lung Cancer (NSCLC)	Approved by FDA for metastatic NSCLC with ALK positive tumours (Yun and Bazhenova 2022) Has shown promising results for metastatic NSCLC with ROS1 positive tumours (Duruisseaux 2019; Facchinetti and Friboulet 2018)
14.	Entrectinib	ROS1 inhibitor	Non-Small Cell Lung	Approved by FDA for metastatic NSCLC with

(continued)

Table 15.5 (continued)

Sl. no.	Name	Description	Types of lung cancer	Comments
			Cancer (NSCLC)	ROS1 positive tumours Also approved by FDA for solid metastatic tumours like lung cancer harbouring Neurotrophic Tyrosine Receptor Kinase (NTRK) gene fusions (Michelotti et al. 2022)
15.	Ceritinib	Anaplastic lymphoma kinase (ALK) inhibitor	Non-Small Cell Lung Cancer (NSCLC)	Approved by FDA for metastatic NSCLC with ALK positive tumours (Khozin et al. 2015)
16.	Alectinib	Anaplastic lymphoma kinase (ALK) inhibitor	Non-Small Cell Lung Cancer (NSCLC)	Approved by FDA for metastatic NSCLC with ALK positive tumours (Larkins et al. 2016)
17.	Brigatinib	Anaplastic lymphoma kinase (ALK) inhibitor	Non-Small Cell Lung Cancer (NSCLC)	Approved by FDA for metastatic NSCLC with ALK positive tumours (Rijavec et al. 2022)
18.	Dabrafenib	BRAF inhibitor	Non-Small Cell Lung Cancer (NSCLC)	FDA approved the combination of dabrafenib and trametinib for metastatic NSCLC patients having V600E mutation in BRAF gene (Khunger et al. 2018; Odogwu et al. 2018)
19.	Trametinib	MEK inhibitor	Non-Small Cell Lung Cancer (NSCLC)	FDA approved the combination of dabrafenib and trametinib for metastatic NSCLC patients having V600E mutation in BRAF gene (Khunger et al. 2018; Odogwu et al. 2018)
20.	Capmatinib	MET inhibitor	Non-Small Cell Lung Cancer (NSCLC)	Approved by FDA for metastatic NSCLC having a mutation that leads to mesenchymal-epithelial transition (MET) exon 14 skipping (Mathieu et al. 2022)
21.	Tepotinib	MET inhibitor	Non-Small Cell Lung Cancer (NSCLC)	Approved by FDA for metastatic NSCLC harbouring alterations that lead to MET exon

(continued)

Table 15.5 (continued)

Sl. no.	Name	Description	Types of lung cancer	Comments
				14 skipping (Mathieu et al. 2022)
22.	Larotrectinib	Inhibitor of tropomyosin kinase receptors (TrkA, TrkB, and TrkC)	Lung cancer	Approved by FDA for solid metastatic tumours harbouring Neurotrophic Tyrosine Receptor Kinase (NTRK) gene fusions (noted in variety of tumours including lung cancer) (Dunn 2020)
23.	Selpercatinib	RET inhibitor	Non-Small Cell Lung Cancer (NSCLC)	Approved by FDA for metastatic RET fusion-positive NSCLC (Bradford et al. 2021)
24.	Pralsetinib	RET inhibitor	Non-Small Cell Lung Cancer (NSCLC)	Approved by FDA for metastatic RET fusion-positive NSCLC (Kim et al. 2021)

This table enlists the different FDA approved biologics and small molecules that are used for targeted drug therapy of lung cancer

References

Adjei AA (2004) Pharmacology and mechanism of action of pemetrexed. Clin Lung Cancer 5 (Suppl 2):S51–S55

Adrianzen Herrera D, Ashai N, Perez-Soler R, Cheng H (2019) Nanoparticle albumin bound-paclitaxel for treatment of advanced non-small cell lung cancer: an evaluation of the clinical evidence. Expert Opin Pharmacother 20(1):95–102

Ahmed N, Bezabeh T, Ijare OB, Myers R, Alomran R, Aliani M, Nugent Z, Banerji S, Kim J, Qing G et al (2016) Metabolic signatures of lung cancer in sputum and exhaled breath condensate detected by (1)H magnetic resonance spectroscopy: a feasibility study. Magn Reson Insights 9: 29–35

Ahmed N, Kidane B, Wang L, Nugent Z, Moldovan N, McElrea A, Shariati-Ievari S, Qing G, Tan L, Buduhan G et al (2022) Metabolic alterations in sputum and exhaled breath condensate of early stage non-small cell lung cancer patients after surgical resection: a pilot study. Front Oncol 12:874964

Akinboro O, Larkins E, Pai-Scherf LH, Mathieu LN, Ren Y, Cheng J, Fiero MH, Fu W, Bi Y, Kalavar S et al (2022) FDA approval summary: pembrolizumab, atezolizumab, and cemiplimab-rwlc as single agents for first-line treatment of advanced/metastatic PD-L1-high NSCLC. Clin Cancer Res 28(11):2221–2228

Alsagaby S, Ahmed AA, Rasheed Z, Althwab SA, Aljohani ASM, Alhumaydhi FA, Alhomaidan HT, Alkhamiss AS, Alkhowailed M, Alaqeel A et al (2022) Association of genetic polymorphisms in DNA repair genes ERCC2 Asp312Asn (rs1799793), ERCC2 Lys 751 Gln (rs13181), XRCC1 Arg399 Gln (rs25487) and XRCC3 Thr 241Met (rs861539) with the susceptibility of lung cancer in Saudi population. Nucleosides Nucleotides Nucleic Acids 41(5–6):530–554

Ansari J, Shackelford RE, El-Osta H (2016) Epigenetics in non-small cell lung cancer: from basics to therapeutics. Transl Lung Cancer Res 5(2):155–171

Bakulski KM, Dou J, Lin N, London SJ, Colacino JA (2019) DNA methylation signature of smoking in lung cancer is enriched for exposure signatures in newborn and adult blood. Sci Rep 9(1):4576

Baldwin EL, Osheroff N (2005) Etoposide, topoisomerase II and cancer. Curr Med Chem Anticancer Agents 5(4):363–372

Baran K, Brzezianska-Lasota E (2021) Proteomic biomarkers of non-small cell lung cancer patients. Adv Respir Med 89(4):419–426

Bekisz S, Geris L (2020) Cancer modeling: from mechanistic to data-driven approaches, and from fundamental insights. J Comput Sci:46

Bhattacharya A, Cui Y (2016) SomamiR 2.0: a database of cancer somatic mutations altering microRNA-ceRNA interactions. Nucleic Acids Res 44(D1):D1005–D1010

Bhattacharya A, Ziebarth JD, Cui Y (2013) SomamiR: a database for somatic mutations impacting microRNA function in cancer. Nucleic Acids Res 41(Database issue):D977–D982

Bilimoria KY, Stewart AK, Winchester DP, Ko CY (2008) The National Cancer Data Base: a powerful initiative to improve cancer care in the United States. Ann Surg Oncol 15(3):683–690

Bishop JF: Carboplatin/etoposide in small cell lung cancer. Oncology 1992, 49 Suppl 1:11–17; discussion 17–18

Bourbonne V, Geier M, Schick U, Lucia F (2022) Multi-omics approaches for the prediction of clinical endpoints after immunotherapy in non-small cell lung cancer: a comprehensive review. Biomedicine 10(6)

Bradford D, Larkins E, Mushti SL, Rodriguez L, Skinner AM, Helms WS, Price LSL, Zirkelbach JF, Li Y, Liu J et al (2021) FDA approval summary: selpercatinib for the treatment of lung and thyroid cancers with RET gene mutations or fusions. Clin Cancer Res 27(8):2130–2135

Bustos A, Pertusa A, Salinas JM, de la Iglesia-Vaya M (2020) PadChest: a large chest X-ray image dataset with multi-label annotated reports. Med Image Anal 66:101797

Byun J, Schwartz AG, Lusk C, Wenzlaff AS, de Andrade M, Mandal D, Gaba C, Yang P, You M, Kupert EY et al (2018) Genome-wide association study of familial lung cancer. Carcinogenesis 39(9):1135–1140

Cai L, Lin S, Girard L, Zhou Y, Yang L, Ci B, Zhou Q, Luo D, Yao B, Tang H et al (2019) LCE: an open web portal to explore gene expression and clinical associations in lung cancer. Oncogene 38(14):2551–2564

Cameron SJS, Lewis KE, Huws SA, Hegarty MJ, Lewis PD, Pachebat JA, Mur LAJ (2017) A pilot study using metagenomic sequencing of the sputum microbiome suggests potential bacterial biomarkers for lung cancer. PLoS One 12(5):e0177062

Cancer Genome Atlas Research N, Weinstein JN, Collisson EA, Mills GB, Shaw KR, Ozenberger BA, Ellrott K, Shmulevich I, Sander C, Stuart JM (2013) The Cancer Genome Atlas Pan-Cancer analysis project. Nat Genet 45(10):1113–1120

Capasso A (2012) Vinorelbine in cancer therapy. Curr Drug Targets 13(8):1065–1071

Carrola J, Rocha CM, Barros AS, Gil AM, Goodfellow BJ, Carreira IM, Bernardo J, Gomes A, Sousa V, Carvalho L et al (2011) Metabolic signatures of lung cancer in biofluids: NMR-based metabonomics of urine. J Proteome Res 10(1):221–230

Castro M, Grau L, Puerta P, Gimenez L, Venditti J, Quadrelli S, Sanchez-Carbayo M (2010) Multiplexed methylation profiles of tumor suppressor genes and clinical outcome in lung cancer. J Transl Med 8:86

Cellina M, Ce M, Irmici G, Ascenti V, Khenkina N, Toto-Brocchi M, Martinenghi C, Papa S, Carrafiello G (2022) Artificial intelligence in lung cancer imaging: unfolding the future. Diagnostics (Basel) 12(11)

Cerami E, Gao J, Dogrusoz U, Gross BE, Sumer SO, Aksoy BA, Jacobsen A, Byrne CJ, Heuer ML, Larsson E et al (2012) The cBio cancer genomics portal: an open platform for exploring multidimensional cancer genomics data. Cancer Discov 2(5):401–404

Charoentong P, Finotello F, Angelova M, Mayer C, Efremova M, Rieder D, Hackl H, Trajanoski Z (2017) Pan-cancer immunogenomic analyses reveal genotype-immunophenotype relationships and predictors of response to checkpoint blockade. Cell Rep 18(1):248–262

Chen R, Hong Q, Jiang J, Chen X, Jiang Z, Wang J, Liu S, Duan S, Shi S (2017) AGTR1 promoter hypermethylation in lung squamous cell carcinoma but not in lung adenocarcinoma. Oncol Lett 14(4):4989–4994

Chen L, Zeng H, Xiang Y, Huang Y, Luo Y, Ma X (2021) Histopathological images and multi-omics integration predict molecular characteristics and survival in lung adenocarcinoma. Front Cell Dev Biol 9:720110

Chen AS, Liu H, Wu Y, Luo S, Patz EF Jr, Glass C, Su L, Du M, Christiani DC, Wei Q (2022) Genetic variants in DDO and PEX5L in peroxisome-related pathways predict non-small cell lung cancer survival. Mol Carcinog 61(7):619–628

Cheung CHY, Juan HF (2017) Quantitative proteomics in lung cancer. J Biomed Sci 24(1):37

Ciccolini J, Serdjebi C, Peters GJ, Giovannetti E (2016) Pharmacokinetics and pharmacogenetics of Gemcitabine as a mainstay in adult and pediatric oncology: an EORTC-PAMM perspective. Cancer Chemother Pharmacol 78(1):1–12

Clark K, Vendt B, Smith K, Freymann J, Kirby J, Koppel P, Moore S, Phillips S, Maffitt D, Pringle M et al (2013) The Cancer Imaging Archive (TCIA): maintaining and operating a public information repository. J Digit Imaging 26(6):1045–1057

Cohen MH, Gootenberg J, Keegan P, Pazdur R (2007) FDA drug approval summary: bevacizumab (Avastin) plus Carboplatin and Paclitaxel as first-line treatment of advanced/metastatic recurrent nonsquamous non-small cell lung cancer. Oncologist 12(6):713–718

Coker EA, Mitsopoulos C, Tym JE, Komianou A, Kannas C, Di Micco P, Villasclaras Fernandez E, Ozer B, Antolin AA, Workman P et al (2019) canSAR: update to the cancer translational research and drug discovery knowledgebase. Nucleic Acids Res 47(D1):D917–D922

Das M, Padda SK, Weiss J, Owonikoko TK (2021) Advances in Treatment of Recurrent Small Cell Lung Cancer (SCLC): insights for optimizing patient outcomes from an expert roundtable discussion. Adv Ther 38(11):5431–5451

de Marinis F, Attili I, Gridelli C, Cecere F, Curcio C, Facciolo F, Spaggiari L (2022) Incorporating atezolizumab in the adjuvant setting of non-small cell lung cancer: key discussion points from an expert multidisciplinary panel by Italian Association of Thoracic Oncology. Front Oncol 12: 971042

de Sousa CL, Monteiro G (2014) Gemcitabine: metabolism and molecular mechanisms of action, sensitivity and chemoresistance in pancreatic cancer. Eur J Pharmacol 741:8–16

Diaz-Lagares A, Mendez-Gonzalez J, Hervas D, Saigi M, Pajares MJ, Garcia D, Crujerias AB, Pio R, Montuenga LM, Zulueta J et al (2016) A novel epigenetic signature for early diagnosis in lung cancer. Clin Cancer Res 22(13):3361–3371

Dingerdissen HM, Torcivia-Rodriguez J, Hu Y, Chang TC, Mazumder R, Kahsay R (2018) BioMuta and BioXpress: mutation and expression knowledgebases for cancer biomarker discovery. Nucleic Acids Res 46(D1):D1128–D1136

Du H, Liu L, Liu H, Luo S, Patz EF Jr, Glass C, Su L, Du M, Christiani DC, Wei Q (2021) Genetic variants of DOCK2, EPHB1 and VAV2 in the natural killer cell-related pathway are associated with non-small cell lung cancer survival. Am J Cancer Res 11(5):2264–2277

Duke ES, Stapleford L, Drezner N, Amatya AK, Mishra-Kalyani PS, Shen YL, Maxfield K, Fourie Zirkelbach J, Bi Y, Liu J et al (2022) FDA approval summary: mobocertinib for metastatic non-small cell lung cancer with EGFR exon 20 insertion mutations. Clin Cancer Res

Dumont-Leblond N, Veillette M, Racine C, Joubert P, Duchaine C (2021) Non-small cell lung cancer microbiota characterization: Prevalence of enteric and potentially pathogenic bacteria in cancer tissues. PLoS One 16(4):e0249832

Dungo RT, Keating GM (2013) Afatinib: first global approval. Drugs 73(13):1503–1515

Dunn DB (2020) Larotrectinib and entrectinib: TRK inhibitors for the treatment of pediatric and adult patients with NTRK gene fusion. J Adv Pract Oncol 11(4):418–423

Duruisseaux M (2019) Lorlatinib: a new treatment option for ROS1-positive lung cancer. Lancet Oncol 20(12):1622–1623

Durvalumab Durvalumab approval expanded to reduce risk of NSCLC progression. Oncol Times 40(6)

Emaminejad N, Qian W, Guan Y, Tan M, Qiu Y, Liu H, Zheng B (2016) Fusion of quantitative image and genomic biomarkers to improve prognosis assessment of early stage lung cancer patients. IEEE Trans Biomed Eng 63(5):1034–1043

Facchinetti F, Friboulet L (2018) Lorlatinib in ALK- and ROS1-positive NSCLC: the future has a start. Transl Lung Cancer Res 7(Suppl 2):S103–S106

Fala L (2016) Portrazza (Necitumumab), an IgG1 monoclonal antibody, FDA approved for advanced squamous non-small-cell lung cancer. Am Health Drug Benefits 9(Spec Feature):119–122

Feng X, Xie X, Zheng B, Peng C, Zhou H, Qin J (2018) The more potential performance of nidogen 2 methylation by tissue or plasma DNA over brichoalveolar lavage DNA in diagnosis of nonsmall cell lung cancer. J Cancer Res Ther 14(Supplement):S341–S346

Fennell DA, Summers Y, Cadranel J, Benepal T, Christoph DC, Lal R, Das M, Maxwell F, Visseren-Grul C, Ferry D (2016) Cisplatin in the modern era: the backbone of first-line chemotherapy for non-small cell lung cancer. Cancer Treat Rev 44:42–50

Firdous S, Ghosh A, Saha S: BCSCdb: a database of biomarkers of cancer stem cells. Database (Oxford) 2022, 2022

Forde PM, Ettinger DS (2013) Targeted therapy for non-small-cell lung cancer: past, present and future. Expert Rev Anticancer Ther 13(6):745–758

Freeberg MA, Fromont LA, D'Altri T, Romero AF, Ciges JI, Jene A, Kerry G, Moldes M, Ariosa R, Bahena S et al (2022) The European Genome-phenome Archive in 2021. Nucleic Acids Res 50 (D1):D980–D987

Fuertes MA, Castilla J, Alonso C, Perez JM (2003) Cisplatin biochemical mechanism of action: from cytotoxicity to induction of cell death through interconnections between apoptotic and necrotic pathways. Curr Med Chem 10(3):257–266

Furuse K (1992) Platinum/oral etoposide therapy in non-small cell lung cancer. Oncology 49(Suppl 1):63–69. discussion 70

Gao J, Aksoy BA, Dogrusoz U, Dresdner G, Gross B, Sumer SO, Sun Y, Jacobsen A, Sinha R, Larsson E et al (2013) Integrative analysis of complex cancer genomics and clinical profiles using the cBioPortal. Sci Signal 6(269):pl1

Georgoulias V (2002) Docetaxel (taxotere) in the treatment of non-small cell lung cancer. Curr Med Chem 9(8):869–877

Gomes S, Cavadas B, Ferreira JC, Marques PI, Monteiro C, Sucena M, Sousa C, Vaz Rodrigues L, Teixeira G, Pinto P et al (2019) Profiling of lung microbiota discloses differences in adenocarcinoma and squamous cell carcinoma. Sci Rep 9(1):12838

Gong J, Fan Y, Lu H (2021) Pulmonary enteric adenocarcinoma. Transl Oncol 14(8):101123

Gonzalez VM, Fuertes MA, Alonso C, Perez JM (2001) Is cisplatin-induced cell death always produced by apoptosis? Mol Pharmacol 59(4):657–663

Green MR, Kreisman H, Doll DC, Lyss AP, Clamon GH, Goutsou M, Perry MC, Propert KJ (1992) Carboplatin in non-small cell lung cancer: an update on the Cancer and Leukemia Group B experience. Semin Oncol 19(1 Suppl 2):44–49

Grutters JP, Pijls-Johannesma M, Ruysscher DD, Peeters A, Reimoser S, Severens JL, Lambin P, Joore MA (2010) The cost-effectiveness of particle therapy in non-small cell lung cancer: exploring decision uncertainty and areas for future research. Cancer Treat Rev 36(6):468–476

Hanauske AR, Chen V, Paoletti P, Niyikiza C (2001) Pemetrexed disodium: a novel antifolate clinically active against multiple solid tumors. Oncologist 6(4):363–373

Harel M, Lahav C, Jacob E, Dahan N, Sela I, Elon Y, Raveh Shoval S, Yahalom G, Kamer I, Zer A et al (2022) Longitudinal plasma proteomic profiling of patients with non-small cell lung cancer undergoing immune checkpoint blockade. J Immunother Cancer 10(6)

Heller G, Weinzierl M, Noll C, Babinsky V, Ziegler B, Altenberger C, Minichsdorfer C, Lang G, Dome B, End-Pfutzenreuter A et al (2012) Genome-wide miRNA expression profiling identifies miR-9-3 and miR-193a as targets for DNA methylation in non-small cell lung cancers. Clin Cancer Res 18(6):1619–1629

Heller G, Babinsky VN, Ziegler B, Weinzierl M, Noll C, Altenberger C, Mullauer L, Dekan G, Grin Y, Lang G et al (2013) Genome-wide CpG island methylation analyses in non-small cell lung cancer patients. Carcinogenesis 34(3):513–521

Hoang PH, Landi MT (2022) DNA methylation in lung cancer: mechanisms and associations with histological subtypes, molecular alterations, and major epidemiological factors. Cancers (Basel) 14(4)

Hollings N, Shaw P (2002) Diagnostic imaging of lung cancer. Eur Respir J 19(4):722–742

Hopkins-Donaldson S, Ziegler A, Kurtz S, Bigosch C, Kandioler D, Ludwig C, Zangemeister-Wittke U, Stahel R (2003) Silencing of death receptor and caspase-8 expression in small cell lung carcinoma cell lines and tumors by DNA methylation. Cell Death Differ 10(3):356–364

Hosgood HD 3rd, Sapkota AR, Rothman N, Rohan T, Hu W, Xu J, Vermeulen R, He X, White JR, Wu G et al (2014) The potential role of lung microbiota in lung cancer attributed to household coal burning exposures. Environ Mol Mutagen 55(8):643–651

Jairam V, Park HS (2019) Strengths and limitations of large databases in lung cancer radiation oncology research. Transl Lung Cancer Res 8(Suppl 2):S172–S183

Jensen MA, Ferretti V, Grossman RL, Staudt LM (2017) The NCI Genomic Data Commons as an engine for precision medicine. Blood 130(4):453–459

Jiang M, Li X, Quan X, Li X, Zhou B (2018) Single nucleotide polymorphisms in HMGB1 correlate with lung cancer risk in the Northeast Chinese Han Population. Molecules 23(4)

Jones GS, Baldwin DR (2018) Recent advances in the management of lung cancer. Clin Med (Lond) 18(Suppl 2):s41–s46

Kadir T, Gleeson F (2018) Lung cancer prediction using machine learning and advanced imaging techniques. Transl Lung Cancer Res 7(3):304–312

Kalari S, Jung M, Kernstine KH, Takahashi T, Pfeifer GP (2013) The DNA methylation landscape of small cell lung cancer suggests a differentiation defect of neuroendocrine cells. Oncogene 32(30):3559–3568

Kazandjian D, Suzman DL, Blumenthal G, Mushti S, He K, Libeg M, Keegan P, Pazdur R (2016a) FDA Approval Summary: Nivolumab for the treatment of metastatic non-small cell lung cancer with progression on or after platinum-based chemotherapy. Oncologist 21(5):634–642

Kazandjian D, Blumenthal GM, Yuan W, He K, Keegan P, Pazdur R (2016b) FDA approval of gefitinib for the treatment of patients with metastatic EGFR mutation-positive non-small cell lung cancer. Clin Cancer Res 22(6):1307–1312

Kazandjian D, Blumenthal GM, Luo L, He K, Fran I, Lemery S, Pazdur R (2016c) Benefit-risk summary of crizotinib for the treatment of patients with ROS1 alteration-positive, metastatic non-small cell lung cancer. Oncologist 21(8):974–980

Khozin S, Blumenthal GM, Jiang X, He K, Boyd K, Murgo A, Justice R, Keegan P, Pazdur R: U.S. (2014) Food and Drug Administration approval summary: Erlotinib for the first-line treatment of metastatic non-small cell lung cancer with epidermal growth factor receptor exon 19 deletions or exon 21 (L858R) substitution mutations. Oncologist 19(7):774–779

Khozin S, Blumenthal GM, Zhang L, Tang S, Brower M, Fox E, Helms W, Leong R, Song P, Pan Y et al (2015) FDA approval: ceritinib for the treatment of metastatic anaplastic lymphoma kinase-positive non-small cell lung cancer. Clin Cancer Res 21(11):2436–2439

Khunger A, Khunger M, Velcheti V (2018) Dabrafenib in combination with trametinib in the treatment of patients with BRAF V600-positive advanced or metastatic non-small cell lung cancer: clinical evidence and experience. Ther Adv Respir Dis 12:1753466618767611

Kim YJ, Sertamo K, Pierrard MA, Mesmin C, Kim SY, Schlesser M, Berchem G, Domon B (2015) Verification of the biomarker candidates for non-small-cell lung cancer using a targeted proteomics approach. J Proteome Res 14(3):1412–1419

Kim YI, Ahn JM, Sung HJ, Na SS, Hwang J, Kim Y, Cho JY (2016) Meta-markers for the differential diagnosis of lung cancer and lung disease. J Proteome 148:36–43

Kim J, Bradford D, Larkins E, Pai-Scherf LH, Chatterjee S, Mishra-Kalyani PS, Wearne E, Helms WS, Ayyoub A, Bi Y et al (2021) FDA approval summary: pralsetinib for the treatment of lung and thyroid cancers with RET gene mutations or fusions. Clin Cancer Res 27(20):5452–5456

Koch AL, Vellanki PJ, Drezner N, Li X, Mishra-Kalyani PS, Shen YL, Xia H, Li Y, Liu J, Zirkelbach JF et al (2021) FDA approval summary: osimertinib for adjuvant treatment of surgically resected non-small cell lung cancer, a collaborative project Orbis review. Clin Cancer Res 27(24):6638–6643

Kollmannsberger C, Mross K, Jakob A, Kanz L, Bokemeyer C (1999) Topotecan – a novel topoisomerase I inhibitor: pharmacology and clinical experience. Oncology 56(1):1–12

Kong J, Chen X, Wang J, Li J, Xu F, Gao S, Yu H, Qian B (2020) Genetic polymorphisms in the Vitamin D pathway and non-small cell lung cancer survival. Pathol Oncol Res 26(3):1709–1715

Kubota K, Hida T, Ishikura S, Mizusawa J, Nishio M, Kawahara M, Yokoyama A, Imamura F, Takeda K, Negoro S et al (2014) Etoposide and cisplatin versus irinotecan and cisplatin in patients with limited-stage small-cell lung cancer treated with etoposide and cisplatin plus concurrent accelerated hyperfractionated thoracic radiotherapy (JCOG0202): a randomised phase 3 study. Lancet Oncol 15(1):106–113

Kumar R, Chaudhary K, Gupta S, Singh H, Kumar S, Gautam A, Kapoor P, Raghava GP (2013) CancerDR: cancer drug resistance database. Sci Rep 3:1445

Lababede O, Meziane MA (2018) The eighth edition of TNM staging of lung cancer: reference chart and diagrams. Oncologist 23(7):844–848

Langevin SM, Kratzke RA, Kelsey KT (2015) Epigenetics of lung cancer. Transl Res 165(1):74–90

Lara PN Jr, Natale R, Crowley J, Lenz HJ, Redman MW, Carleton JE, Jett J, Langer CJ, Kuebler JP, Dakhil SR et al (2009) Phase III trial of irinotecan/cisplatin compared with etoposide/cisplatin in extensive-stage small-cell lung cancer: clinical and pharmacogenomic results from SWOG S0124. J Clin Oncol 27(15):2530–2535

Larkins E, Blumenthal GM, Chen H, He K, Agarwal R, Gieser G, Stephens O, Zahalka E, Ringgold K, Helms W et al (2016) FDA approval: alectinib for the treatment of metastatic, ALK-positive non-small cell lung cancer following crizotinib. Clin Cancer Res 22(21): 5171–5176

Laroumagne S, Salinas-Pineda A, Hermant C, Murris M, Gourraud PA, Do C, Segonds C, Didier A, Mazieres J (2011) Incidence and characteristics of bronchial colonisation in patient with lung cancer: a retrospective study of 388 cases. Rev Mal Respir 28(3):328–335

Lawi ZK, Al-Shuhaib MBS, Amara IB (2022) The rs1801280 SNP is associated with non-small cell lung carcinoma by exhibiting a highly deleterious effect on N-acetyltransferase 2. J Cancer Res Clin Oncol

Lebanony D, Benjamin H, Gilad S, Ezagouri M, Dov A, Ashkenazi K, Gefen N, Izraeli S, Rechavi G, Pass H et al (2009) Diagnostic assay based on hsa-miR-205 expression distinguishes squamous from nonsquamous non-small-cell lung carcinoma. J Clin Oncol 27(12):2030–2037

Lee SH, Sung JY, Yong D, Chun J, Kim SY, Song JH, Chung KS, Kim EY, Jung JY, Kang YA et al (2016) Characterization of microbiome in bronchoalveolar lavage fluid of patients with lung cancer comparing with benign mass like lesions. Lung Cancer 102:89–95

Lee G, Park H, Bak SH, Lee HY (2020) Radiomics in lung cancer from basic to advanced: current status and future directions. Korean J Radiol 21(2):159–171

Li C, Lu H (2018) Adenosquamous carcinoma of the lung. Onco Targets Ther 11:4829–4835

Li LS, Kim H, Rhee H, Kim SH, Shin DH, Chung KY, Park KS, Paik YK, Chang J, Kim H (2004) Proteomic analysis distinguishes basaloid carcinoma as a distinct subtype of nonsmall cell lung carcinoma. Proteomics 4(11):3394–3400

Li Y, Zhang Y, Qiu F, Qiu Z (2011) Proteomic identification of exosomal LRG1: a potential urinary biomarker for detecting NSCLC. Electrophoresis 32(15):1976–1983

Li X, Li X, Yin Z, Jiang M, Tian W, Tang M, Zhou B (2019) Polymorphisms of rs4787050 and rs8045980 are associated with lung cancer risk in northeast Chinese female nonsmokers. Biomark Med 13(13):1119–1128

Li W, Liu B, Wang W, Sun C, Che J, Yuan X, Zhai C (2022) Lung cancer stage prediction using multi-omics data. Comput Math Methods Med 2022:2279044

Liu WB, Han F, Jiang X, Yang LJ, Li YH, Liu Y, Chen HQ, Ao L, Cui ZH, Cao J et al (2012) ANKRD18A as a novel epigenetic regulation gene in lung cancer. Biochem Biophys Res Commun 429(3–4):180–185

Liu SH, Shen PC, Chen CY, Hsu AN, Cho YC, Lai YL, Chen FH, Li CY, Wang SC, Chen M et al (2020) DriverDBv3: a multi-omics database for cancer driver gene research. Nucleic Acids Res 48(D1):D863–D870

Liu Y, Liang X, Zhang H, Dong J, Zhang Y, Wang J, Li C, Xin X, Li Y (2022) ER Stress-related genes EIF2AK3, HSPA5, and DDIT3 polymorphisms are associated with risk of lung cancer. Front Genet 13:938787

Loehrer PJ Sr, Einhorn LH, Greco FA (1988) Cisplatin plus etoposide in small cell lung cancer. Semin Oncol 15(3 Suppl 3):2–8

Lopez-Sanchez LM, Jurado-Gamez B, Feu-Collado N, Valverde A, Canas A, Fernandez-Rueda JL, Aranda E, Rodriguez-Ariza A (2017) Exhaled breath condensate biomarkers for the early diagnosis of lung cancer using proteomics. Am J Physiol Lung Cell Mol Physiol 313(4): L664–L676

Louis E, Adriaensens P, Guedens W, Bigirumurame T, Baeten K, Vanhove K, Vandeurzen K, Darquennes K, Vansteenkiste J, Dooms C et al (2016) Detection of lung cancer through metabolic changes measured in blood plasma. J Thorac Oncol 11(4):516–523

Lu M, Zhan X (2018) The crucial role of multiomic approach in cancer research and clinically relevant outcomes. EPMA J 9(1):77–102

Lu C, Onn A, Vaporciyan AA (2017) Chapter 84: cancer of the Lung. In: Holland-Frei cancer medicine, 9th edn. Wiley Blackwell. ISBN 9781119000846

Ma L, Muscat JE, Sinha R, Sun D, Xiu G (2021a) Proteomics of exhaled breath condensate in lung cancer and controls using data-independent acquisition (DIA): a pilot study. J Breath Res 15(2)

Ma Y, Qiu M, Wang S, Meng S, Yang F, Jiang G (2021b) Distinct tumor bacterial microbiome in lung adenocarcinomas manifested as radiological subsolid nodules. Transl Oncol 14(6):101050

Madama D, Martins R, Pires AS, Botelho MF, Alves MG, Abrantes AM, Cordeiro CR (2021) Metabolomic profiling in lung cancer: a systematic review. Meta 11(9)

Majumdar S, Saha S (2020) Systems immunology approach in understanding the association of allergy and cancer. In: Singh S (ed) Systems and synthetic immunology. Springer, Singapore

Martinez-Jimenez F, Muinos F, Sentis I, Deu-Pons J, Reyes-Salazar I, Arnedo-Pac C, Mularoni L, Pich O, Bonet J, Kranas H et al (2020) A compendium of mutational cancer driver genes. Nat Rev Cancer 20(10):555–572

Mathieu L, Shah S, Pai-Scherf L, Larkins E, Vallejo J, Li X, Rodriguez L, Mishra-Kalyani P, Goldberg KB, Kluetz PG et al (2021) FDA approval summary: atezolizumab and durvalumab in combination with platinum-based chemotherapy in extensive stage small cell lung cancer. Oncologist 26(5):433–438

Mathieu LN, Larkins E, Akinboro O, Roy P, Amatya AK, Fiero MH, Mishra-Kalyani PS, Helms WS, Myers CE, Skinner AM et al (2022) FDA approval summary: capmatinib and tepotinib for the treatment of metastatic NSCLC harboring MET exon 14 skipping mutations or alterations. Clin Cancer Res 28(2):249–254

McNitt-Gray MF, Armato SG 3rd, Meyer CR, Reeves AP, McLennan G, Pais RC, Freymann J, Brown MS, Engelmann RM, Bland PH et al (2007) The Lung Image Database Consortium (LIDC) data collection process for nodule detection and annotation. Acad Radiol 14(12): 1464–1474

Meng W, Ye Z, Cui R, Perry J, Dedousi-Huebner V, Huebner A, Wang Y, Li B, Volinia S, Nakanishi H et al (2013) MicroRNA-31 predicts the presence of lymph node metastases and survival in patients with lung adenocarcinoma. Clin Cancer Res 19(19):5423–5433

Meng YQ, Shi Q, Zhu SJ, He GH, Zhang SJ, Sun C (2021) Asbestos-induced lung cancer-associated SNP rs13383928 regulates PTH2R expression in lung tissue. Hum Hered:1–5

Michelotti A, de Scordilli M, Bertoli E, De Carlo E, Del Conte A, Bearz A (2022) NSCLC as the paradigm of precision medicine at its finest: the rise of new druggable molecular targets for advanced disease. Int J Mol Sci 23(12)

Miele E, Spinelli GP, Miele E, Tomao F, Tomao S (2009) Albumin-bound formulation of paclitaxel (Abraxane ABI-007) in the treatment of breast cancer. Int J Nanomedicine 4:99–105

Minguet J, Smith KH, Bramlage P (2016) Targeted therapies for treatment of non-small cell lung cancer--recent advances and future perspectives. Int J Cancer 138(11):2549–2561

Mur LA, Huws SA, Cameron SJ, Lewis PD, Lewis KE (2018) Lung cancer: a new frontier for microbiome research and clinical translation. Ecancermedicalscience 12:866

Najafi Z, Mohamadnia A, Ahmadi R, Mahmoudi M, Bahrami N, Khosravi A, Jamaati H, Tabarsi P, Kazem Pour Dizaji M, Shirian S (2020) Proteomic and genomic biomarkers for non-small cell lung cancer: peroxiredoxin, haptoglobin, and alpha-1 antitrypsin. Cancer Med 9(11):3974–3982

Nakagawa K, Garon EB, Seto T, Nishio M, Ponce Aix S, Paz-Ares L, Chiu CH, Park K, Novello S, Nadal E et al (2019) Ramucirumab plus erlotinib in patients with untreated, EGFR-mutated, advanced non-small-cell lung cancer (RELAY): a randomised, double-blind, placebo-controlled, phase 3 trial. Lancet Oncol 20(12):1655–1669

Nakajima EC, Drezner N, Li X, Mishra-Kalyani PS, Liu Y, Zhao H, Bi Y, Liu J, Rahman A, Wearne E et al (2022) FDA approval summary: Sotorasib for KRAS G12C-mutated metastatic NSCLC. Clin Cancer Res 28(8):1482–1486

Nicholson AG, Tsao MS, Beasley MB, Borczuk AC, Brambilla E, Cooper WA, Dacic S, Jain D, Kerr KM, Lantuejoul S et al (2022) The 2021 WHO classification of lung tumors: impact of advances since 2015. J Thorac Oncol 17(3):362–387

Nobili S, Lavacchi D, Perrone G, Vicini G, Tassi R, Landini I, Grosso A, Roviello G, Mazzanti R, Santomaggio C et al (2020) Vinorelbine in non-small cell lung cancer: real-world data from a single-institution experience. Oncol Res 28(3):237–248

Nooreldeen R, Bach H (2021) Current and future development in lung cancer diagnosis. Int J Mol Sci 22(16)

Odogwu L, Mathieu L, Blumenthal G, Larkins E, Goldberg KB, Griffin N, Bijwaard K, Lee EY, Philip R, Jiang X et al (2018) FDA approval summary: dabrafenib and trametinib for the treatment of metastatic non-small cell lung cancers harboring BRAF V600E mutations. Oncologist 23(6):740–745

Olivier T, Prasad V (2022) Amivantamab and mobocertinib in exon 20 insertions EGFR mutant lung cancer, challenge to the current guidelines. Transl Oncol 23:101475

Ongenaert M, Van Neste L, De Meyer T, Menschaert G, Bekaert S, Van Criekinge W (2008) PubMeth: a cancer methylation database combining text-mining and expert annotation. Nucleic Acids Res 36(Database issue):D842–D846

Pai-Scherf L, Blumenthal GM, Li H, Subramaniam S, Mishra-Kalyani PS, He K, Zhao H, Yu J, Paciga M, Goldberg KB et al (2017) FDA approval summary: pembrolizumab for treatment of metastatic non-small cell lung cancer: first-line therapy and beyond. Oncologist 22(11): 1392–1399

Panunzio A, Sartori P (2020) Lung cancer and radiological imaging. Curr Radiopharm 13(3): 238–242

Parashar B, Arora S, Wernicke AG (2013) Radiation therapy for early stage lung cancer. Semin Intervent Radiol 30(2):185–190

Patel S, Petty WJ, Sands JM (2021) An overview of lurbinectedin as a new second-line treatment option for small cell lung cancer. Ther Adv Med Oncol 13:17588359211020529

Pedersen S, Hansen JB, Maltesen RG, Szejniuk WM, Andreassen T, Falkmer U, Kristensen SR (2021) Identifying metabolic alterations in newly diagnosed small cell lung cancer patients. Metabol Open 12:100127

Pircher A, Wasle IK, Mian M, Gamerith G, Ulsperger E, Studnicka M, Mohn-Staudner A, Hilbe W, Fiegl M (2013) Docetaxel in the treatment of non-small cell lung cancer (NSCLC) -- an observational study focusing on symptom improvement. Anticancer Res 33(9):3831–3836

Purandare NC, Rangarajan V (2015) Imaging of lung cancer: implications on staging and management. Indian J Radiol Imaging 25(2):109–120

Ren Y, Zhao S, Jiang D, Feng X, Zhang Y, Wei Z, Wang Z, Zhang W, Zhou QF, Li Y et al (2018) Proteomic biomarkers for lung cancer progression. Biomark Med 12(3):205–215

Repana D, Nulsen J, Dressler L, Bortolomeazzi M, Venkata SK, Tourna A, Yakovleva A, Palmieri T, Ciccarelli FD (2019) The Network of Cancer Genes (NCG): a comprehensive catalogue of known and candidate cancer genes from cancer sequencing screens. Genome Biol 20(1):1

Rho JH, Roehrl MH, Wang JY (2009) Tissue proteomics reveals differential and compartment-specific expression of the homologs transgelin and transgelin-2 in lung adenocarcinoma and its stroma. J Proteome Res 8(12):5610–5618

Rhodes DR, Yu J, Shanker K, Deshpande N, Varambally R, Ghosh D, Barrette T, Pandey A, Chinnaiyan AM (2004) ONCOMINE: a cancer microarray database and integrated data-mining platform. Neoplasia 6(1):1–6

Richardson CJ, Gao Q, Mitsopoulous C, Zvelebil M, Pearl LH, Pearl FM (2009) MoKCa database--mutations of kinases in cancer. Nucleic Acids Res 37(Database issue):D824–D831

Rijavec E, Biello F, Indini A, Grossi F, Genova C (2022) Current insights on the treatment of anaplastic lymphoma kinase-positive metastatic non-small cell lung cancer: focus on brigatinib. Clin Pharmacol 14:1–9

Rocha CM, Carrola J, Barros AS, Gil AM, Goodfellow BJ, Carreira IM, Bernardo J, Gomes A, Sousa V, Carvalho L et al (2011) Metabolic signatures of lung cancer in biofluids: NMR-based metabonomics of blood plasma. J Proteome Res 10(9):4314–4324

Ruan X, Ye Y, Cheng W, Xu L, Huang M, Chen Y, Zhu J, Lu X, Yan F (2022) Multi-omics integrative analysis of lung adenocarcinoma: an in silico profiling for precise medicine. Front Med (Lausanne) 9:894338

Rutledge MR, Waddell JA, Solimando DA Jr (2014) Irinotecan and Carboplatin (IC) regimen for small-cell lung cancer and non-small-cell lung cancer. Hosp Pharm 49(7):603–611

Santarelli L, Gaetani S, Monaco F, Bracci M, Valentino M, Amati M, Rubini C, Sabbatini A, Pasquini E, Zanotta N et al (2019) Four-miRNA signature to identify asbestos-related lung malignancies. Cancer Epidemiol Biomark Prev 28(1):119–126

Schmidt B, Liebenberg V, Dietrich D, Schlegel T, Kneip C, Seegebarth A, Flemming N, Seemann S, Distler J, Lewin J et al (2010) SHOX2 DNA methylation is a biomarker for the diagnosis of lung cancer based on bronchial aspirates. BMC Cancer 10:600

Scott LJ (2018) Osimertinib as first-line therapy in advanced NSCLC: a profile of its use. Drugs Ther Perspect 34(8):351–357

Scott A, Salgia R (2008) Biomarkers in lung cancer: from early detection to novel therapeutics and decision making. Biomark Med 2(6):577–586

Sellers K, Allen TD, Bousamra M 2nd, Tan J, Mendez-Lucas A, Lin W, Bah N, Chernyavskaya Y, MacRae JI, Higashi RM et al (2019) Metabolic reprogramming and Notch activity distinguish between non-small cell lung cancer subtypes. Br J Cancer 121(1):51–64

Shaikh FJ, Rao DS (2022) Prediction of cancer disease using machine learning approach. Mater Today Proc 50:40–47

Shen Y, Yao H, Li A, Wang M (2016) CSCdb: a cancer stem cells portal for markers, related genes and functional information. Database (Oxford) 2016

Shen N, Du J, Zhou H, Chen N, Pan Y, Hoheisel JD, Jiang Z, Xiao L, Tao Y, Mo X (2019) A Diagnostic Panel of DNA Methylation Biomarkers for Lung Adenocarcinoma. Front Oncol 9: 1281

Shi YX, Wang Y, Li X, Zhang W, Zhou HH, Yin JY, Liu ZQ (2017) Genome-wide DNA methylation profiling reveals novel epigenetic signatures in squamous cell lung cancer. BMC Genomics 18(1):901

Shirley M (2018) Dacomitinib: first global approval. Drugs 78(18):1947–1953

Sim Y, Chung MJ, Kotter E, Yune S, Kim M, Do S, Han K, Kim H, Yang S, Lee DJ et al (2020) Deep convolutional neural network-based software improves radiologist detection of malignant lung nodules on chest radiographs. Radiology 294(1):199–209

Su L, Huang H, Gao M, Li Y, Shi R, Chen C, Li X, Zhu G, Liu H, Chen J (2022) Association of CDH1, FANCB and APC gene polymorphisms with lung cancer susceptibility in Chinese Population. Zhongguo Fei Ai Za Zhi 25(9):658–664

Sul J, Blumenthal GM, Jiang X, He K, Keegan P, Pazdur R (2016) FDA approval summary: pembrolizumab for the treatment of patients with metastatic non-small cell lung cancer whose tumors express programmed death-ligand 1. Oncologist 21(5):643–650

Sulewska A, Niklinski J, Charkiewicz R, Karabowicz P, Biecek P, Baniecki H, Kowalczuk O, Kozlowski M, Modzelewska P, Majewski P et al (2022) A signature of 14 long non-coding RNAs (lncRNAs) as a step towards precision diagnosis for NSCLC. Cancers (Basel) 14(2)

Sun X, Yi J, Yang J, Han Y, Qian X, Liu Y, Li J, Lu B, Zhang J, Pan X et al (2021) An integrated epigenomic-transcriptomic landscape of lung cancer reveals novel methylation driver genes of diagnostic and therapeutic relevance. Theranostics 11(11):5346–5364

Szczyrek M, Mlak R, Szudy-Szczyrek A, Wojas-Krawczyk K, Kedziora K, Milanowski J (2022) Polymorphism of Baculoviral inhibitor of apoptosis repeat-containing 5 (BIRC5) can be associated with clinical outcome of non-small cell lung cancer. Cell 11(6)

Takahashi S, Asada K, Takasawa K, Shimoyama R, Sakai A, Bolatkan A, Shinkai N, Kobayashi K, Komatsu M, Kaneko S et al (2020) Predicting deep learning based multi-omics parallel integration survival subtypes in lung cancer using reverse phase protein array data. Biomol Ther 10(10)

Tam M, Dyer T, Dissez G, Morgan TN, Hughes M, Illes J, Rasalingham R, Rasalingham S (2021) Augmenting lung cancer diagnosis on chest radiographs: positioning artificial intelligence to improve radiologist performance. Clin Radiol 76(8):607–614

Tan X, Li D, Huang P, Jian X, Wan H, Wang G, Li Y, Ouyang J, Lin Y, Xie L (2020) dbPepNeo: a manually curated database for human tumor neoantigen peptides. Database (Oxford) 2020

Tate JG, Bamford S, Jubb HC, Sondka Z, Beare DM, Bindal N, Boutselakis H, Cole CG, Creatore C, Dawson E et al (2019) COSMIC: the catalogue of somatic mutations in cancer. Nucleic Acids Res 47(D1):D941–D947

Vaddepally RK, Kharel P, Pandey R, Garje R, Chandra AB (2020) Review of Indications of FDA-approved immune checkpoint inhibitors per NCCN guidelines with the level of evidence. Cancers (Basel) 12(3)

Valk K, Vooder T, Kolde R, Reintam MA, Petzold C, Vilo J, Metspalu A (2010) Gene expression profiles of non-small cell lung cancer: survival prediction and new biomarkers. Oncology 79(3–4):283–292

Vellanki PJ, Mulkey F, Jaigirdar AA, Rodriguez L, Wang Y, Xu Y, Zhao H, Liu J, Howe G, Wang J et al (2021) FDA approval summary: nivolumab with ipilimumab and chemotherapy for metastatic non-small cell lung cancer, a collaborative project orbis review. Clin Cancer Res 27(13):3522–3527

Wang S, Zhang H, Cheng L, Evans C, Pan CX (2010) Analysis of the cytotoxic activity of carboplatin and gemcitabine combination. Anticancer Res 30(11):4573–4578

Wang R, Feng N, Wang Y, Gao S, Zhang F, Qian Y, Gao M, Yu H, Zhou B, Qian B (2019) SNPs in LncRNA genes are associated with non-small cell lung cancer in a Chinese population. J Clin Lab Anal 33(4):e22858

Wang C, Yang D, Zhang X, Zhang X, Yang L, Wang P, Zhou W, Li H, Li Y, Nie H et al (2020) Association of PTEN gene SNPs rs2299939 with PFS in patients with small cell lung cancer treated with early radiotherapy. Front Genet 11:298

Watanabe K, Emoto N, Hamano E, Sunohara M, Kawakami M, Kage H, Kitano K, Nakajima J, Goto A, Fukayama M et al (2012) Genome structure-based screening identified epigenetically silenced microRNA associated with invasiveness in non-small-cell lung cancer. Int J Cancer 130(11):2580–2590

Weissfeld JL, Lin Y, Lin HM, Kurland BF, Wilson DO, Fuhrman CR, Pennathur A, Romkes M, Nukui T, Yuan JM et al (2015) Lung cancer risk prediction using common SNPs located in GWAS-identified susceptibility regions. J Thorac Oncol 10(11):1538–1545
Wu C, Jin X, Tsueng G, Afrasiabi C, Su AI (2016) BioGPS: building your own mash-up of gene annotations and expression profiles. Nucleic Acids Res 44(D1):D313–D316
Xi S, Xu H, Shan J, Tao Y, Hong JA, Inchauste S, Zhang M, Kunst TF, Mercedes L, Schrump DS (2013) Cigarette smoke mediates epigenetic repression of miR-487b during pulmonary carcinogenesis. J Clin Invest 123(3):1241–1261
Xiao S, Sun S, Long W, Kuang S, Liu Y, Huang H, Zhou J, Zhou Y, Lu X (2018) A meta-analytic review of the association between two common SNPs in miRNAs and lung cancer susceptibility. Onco Targets Ther 11:2419–2427
Xie B, Ding Q, Han H, Wu D (2013) miRCancer: a microRNA-cancer association database constructed by text mining on literature. Bioinformatics 29(5):638–644
Xu H, Ma J, Wu J, Chen L, Sun F, Qu C, Zheng D, Xu S (2016) Gene expression profiling analysis of lung adenocarcinoma. Braz J Med Biol Res 49(3)
Yan X, Yang M, Liu J, Gao R, Hu J, Li J, Zhang L, Shi Y, Guo H, Cheng J et al (2015) Discovery and validation of potential bacterial biomarkers for lung cancer. Am J Cancer Res 5(10): 3111–3122
Yang Y, Zhao S, Fan Y, Zhao F, Liu Q, Hu W, Liu D, Fan K, Wang J, Wang J (2009) Detection and identification of potential biomarkers of non-small cell lung cancer. Technol Cancer Res Treat 8(6):455–466
Yang W, Soares J, Greninger P, Edelman EJ, Lightfoot H, Forbes S, Bindal N, Beare D, Smith JA, Thompson IR et al (2013) Genomics of Drug Sensitivity in Cancer (GDSC): a resource for therapeutic biomarker discovery in cancer cells. Nucleic Acids Res 41(Database issue): D955-961
Yang D, Yang X, Li Y, Zhao P, Fu R, Ren T, Hu P, Wu Y, Yang H, Guo N (2020) Clinical significance of circulating tumor cells and metabolic signatures in lung cancer after surgical removal. J Transl Med 18(1):243
Yardley DA (2013) nab-Paclitaxel mechanisms of action and delivery. J Control Release 170(3): 365–372
Yin J, Hou W, Vogel U, Ma Y, Wang C, Wang H, Sun Z (2020) Interaction between common variants of MDM2 and PPP1R13L and CD3EAP and TP53 SNPs in relation to lung cancer risk among Chinese. Ann Transl Med 8(15):934
Yun KM, Bazhenova LA (2022) Update on Lorlatinib: role in reducing the risk of disease progression in ALK-positive NSCLC. Cancer Manag Res 14:843–850
Yun Y, Wang Y, Yang E, Jing X (2022) Cuproptosis-Related Gene - SLC31A1, FDX1 and ATP7B - polymorphisms are associated with risk of lung cancer. Pharmgenomics Pers Med 15:733–742
Zappa C, Mousa SA (2016) Non-small cell lung cancer: current treatment and future advances. Transl Lung Cancer Res 5(3):288–300
Zeng GQ, Zhang PF, Deng X, Yu FL, Li C, Xu Y, Yi H, Li MY, Hu R, Zuo JH et al (2012) Identification of candidate biomarkers for early detection of human lung squamous cell cancer by quantitative proteomics. Mol Cell Proteomics 11(6):M111 013946
Zhang J, Baran J, Cros A, Guberman JM, Haider S, Hsu J, Liang Y, Rivkin E, Wang J, Whitty B et al (2011) International Cancer Genome Consortium Data Portal--a one-stop shop for cancer genomics data. Database (Oxford) 2011:bar026
Zhang XY, Zhang G, Jiang Y, Liu D, Li MZ, Zhong Q, Zeng SQ, Liu WL, Zeng MS (2015) The prognostic value of serum C-reactive protein-bound serum amyloid A in early-stage lung cancer. Chin J Cancer 34(8):335–349
Zhang Y, Zhang L, Li R, Chang DW, Ye Y, Minna JD, Roth JA, Han B, Wu X (2017) Genetic variations in cancer-related significantly mutated genes and lung cancer susceptibility. Ann Oncol 28(7):1625–1630

Zhang C, Leng W, Sun C, Lu T, Chen Z, Men X, Wang Y, Wang G, Zhen B, Qin J (2018) Urine proteome profiling predicts lung cancer from control cases and other tumors. EBioMedicine 30: 120–128

Zhang H, Li Y, Guo S, Wang Y, Wang H, Lu D, Wang J, Jin L, Jiang G, Wu J et al (2020a) Effect of ERCC2 rs13181 and rs1799793 polymorphisms and environmental factors on the prognosis of patients with lung cancer. Am J Transl Res 12(10):6941–6953

Zhang L, Zheng J, Ahmed R, Huang G, Reid J, Mandal R, Maksymuik A, Sitar DS, Tappia PS, Ramjiawan B et al (2020b) A high-performing plasma metabolite panel for early-stage lung cancer detection. Cancers (Basel) 12(3)

Zhao Z, Fei K, Bai H, Wang Z, Duan J, Wang J (2021a) Metagenome association study of the gut microbiome revealed biomarkers linked to chemotherapy outcomes in locally advanced and advanced lung cancer. Thorac Cancer 12(1):66–78

Zhao Y, Gao Y, Xu X, Zhou J, Wang H (2021b) Multi-omics analysis of genomics, epigenomics and transcriptomics for molecular subtypes and core genes for lung adenocarcinoma. BMC Cancer 21(1):257

Zhu L, Chen L (2019) Progress in research on paclitaxel and tumor immunotherapy. Cell Mol Biol Lett 24:40

Zhu X, Luo H, Xu Y (2019) Transcriptome analysis reveals an important candidate gene involved in both nodal metastasis and prognosis in lung adenocarcinoma. Cell Biosci 9:92